Handbook of Soil Science

Handbook of Soil Science

Edited by **Brian Bechdal**

R CALLISTO REFERENCE

New York

Published by Callisto Reference,
106 Park Avenue, Suite 200,
New York, NY 10016, USA
www.callistoreference.com

Handbook of Soil Science
Edited by Brian Bechdal

International Standard Book Number: 978-1-63239-412-5 (Hardback)

Printed in the United States of America.

Contents

Preface

Soil science is critical for the provision of adequate food, fiber and renewable natural resources (fuel, wood etc.). Moreover, in the present times, soil also plays a crucial role in maintaining the water table. In the developing world, soil science is linked to the economic well-being of many farm families. Poor soils mean poor harvest resulting in low returns for over 60% of the population. Most families in developing countries rely on agriculture for survival. Use of inorganic and organic fertilizers, in addition to biological processes can be used to improve the science of the soil. Soil Science finds its applications in areas such as wetland studies, ecosystem studies, climate change, nutrient management and soil survey among numerous others.

A Russian geologist, Vasily Dokuchaev has been credited for identifying soil as a crucial resource for the use and development of mankind. There are many locally available resources that can be collected, improved and used to improve the soil science. However, it requires good understanding of their practical use and management. The effect of these materials on the environment and their economic returns need to be understood. There are various areas of specializations in the field of soil science. The five popular ones in which most soil scientists undertake research are microbiology, pedology, edaphology, physics and chemistry.

We especially wish to acknowledge the contributing authors, without whom a work of this magnitude would clearly not have been realizable. We thank all our authors for allocating much of their scarce time to this project. Not only do we appreciate their participation, but also their adherence as a group to the time parameters set for this publication.

Editor

Short-Term Effects of Land Leveling on Irrigation-Related Some Soil Properties in a Clay Loam Soil

Tekin Öztekin

Department of Biosystem Engineering, Agricultural Faculty, Gaziosmanpasa University, Taslicifthk Campus, 60250 Tokat, Turkey

Correspondence should be addressed to Tekin Öztekin; tekin.oztekin@gop.edu.tr

Academic Editors: F. Darve, H. Freitas, and A. Moldes

There are few studies conducted on the short-term effects of land leveling on soil water holding capacity. The objectives of this study were to analyze the short-term effects of land leveling on the magnitudes, variances, spatial variability, and distributions of surface (0–20 cm) and subsurface (20–40 cm) soil properties of bulk density, field capacity, permanent wilting point, water holding capacity and particle size fractions. The study was conducted in a 1.2 ha field with clay loam soil located on the low terraces of Yesilirmak River, Tokat, Turkey. According to the paired t-test results, water holding capacity, and bulk density significantly increased, while permanent wilting point ($P \leq 0.001$) and field capacity ($P \leq 0.05$) significantly decreased for surface soil due to land leveling. The reasons for the increases in WHC values in both cut and fill areas (29%, and 12%, resp.) of surface soil are look like the much more decreases in PWP values than those of FC values and the increases in BD values. The moderate positive linear relationship between the surface soil clay contents and cut depths through cut areas ($r = 0.64$) was also determined in this study.

1. Introduction

In Turkey, land consolidation projects have gained a speedy acceleration during the last decade. This trend will likely continue into the near future when we consider the government common agricultural policy. One of the main purposes of land consolidation projects is to provide the land a uniform grade compatible for both drainage and surface irrigation methods. In order to create a slight but uniform slope across a field with land leveling, most commonly, topsoil from spots of relatively high elevations are simply scrapped away (i.e., cut) and deposited (i.e., filled) in areas of relatively low elevations [1]. In other words, land leveling is the operation of shaping the surface of land to predetermined grades so that each surface slopes to a drain or is configured for efficient application of irrigation water.

Land leveling alters the depth to soil horizons relative to the original soil surface and, thus, may cause unfavorable surface conditions after land leveling, depending on the nature of the surface horizons [2]. Deep tillage with the application of N- and P-rich fertilizers may be required following land leveling to help alleviate the poor soil physical

conditions. In addition, land leveling is to be maintained routinely because tillage or inappropriate erosive discharges disturb land leveling and the uniformity of water distribution.

The major problem with land leveling is the effect of removing the topsoil and its subsequent influence on plant growth. Reduced growth may occur on the fill areas, although the exposure of subsoil in the cuts is usually a more serious problem [3]. In some less common instances, all of the topsoils in the area to be graded are scrapped off and piled, and topographic variations at the top of the subsoil are smoothed out, and the topsoils spread back on the manipulated area [1, 4]. In this case the cost can be justified.

Land leveling improves uniformity of crop growth and yield. It conserves soil and water by creating slight, but uniform, slope gradients to improve drainage, to drive irrigation water across the field with a suitable flow velocity without erosion, to facilitate more even distributions of irrigation water, to improve the effectiveness of surface irrigation, and so forth. Besides some of the mentioned benefits, land leveling also has some disadvantages. First of all, land leveling is a severe soil disturbance that disrupts or alters the entire equilibrium among near surface soil properties.

The spatial variability, distributions, changes, and relationships in soil physical, chemical, and biological properties as a result of land leveling were reported mostly by Brye et al. in a few studies [1, 2, 4–9]. Bulk density, sand, and clay contents significantly increased, while silt content significantly decreased, and the variances of soil physical properties were unaffected due to land leveling in a Stuttgart silt loam soil [1]. However, in another study by Brye et al. [8] at the same soil, increases in bulk density and clay percentage and decreases in sand and silt percentages were found. In addition, soil pH and EC values and the variance of soil pH in the top 10 cm increased due to land leveling [9]. In another study by Brye et al. [7], there was a significant increase in the EC values in the top 10 cm, while, significant decrease in the soil pH values due to land leveling was reported. Furthermore, Ramos et al. [10] reported that the changes in the particle size distribution of the fine fraction due to land terracing affected hydraulic conductivity, water retention capacity, and aggregate stability, as well as the relationships between all these variables.

To our knowledge, few studies have reported the effects of land leveling on the soil physical properties of field capacity, permanent wilting point, and water holding capacity in clay loam soils. Evaluations about the changes and variability in soil properties in this study will improve management capabilities to ensure maximum production from the graded fields at clay loam soils. Furthermore, we are going to try to relate the changes in soil properties to the process of grading through analyses of cut and fill areas separately. This may help engineers decide whether the designed leveling projects with determined depths of cut and fill are also appropriate in the sense of soil physical properties, and readers generalize the results to other fields and types of leveling.

The objective of this study was to analyze the short-term effects of land leveling on the magnitudes, variances, spatial variability, and distributions of surface (0–20 cm) and subsurface (20–40 cm) soil properties of bulk density (BD), field capacity (FC), permanent wilting point (PWP), water holding capacity (WHC), particle size fractions, pH, and soil electrical conductivity (EC) in a clay loam soil.

2. Materials and Methods

2.1. Site Description and Experimental Design. A 1.2 ha (120 × 100 m) field, previously cropped to wheat, on Channel clay loam soil [11] at the Agricultural Research and Application Center of Gaziosmanpaşa University in Tokat, Turkey (40°20′N lat., 36°28′E long., elevation = 600 m from mean sea level) was chosen as the study site in Fall 2010. The study area is located at a junction between colluvial and alluvial materials. The channel clay loam has a deep soil profile, low permeability, and poor drainage. This soil is typically located on the low terraces of Yeşilırmak River and classified as an ustorthent and ustifluvent according to the soil taxonomy [12] for surface and subsurface soils, respectively. Before land leveling, the research field was slightly nonuniform sloped (<3%) from north to south, and almost leveled (<0.3%) in the west-east direction. In general, the research area has been used for the research experiments of Agricultural Faculty of Gaziosmanpaşa University to grow forage and cereal plants irrigated by surface irrigation methods. In order to facilitate statistical evaluation of the effects of land leveling on the magnitude, variability, and spatial distribution associated with soil properties, a 120 m wide by 100 m long study area covering a 30-point grid was established with sampling points spaced evenly 20 m apart just before the land leveling. The grid-sampling approach used by previous studies by Brye et al. [1, 4, 7, 8] was employed in this study to characterize the short-term impacts of land leveling. The average annual rainfall of Tokat city is 443 mm, mainly distributed in winter, spring, and autumn; and the average annual temperature is 12.5°C.

Considering both irrigation and drainage methods to be applied on the study area in the near future, the plane method [3] to determine the amounts in depths of cut or fill at each grid was employed. The slopes of the study area in both north-south and east-west directions in this method are determined by the least-squares procedure presented by Chugg [13]. Figure 1 shows the determined cut and fill grid square areas with the computed depth values of cuts and fills in centimeter. From Figure 1, the cut areas or grids were located along the South (except south-east corner), East (almost two lines along north-south direction), and west borders of the study area. While total 13 grids of 30 were determined as fill areas, 16 grids were determined as cut areas. The fill depths were ranged between 1.0 and 48.5 cm, and the cut depths were ranged between 0.5 and 50.0 cm. According to the cut and fill depths given in Figure 1 for each grid, the computed cut-fill ratio for the study area is 1.25. This computed ratio of the sum of the cuts to the sum of the fills is in the limits (1.2–1.5) recommended by James [14] and Schwab et al. [3]. After the land leveling by heavy carrier-type scraper powered with rubber tires, a 2.9% uniform slope from north to south, and 0.29% uniform slope from west to east at the study area were obtained.

2.2. Soil Sampling and Measurements. Before the leveling, the 30 grid centers (20 m apart) at the study area were setup with stakes by using a theodolite, level rods, and steel tapes in the mid of November 2010. The elevations at these centers were also measured by using these theodolite and level rods. The preleveling soil samples from both surface (0–20 cm) and subsurface soils (20–40 cm) at each of the 30 grid centers in the study area were obtained just after setting up the grids just before the land leveling activities. The land leveling activities were performed over a 2-day period (December 5-6, 2010). To characterize the changes in soil properties as a result of the land leveling, the post-leveling soil samples from both surface and subsurface soils at each of the 30 grid centers were obtained the mid of December, 2010.

Bulk soil samples to determine the particle size distributions, FC, PWP, WHC, pH, EC, and 100 cm^3 (5 cm high by 5.1 cm diameter) undisturbed core samples to determine bulk densities (BDs) of surface and subsurface soils at each grid centers were used in this study. The core samples were taken vertically with a double-cylinder, hammer-driven core sampler. Bulk densities were determined as described by Blake and Hartge [15]. Bulk samples were air-dried, manually

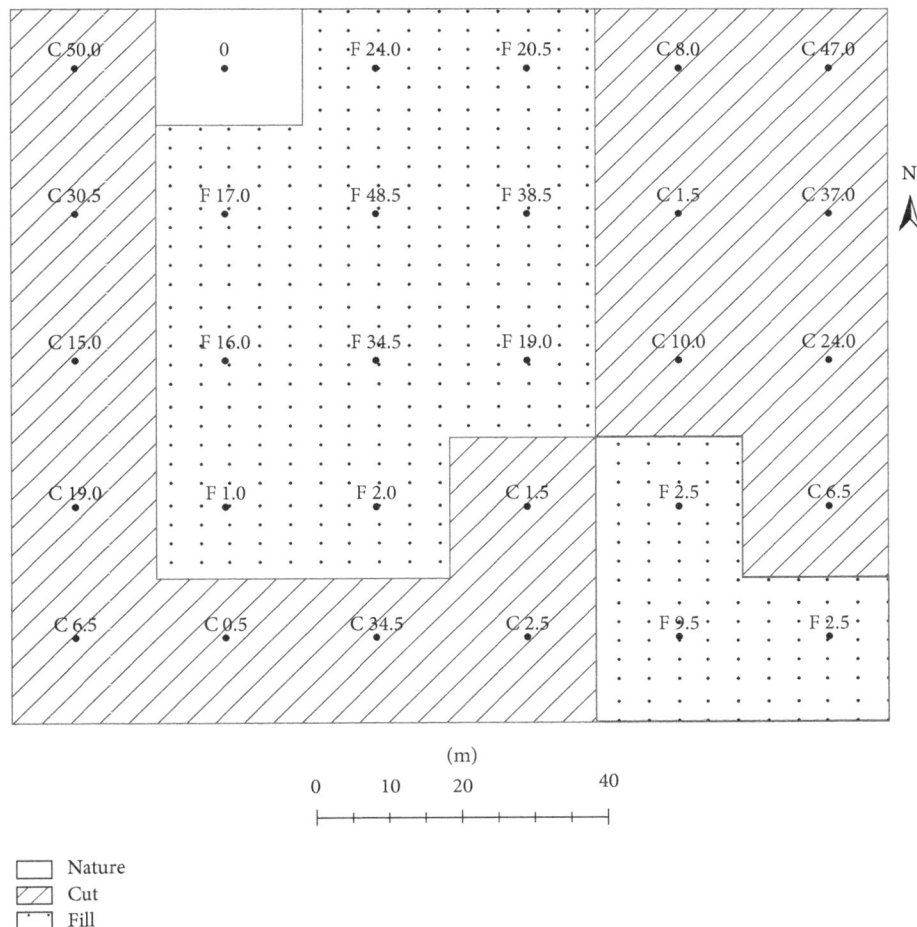

FIGURE 1: The grids of cuts or fills of the study area for land leveling (the unit of cut or fill depth in each grid is in centimeter).

grounded, and sieved to remove coarse fragments >2 mm. The analysis of particle size distribution on the soil samples passing through 2 mm mesh screen was performed using the hydrometer method [16]. Soil water contents at the suctions of field capacity (FC) (1/3 atm.) and permanent wilting point (PWP) (15 atm.) were determined in the laboratory by ceramic plates following the procedures by Klute [17]. Soil pH and EC were determined with an electrode on a 1 : 2 soil/water solution. To determine the water holding capacity (WHC) for each soil sample, (1) was used:

$$\text{WHC} = (\text{FC} - \text{PWP}) \times \text{BD} \times D, \qquad (1)$$

where WHC is the water holding capacity of 2 dm depth of soil layer (mm), FC is the soil water content at the suction of field capacity (1/3 atm.) (% by weight), PWP is the soil water content at the suction of permanent wilting point (15 atm.) (% by weight), BD is the soil bulk density (g cm^{-3}), and D is the soil depth (=2 dm).

2.3. Data Analysis. Analysis of variance and subsequent statistical testing have been a common methods for comparing variations for land leveled fields. The surface subsurface soil properties (BD, sand, silt, and clay contents, pH, EC, FC,

PWP, and WHC) belonging to the study area just before and after land leveling were reported as mean, standard error (SE), coefficient of variation (CV), and variance.

Paired t-tests were performed to determine whether differences in the soil properties before and after land leveling are significant or not. Therefore, the results of this test shows the overall short-term effects of land leveling on the soil properties considered in this study. Employing the paired t-test, the means of soil properties were separated by least significant differences (LSDs) at the 0.001, 0.01, and 0.05 levels. For paired t-test, Minitab package program (Minitab version 12.1, Minitab, Inc., State College, PA, USA) was used. In this analysis, it was assumed that the paired differences follow a normal distribution. The effects of land leveling on the variability of the studied soil properties were also determined by Levene's test [18]. Levene's test is an inferential statistic used to assess the equality of variances in different samples. In this study, it was tested that the pre- and postleveling variances of soil properties are equal (called homogeneity of variance). Therefore, to determine the effect of land leveling on the sample variance, homogeneity of variance was evaluated using Levene's test. Variances were separated by LSD at the 0.05 level. For Levene's test, also Minitab package program was used.

The effects of land leveling on the spatial variability of the studied soil properties were determined by geostatistical analyses. In order to determine the effects of land leveling on statistical distribution of soil properties and on the spatial relationships among soil properties before and after land leveling, geostatistical analysis can give useful information. Geostatistical analyses were conducted using GS+ (version 7.0, Gamma Design Software, Plainwell, MI). Only isotropic semivariograms were considered, and semivariance parameters of the best-fitting (i.e., smallest residual sum of squares, and the highest coefficient of determination-r^2) Gaussian, linear, spherical, or exponential models were reported for pre- and postleveled soil properties. The effects of land leveling on the spatial distributions and variability of soil properties (BD, sand, silt, clay contents, pH, EC, FC, PWP, and WHC) were analyzed by mapping pre- and post-leveling values of these properties for both surface and subsurface soils.

In order to explain the effects of leveling on the water holding capacities of both surface and subsurface soils of both cut and fill areas, the relationships between the changing amounts of the soil physical properties and the depths of cuts and fills through cut and fill areas were also analyzed separately by employing simple linear regression analysis.

3. Results and Discussion

In this study, the land leveling resulted in an average surface elevation change of −2.5 cm (i.e., an overall cut), ranging from +18.1 cm (i.e., average fill) to −18.4 cm (i.e., average cut) across the 1.2 ha study area (Figure 1). According to the cut volume per ha area (1220 m³ ha⁻¹), the performed land leveling operation can be classified as a heavy leveling [19].

3.1. Changes in Soil Characteristics. According to the paired t-test results, land leveling significantly affected the magnitudes of soil parameters of BD, pH, FC, PWP, and WHC for surface soil (Table 1). Land leveling also significantly affected the magnitudes of BD, clay content, pH, EC, PWP, and WHC for subsurface soil. Before the land leveling, the bulk density of surface soil ranged from 1.14 to 1.67 g cm⁻³ and averaged 1.40 (SE = 0.03 and CV = 11.01) g cm⁻³ over the 1.2 ha sampling area (Table 1). After leveling, it ranged from 1.44 to 1.85 g cm⁻³ and averaged 1.68 (SE = 0.02 and CV = 5.47) g cm⁻³. The increase in bulk density due to leveling is significant at 0.001 level. These increases in bulk density are probably due to exposure of subsoil (alluvial material) with high sand contents thereby having inherently high bulk density, soil compaction by the scraper, and arriving to a hardpan. Furthermore, land leveling decreased the overall variability of bulk density (ΔCV = −50%) in the surface soil (Table 1). Similar results with an increase of 0.08 g cm⁻³ in mean value of bulk density and a decrease of 38% in ΔCV after land leveling were also observed for subsurface soil. These kinds of increases for bulk density were also reported in the studies by Brye et al. [1, 8]. A deep tillage may be

required following land leveling to help alleviate this poor soil condition.

When the mean values of sand, silt, and clay contents in Table 1 were used to determine soil textural classes of surface and subsurface soils before and after land leveling, the preleveled data of both surface and subsurface, and postleveled surface soils were classified as clay loam. However, the postleveled subsurface soil was classified as sandy clay loam. This change is due to probably increases of sand contents and decreases of clay contents in the deep horizons of study site soil. Furthermore, while any significant changes in soil particle size distributions of surface soil were not determined, a small decrease in clay percentage (2%) for subsurface soil was found significant ($P \leq 0.05$). The mean sand content of subsurface soil also increased (about 3%) due to land leveling; however, this increase was not found to be significant. Textural class change of subsurface soil due to land leveling was further investigated by illustrating the sand, silt, and clay contents at each grid center on the soil texture triangle in Figure 2. From this figure, it is clear that especially four yellow-colored dots on the left bottom corner of triangle for postleveled soil are located apart from the rest of groups. The sand contents of these samples are higher, and the clay contents are lower than those of the rest of samples. Most probably the changes in textural classes of subsurface soil due to land leveling are because of these data of these samples. These samples were located in the cut areas.

Increases in the pH values due to land leveling for both surface and subsurface soils are also significant ($P \leq 0.001$) (Table 1). This result is different than that by Brye et al. [7]. But these increases (0.2 and 0.3 unit for surface and subsurface soils, resp.) can be considered relatively small and agronomically not significant. Furthermore, variability in the pH values was increased due to land leveling. In addition, the decrease (0.03 dS m⁻¹) of EC mean for subsurface soil was also found significant ($P \leq 0.05$) (Table 1).

According to the analysis of paired t-test, the decreases in FC and PWP, the increase in WHC values for surface soil, the decrease in PWP, and the increase in WHC values for subsurface soil were found to be significant due to land leveling. Before land leveling, FC values of surface soil ranged from 21.8 to 31.0% and averaged 26% (SE = 0.37 and CV = 7.77) (Table 1). After leveling, FC values for surface soil ranged from 19.8 to 30% and averaged 24.9% (SE = 0.53 and CV = 11.56). The decrease (1.1%) in FC due to the leveling was found significant ($P \leq 0.05$) according to the analysis of paired t-test. Furthermore, the land leveling increased the overall variability of FC (ΔCV = 49%) in the surface soil. When we consider the mean values of FC after land leveling, the FC values for both surface and subsurface soils became close to each other (difference 0.5%). On the other hand, the PWP values of preleveling surface soil ranged from 14.2 to 23.5% and averaged 17.8% (SE = 0.45 and CV = 13.96). After the land leveling, the PWP values for the surface soil ranged from 10.9 to 20.3% and averaged 15% (SE = 0.35 and CV = 12.78). The decrease (2.8%) in PWP due to the leveling is significant at 0.001 level. The land leveling decreased the overall variability of PWP (ΔCV = −8.5%) in the surface soil. For subsurface

TABLE 1: The effect of land leveling on some selected soil properties from both 0–20 cm (surface soil) and 20–40 cm (subsurface soil). Means, standard errors, coefficients of variation, and the changes in coefficients of variation (ΔCV) of pre- and postleveling soil properties for both surface and subsurface soils were reported. Asterisks next to postleveling means represent significant differences between pre- and postleveling measurements caused by land leveling. These asterisks were determined after paired t-test ($n = 30$).

| Soil parameter | Surface soil | | | | | | | Subsurface soil | | | | | | |
| | Preleveling | | | Postleveling | | | ΔCV[+] (%) | Preleveling | | | Postleveling | | | ΔCV (%) |
	Mean	S. error	CV_{pre}	Mean	S. error	CV_{post}		Mean	S. error	CV_{pre}	Mean	S. error	CV_{post}	
BD (g cm^{-3})	1.40	0.03	11.01	1.68[***]	0.02	5.47	−50	1.59	0.02	7.67	1.67[***]	0.01	4.73	−38
Sand (%)	43.88	1.37	17.16	44.36	1.17	14.50	−16	44.00	1.42	17.73	46.88	2.24	26.22	48
Silt (%)	28.33	0.98	18.87	28.33	0.59	11.39	−40	28.02	0.87	17.07	27.44	1.19	23.83	40
Clay (%)	27.79	0.94	18.46	27.31	0.72	14.36	−22	27.72	0.85	16.74	25.67[*]	1.29	27.54	65
pH	8.02	0.01	0.74	8.21[***]	0.02	1.08	45	7.92	0.02	1.12	8.21[***]	0.03	1.82	63
EC (dS m^{-1})	0.187	0.01	18.66	0.198	0.01	21.90	17	0.246	0.01	23.49	0.216[*]	0.01	27.37	17
FC (%wt)	26.0	0.37	7.77	24.9[**]	0.53	11.56	49	25.1	0.46	9.99	24.4	0.80	17.94	80
PWP (%wt)	17.8	0.45	13.96	15.0[***]	0.35	12.78	−8.5	16.8	0.48	15.80	15.0[**]	0.64	23.33	48
WHC (mm)	22.7	0.76	18.40	33.1[***]	1.19	19.66	6.8	26.2	0.87	18.25	31.4[***]	1.06	18.45	1.1

+: ΔCV $= 100 \times [CV_{post} - CV_{pre}]/CV_{pre}$.
*: $0.01 < P \leq 0.05$.
**: $0.001 < P \leq 0.01$.
* * *: $P \leq 0.001$.

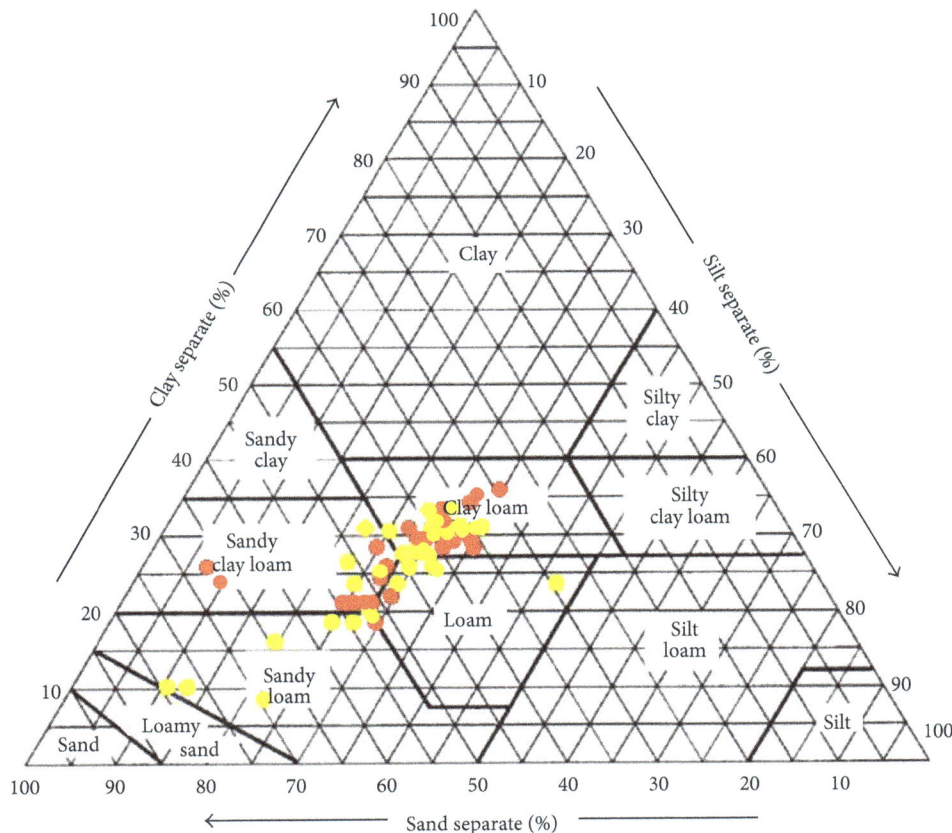

FIGURE 2: Distributions of the subsurface soil texture of study area before (red dot) and after (yellow dot) land leveling.

soil, preleveling PWP values ranged from 10.7 to 22.7% and averaged 16.8% (SE = 0.48 and CV = 15.80). After the land leveling, the PWP values of subsurface soil ranged from 7.1 to 20.7% and averaged 15% (SE = 0.64 and CV = 23.33). The decrease (1.8%) in the PWP values of subsurface soil due to the leveling was also found significant ($P \leq 0.05$). In contrast to the result for the surface soil, land leveling increased the overall variability of PWP (ΔCV = 48%) in the subsurface soil. Similar to the FC, after the leveling, the mean PWP values of surface and subsurface soils became equal. When we consider

TABLE 2: The effects of land leveling on the sample variances of some selected soil properties from both 0–20 cm (surface soil) and 20–40 cm (subsurface soil). Asterisks next to postleveling variance values represent significant differences in the sample variances of pre- and postleveling soil properties caused by land leveling. The asterisks were obtained after employing Levene's test ($n = 30$).

Soil parameter	Surface soil		Subsurface soil	
	Preleveling	Postleveling	Preleveling	Postleveling
BD ($g\,cm^{-3}$)	0.024	0.008*	0.015	0.006*
Sand (%)	56.7	41.3	60.8	151.0
Silt (%)	28.6	10.4	22.9	42.8
Clay (%)	26.3	15.4	21.5	50.0
pH	0.004	0.008*	0.008	0.022*
EC ($dS\,m^{-1}$)	0.001	0.002	0.003	0.003
FC (%wt)	4.07	8.30*	6.26	19.13
PWP (%wt)	6.17	13.69	7.06	12.20
WHC (mm)	17.42	42.25*	22.89	35.52

*: $P \le 0.05$.

TABLE 3: Summary of geostatistical parameters for surface soil (0–20 cm) properties measured before and after land leveling.

Soil parameter		Preleveling						Postleveling				
	Model	Nugget (C_0)[†]	Sill ($C_0 + C$)[†]	Range (m)	$C/(C_0 + C)$	r^2	Model	Nugget (C_0)	Sill ($C_0 + C$)	Range (m)	$C/(C_0 + C)$	r^2
BD ($g\,cm^{-3}$)	Gaussian	0.0092	0.0262	58.5	0.651	0.98	Linear	0.0085	0.0085	74.6	0.000	0.46
Sand (%)	Gaussian	38.90	98.95	183.4	0.607	0.94	Linear	31.03	40.12	74.6	0.227	0.46
Silt (%)	Linear	25.45	32.03	74.6	0.206	0.42	Linear	8.36	10.72	74.6	0.220	0.30
Clay (%)	Spherical	3.44	30.23	83.0	0.886	0.99	Spherical	0.20	14.13	28.0	0.986	0.39
pH	Linear	0.004	0.004	74.6	0.000	0.00	Spherical	0.00	0.0077	23.3	0.999	0.02
EC ($dS\,m^{-1}$)	Spherical	0.0005	0.0014	69.4	0.609	0.84	Linear	0.00	0.0016	20.0	0.000	0.06
FC (%wt)	Gaussian	1.66	5.015	73.3	0.668	0.99	Gaussian	0.01	8.35	35.2	0.999	0.87
PWP (%wt)	Spherical	0.01	8.09	80.2	0.999	0.99	Spherical	0.01	3.54	31.5	0.997	0.56
WHC (mm)	Gaussian	8.86	36.15	158.7	0.755	0.93	Gaussian	10.7	68.78	98.0	0.844	0.99

†: C_0 represents the inherent nonspatially related variability in the data, and C represents the variability explained by any spatial component in the data.

the WHC values which are functions of FC, PWP, and BD, the increases (10.4 and 5.2 mm for surface and subsurface soils, resp.) due to the land leveling were found to be significant at 0.001 level. The land leveling did not make a big change on the overall variability of WHC ($\Delta CV = 6.8\%$ and $\Delta CV = 1.1\%$ for surface and subsurface soils, resp.).

In order to measure variability due to the land leveling, the variances of soil groups before and after land leveling were employed. As indicated, the significant changes for pre- and postleveling mean values of BD, pH, FC, PWP, and WHC for the surface soil according to the paired t-test (Table 1), the sample variance (σ) values of, same properties except PWP, were also significantly affected ($P \le 0.05$) by the land leveling (Table 2) according to the Levene's test. While land leveling decreased the overall variability of surface soil BD, it increased the overall variability of soil pH, FC, and WHC. Among BD, clay content, pH, EC, PWP, and WHC of the subsurface soil, only the variance values of BD and pH were significantly affected by land leveling. Similar to the results for the surface soil, the land leveling decreased the overall variability of BD of subsurface soil and it increased the overall variability of soil pH. The increases in the sample variance of pH were also reported by Brye et al. [7] and Brye [9].

As a result, according to the Levene's test, land leveling significantly affected the variance of BD, pH, FC, and WHC of the surface soil and BD and pH of the subsurface soil in this study.

3.2. Changes in Spatial Variability and Distributions. Based on the geostatistical analyses, the best-fit semivariogram models of the surface soil were changed for BD, sand content, and EC due to the land leveling (Table 3). Before the leveling, the best-fit models for surface soil parameters, except silt content, tended to have high predictive capability, where all r^2 values were ≥ 0.84. The best-fitting model efficiencies (r^2 values) decreased for all soil parameters (except WHC) due to leveling. After the leveling, except the FC, PWP, and WHC, the best-fit models also tended to have low predictive capability ($r^2 \le 0.46$). The Gaussian and spherical models best characterized the structures of semivariograms for surface soil FC, WHC, PWP, respectively (Table 3). Both before and after the land leveling, the model fit was very poor for the pH of the surface soil ($r^2 \le 0.02$). In addition, the model fit for the EC of postleveled surface soil was also very poor.

Table 4: Summary of geostatistical parameters for subsurface soil (20–40 cm) properties measured before and after land leveling.

Soil parameter	Preleveling						Postleveling					
	Model	Nugget $(C_0)^\dagger$	Sill $(C_0 + C)^\dagger$	Range (m)	$C/(C_0 + C)$	r^2	Model	Nugget (C_0)	Sill $(C_0 + C)$	Range (m)	$C/(C_0 + C)$	r^2
BD (g cm^{-3})	Gaussian	0.0054	0.019	69.9	0.718	0.97	Linear	0.006	0.0063	74.6	0.051	0.01
Sand (%)	Spherical	23.50	65.82	68.8	0.643	0.91	Linear	92.59	152.9	74.6	0.394	0.67
Silt (%)	Linear	16.87	25.81	74.6	0.347	0.83	Linear	27.4	48.02	74.6	0.435	0.88
Clay (%)	Spherical	5.02	21.71	55.0	0.769	0.84	Linear	33.31	46.27	74.6	0.280	0.43
pH	Spherical	0.003	0.009	75.6	0.670	0.96	Linear	0.0232	0.0232	74.6	0.000	0.45
EC (dS m^{-1})	Gaussian	0.0018	0.0066	151.6	0.723	0.99	Linear	0.0038	0.0038	74.6	0.000	0.93
FC (%wt)	Gaussian	2.44	9.339	90.6	0.739	0.99	Linear	11.52	20.16	74.6	0.429	0.80
PWP (%wt)	Gaussian	1.50	11.21	92.5	0.866	0.99	Spherical	4.79	12.60	73.1	0.620	0.81
WHC (mm)	Gaussian	11.66	26.82	93.4	0.565	0.95	Linear	25.30	36.28	74.6	0.302	0.57

†: C_0 represents the inherent nonspatially related variability in the data, and C represents the variability explained by any spatial component in the data.

The obtained high nugget values (C_0) for the sand and silt contents and WHC (Table 3) indicate microscale effects and some sampling and measurement errors. As indicated by Isaaks and Srivastava [20], the sill values ($C_0 + C$) are also the variance values (Table 2) of random soil properties. Therefore, the sill values (Table 3) and variance values (Table 2) of soil properties are parallel to each other. In general, the range parameters from the best-fit semivariogram models for the surface soil properties were large (>58 m) before the leveling, indicating spatial autocorrelation among the sampling points at the 20 m spacing, and the data were not truly independent within the study area (Table 3). After the leveling, the range parameter for bulk density increased (28%). In other words, the homogeneity of bulk density across the study area increased after the land leveling. This means that after the land leveling, the bulk density is not changing more rapidly within the study area than it did before the leveling. A similar result about increase of bulk density after leveling was also stated by Brye et al. [8]. Furthermore, the range parameter for silt did not change after the leveling (Table 3). The range parameters decreased for other soil properties after the leveling. The highest decreases or changes in the range values were seen for the EC, pH, PWP, clay, and sand contents (>58%). Though the best-fit semivariogram models did not change after the leveling, the range parameters decreased about 52% and 61% for FC and PWP, respectively. The clay content, pH, EC, FC, and PWP achieved spatial independence (i.e., the range parameter < 36 m) within the study area due to land leveling (Table 3). These changes may be the causes of immediate disruption of a previous quasi-equilibrium, in which land leveling activities imparted on the soil can have lasting negative effects on soil properties [1, 4, 7, 8].

For surface soil, while the PWP had a significant spatial component (C) before land leveling, PWP also had a significant spatial component with a little bit decreasing significance order after land leveling (Table 3). Another significant spatial component was observed for the FC of surface soil after the land leveling. The proportion $C/(C_0 + C)$, representing inherent variability in the data,

explained the highest variability (99.9%) in PWP and no variability in pH (0%) of the surface soil before the leveling. The propor-tion values of other preleveling surface soil properties, except the silt content, are high (>60%). These results are parallel to the values of r^2 (predictive capability of the best-fit model). After leveling, the 20 m spacing appeared too large to ascertain a spatial dependency for the BD and EC within the sampling area. Furthermore, after the leveling, the inherent variability values explained also large fractions of the total variability in the postleveled soil properties as it did for the soil properties of clay content, pH, FC, PWP, and WHC before the leveling.

The best-fit semivariogram models of subsurface soil were changed for all soil properties, except silt content, due to land leveling (Table 4). After leveling, the best-fit semivariogram models were linear for all the soil properties, except PWP (spherical). The best-fit models for the subsurface soil properties tended to have high predictive capability ($r^2 \geq 0.83$) before leveling. The best-fitting model efficiencies (r^2 values) decreased for the soil parameters (except silt content) due to leveling. The model fit for the BD of postleveled subsurface soil was very poor ($r^2 = 0.01$). After leveling, except the clay content, and pH, the best-fit models also tended to have high predictive capability ($r^2 \geq 0.57$). The model fits were poor for the clay content and pH of the postleveled subsurface soil ($r^2 \leq 0.45$). While the Gaussian model best characterized the structures of semivariograms for the subsurface soil FC, WHC, and PWP before leveling, the linear and spherical models best characterized the structures of semivariograms for the postleveled subsurface soil FC, WHC, and PWP, respectively.

Similar to those of the surface soil, the high nugget values (C_0) for the sand and silt contents and WHC of subsurface soil before leveling; for the sand, silt and clay contents FC, and WHC of the postleveled subsurface soil were also obtained (Table 4). Both pre- and post-leveling range values from the best-fit semivariogram models for the subsurface soil properties are large (≥ 55 m). Unlike to the surface soil, the land leveling did not cause much spatial independence to the subsurface soil properties. After leveling, small increases in

Preleveling   Postleveling

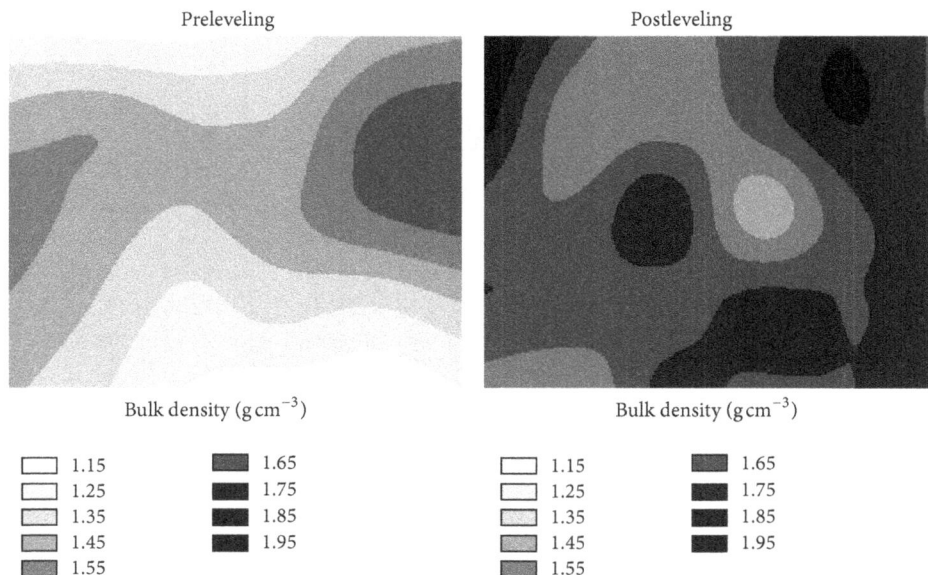

Bulk density (g cm^{-3})   Bulk density (g cm^{-3})

☐ 1.15	■ 1.65
☐ 1.25	■ 1.75
☐ 1.35	■ 1.85
☐ 1.45	■ 1.95
☐ 1.55	

FIGURE 3: Pre- and post-leveling spatial distributions of soil bulk density for surface soil (0–20 cm).

the range values of BD and sand and silt contents and small decreases in the range values of FC, PWP, and WHC were obtained. The decrease in the range value of EC is high (51%). The range parameters for the silt and pH did not change after leveling (Table 4).

The properties of subsurface soil did not have any significant spatial component (C). Before leveling and the proportion values, $C/(C_0 + C)$, explained between 56 and 87% of the variation in soil properties, except silt content (35%) (Table 4). In contrast, the 20 m spacing appeared too large to ascertain a spatial dependency for the BD, pH, and EC of the postleveled subsurface soil within the sampling area. Due to leveling, the amount of variation in the FC and WHC, explained by the proportion column, decreased to 43 and 30%, respectively. The amount of variation explained by the proportional spatial component following leveling also decreased by 28% for the PWP.

The spatial distributions of all soil properties measured in the both surface (0–20 cm) and subsurface (20–40 cm) soils were altered by land leveling. While the areas that were initially cut and the areas that were initially fill were given in Figure 1, they also can be observable from the figures of spatial distributions of soil properties with the help of Figure 1. However, it will be difficult to associate exactly the changes according to the initial cut and fill areas.

Almost the all locations within the study area with the low bulk densities before leveling had noticeably high bulk densities after leveling (Figure 3). In general, the cut areas before and after leveling had higher bulk densities than that of fill areas due to most probably exposure of the subsoil that had higher bulk density than that of the original surface soil, soil compaction by machinery, and arriving to a hardpan. Similar results are valid for the bulk densities of subsurface soil (spatial distribution map of bulk density for subsurface soil was not given in order to save space in the paper).

Changes in the spatial distributions of sand, silt, and clay contents of the surface soil also occurred (Figure 4). Some locations with high sand, silt, and clay contents became with slightly low sand, silt, and clay contents, or vice versa. Especially the sand content increases are high in the fill areas. While the silt contents of the couple of locations of study area decreased after leveling, in general, small increases in silt contents at the majority of other locations were seen. After leveling, the spatial distribution of silt content of the study area became more homogeneous as it was also indicated by the CV values in Table 1. Furthermore, a small increase in the homogeneity of clay content spatial distribution over the study area after leveling was also noticed. In general, the clay contents of the fill areas decreased (Figure 4). As a result, small, not significant alterations on the soil particle size fractions in the surface soil were noticed in this study. This is most probably due to exposure of and mixing with subsoil with similar particle size fractions of surface soil. This result is different to those of Brye et al. [1, 8]. Increases in the sand contents especially on three grid centers of the cut areas and not much change of the sand contents in the fill areas were observed for the subsurface soil after leveling (not shown). After leveling, small decreases in the clay contents of the subsurface soil at the cut areas were also noticed.

The spatial distribution maps of pH and EC of the surface soil were altered to some degree by the land leveling (Figure 5). The pH values of the postleveled surface soil for almost all locations increased. In addition, the spatial distribution of pH values at the middle part of the fill areas seems to have the lowest pH values after the leveling. Similar increases of the pH values after the leveling for the subsurface soil were also observed (not shown). As a result of the land leveling, the soil of the study area became slightly more alkaline. The increases or decreases of pH for surface soil after leveling were also stated by Brye et al. [7, 8]. [The authors reported that exposing and mixing alkaline surface soil with

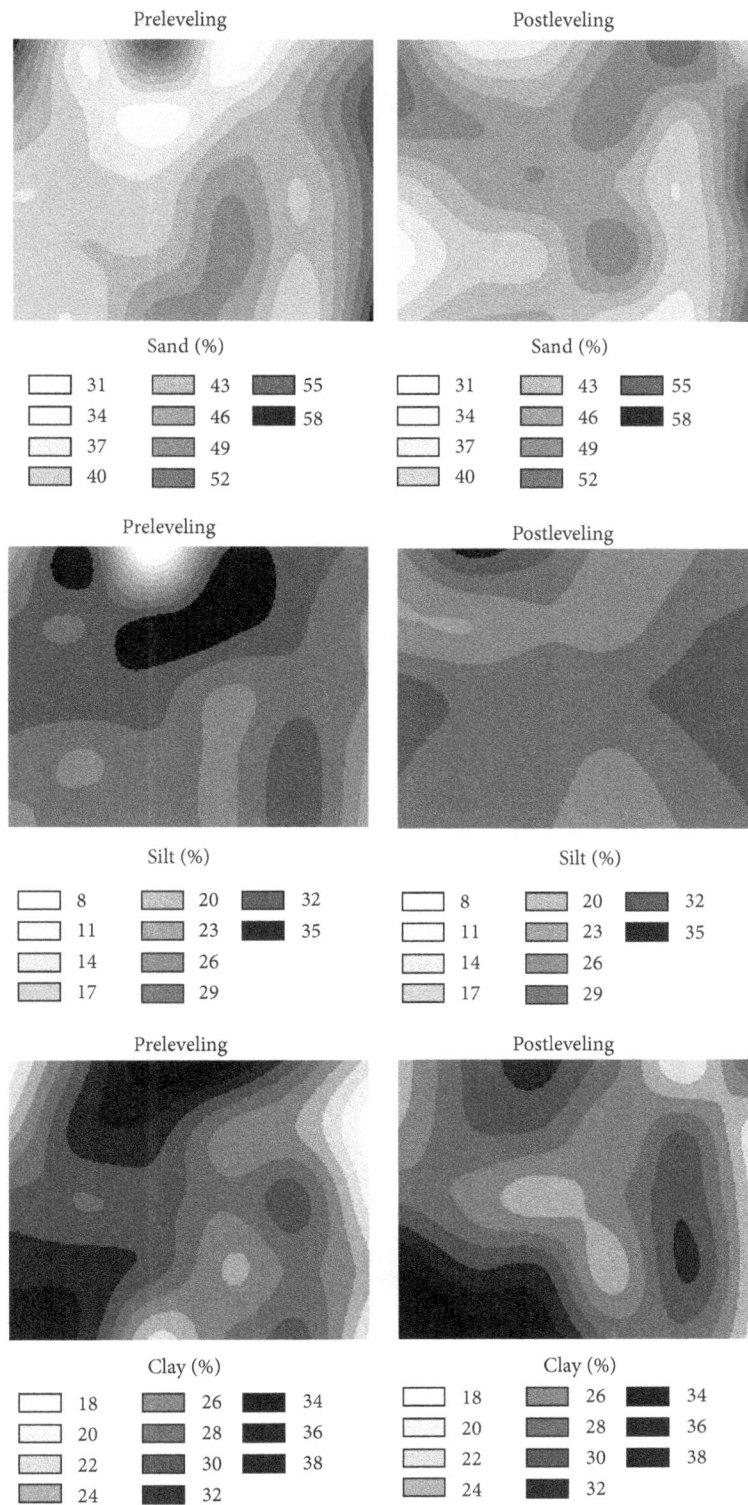

FIGURE 4: Pre- and post-leveling spatial distributions of soil particle-size fractions in surface soil (0–20 cm).

Preleveling   Postleveling

pH (%)   pH (%)

☐ 7.75   ▨ 8.00   ■ 8.25   ☐ 7.75   ▨ 8.00   ■ 8.25
☐ 7.80   ▨ 8.05   ■ 8.30   ☐ 7.80   ▨ 8.05   ■ 8.30
☐ 7.85   ▨ 8.10   ■ 8.35   ☐ 7.85   ▨ 8.10   ■ 8.35
☐ 7.90   ▨ 8.15   ■ 8.40   ☐ 7.90   ▨ 8.15   ■ 8.40
☐ 7.95   ▨ 8.20   ■ 8.45   ☐ 7.95   ▨ 8.20   ■ 8.45

Preleveling   Postleveling

EC (dS m^{-1})   EC (dS m^{-1})

☐ 0.12   ▨ 0.22   ■ 0.32   ☐ 0.12   ▨ 0.22   ■ 0.32
☐ 0.14   ▨ 0.24   ■ 0.34   ☐ 0.14   ▨ 0.24   ■ 0.34
☐ 0.16   ▨ 0.26   ■ 0.36   ☐ 0.16   ▨ 0.26   ■ 0.36
☐ 0.18   ■ 0.28   ☐ 0.18   ■ 0.28
☐ 0.20   ■ 0.30   ☐ 0.20   ■ 0.30

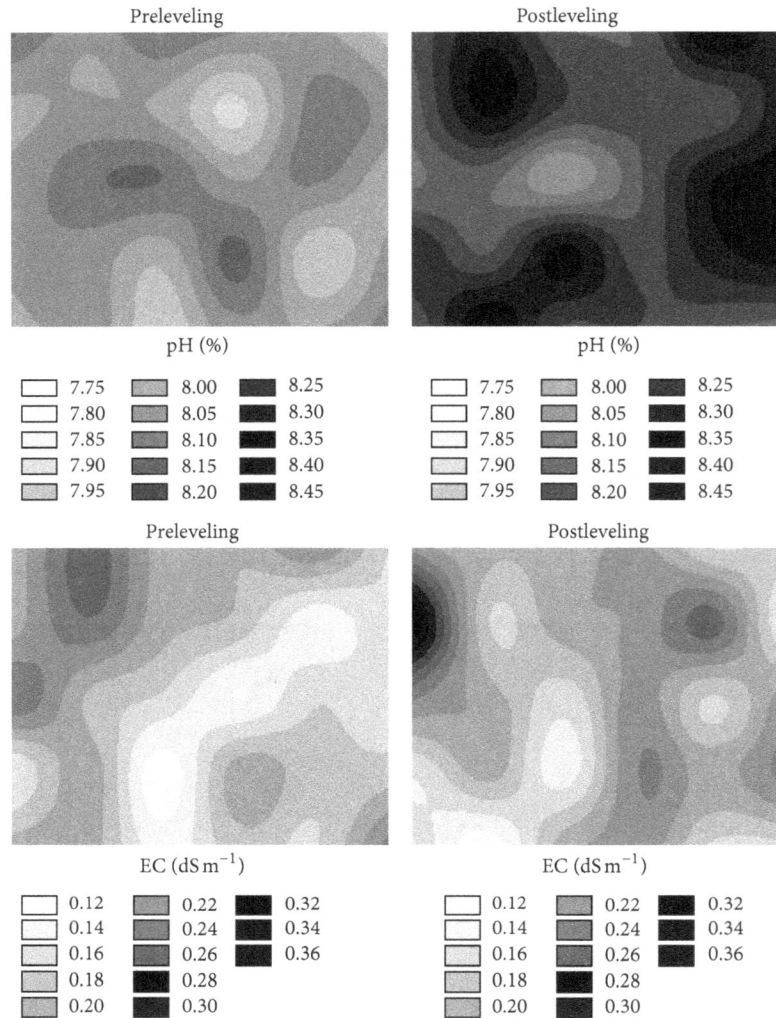

FIGURE 5: Pre- and post-leveling spatial distributions of soil pH, electrical conductivity (EC), and available water holding capacity (AWHC) in surface soil (0–20 cm).

the typically acidic or alkaline subsoil as reasons for these changes on the pH.] On the other hand, small increases or decreases in the EC values on some locations of the study area caused differences in the spatial distribution of the EC values for the surface soil. In general, the low EC values were located in the fill areas. On the other hand, in general, the EC values of the subsurface soil over the study area, except three locations or grids, decreased after the leveling (not shown). These results of EC changes after the leveling are different from those by Brye et al. [7, 8].

The spatial distributions of the FC, PWP, and WHC of both the surface and subsurface soils also changed as a result of land leveling (Figures 6 and 7). The distribution of FC values over the study area before leveling was much more homogeneous than that after leveling (Figure 6). After the leveling, the FC values for almost all locations, except a couple of locations on the south-west portion of the study area, slightly decreased. It seems that the decreases in the fill areas were higher than those from the cut areas. After leveling, the PWP values of the surface soil over almost all

locations of the study area decreased (Figure 6). The reasons for these decreases of both FC and PWP values can be the slight decreases of clay contents and slight increases of sand contents of surface soil after leveling. In contrast to the FC and PWP values, the WHC values increased over almost all grids of the research area after leveling (Figure 6). It looks like the fill areas located in the middle part of the research area had the lowest WHC values after leveling. As it can be inferred from (1), WHC is a function of FC, PWP, and BD. Therefore, the increases of WHC are due to high differences between the FC and PWP values with the high BD values. While we expected a more uniform spatial distribution of WHC values over the research area after leveling, it did not happen. Therefore, this kind of area may need much more attention especially for irrigation planning and management. Maybe, site specific irrigation planning and management can be an alternative for sustainable agriculture or research on these kinds of areas.

After the land leveling, decreases of the FC values of subsurface soil (20–40 cm) over almost all locations of the

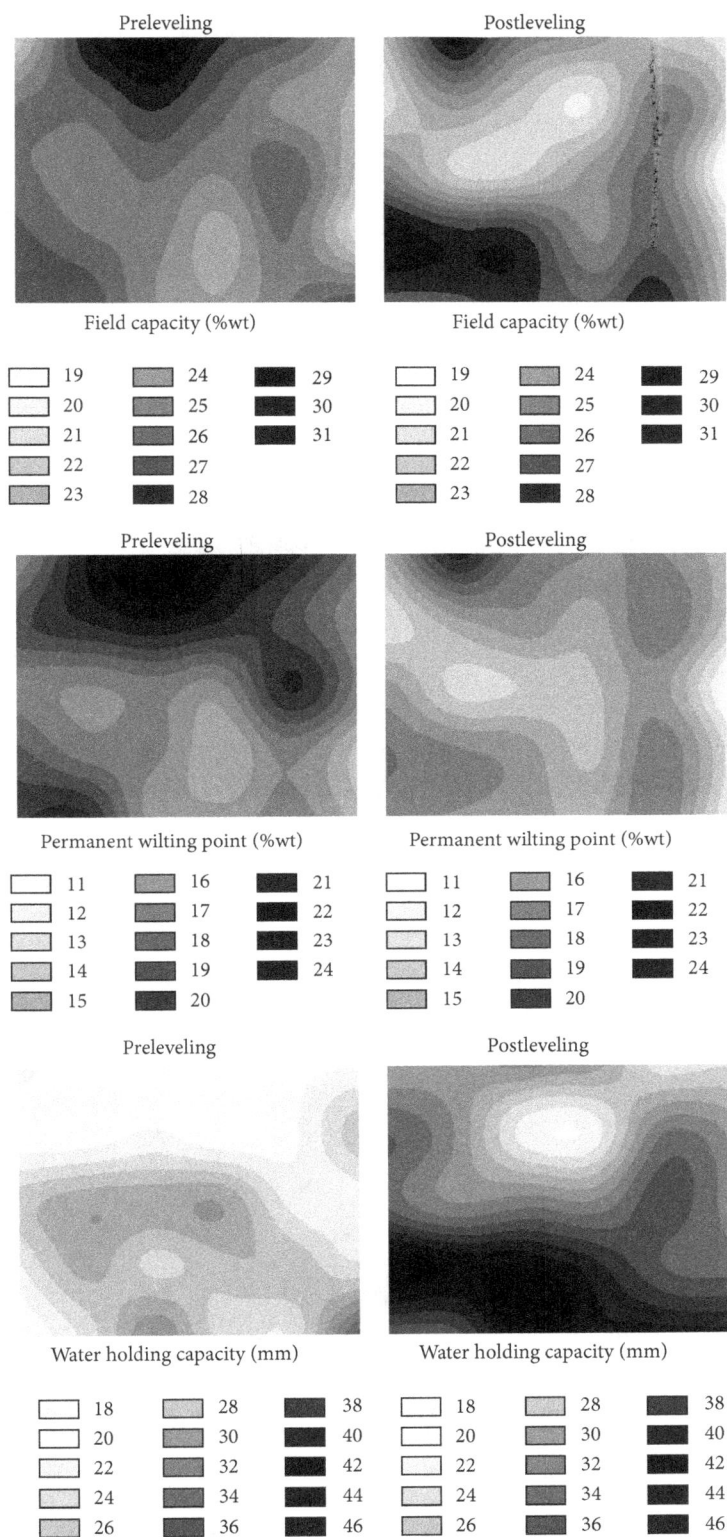

FIGURE 6: Pre- and post-leveling spatial distributions of soil field capacity, permanent wilting point, and water holding capacity for surface soil (0–20 cm).

Preleveling

Postleveling

Field capacity (%wt)

	13		23
	15		25
	17		27
	19		29
	21		31

Field capacity (%wt)

	13		23
	15		25
	17		27
	19		29
	21		31

Preleveling

Postleveling

Permanent wilting point (%wt)

Permanent wilting point (%wt)

	8		13		18		23
	9		14		19		24
	10		15		20		
	11		16		21		
	12		17		22		

	8		13		18		23
	9		14		19		24
	10		15		20		
	11		16		21		
	12		17		22		

Preleveling

Postleveling

Water holding capacity (mm)

Water holding capacity (mm)

	18		28		38
	20		30		40
	22		32		42
	24		34		44
	26		36		46

	18		28		38
	20		30		40
	22		32		42
	24		34		44
	26		36		46

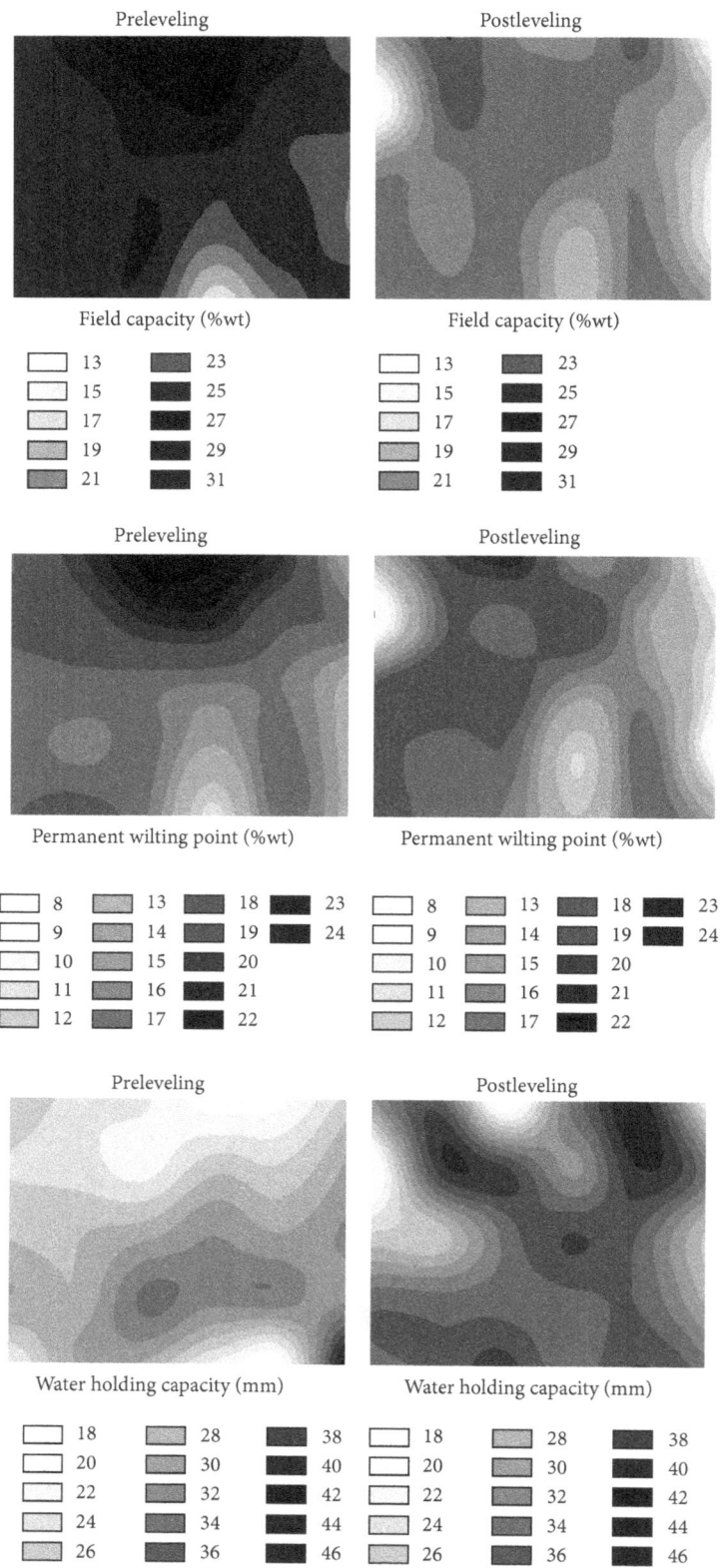

FIGURE 7: Pre- and post-leveling spatial distributions of soil field capacity, permanent wilting point, and water holding capacity for subsurface soil (20–40 cm).

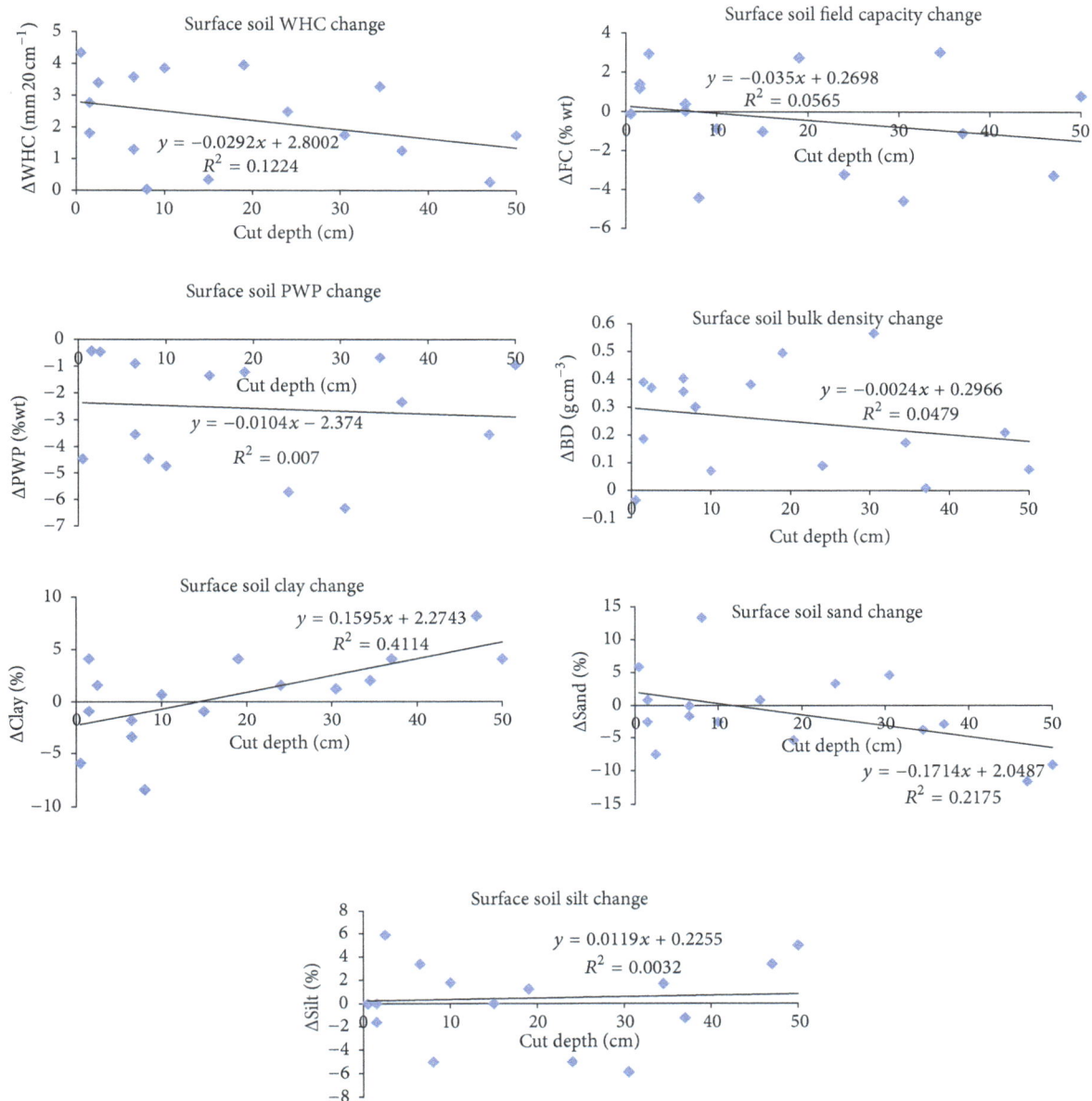

FIGURE 8: The relationships between the amounts of change of surface soil (0–20 cm) physical properties that occurred as a result of leveling and cut depths through cut areas.

study area can be seen in Figure 7. Furthermore, it seems that the spatial distribution of the FC values before leveling was more homogeneous than that after leveling. In addition, both before and after leveling, the FC values in the fill areas are higher than those of the cut areas. The decreases of FC values in the cut locations may be due to exposure of subsoil with the high sand contents. The PWP values of the subsurface soil showed similar spatial distribution to that of the FC of subsurface soil for both before and after levelings (Figure 7). In general, the locations had the high FC values also had the high PWP values; and the locations that had the low FC values also had the low PWP values. Similar to the FC values, the PWP values of subsurface soil (20–40 cm) over almost all

locations of the study area decreased after leveling (Figure 7). The WHC values of subsurface soil also increased as the WHC values of surface soil as a result of the land leveling (Figure 7).

3.3. Relationships between the Amounts of Changes in Soil Physical Properties and Depths of Cut and Fill. The mean values of changes for the soil properties of cut and fill areas were given in Table 5 separately. The result of the overall decreases in both FC and PWP values and the increases in BD values due to leveling, which caused the increases in WHC values, can also be induced from the given values in Table 5. As stated earlier, the reasons for the increases in WHC values

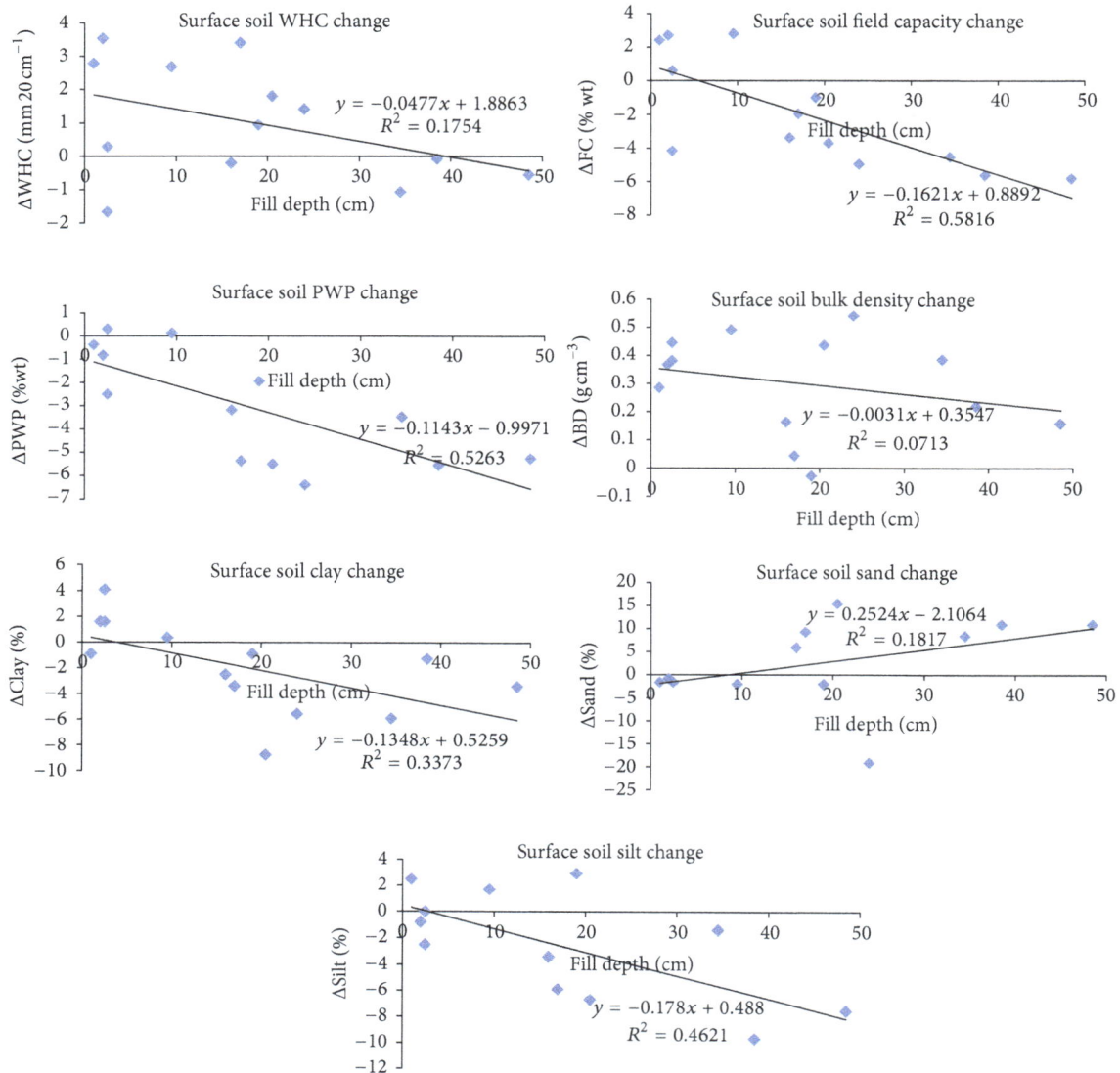

FIGURE 9: The relationships between the amounts of change of surface soil (0–20 cm) physical properties that occurred as a result of leveling and fill depths through fill areas.

due to leveling in both cut and fill areas of both surface and subsurface soils [are look like the much more decreases] in PWP values than those of FC values and the increases in BD values. The overall WHC increase in the cut areas (29%) occurred more than twice of that in the fill areas (12%) for the surface soil of the study area. However, the overall WHC increase in the cut areas (13%) is close to that in the fill areas (15%) for the subsurface soil of the study area. In addition, the reasons that caused the decreases in both FC and PWP and the increases in BD may be mainly due to the decreases of clay and increases of sand contents of both cut and fill subsurface and fill surface soils (Table 5). The values of increases or decreases (change amounts) against the cut depths through cut areas and the fill depths through fill areas are presented in Figures 8 and 9, respectively. The situation of overall decreases of clay and increases of sand contents [looks like did not resulted] for the surface soils through

cut areas mainly because of the high increases of clay and high decreases of sand contents when the cut depth is greater than 30 cm (Figure 8). The WHC values in all grids of cut area (Figure 8) and in 2/3rd of grids of fill area (Figure 9) of surface soil increased (above horizontal axes) due to leveling. In addition, increases of the WHC values in 2/3rd of grids of both cut and fill areas (not given) of subsurface soil were also observed. The decreases of PWP and increases of BD values of surface soil in almost all grids of both cut and fill areas are also clear from these figures.

The results of linear relationships with the values of their importance (R^2) between the amounts of changes in the soil physical properties due to the leveling and the depths of cut and fill are also presented in Figures 8 and 9 for surface soils through the cut and fill areas separately. Some important results from this analysis are as follows: the surface soil clay contents through cut areas increased as cut depths increased

TABLE 5: Mean changes in surface and subsurface soil properties of cut and fill areas separately that occurred as a result of land leveling.

Area	Soil depth (cm)	WHC (mm)	FC (%wt)	PWP (%wt)	BD (g cm^{-3})	Clay (%)	Sand (%)	Silt (%)	EC (dS m^{-1})	pH
Cut	0–20	2.26	−0.37	−2.57	0.25	0.66	−1.10	0.44	22.81	0.21
	20–40	1.19	−1.51	−2.70	0.04	−2.77	5.00	−2.25	−16.81	0.28
Fill	0–20	1.02	−2.05	−3.07	0.30	−1.92	2.47	−0.55	2.47	0.16
	20–40	1.06	0.17	−0.89	0.12	−1.52	0.64	0.82	−34.73	0.30

(r = 0.64) (Figure 8), the surface soil FC and PWP values through fill areas decreased as fill depths increased (r = 0.76, and r = 0.73, resp.) (Figure 9), and the surface soil clay and silt contents through fill areas decreased as fill depths increased (r = 0.58 and r = 0.68, resp.) (Figure 9). We did not determine any moderate or strong relationships between the changing amounts of the subsurface soil physical properties due to leveling and the depths of cut and fill.

4. Conclusions

When comparing some irrigation planning and management related soil properties (bulk density, particle size distributions, pH, electrical conductivity, field capacity, permanent wilting point, and water holding capacity) of a clay loam soil before and just after land leveling, we saw that the practice of land leveling significantly affected the magnitude, spatial variability, and spatial distribution of these soil properties in both surface (0–20 cm) and subsurface (20–40 cm) clay loam soils. Increased bulk density, pH, and water holding capacity and decreased permanent wilting point due to land leveling are some of the results of this study. The moderate negative relationships between the surface soil FC and PWP values and fill depths and between the clay and silt contents and fill depths through fill areas were found. Further research about these kinds of relationships may be needed in larger study areas.

The study showed that while land leveling is an agricultural practice to facilitate more uniform distribution of irrigation water, if leveling is poor, then changes in soil properties are notable and may be detrimental when observed at a short time after the operation. Applying equal amount of irrigation water on overall land with heterogenic soil properties alone will not be enough for productivity. After leveling, some other site specific cautions for defining and refining management practices to regain productivity and for improving homogeneity in soil properties are needed. In addition, research about long-term effects of land leveling with an adapted land reclamation policy to improve homogeneity in soil properties may be needed.

Acknowledgments

The author wishes to thank Selma Öztekin, Gülay Karahan, and Buket Yetkin Uz for their assistances in the laboratory. In addition, the author would like to thank the Agricultural Faculty of Gaziosmanpaşa University for equipment, laboratory, and financial supports.

References

[1] K. R. Brye, N. A. Slaton, M. C. Savin, R. J. Norman, and D. M. Miller, "Short-term effects of land leveling on soil physical properties and microbial biomass," *Soil Science Society of America Journal*, vol. 67, no. 5, pp. 1405–1417, 2003.

[2] P. W. Unger, L. J. Fulton, and O. R. Jones, "Land-leveling effects on soil texture, organic matter content, and aggregate stability," *Journal of Soil & Water Conservation*, vol. 45, no. 3, pp. 412–415, 1990.

[3] G. O. Schwab, D. D. Fangmeier, W. J. Elliot, and R. K. Frevert, *Soil and Water Conservation Engineering*, John Wiley & Sons, New York, NY, USA, 1993.

[4] K. R. Brye, N. A. Slaton, and R. J. Norman, "Penetration resistance as affected by shallow-cut land leveling and cropping," *Soil and Tillage Research*, vol. 81, no. 1, pp. 1–13, 2005.

[5] H. V. Eck, "Characteristics of exposed subsoil-at exposure and 23 years later," *Agronomy Journal*, vol. 79, no. 6, pp. 1067–1073, 1987.

[6] T. W. Walker, W. L. Kingery, J. E. Street et al., "Rice yield and soil chemical properties as affected by precision land leveling in alluvial soils," *Agronomy Journal*, vol. 95, no. 6, pp. 1483–1488, 2003.

[7] K. R. Brye, N. A. Slaton, M. Mozaffari, M. C. Savin, R. J. Norman, and D. M. Miller, "Short-term effects of land leveling on soil chemical properties and their relationships with microbial biomass," *Soil Science Society of America Journal*, vol. 68, no. 3, pp. 924–934, 2004.

[8] K. R. Brye, N. A. Slaton, and R. J. Norman, "Soil physical and biological properties as affected by land leveling in a clayey aquert," *Soil Science Society of America Journal*, vol. 70, no. 2, pp. 631–642, 2006.

[9] K. R. Brye, "Soil biochemical properties as affected by land leveling in a clayey aquert," *Soil Science Society of America Journal*, vol. 70, no. 4, pp. 1129–1139, 2006.

[10] M. C. Ramos, R. Cots-Folch, and J. A. Martínez-Casasnovas, "Effects of land terracing on soil properties in the Priorat region in Northeastern Spain: a multivariate analysis," *Geoderma*, vol. 142, no. 3-4, pp. 251–261, 2007.

[11] H. Taşova, *Tokat Ziraat Fakültesi Yerleşim Alanının Toprak Etüt, Haritalanması ve Sınıflandırılması [M.S. thesis]*, Cumhuriyet University, Sivas, Turkey, 1992.

[12] Soil Survey Staff, *Keys to Soil Taxonomyedition*, U.S. Department of Agriculture-Soil Conservation Service, Washington, DC, USA, 8th edition, 1999.

[13] G. E. Chugg, "Calculations for land gradation," *Agricultural Engineering*, vol. 28, no. 10, pp. 461–463, 1947.

[14] L. G. James, , *Principles of Farm Irrigation System Design*, Krieger, Malabar, Fla, USA, 1993.

[15] G. R. Blake and K. H. Hartge, "Bulk density," in *Methods of Soil Analysis, Part 1, Agronomy*, A. Klute, Ed., vol. 9, pp. 377–382, American Society of Agronomy, Madison, Wis, USA, 1994.

[16] G. W. Gee and J. W. Bauder, "Particle-size analysis," in *Methods of Soil Analysis, Part 1, Agronomy*, A. Klute, Ed., vol. 9, pp. 383–411, American Society of Agronomy, Madison, Wis, USA, 1994.

[17] A. Klute, "Water retention: laboratory methods," in *Methods of Soil Analysis, Part 1. Agronomy*, A. Klute, Ed., vol. 9, pp. 635–662, American Society of Agronomy, Madison, Wis, USA, 1994.

[18] H. Levene, *Contributions to Probability and Statistics*, Stanford University Press, Stanford, Calif, USA, 1960.

[19] Y. Güngör and O. Yıldırım, *Tarla Sulama Sistemleri*, Ankara Üniversitesi Ziraat Fakültesi Yayınları, Ankara, Turkey, 1989.

[20] E. H. Isaaks and R. M. Srivastava, *Applied Geostatistics*, Oxford University Press, Oxford, UK, 1989.

2

Determination of the Proportion of Total Soil Extracellular Acid Phosphomonoesterase (E.C. 3.1.3.2) Activity Represented by Roots in the Soil of Different Forest Ecosystems

Klement Rejsek, Valerie Vranova, and Pavel Formanek

Department of Geology and Soil Science, Mendel University in Brno, Zemedelska 3, 613 00, Brno, Czech Republic

Correspondence should be addressed to Pavel Formanek, formanek@mendelu.cz

Academic Editors: F. Darve and U. Feller

The aim of this study is to present a new method for determining the root-derived extracellular acid phosphomonoesterase (EAPM) activity fraction within the total EAPM activity of soil. EAPM activity was determined for roots, organic and mineral soil. Samples were collected using paired PVC cylinders, inserted to a depth of 15 cm, within seven selected forest stands. Root-derived EAPM formed between 4 and 18% of the total EAPM activity of soil from forests of differing maturity. A new approach, presented in this work, enables separation of root-derived EAPM activity from total soil EAPM. Separation of root-derived EAPM from soil provides a better understanding of its role in P-cycling in terrestrial ecosystems. The method presented in this work is a first step towards the separation of root- and microbe-derived EAPM in soils, which are thought to possess different kinetic properties and different sensitivity to environmental change.

1. Introduction

Extracellular acid phosphomonoesterase (EAPM) (orthophosphoric monoester phosphohydrolase, E.C. 3.1.3.2) plays an important role in the mineralization of soil organic phosphorus in a range of terrestrial ecosystems [1, 2]. This enzyme may be produced in soil by microorganisms including bacteria, protozoa, and mycorrhizal or saprophytic fungi and by plant roots [3–5]. Root-derived EAPM is bound onto root surfaces or released to external media as a part of the root exudates [6].

Different ecosystems are thought to have either plant- or microbe-derived EAPM prevailing in soil [7, 8]. Nevertheless, there were no available data indicating the significance of plant roots *versus* soil microorganisms in the production of EAPM and thus their relative importance for P-cycling. EAPM from roots is known to possess different kinetic properties and sensitivity to other factors of environment compared to that derived from microorganisms [9]. Consequently, these two fractions of total soil EAPM may respond differently to climate change and other environmental perturbations [6].

Separation of plant root- and microbe-derived EAPM in soil is difficult. Hence, we have developed a new approach, focusing on the activity of root-derived EAPM as a part of the total EAPM activity of soil in different forest ecosystems.

2. Material and Methods

2.1. Site and Soil Sampling. In total, seven forest stands were selected for this study. These included young (19 years) and old (207 years) beech (*Fagus sylvatica* L.) stands (480 m asl, N 49°16′54″, E 16°37′52″), young (33 years, oak 60%, hornbeam 30%, beech 10%) and old (133 years, oak 87%, Douglas fir 6%, beech 4%, and larch 3%) oak (*Quercus robur* L.) stands (460 m asl, N 49°32′16″, E 16°79′75″), and young (15 years, spruce 100%), middle-aged (51 years, spruce 68%, larch 32%), and old (94 years, spruce 92%, larch 6%, beech 2%) spruce (*Picea abies* L.) stands (500 m asl, N 49°32′19″, E 16°78′54″). Soils within the studied stands were Dystric Luvisol (young beech and old oak), Haplic Cambisol (old beech and young oak), Dystric Cambisol (old spruce), Gleyic Cambisol (young spruce), and Leptic Cambisol (middle aged spruce) [10].

Five pairs of PVC cylinders (15 cm long, 5.9 cm dia) were randomly inserted in every stand; cylinders within the same pair were always inserted side by side, to ensure similarity. After transportation to the laboratory in plastic bags, the litter layer in all cylinders was removed. One cylinder in each pair was used for separation of all the roots, which were washed in tap water and then in demineralized water. The soil from the second cylinder was separated into organic (F + H horizons) and mineral part to ensure consequent determination of EAPM activity of naturally developed soil layers without their artificial mixing together both parts were separately sieved through a 5 mm mesh, homogenised, and weighed.

2.2. Root Analysis. All roots from each of the cylinders were incubated, in succinate-borate buffer (pH 4.8) at 37°C for 30 min., with p-nitrophenyl phosphate (p-NPP) as a substrate [11]. The substrate dissolved in succinate-borate buffer was applied in ratio 12 mL per 0.5 g of fresh roots.

2.3. Soil Analysis. EAPM was measured separately in organic and mineral soil. Fresh soil (1 g) was incubated, in 12 mL of succinate-borate buffer (pH 4.8) at 37°C for 1 h, with p-NPP as a substrate [11]. EAPM activity was consequently calculated per the total amount of organic and mineral soil of every cylinder (data were pooled together), and, further, EAPM of roots of the same cylinders was added to obtain the total EAPM of the whole cylinder. Results were consequently calculated per 100 cm^2 of soil surface.

2.4. Statistical Treatment. Values are given as means of five replicates with standard errors (SE). Significant differences were calculated using one-way ANOVA plus Fisher's LSD test.

3. Results and Discussion

The total soil activity of EAPM, including roots, was significantly ($P < 0.05$) higher in young oak, old oak, and young spruce than that in other forest stands (Figure 1). Significantly ($P < 0.05$), the lowest total EAPM activity was found in the soil from the old beech forest stand. From the total EAPM activity of soil, up to 18% was derived from roots (Figure 2). The proportion of root-derived EAPM was higher for all spruce stands (at average >12%) than for beech or oak stands, due to higher EAPM related to unit fresh root mass (Table 1).

Historically, different approaches have been tested to separate acid phosphomonoesterase activity in soils. These have included separation of the intra- and extracellular APM pool [12–15], assessment of APM in rhizosphere *versus* bulk soil [3, 16, 17] or within particle-size fractions [18–20], soluble *versus* immobilized soil APM fractions [21–24], or phosphatase bonded to humic substances [25]. In addition to these, fractions of APM derived from plant roots have been studied in intact roots, external-root solution (as a part of rhizodeposition), root apoplastic sap, total root, and root segment extracts. Anatomical-physiological studies

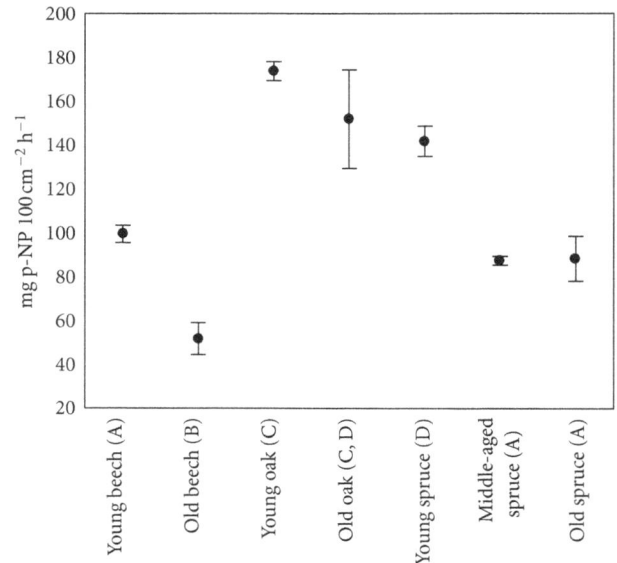

FIGURE 1: Total soil activity of EAPM, including roots (up to 15 cm depth), from seven forest stands (Mean ± SE). Different letters (in brackets) mark significant differences ($P < 0.05$).

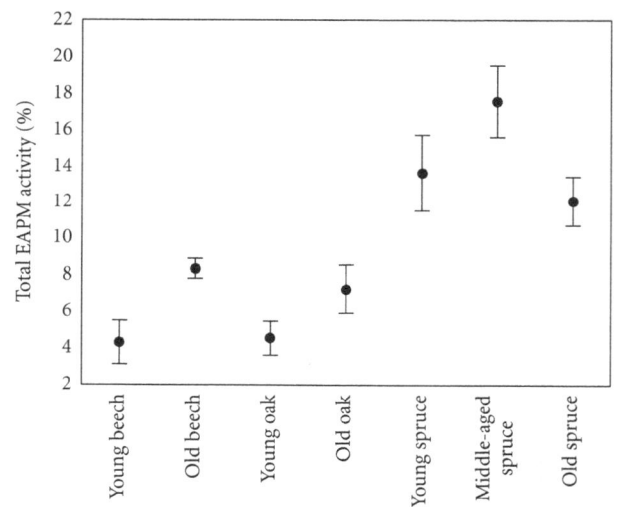

FIGURE 2: Proportion of root-derived EAPM within total soil EAPM up to 15 cm depth (Mean ± SE).

of surface-bound phosphomonoesterase activity in cross sections of roots and mycorrhizal associations have also been carried out [3, 6, 26–29].

As APM from roots and microorganisms is known to possess different kinetic properties, separation of APM sources in soil components allows us to better understand the response of P-transformation in soil in different conditions. The new approach presented in this work does not enable us to distinguish between root- and microbe-derived EAPM in soil, nor can we determine if the studied forest ecosystems have either plant- or microbe-derived EAPM prevalent in the soil. Nevertheless, the presented approach enables separation of root-derived EAPM activity from EAPM of the soil which

Determination of the Proportion of Total Soil Extracellular Acid Phosphomonoesterase (E.C. 3.1.3.2) Activity Represented by Roots in the Soil of Different Forest Ecosystems

19

TABLE 1: Total Root Mass and EAPM Activity of Roots in Seven Forest Stands (Mean ± SE). Different Letters Mark Significant Differences ($P < 0.05$).

Forest	g fresh roots	mg p-NP g^{-1} fresh roots h^{-1}
Young beech	3.98 ± 0.85^a	0.27 ± 0.03^a
Old beech	6.79 ± 1.10^{ab}	0.17 ± 0.03^a
Young oak	7.49 ± 1.27^{bc}	0.26 ± 0.02^a
Old oak	9.77 ± 1.00^{cd}	0.27 ± 0.02^a
Young spruce	4.18 ± 0.52^{ae}	1.18 ± 0.10^b
Middle-aged spruce	6.95 ± 0.86^{bde}	0.58 ± 0.08^c
Old spruce	4.02 ± 1.11^a	0.81 ± 0.13^d

may originate from both microorganisms and roots. The results presented in this work showed up to 18% of EAPM in soil to be root-derived when mycorrhizal status of roots was not considered. This work represents a first step in research leading to separation of root- and microbe-derived EAPM in soils.

Origin of phosphomonoesterase was shown to affect its Michaelis-Menten characteristics (K_m and V_{max} values) and response to pollutants (e.g., Cu) and other compounds in soil [30, 31]. Also, Gould et al. [9] reported that properties of microbe- and root-derived EAPM were different including their kinetic parameters and temperature sensitivity. Further research is necessary to separate the importance of root- and microbe-derived sources of EAPM in the soils of different ecosystems in order to better understand their importance in P-cycling and to evaluate their sensitivity to climate change and other types of environmental perturbations.

In conclusion, root-derived EAPM forms a lesser part of the total EAPM activity of soils in forest ecosystems. These findings can be generalized for acid forest soils where EAPM is of microbial and root origin. Alkaline soils with dominance of alkaline phosphomonoesterase of microbial origin are hypothesised to have especially plant root-derived EAPM activity; however, it still remains to be experimentally determined.

Acknowledgments

This paper was created within the framework of the Grant TA02020867 and the IGA Projects 47/2010–2012.

References

[1] P. Nannipieri, B. Ceccanti, C. Conti, and D. Bianchi, "Hydrolases extracted from soil: their properties and activities," *Soil Biology and Biochemistry*, vol. 14, no. 3, pp. 257–263, 1982.

[2] B. Sarapatka, *Phosphatase Activities (ACP, ALP) in Agroecosystem Soils*, Swedish University of Agricultural Sciences, Uppsala, Sweden, 2003.

[3] T. S. George, P. J. Gregory, M. Wood, D. Read, and R. J. Buresh, "Phosphatase activity and organic acids in the rhizosphere of potential agroforestry species and maize," *Soil Biology and Biochemistry*, vol. 34, no. 10, pp. 1487–1494, 2002.

[4] S. M. Martin and T. J. Byers, "Acid hydrolase activity during growth and encystment in Acanthamoeba castellanii," *Journal of Protozoology*, vol. 23, no. 4, pp. 608–613, 1976.

[5] G. Shaw and D. J. Read, "The biology of mycorrhiza in the Ericaceae. XIV. Effects of iron and aluminium on the activity of acid phosphatase in the ericoid endophyte *Hymenoscyphus ericae* (Read) Korf and Kernan," *New Phytologist*, vol. 113, pp. 529–533, 1989.

[6] F. Asmar and G. Gissel-Nielsen, "Extracellular phosphomono- and phosphodiesterase associated with the released by the roots of barley genotypes: a non-destructive method for the measurement of the extracellular enzymes of roots," *Biology and Fertility of Soils*, vol. 25, no. 2, pp. 117–122, 1997.

[7] S. R. Colvan, J. K. Syers, and A. G. O'Donnell, "Effect of long-term fertiliser use on acid and alkaline phosphomonoesterase and phosphodiesterase activities in managed grassland," *Biology and Fertility of Soils*, vol. 34, no. 4, pp. 258–263, 2001.

[8] H. Rodríguez, R. Fraga, T. Gonzalez, and Y. Bashan, "Genetics of phosphate solubilization and its potential applications for improving plant growth-promoting bacteria," *Plant and Soil*, vol. 287, no. 1-2, pp. 15–21, 2006.

[9] W. D. Gould, D. C. Coleman, and A. J. Rubink, "Effect of bacteria and amoebae on rhizosphere phosphatase activity," *Applied and Environmental Microbiology*, vol. 37, no. 5, pp. 943–946, 1979.

[10] D. T. L. Shek, H. K. Ma, and J. Merrick, Eds., *Positive Youth Development: Development of A Pioneering Program in a Chinese Context*, Freund Publishing Company, London, UK, 2002.

[11] K. Rejšek, "Acid phosphomonoesterase activity of ectomycorrhizal roots in norway spruce pure stands exposed to pollution," *Soil Biology and Biochemistry*, vol. 23, no. 7, pp. 667–671, 1991.

[12] G. Renella, L. Landi, and P. Nannipieri, "Hydrolase activities during and after the chloroform fumigation of soil as affected by protease activity," *Soil Biology and Biochemistry*, vol. 34, no. 1, pp. 51–60, 2002.

[13] M. E. Stromberger, S. Klose, H. Ajwa, T. Trout, and S. Fennimore, "Microbial populations and enzyme activities in soils fumigated with methyl bromide alternatives," *Soil Science Society of America Journal*, vol. 69, no. 6, pp. 1987–1999, 2005.

[14] S. Klose, V. Acosta-Martínez, and H. A. Ajwa, "Microbial community composition and enzyme activities in a sandy loam soil after fumigation with methyl bromide or alternative biocides," *Soil Biology and Biochemistry*, vol. 38, no. 6, pp. 1243–1254, 2006.

[15] A. Margon and F. Fornasier, "Determining soil enzyme location and related kinetics using rapid fumigation and high-yield extraction," *Soil Biology and Biochemistry*, vol. 40, no. 9, pp. 2178–2181, 2008.

[16] A. Hernesmaa, K. Björklöf, O. Kiikkilä, H. Fritze, K. Haahtela, and M. Romantschuk, "Structure and function of microbial communities in the rhizosphere of Scots pine after tree-felling," *Soil Biology and Biochemistry*, vol. 37, no. 4, pp. 777–785, 2005.

[17] Q. Zhao, D. H. Zeng, D. K. Lee, X. Y. He, Z. P. Fan, and Y. H. Jin, "Effects of Pinus sylvestris var. mongolica afforestation on soil phosphorus status of the Keerqin Sandy Lands in China," *Journal of Arid Environments*, vol. 69, no. 4, pp. 569–582, 2007.

[18] M. J. Rojo, S. G. Carcedo, and M. P. Mateos, "Distribution and characterization of phosphatase and organic phosphorus in soil fractions," *Soil Biology and Biochemistry*, vol. 22, no. 2, pp. 169–174, 1990.

[19] E. Kandeler, S. Palli, M. Stemmer, and M. H. Gerzabek, "Tillage changes microbial biomass and enzyme activities in particle-size fractions of a Haplic Chernozem," *Soil Biology and Biochemistry*, vol. 31, no. 9, pp. 1253–1264, 1999.

[20] M. C. Marx, E. Kandeler, M. Wood, N. Wermbter, and S. C. Jarvis, "Exploring the enzymatic landscape: distribution and kinetics of hydrolytic enzymes in soil particle-size fractions," *Soil Biology and Biochemistry*, vol. 37, no. 1, pp. 35–48, 2005.

[21] P. Nannipieri, B. Ceccanti, S. Cervelli, and E. Matarese, "Extraction of phosphatase, urease, proteases, organic carbon and nitrogen from soil," *Soil Science Society of America Journal*, vol. 44, pp. 1011–1016, 1980.

[22] J. A. Pascual, J. L. Moreno, T. Hernández, and C. García, "Persistence of immobilised and total urease and phosphatase activities in a soil amended with organic wastes," *Bioresource Technology*, vol. 82, no. 1, pp. 73–78, 2002.

[23] S. Criquet, E. Ferre, A. M. Farnet, and J. Le Petit, "Annual dynamics of phosphatase activities in an evergreen oak litter: influence of biotic and abiotic factors," *Soil Biology and Biochemistry*, vol. 36, no. 7, pp. 1111–1118, 2004.

[24] F. Fornasier and A. Margon, "Bovine serum albumin and Triton X-100 greatly increase phosphomonoesterases and aryl-sulphatase extraction yield from soil," *Soil Biology and Biochemistry*, vol. 39, no. 10, pp. 2682–2684, 2007.

[25] U. Gosewinkel and F. E. Broadbent, "Decomplexation of phosphatase from extracted soil humic substances with electron donating reagents," *Soil Science*, vol. 141, pp. 261–267, 1986.

[26] A. E. Richardson, P. A. Hadobas, and J. E. Hayes, "Acid phosphomonoesterase and phytase activities of wheat (*Triticum aestivum* L.) roots and utilization of organic phosphorus substrates by seedlings grown in sterile culture," *Plant, Cell and Environment*, vol. 23, no. 4, pp. 397–405, 2000.

[27] M. Alvarez, R. Godoy, W. Heyser, and S. Härtel, "Anatomical-physiological determination of surface bound phosphatase activity in ectomycorrhizae of *Nothofagus obliqua*," *Soil Biology and Biochemistry*, vol. 37, no. 1, pp. 125–132, 2005.

[28] L. Tamas, J. Dudikova, K. Durceková, J. Huttova, I. Mistrik, and V. Zelinova, "The impact of heavy metals on the activity of some enzymes along the barley root," *Environmental and Experimental Botany*, vol. 62, pp. 86–91, 2008.

[29] P. Priya and S. V. Sahi, "Influence of phosphorus nutrition on growth and metabolism of Duo grass (*Duo festulolium*)," *Plant Physiology and Biochemistry*, vol. 47, no. 1, pp. 31–36, 2009.

[30] B. R. Gibson and D. T. Mitchell, "Phosphatases of ericoid mycorrhizal fungi: kinetic properties and the effect of copper on activity," *Mycological Research*, vol. 109, no. 4, pp. 478–486, 2005.

[31] M. M. Goil and R. P. Harpur, "A comparison of the non-specific acid phosphomonoesterase activity in the larva of *Phocanema decipiens* (nematoda) with that of the muscle of its host the codfish (*Gadus morhua*)," *Zeitschrift fur Parasitenkunde*, vol. 60, no. 2, pp. 177–183, 1979.

3

Effect of Long-Term Paddy-Upland Yearly Rotations on Rice (*Oryza sativa*) Yield, Soil Properties, and Bacteria Community Diversity

Song Chen,[1] Xi Zheng,[2] Dangying Wang,[1] Liping Chen,[1] Chunmei Xu,[1] and Xiufu Zhang[1]

[1] *China National Rice Research Institute, Chinese Academy of Agricultural Sciences, Zhejiang, Hangzhou 310006, China*
[2] *985-Institute of Agrobiology and Environmental Sciences, Zhejiang University, Hangzhou 310029, China*

Correspondence should be addressed to Song Chen, chais.zju@gmail.com

Academic Editors: H. P. Bais, L. E. Parent, and A. R. Garrigós

A 10-year-long field trial (between 2001 and 2010) was conducted to investigate the effect of paddy-upland rotation on rice yield, soil properties, and bacteria community diversity. Six types of paddy-upland crop rotations were evaluated: rice-fallow (control; CK), rice-rye grass (RR), rice-potato with rice straw mulches (RP), rice-rapeseed with straw incorporated into soil at flowering (ROF), rice-rapeseed incorporated in soil after harvest (ROM), and rice-Chinese milk vetch (RC). Analysis of terminal restriction fragment length polymorphism (T-RFLP) was used to determine microbial diversity among rotations. Rice yield increased for upland crops planted during the winter. RC had the highest average yield of 7.74 t/ha, followed by RR, RP, ROM, and ROF. Soil quality differences among rotations were found. RC and RP improved the soil mean weight diameter (MWD), which suggested that rice rotated with milk vetch and potato might improve the paddy soil structure. Improved total nitrogen (TN) and soil organic matter (SOM) were also found in RC and RP. The positive relationship between yield and TN/SOM might provide evidence for the effect of RC rotation on rice yield. A strong time dependency of soil bacterial community diversity was also found.

1. Introduction

In China and other Asian countries, continuous rice planting has had a negative impact on soil properties, such as reduced soil nitrogen supply and organic carbon content [1, 2]. Paddy-rice-upland crop rotations have been recommended and used to improve soil quality and reduce input [3–8].

In conventional paddy-upland rotation systems, farmers drain the fields after harvesting rice and then plant an upland crop, such as milk vetch, wheat, or oilseed [9–11]; However, the growth conditions required by rice are quite different from those required by upland crops. Rice will grow best under puddled, reduced, and anaerobic soil conditions, whereas upland crops require unpuddled, aerobic and oxidized soil conditions. Paddy soils show a large difference from upland soils in physical, chemical, and biological properties [12]. Furthermore, because of long-term submergence and mineral fertilizer application, paddy soils experience degradation of soil quality, such as breakdown of stable aggregation and deterioration of soil organic matter (SOM), which negatively affects agricultural sustainability [13, 14].

Soil quality is a term used to describe the health of agricultural soils. It has been suggested as an indicator for evaluating sustainability of soil and crop management practices [15–17]. Many soil attributes have been proposed to describe soil quality, but evaluation of pH, soil organic matter (SOM), and total nitrogen content (TN) of soil have been considered essential for assessing the chemical aspects of soil quality [15, 17]. These chemical traits are so important because they provide a measure of the ability of soil to supply nutrients and to buffer against chemical additives [18–20].

Soil physical properties are indicators of the impact of soil and crop management practices. Soil size distribution and water stability of soil aggregates would be influenced by crop types as well as soil management practices [21]. Furthermore, microbial populations in soil interact with each other and with soil. These interactions, in turn, affect major environmental processes, including biogeochemical cycling of nutrients, plant health, and soil quality [22]. Most microbial interactions in soil occur near the plant roots and the root-soil interface, called the rhizosphere [23, 24]. It is not surprising that microbial communities in the rhizosphere depend on plant species [25, 26]. Although the relationships between soil microbial diversity and function and sustainability (or stability) of agricultural ecosystems are still unclear [17, 27, 28], it has been documented that diversity of soil biota is important to the beneficial function of agro-ecosystems [29, 30].

In China, paddy-upland crop rotation is a major cropping system utilized along the Yangtze River basin [31–33]. Traditionally, the main patterns of rice and upland crop are rice-wheat, rice-oilseed, rice- milk vetch, and rice-ryegrass [4]. However, some practices may not be economical because of high input and increased risk of financial loss for the rice crop. Currently, new patterns including rice-oilseed rape (*Brassica napus* L.) and rice-potato (*Solanum tuberosum*) are widely used. However, few studies have been done to determine the effect of these paddy-upland crop rotations on soil physical and chemical properties. The present study was conducted to determine the effect of long-term cropping system on (i) rice and other crop yields, (ii) soil quality attributes (chemical and physical), and (iii) the diversity of soil bacterial community. The results offer helpful insight into the effect of various rice-upland combinations on crop productivity and soil properties.

2. Materials and Methods

2.1. Soil and Site. The study was conducted over a period of ten years at the experimental farm of the China Rice Research Institute (120.2 E, 30.3 N, and 11 m above sea level) located in Fuyang, Hangzhou, China. The long-term field experiment has been carried out in an irrigated rice paddy starting in 2001. The historical cropping background was monoculture rice cropping before this time. The area is characterized by a subtropical monsoon climate with an annual mean temperature of 13–20°C, ranging from 2°C in January to 35°C in July and mean precipitation of 1200–1600 mm per year, with about 80% falling between April and September. Soil is classified as Ferric-accumulic Stagnic Anthrosols [34] and entic Halpudept [35]. Further details about the experimental site can be found in the study of Fu et al. [36].

2.2. Experimental Design and Crop Cultivation. The long-term field experiment included six different types of cropping rotation: continuous monoculture rice-fallow (CK), rice-ryegrass (*Lolium perenne*) (RR), rice-potato (*Solanum tuberosum*) (RP), rice-Chinese milk vetch (*Astragalus sinicus* L.) (RC), rice-oilseed rape (*Brassica napus* L.) with burned straw returned to the soil after harvest (ROM), and rice-oilseed rape with fresh oilseed straw returned as mulch at flowering (ROF). Primary tillage was followed by a pass of a stubble crushing machine to a depth of 0–20 cm about three days before rice seedling transplanting in early June every

year. During the rice season, the rice species used was Guodao 6, an elite *Japonica* hybrid widely planted in Southeast China. Direct sowing was applied for seedling establishment, with a seeding rate of 15 kg/ha. The amount of fertilizer application was set according to local agronomic practices. It is noteworthy that there was a significant reduction in nitrogen input starting in 2005 due to the serious rice lodging of several rotations. The rate and timing of fertilizer was set as follows: basal application, including total phosphorous, 50% nitrogen, and 50% potassium, was conducted a week before transplantation in the form of compound fertilizer; 25% nitrogen was applied as topdressing at mid-tillering in the form of urea, and 25% nitrogen and 50% potassium was applied as topdressing at panicle initiation in the form of urea and potassium chloride, respectively. Weeds, insects, and diseases were controlled as required to avoid yield loss. More details of the agronomic practice are presented in Table 1.

The field experiment was planned with a large plot area (20 m long, 20 m wide) but without treatment replicates due to practical reasons. Although this design could lead to statistical problems, the cropping system has been used for decades with uniform fertilizer management. The small variability (CV < 5%) of the representative parameters of soil fertility in the treatment plots (Table 2) indicates a low spatial heterogeneity of the field. We therefore believe it is reasonable to assume that any significant differences later observed between plots were caused by the different rotations.

2.3. Sampling and Measurement. In 2010, a survey of rice yield was carried out as follows: hills were harvested from the uniform part of each plot of 5 m^2 with 4 subsamples. Unhulled (rough) rice was obtained after reaping, threshing, and wind selection. Rough rice from 80 hills was hulled and then put through a 1.8 mm sieve to remove any immature kernels. The weight of hulled rice (brown rice) was adjusted to a moisture content of 14%. Upland crop biomass was sampled from 1 m^2 of four sub-sample plots. For RC, RP, ROM, and ROF, plants were sampled at harvest. For RR, sampling was carried on during the growth season. All samples were oven-dried at 70°C to a constant weight to determine the dry weight.

Soil samples from 0 to 20 cm depth at pretransplanting in early June 2010 were used for soil physical quality analysis. Bulk density was determined for undisturbed soil samples using a steel cylinder of 100 cm^3 volume (5 cm in diameter, and 5.1 cm in height) [37] Soil aggregation was determined by the wet sieving method [38]. The mean weight diameter was calculated as follows:

$$\text{MWD} = \sum_{i=1}^{n} X_i W_i, \tag{1}$$

where MWD is the mean weight diameter of water-stable aggregates, X_i is the mean diameter of each size fraction (mm), and W_i is the proportion of the total sample mass in the corresponding size fraction after the mass of stones was deducted (upon dispersion and passing through the same sieve).

Effect of Long-Term Paddy-Upland Yearly Rotations on Rice (Oryza sativa) Yield, Soil Properties, and Bacteria Community Diversity

23

TABLE 1: Experimental design of six paddy-upland crop rotation systems.

Label	Cropping system	Year of start of experiment	Rice season Sowing	Rice season TP	Rice season Harvest	Upland crop season Sowing	Upland crop season Harvest	Herbicides	Annual N-P-K input Kg/ha Rice	Annual N-P-K input Kg/ha Upland crop season
CK	Rice-fallow phase	Continues				—	—	Yes		No
RR	Rice-ryegrass	2001	Early May	Early June	Late October	November	Harvest 2-3 times	Yes	[b]2001–2005: 180-50-180	No
RP	Rice-potato, with rice straw as mulch	2001				December	March	No		No
RC	Rice-Chinese milk vetch	2001				November	—	No	2005–by now: 120-50-120	No
ROM	Rice-oilseed rape with burned straw return after harvest	2001				November	April	No		[c]20-20-20
[a]ROF	Rice-oilseed rape with fresh straw return as green manure at flowering	2003				November	March	No		No

[a] Oilseed rape was considered as manure crops but economic crops in ROF. This is a new cropping rotation under evaluating.
[b] Fertilizer input was reduced due to the lodging problem of rice season in RP and RC treatment.
[c] Fertilizer applied at bolting stage for oilseed rape.

TABLE 2: General soil properties before the experiment started in 2001 (0–20 cm soil depth).

Treatment	pH	SOM (g/kg)	TN (g/kg)	TP (g/kg)	[a]Available K (g/kg)	Bulk density (g/cm³)
CK	6.51	26.90	2.53	0.62	0.22	1.17
RP	6.69	24.50	2.49	0.65	0.24	1.21
RR	6.82	25.60	2.50	0.63	0.23	1.12
RC	6.62	29.10	2.46	0.66	0.23	1.24
ROM	6.43	27.50	2.69	0.61	0.22	1.11
ROF	—	—	—	—	—	—
Means	6.61	26.72	2.53	0.64	0.23	1.17
[b]CV%	2.30	4.81	3.32	3.08	3.61	4.85

[a] Available K was extracted with 1 mol NH$_4$AC.
[b] C.V. is coefficient of variation.
Abbreviations: CK: rice-fallow phase; RR: rice-ryegrass; RP: rice-potato, with rice straw as mulch; RC: rice-Chinese milk vetch; ROM: rice-oilseed rape with burned straw return after harvest; ROF: rice-oilseed rape with fresh straw return as mulch at flowering.

Soil samples for chemical analysis were collected at pretransplanting in early June 2010 from two depths (0–10 and 10–20 cm). The field-moist soil samples were passed through an 8 mm sieve and air-dried. Cation exchange capacity (CEC) was measured according to the procedure used by Hendershot et al. [39]. Electrical conductivity was measured using a digital conductivity meter. Soil total N (TN) and total P were estimated by the methods given by Bremner and Mulvaney [40] and Sparling et al. [41], respectively. Soil water content was determined gravimetrically (105°C, 24 h). Results were expressed based on the dry weight of the soil. Soil pH was determined (1 : 5 water suspension) by pH meter. Soil available potassium was determined by extraction with 1 mol NH$_4$AC. Soil organic matter (SOM) was determined by a dichromate oxidation procedure. Multiplying the soil organic carbon by 1.72 resulted in the SOM [42]. The microbial biomass C (MBC) was determined using the chloroform fumigation-extraction method on fresh soils [43]. Each replicate was divided into two equivalent portions: one was fumigated for 24 h with ethanol-free chloroform and the other was not fumigated as a control. Both fumigated and unfumigated soils were shaken for 30 min with 0.5 M K$_2$SO$_4$ (1 : 4 soil : extraction ratio) and centrifuged and filtered. Extracts were analyzed for DOC on a total organic C analyzer (TOC-V CPH, Shimadzu).

Soil samples for T-RFLP analysis were collected from each plot using a soil auger (5 cm in diameter) at pretransplanting in early June 2010 (BR) and after harvest in late October 2010 (AR) from 0–20 cm soil depth. Samples were packed in sterile plastic bags and sent to the laboratory, then air-dried until the water content was about 75%. Later, the moist soil was passed through a 2 mm sieve and stored at 4°C for DNA extraction.

2.4. DNA Extraction and T-RFLP Analysis. Genomic DNA of the soil samples was isolated using a SDS-hyperhaline buffer solution as used in Zhou et al. [44]. Approximately 1 g of dry soil was suspended in 2.7 mL of extraction buffer (100 mM Tris-HCl, 100 mM EDTA, 100 mM Na$_3$PO$_4$, 1.5 mM NaCl,

1% CTAB, pH 8.0), proteinase K (20 μL) was added, and the mixture was shaken at 225 rpm at 37°C for 30 min. The suspension was further incubated in 20% SDS at 65°C for 2 h. During incubation, the tubes were gently frozen-thawed with liquid nitrogen for 20 min and run for three cycles. After that, the soil samples were centrifuged at 8,000 rpm for 10 min at 4°C, followed by the extraction of supernatant with 2.6 mL chloroform-isoamyl alcohol (24 : 1) and centrifuged at 5,000 rpm for 5 min. The supernatant was precipitated for 2 h with 2 mL isopropanol before recovering the DNA with 10,000 rpm centrifugation for 10 min. The resulting pellet was washed with 2 mL of 70% (v/v) ice-cold ethanol, dried, and dissolved in 200 μL of sterile distilled water. The purified DNA was stored at 4°C for at least 1 day before PCR amplification. Preliminary analysis showed that the heterogeneity of the profiles obtained from independent preparations of the same soil sample decreased after storing of the fresh DNA.

The eubacterial primers 8f (5′-AGAGTTTGATCCTGG-CTCAG-3′) labeled at the 5′ end with 6-carboxyfluorescein (6-FAM) and 926r (5′-CCGTCAATTCCTTTRAGTTT-3′) were used to amplify approximately 920 bp of the 16S rRNA gene [45]. Each PCR reaction mixture (25 μL) contained 2.5 μL 10 × *TransTaq* HiFi Buffer I (200 mM Tris-HCl (pH 8.4), 200 mM KCl, 100 mM (NH$_4$)$_2$SO$_4$, 20 mM MgCl$_2$), 2 μL 2.5 mM dNTPs, 2 μL 10 μM primers, 0.5 μL 5 unit μL^{-1} of *TransTaq* polymerase High Fidelity (Beijing TransGen Biotech Co., Ltd. China), and 2 μL genomic DNA temples in a final volume of 25 μL. All amplifications were performed on a DNA Engine Dyad thermal cycler (Bio-Rad, Inc., USA) using the following program: a 5-min hot start at 94°C, followed by 39 cycles consisting of denaturation (1 s at 94°C), annealing (45 s at 50°C and 60°C, resp.), and extension (1 min at 72°C), with a final extension at 72°C for 10 min. PCR products were detected using 1.0% agarose gel electrophoresis in a 1 × TAE buffer. Fluorescently-labeled PCR products (200 μL) were purified with a UNIQ-10 DNA purification kit (Sangon Biotech Co., Ltd., China). Approximately 50 ng of amplified 16S rRNA gene fragments

Effect of Long-Term Paddy-Upland Yearly Rotations on Rice (Oryza sativa) Yield, Soil Properties, and Bacteria Community Diversity

25

was digested with HaeIII (GG′CC), HhaI (GCG′C), and HinfI (G′ANTC) for 4 h at 37°C and precipitated with iso-propanol. The precipitated DNA was washed twice with 70% isopropanol, vacuum dried, and resuspended in 20 mL water. In the preliminary experiments, seven restriction enzymes (HaeIII, HhaI, HinfI, MspI, AluI, and TaqI) were tested; the enzymes HhaI, HinfI, and HaeIII had the higher yield. The sample was mixed with 0.1 mL of GeneScan 1000 Rox size standard, denatured at 95°C for 2 min, immediately placed on ice, and evaluated following electrophoresis in POP6 polymer with an automated DNA sequencer (ABI 3100, Applied Biosystems Instruments, California, USA). Terminal fragment sizes between 29 and 940 bp were determined using Gene-Mapper v3.7 software (Perkin-Elmer, California, USA).

2.5. Data Analysis. Data were analyzed by using Microsoft Excel 2003 and SAS 8.0 (2003). Means and standard deviations/standard errors are reported for each of the measurements. One-way analysis of variance (ANOVA) of Tukey's test was used to compare the effects of rotations on soil properties determined for the two soil depths of 0–10 cm and 10–20 cm separately.

All T-RFLP community profiles were labeled for statistical analyses by rotation (CK, RR, RP, RC, ROM, or ROF), sampling time (BR or AR), restriction enzyme (HaeIII, HhaI, or HinfI), and field plot replicate (1, 2, or 3). T-RF peaks between 35 and 500 bp and peak heights of <50 fluorescence units were included in the analysis according to the range of the size marker. Generally, the error for determining fragment sizes with our automated DNA sequencer was less than 1 bp; however, in some cases, a higher variation was found. Therefore, T-RFs that differed by less than 1.5 bp were clustered unless individual peaks were detected in a reproducible manner. Three replicate samples of all rotations and particle sizes were analyzed individually, or a representative sample profile was determined in a way similar to that suggested by Dunbar et al. [46]. Essentially, the sum of the peak heights in each replicate profile was calculated and used to indicate the total DNA quantity. Total fluorescence was adjusted to the medium DNA quantity by calculating a correction factor. For example, three replicate profiles had total fluorescence values of 4,500, 4,700, and 4,900, and then each peak in the latter profile was multiplied by a factor of 0.96 (i.e., the quotient of 4,700/4,900), and peaks in the first profile were multiplied by a factor of 1.04 (i.e., a quotient of 4,700/4,500). After adjustment, only peaks of >50 fluorescence units were considered. In addition, T-RFs were scored as positive only when they were present in at least two of the three replicates.

In order to determine similarities between T-RFLP profiles, a binary matrix that recorded the absence and presence of aligned fragments was generated. The distance matrix of fragments was generated according to the Jaccard index (1908) using NTSYS version 2.10e software for PC (Applied Biostatistics). Based on the distance matrix, cluster analysis was performed utilizing an unweighted pair group method with arithmetic average (UPGMA).

TABLE 3: Soil physical properties in 0–20 cm depth.

Rotation	^aSand (g/kg)	Slit (g/kg)	Clay (g/kg)	MWD (mm)	Bulk density (g/cm³)
CK	40.9b	650.9	300.0b	0.07b	1.20b
RC	89.4a	574.5	336.1b	0.11a	1.39a
RR	51.4b	559.2	389.4a	0.08b	1.26ab
RP	76.2a	586.2	337.5b	0.10a	1.21b
ROM	54.6b	551.5	393.9a	0.08b	1.21b
ROF	42.6b	531.4	426.0a	0.06b	1.35a

^aSand 2–0.05 mm, silt 0.05–0.002 mm, clay < 0.002 mm. Means on the same column and for the same sampling time followed by the same letter (or none) are not significantly different at $P < 0.05$ by Tukey Test. Abbreviations: CK: rice-fallow phase; RR: rice-ryegrass; RP: rice-potato, with rice straw as mulch; RC: rice-Chinese milk vetch; ROM: rice-oilseed rape with burned straw return after harvest; ROF: rice-oilseed rape with fresh straw return as mulch at flowering.

3. Results

3.1. Rice Yield and Upland Crops Biomass Production. Rice yield increased in the plots with upland crops applied during the winter season (Figure 1(a)). RC had the highest average yield of 7.74 t/ha, which was 27.8% significantly higher than CK. No significant difference was observed among RR, RP, ROM, and ROF, although their average yields were also higher than CK, with average yield increase ranging from 14.2–17.8%. Upland crop biomass production was estimated and is shown in Figure 1(b). RP had the highest average biomass production of 22.3 t/ha, followed by ROF, RC, RR, and ROM. In CK, weeds grew and died during the fallow phase, making it difficult to estimate the biomass production due to the uneven growth of the weeds. As a result, no data are shown here to describe the biomass production in CK. Different from RP, RC, ROM, and ROF, biomass produced in RR was removed from the field as pasture crops. But all the plant residues in ROF, RP, and RC were returned to the field as cover crops and incorporated into soil before rice season. For ROM, the straw was burned after harvest and the ash was incorporated into soil by tillage. Generally speaking, RP had the highest value of biomass C return, followed by RC and ROF. ROM and RR might have the lowest values next to CK.

3.2. Effect of Paddy-Upland Rotation System on Soil Properties. The soil bulk density, soil aggregation, and mean weight diameter (MWD) of different paddy-upland crop rotations are presented in Table 3. The bulk density was significantly greater in RC and ROF than in the others, and RP and RC had relatively higher values of MWD compared with other rotations.

There was a strong depth-dependency of soil pH value, total soil nitrogen (TN), total soil phosphorus (TP), available potassium (K), and cation exchange capacity (CEC) in all rotations (see Table 4). Soil was acid in all six rotations from 0–10 cm and 10–20 cm depth. However, the average pH value was significantly greater in 10–20 cm (5.90) than in 0–10 cm (5.44). In comparison with CK, the pH values in RC, ROF,

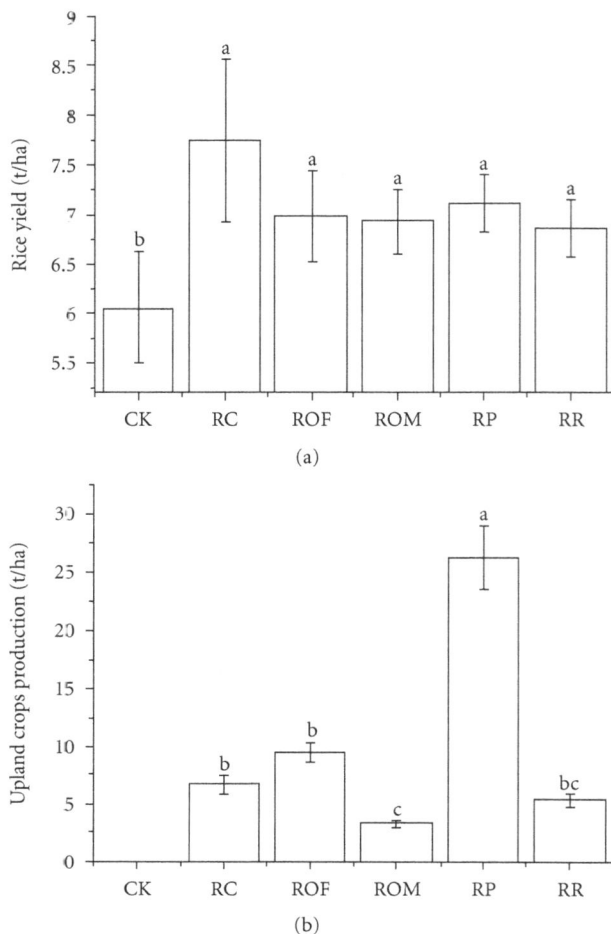

FIGURE 1: Effect of different paddy-upland crops rotations on the rice yield (a) and upland crops biomass production (b) during 2010-2011. Error bars represent standard deviation, $n = 3$. Values followed by different lowercase letters within depth are significantly different between treatment by Tukey's test ($P < 0.05$). Abbreviations: CK, rice-Fallow phase; RR, rice-ryegrass; RP, rice-potato, with rice straw as mulch; RC, rice-Chinese milk vetch; ROM, rice-oilseed rape with burned straw return after harvest; ROF, rice-oilseed rape with fresh straw return as mulch at flowering.

RP, and RR decreased in 0–10 cm depth, but increased in the 10–20 cm depth for ROF and ROM.

For TN and TP, remarkable differences were found between depths, with 0–10 cm greater than 10–20 cm. The average values of TN/TP were 2.73/0.63 (g/kg) in 0–10 cm and 2.51/0.51 (g/kg) in 10–20 cm. However, the difference of available K between depths was not statistically significant. The rotation effect on soil TN, TP, and available K was significant. In comparison with CK, the values of TN were significantly increased in RC, ROF, and RP in 0–10 cm, with average increments of 18.7%, 14.6%, and 20.3%, respectively. However, similar results in 1–20 cm were only found in RC and RP, with average increments of 10.6% and 14.4%, respectively. There was a variation of rotation effects on TP in 0–10 cm, with ROF (0.70 g/kg), RP (0.72 g/kg), and

RR (0.67 g/kg) significantly greater than CK (0.53 g/kg). However, little difference in TP was found in 10–20 cm depth among all rotations. The available K was greater in ROF, ROM, and RP compared with CK in 0–10 cm depth. However, little difference was found among rotations in 10–20 cm depth, except for ROM, which had the highest value among the six rotations. The rotation effect on soil CEC was variable with soil depth. For 0–10 cm depth, CEC in RC and ROM was significantly greater than that in CK, but only RC had significantly different results in 10–20 cm depth.

Soil organic matter (SOM), soil dissolved organic carbon content (DOC), and soil microbial biomass carbon content (MBC) of the six rotations in two soil depths were analyzed and are shown in Table 5. Significant difference was found in SOM between soil depths, with values in 0–10 cm greater than those in 10–20 cm. SOM of rotations ranged from 18.9–25.8 g/kg in 0–10 cm to 16.6–23.4 g/kg in 10–20 cm depth. Moreover, RP rotation was greater than CK in both depths. However, SOM in ROF and ROM decreased compared with CK in both depths. Small differences for DOC and MBC between depths were found, but the difference among rotations was significant. DOC in CK was the lowest among rotations in both depths. Compared with CK, DOC and MBC increased in RC, RP, and ROF, but decreased in ROM and RP.

3.3. Diversity of Bacterial Communities in Soils from Different Paddy-Upland Crop Rotation Systems. Terminal restriction fragment length polymorphism analysis of 16S rRNA gene fragments amplified from community DNA was applied to compare the bacterial community structure in the field sites described above. Consistent T-RFLP profiles were obtained from three sampling points of the same field site, as shown by respective replications of the six different rotations (Figure 2).

As shown by cluster analysis, a total of eight rotations formed mainly two major separate branches: pretransplanting (BR) and postharvest (AR) (Figure 3). Within each branch, the lowest similarity was found in CK (followed by RP) regardless of sample time, with a mean similarity to others of 62.5% for BR and 71.3% for AR. Further, four out of the other five rotations were grouped together with a mean similarity to RP of 69.8% for BR and 75.8% for AR. These results indicate remarkable differences in the effect of long-term paddy-upland rotations on soil bacterial community construction.

4. Discussion

4.1. Paddy-Upland Crop Rotation Effect on Rice Yield and Upland Crop Biomass Production. Rice yield increased when upland crops were applied during the winter season (Figure 1(a)). The increases were slight but positive, indicating the yield benefit of long-term application of paddy-upland crop rotations. Similar results were reported by Ghoshal and Singh [47] and Kim et al. [48]. However, in terms of different agronomic practice, such as upland crop species and the amount of organic and chemical fertilizer input, the magnitudes of these increases were variable. RC

Effect of Long-Term Paddy-Upland Yearly Rotations on Rice (Oryza sativa) Yield, Soil Properties, and Bacteria
Community Diversity

27

TABLE 4: Soil chemical properties in 0–10 cm depth.

Depth	Rotation	pH (1 : 2.5 H$_2$O)	Total N (g/kg)	Total P (g/kg)	Available K (g/kg)	CEC (Cmol/kg)
0–10 cm	CK	5.78a	2.46b	0.53b	0.25b	10.73b
	RC	5.48b	2.92a	0.59b	0.28ab	12.23a
	ROF	5.21c	2.82a	0.70a	0.30a	10.97b
	ROM	5.64a	2.56b	0.59b	0.33a	11.92a
	RP	5.15c	2.96a	0.72a	0.31a	10.80b
	RR	5.40b	2.67b	0.67a	0.24b	11.23b
10–20 cm	CK	5.86b	2.36b	0.47	0.22b	10.26c
	RC	5.82b	2.61ab	0.49	0.23b	13.50a
	ROF	6.00a	2.52b	0.54	0.23b	11.41b
	ROM	5.96a	2.41b	0.48	0.30a	11.43b
	RP	5.84b	2.70a	0.53	0.25b	11.01b
	RR	5.89b	2.48b	0.54	0.21b	10.34c
Comparison of depth	0–10 cm	5.44b	2.73a	0.63a	0.29	11.31
	10–20 cm	5.90a	2.51b	0.51b	0.24	11.33
ANOVA	Rotation	**	**	**	**	**
	Depth	**	**	**	ns	ns
	R × D	*	*	*	**	**

Means on the same column and for the same sampling time followed by the same letter (or none) are not significantly different at $P < 0.05$ by Tukey Test. ns: not significant; *$P < 0.05$; **$P < 0.01$. Abbreviations: CK: rice-fallow phase; RR: rice-ryegrass; RP: rice-potato, with rice straw as mulch; RC: rice-Chinese milk vetch; ROM: rice-oilseed rape with burned straw return after harvest; ROF: rice-oilseed rape with fresh straw return as mulch at flowering.

TABLE 5: Soil organic carbon content, dissolved organic content, and soil microbial organic content in 0–10 and 10–20 cm depth

Rotation	SOM (g/kg)	DOC (g/kg)	MBC (g/kg)	SOM (g/kg)	DOC (mg/kg)	MBC (g/kg)
		0–10 cm			10–20 cm	
CK	21.6b	0.08b	0.76b	22.2ab	0.07b	0.64b
RC	25.8a	0.16a	1.08a	21.4b	0.13a	1.11a
ROF	20.3bc	0.15a	1.22a	16.6c	0.12a	1.08a
ROM	18.9c	0.09b	0.51b	17.1c	0.15a	0.44b
RP	23.7a	0.20a	0.57b	23.4a	0.08b	0.60b
RR	21.3b	0.19a	1.09a	18.6c	0.14a	1.04a
Comparison of depth						
0–10 cm	21.93a	0.15	0.87			
10–20 cm	19.88b	0.11	0.82			
ANOVA						
Rotation	**	**	**			
Depth	**	ns	ns			
R × D	**	**	**			

Means on the same column and for the same sampling time followed by the same letter (or none) are not significantly different at $P < 0.05$ by Tukey Test; ns: not significant; *$P < 0.05$; **$P < 0.01$. Abbreviations: CK: rice-fallow phase; RR: rice-ryegrass; RP: rice-potato, with rice straw as mulch; RC: rice-Chinese milk vetch; ROM: rice-oilseed rape with burned straw return after harvest; ROF: rice-oilseed rape with fresh straw return as mulch at flowering.

had the highest average yield of 7.74 t/ha, which was 27.8% higher than CK. The yield increase might be attributed to the high nitrogen fix capabilities and biomass accumulation in RC [49, 50] However, the capacity of green manure to sufficiently supply soil nutrients is still variable and depends on biomass production and soil management. Slight increases were found in RR, RP, ROM, and ROF compared with CK, although the differences were not statistically

significant. The rotation of rice-potato produced the highest biomass production, followed by ryegrass, rapeseed, and Chinese milk vetch (Figure 1). However, considering the harvest of RR for pasture and the burned straw of ROM, the amount of plant residues returned to the soil was limited. Therefore, only three rotations of RP, RC, and ROF had the organic material returned. RP had the highest value of returned organic material, but the yield improvement

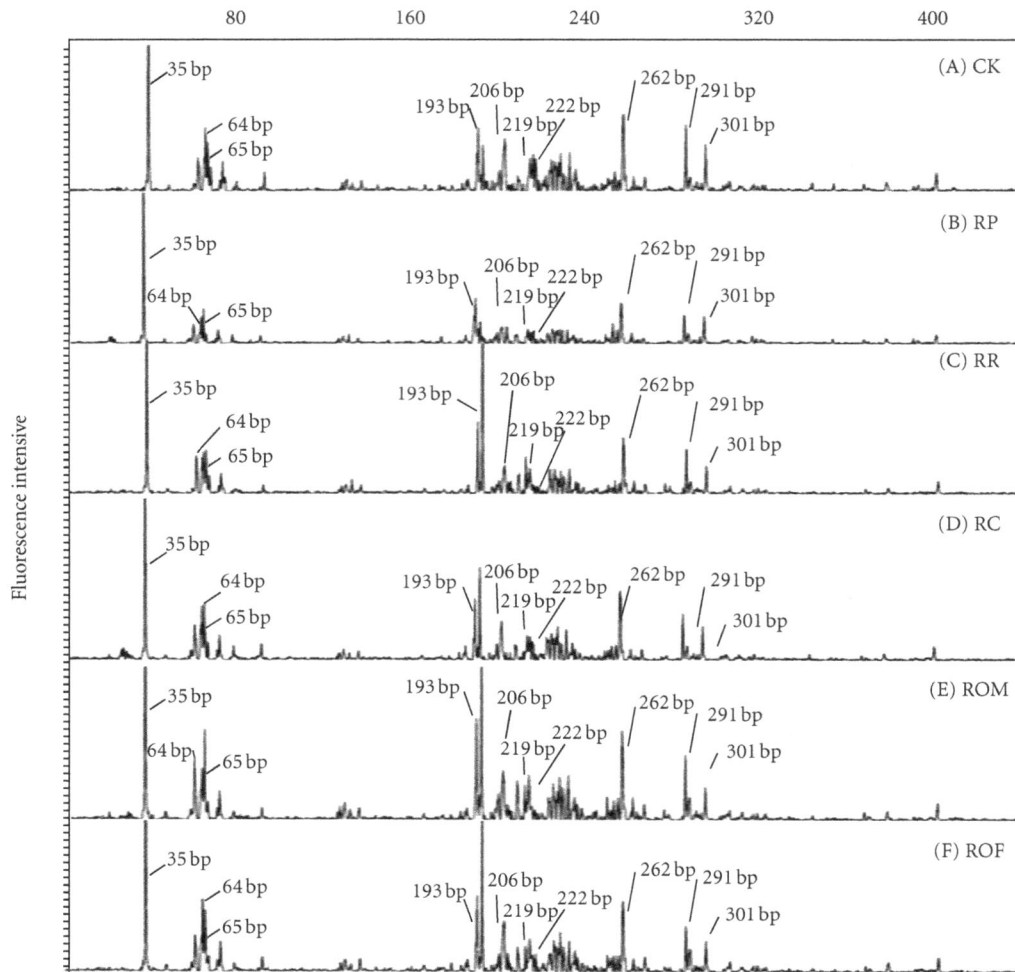

FIGURE 2: Terminal restriction fragment length polymorphism profiles of soil bacterial communities derived from six different rotation fields before transplanting (BR): CK (rice-fallow phase), RR (rice-ryegrass), RP (rice-potato, with rice straw as mulch), RC (rice-Chinese milk vetch), ROM (rice-oilseed rape with burned straw return after harvest), and ROF (rice-oilseed rape with fresh straw return as mulch at flowering) are shown. Terminal fragments were generated by a HaeIII digestion of 16S rRNA gene fragments amplified from total community DNA. Selected terminal restriction fragments differing in their relative abundance between the studied sites are indicated.

was limited. Ghoshal and Singh [47] suggested that crop biomass and grain yield were improved when straw was returned. However, Henderson [51] remarked that most of the studies did not provide an insight into how the procedure influenced crop yield. The results obtained were often site- and year-specific and often contradictory and inconclusive due to variability in soil type, cropping systems, and climate [52, 53]. Therefore, further research is needed to investigate the influence of organic return on rice yield in different paddy-upland rotations.

4.2. Paddy-Upland Crop Rotation Effects on Soil Quality Attributes. In paddy-upland crop rotations, soil puddling in advance of transplanting can foster high productivity [54]. This procedure involves plowing the soil when wet, puddling it, and keeping the area flooded for the duration of rice growth. Puddling breaks down and disperses soil aggregates

into microaggregates and individual particles [12]. However, continuous use of this method of rice cropping will destroy soil structure and create a poor physical condition [13, 14]. In this study, RC and RP improved the soil MWD compared to CK. In addition, RC and ROF increased the soil bulk density compared with CK (Table 3). These results suggest that rice rotated with milk vetch might improve the paddy soil structure by increasing the soil MWD.

Significant soil chemical quality diversity was found among rotations. Soil pH value was acid in the paddy field, which was consistent in our results. Application of ROM increased the pH value as well as available K compared with CK. This might be due to the use of rapeseed straw burned into ash and applied as fertilizer. Biomass ash is considered as a potassium fertilizer in China. Furthermore, improved TN and SOM were found in RC and RP. The positive relationship between yield and TN and SOM might provide the evidence for the positive effect of RC rotation on rice yield.

Effect of Long-Term Paddy-Upland Yearly Rotations on Rice (Oryza sativa) Yield, Soil Properties, and Bacteria
Community Diversity

29

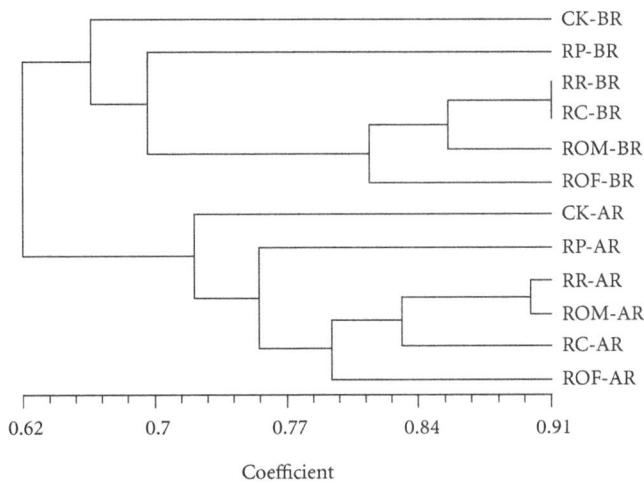

FIGURE 3: UPGMA dendrogram generated from all representative T-RFLP sample profiles. The scale indicated the coefficient between soil from paddy-upland crop rotation systems sampled at postharvest (AR) and pretransplanting (BR); Abbreviations: CK, rice-fallow phase; RR, rice-ryegrass; RP, rice-potato, with rice straw as mulch; RC, rice-Chinese milk vetch; ROM, rice-oilseed rape with burned straw return after harvest; ROF, rice-oilseed rape with fresh straw return as mulch at flowering.

Soil quality is different for different crop species [55–57] as well as the proper management of crop residues in terms of improving soil organic matter dynamics and nutrient cycling [58, 59]. In our experiment, soil quality was influenced by both crop species and residue management, and the integrated results were shown by soil quality indicators. Green manure crop (milk vetch) cultivation of upland plant species during the fallow seasons in the monorice cultivation system was found to improve soil fertility because of its higher nitrogen fixing capabilities and rapid biomass accumulation characteristics [49, 50] We found that MWD, TN, CEC, SOM, DOC, and MBC improved in rice-milk vetch rotations compared with CK. However, increased bulk density was also found in the rice-milk vetch rotation; this has a generally negative effect on soil quality. Rapeseed and winter rye can be sown after rice and harvested before rice transplanting. These systems can maximize benefits of the rotation as well as availability and resources [60]. *Brassica* species are important oil seed crops in China and can aid in controlling pests and weeds because of the allelochemicals they release. In this study, two different rapeseed agronomic practices were introduced in terms of different harvest dates. Unlike the regular harvest practice for oil seed, the rapeseed was harvested artificially at flowering, when its glucosinolate content is relatively high [61, 62]. In addition, rapeseed also has potential for use as a green manure crop, which may prevent soil erosion and reclaim leachable nutrients [63, 64]. We found that ROM and ROF had the most soil CEC, TP, and available K in 10–20 cm depth (Table 4) and that ROF has the highest value of all the rotations in MBC in 10–20 cm depth (Table 5).

4.3. Paddy-Upland Crop Rotations on Soil Bacterial Communities' Diversity.
Soil bacterial communities were significantly affected by soil type and plant species as well as environmental factors. As shown in Figure 3, two major separate branches were found among eight rotations: pretransplanting (BR) and postharvest (AR). Within rotations, CK was totally different from the others; it had the lowest similarity value regardless of sample time, with a mean similarity of 62.5% at BR and 71.3% at AR to others. Further, four out of the other five rotations were grouped together with a mean similarity to RP of 69.8% at BR and 75.8% at AR. These results indicate remarkably different long-term paddy-upland rotation effects on soil bacterial community construction. However, further analysis is needed to explore the details of these bacterial communities.

5. Conclusions

Significant differences in soil chemicals (i.e., soil pH, TN, CEC, SOM, and MBC), physical properties (soil bulk density), and soil bacterial communities were detected between cropping seasons within the year (rice and upland crops season), irrespective of different winter upland crop species. Rice-Chinese milk vetch and rice-rapeseed rotations improved the soil quality to some extent, which might result in the greatest yield performance in rice-Chinese milk vetch rotations among the tested rotations. Soil bacterial communities in CK and RP were remarkably different from those in the other rotations according to the T-RFLP of 16S rRNA genes.

Acknowledgments

This research was supported by grants from the "Five-twelfth" National Science and Technology Support Program (2011BAD16B14) and the National Natural Science Foundation of China (31171502).

References

[1] K. G. Cassman, S. K. De Datta, D. C. Olk et al., "Yield decline and the nitrogen economy of long-term experiments on continuous, irrigated rice systems in the tropics," in *Soil Management, Experimental Basis For Sustainability and Environmental Quality*, R. Lal and B. A. Stewart, Eds., pp. 181–222, Lewis, CRC, Boca Raton, Fla, USA, 1995.

[2] A. Dobermann and C. Witt, "The potential impact of crop intensification on carbon and nitrogen cycling in intensive rice systems," in *Carbon and Nitrogen Dynamics in Flooded Soils*, G. D. Kirk and D. Olk, Eds., pp. 1–25, International Rice Research Institute, Laguna, Philippines, 2000.

[3] K. Yanagisawa and Y. Muramatsu, "Transfer of technetium from soil to paddy and upland rice," *Journal of Radiation Research*, vol. 36, no. 3, pp. 171–178, 1995.

[4] C. Witt, K. G. Cassman, D. C. Olk et al., "Crop rotation and residue management effects on carbon sequestration, nitrogen cycling and productivity of irrigated rice systems," *Plant and Soil*, vol. 225, no. 1-2, pp. 263–278, 2000.

[5] S. W. Chang Chien, M. C. Wang, J. H. Hsu, and K. Seshaiah, "Influence of fertilizers applied to a paddy-upland rotation on

characteristics of soil organic carbon and humic acids," *Journal of Agricultural and Food Chemistry*, vol. 54, no. 18, pp. 6790–6799, 2006.

[6] S. Nishimura, S. Yonemura, T. Sawamoto et al., "Effect of land use change from paddy rice cultivation to upland crop cultivation on soil carbon budget of a cropland in Japan," *Agriculture, Ecosystems and Environment*, vol. 125, no. 1–4, pp. 9–20, 2008.

[7] S. Huang, W. Y. Rui, X. X. Peng, W. R. Liu, and W. J. Zhang, "Responses of soil organic carbon content and fractions to land-use conversion from paddy field to upland," *Chinese Journal of Environmental Science*, vol. 30, no. 4, pp. 1146–1151, 2009.

[8] N. Yamaguchi, A. Kawasaki, and I. Iiyama, "Distribution of uranium in soil components of agricultural fields after long-term application of phosphate fertilizers," *Science of the Total Environment*, vol. 407, no. 4, pp. 1383–1390, 2009.

[9] C. K. Li, *Paddy Soils of China*, Science Press, Beijing, China, 1992.

[10] D. T. Xie and S. L. Chen, *Theory and Technique of Paddy Field Under Soil Virginization*, Southwest Normal University Press, Chongqing, China, 2002.

[11] L. Qunhua, K. Xin, C. Changzhi et al., "New irrigation methods sustain malaria control in Sichuan Province, China," *Acta Tropica*, vol. 89, no. 2, pp. 241–247, 2004.

[12] G. Kirchhof, H. B. So, T. Adisarwanto et al., "Growth and yield response of grain legumes to different soil management practices after rainfed lowland rice," *Soil and Tillage Research*, vol. 56, no. 1-2, pp. 51–66, 2000.

[13] B. S. Boparai, Yadvinder-Singh, and B. D. Sharma, "Effect of green manuring with *Sesbania aculeata* on physical properties of soil and on growth of wheat in rice-wheat and maize-wheat cropping systems in a semiarid region of India," *Arid Soil Research & Rehabilitation*, vol. 6, no. 2, pp. 135–143, 1992.

[14] M. Mohanty and D. K. Painuli, "Land preparatory tillage effect on soil physical environment and growth and yield of rice in a Vertisol," *Journal of the Indian Society of Soil Science*, vol. 51, no. 3, pp. 223–228, 2003.

[15] E. G. Gregorich, C. M. Monreal, M. R. Carter, D. A. Angers, and B. H. Ellert, "Towards a minimum data set to assess soil organic matter quality in agricultural soils," *Canadian Journal of Soil Science*, vol. 74, no. 4, pp. 367–385, 1994.

[16] I. Hussain, K. R. Olson, M. M. Wander, and D. L. Karlen, "Adaptation of soil quality indices and application to three tillage systems in Southern Illinois," *Soil and Tillage Research*, vol. 50, no. 3-4, pp. 237–249, 1999.

[17] J. W. Doran and M. R. Zeiss, "Soil health and sustainability: managing the biotic component of soil quality," *Applied Soil Ecology*, vol. 15, no. 1, pp. 3–11, 2000.

[18] J. M. Tisdall and J. M. Oades, "Organic matter and water-stable aggregates in soils," *Journal of Soil Science*, vol. 33, no. 2, pp. 141–163, 1982.

[19] G. J. Churchman and K. R. Tate, "Stability of aggregates of different size grades in allophanic soils from volcanic ash in New Zealand," *Journal of Soil Science*, vol. 38, no. 1, pp. 19–27, 1987.

[20] C. A. Campbell and R. P. Zentner, "Soil organic matter as influenced by crop rotations and fertilization," *Soil Science Society of America Journal*, vol. 57, no. 4, pp. 1034–1040, 1993.

[21] M. A. Arshad and G. M. Coen, "Characterization of soil quality: physical and chemical criteria," *American Journal of Alternative Agriculture*, vol. 7, no. 1-2, pp. 25–31, 1992.

[22] B. Giri, P. H. Giang, R. Kumari, R. Prasad, and A. Varma, "Microbial diversity in soils," in *Micro-Organisms in Soils,* *Roles in Genesis and Functions*, F. Buscot and S. Varma, Eds., pp. 195–212, Springer, Heidelberg, Germany, 2005.

[23] H. P. Bais, T. L. Weir, L. G. Perry, S. Gilroy, and J. M. Vivanco, "The role of root exudates in rhizosphere interactions with plants and other organisms," *Annual Review of Plant Biology*, vol. 57, pp. 233–266, 2006.

[24] B. Prithiviraj, M. W. Paschke, and J. M. Vivanco, "Root communication, the role of root exudates," *Encyclopedia of Plant and Crop Science*, vol. 1, pp. 1–4, 2007.

[25] L. Innes, P. J. Hobbs, and R. D. Bardgett, "The impacts of individual plant species on rhizosphere microbial communities in soils of different fertility," *Biology and Fertility of Soils*, vol. 40, no. 1, pp. 7–13, 2004.

[26] K. M. Batten, K. M. Scow, K. F. Davies, and S. P. Harrison, "Two invasive plants alter soil microbial community composition in serpentine grasslands," *Biological Invasions*, vol. 8, no. 2, pp. 217–230, 2006.

[27] A. C. Kennedy and K. L. Smith, "Soil microbial diversity and the sustainability of agricultural soils," *Plant and Soil*, vol. 170, no. 1, pp. 75–86, 1995.

[28] K. E. Giller, M. H. Beare, P. Lavelle, A. M. N. Izac, and M. J. Swift, "Agricultural intensification, soil biodiversity and agroecosystem function," *Applied Soil Ecology*, vol. 6, no. 1, pp. 3–16, 1997.

[29] M. J. Swift and J. M. Anderson, "Biodiversity and ecosystem function in agricultural systems," *Biodiversity and Ecosystem Function*, vol. 99, no. 6, pp. 15–41, 1993.

[30] R. G. Joergensen and F. Wichern, "Quantitative assessment of the fungal contribution to microbial tissue in soil," *Soil Biology and Biochemistry*, vol. 40, no. 12, pp. 2977–2991, 2008.

[31] L. Rukun, Zheng, and S. Jian S, "Nutrient balance of agroecosystem in six provinces in Southern China," *Scientia Agricultura Sinica*, vol. 33, no. 2, pp. 63–67, 2000.

[32] Q. Yusheng, T. Shihua, F. Wenqiang, and S. Xifa, "Spatial variability of soil nutrient characteristics under paddy-upland crop rotation in chengdu plain," *Acta Pedologica Sinica*, vol. 45, no. 2, pp. 354–359, 2008.

[33] H. Chu, S. Morimoto, K. Fujii, K. Yagi, and S. Nishimura, "Soil ammonia-oxidizing bacterial communities in paddy rice fields as affected by upland conversion history," *Soil Science Society of America Journal*, vol. 73, no. 6, pp. 2026–2031, 2009.

[34] Z. Gong, G. Zhang, and G. Luo, "Versify of Anthrosols in China," Pedosphere. 9, 1999.

[35] S. S. Staff, *Keys To Soil Taxonomy*, U.S.D. A Soil Conservation Service, Blacksburg, Va, USA, 7th edition, 1996.

[36] G. Fu, D. Wang, C. Xu, and X. Zhang, "Effect of winter conservation tillage in paddy field on soil enzyme activities and grain quality," *Plant Nutrition and Fertilizer Science*, vol. 15, no. 3, pp. 618–624, 2009.

[37] R. L. Parfitt, C. Ross, L. A. Schipper, J. J. Claydon, W. T. Baisden, and G. Arnold, "Correcting bulk density measurements made with driving hammer equipment," *Geoderma*, vol. 157, no. 1-2, pp. 46–50, 2010.

[38] J. Six, K. Paustian, E. T. Elliott, and C. Combrink, "Soil structure and organic matter: I. Distribution of aggregate-size classes and aggregate-associated carbon," *Soil Science Society of America Journal*, vol. 64, no. 2, pp. 681–689, 2000.

[39] W. H. Hendershot, H. Lalande, and M. Duquette, "Ion exchange and exchangeable cations," in *Soil Sampling and Methods of Analysis*, M. R. Carter, Ed., pp. 167–175, Lewis, Boca Raton, Fla, USA, 1993.

[40] J. M. Bremner and C. S. Mulvaney, "Nitrogen-total," in *Methods of Soil Analysis*, Agronomy Monograph 9, Part 2, pp.

Effect of Long-Term Paddy-Upland Yearly Rotations on Rice (Oryza sativa) Yield, Soil Properties, and Bacteria
Community Diversity

31

595–624, American Society of Agronomy, Madison, Wis, USA, 2nd edition, 1982.

[41] G. P. Sparling, K. N. Whale, and A. J. Ramsay, "Quantifying the contribution from the soil microbial biomass to the extractable P levels of fresh and air-dried soils," *Australian Journal of Soil Research*, vol. 23, no. 4, pp. 613–621, 1985.

[42] D. Nelson and L. Sommers, "Total carbon, organic carbon and organic matter," in *Methods of Soil Analysis SSSA Book*, Series No. 7. SSSA, Part 2, pp. 539–579, SSSA, Madison, Wis, USA, 1996.

[43] E. D. Vance, P. C. Brookes, and D. S. Jenkinson, "Microbial biomass measurements in forest soils: the use of the chloroform fumigation-incubation method in strongly acid soils," *Soil Biology and Biochemistry*, vol. 19, no. 6, pp. 697–702, 1987.

[44] J. Zhou, M. A. Bruns, and J. M. Tiedje, "DNA recovery from soils of diverse composition," *Applied and Environmental Microbiology*, vol. 62, no. 2, pp. 316–322, 1996.

[45] W. T. Liu, T. L. Marsh, H. Cheng, and L. J. Forney, "Characterization of microbial diversity by determining terminal restriction fragment length polymorphisms of genes encoding 16S rRNA," *Applied and Environmental Microbiology*, vol. 63, no. 11, pp. 4516–4522, 1997.

[46] J. Dunbar, L. O. Ticknor, and C. R. Kuske, "Phylogenetic specificity and reproducibility and new method for analysis of terminal restriction fragment profiles of 16S rRNA genes from bacterial communities," *Applied and Environmental Microbiology*, vol. 67, no. 1, pp. 190–197, 2001.

[47] N. Ghoshal and K. P. Singh, "Effects of farmyard manure and inorganic fertilizer on the dynamics of soil microbial biomass in a tropical dryland agroecosystem," *Biology and Fertility of Soils*, vol. 19, no. 2-3, pp. 231–238, 1995.

[48] S. Y. Kim, S. H. Oh, W. H. Hwang, K. Y. Choi, and B. G. Oh, "Optimum soil incorporation time of Chinese milk vetch (*Astralagus sinicus* L.) for its natural re-seeding and green manuring of rice in Gyeongnam province Korea," *Journal of Crop Science and Biotechnology*, vol. 11, pp. 193–198, 2008.

[49] N. K. Fageria, "Green manuring in crop production," *Journal of Plant Nutrition*, vol. 30, no. 5, pp. 691–719, 2007.

[50] C. H. Lee, K. D. Park, K. Y. Jung et al., "Effect of Chinese milk vetch (*Astragalus sinicus* L.) as a green manure on rice productivity and methane emission in paddy soil," *Agriculture, Ecosystems and Environment*, vol. 138, no. 3-4, pp. 343–347, 2010.

[51] D. W. Henderson, "Soil water management in semiarid environments," in *Agriculture in the Semi Arid Environments*, A. E. Hall, G. H. Cannel, and H. W. Lawton, Eds., pp. 225–237, Springer, Berlin, Germany, 1979.

[52] C. J. Wright and D. C. Coleman, "The effects of disturbance events on labile phosphorus fractions and total organic phosphorus in the Southern Appalachians," *Soil Science*, vol. 164, no. 6, pp. 391–402, 1999.

[53] A. C. Guzha, "Effects of tillage on soil microrelief, surface depression storage and soil water storage," *Soil and Tillage Research*, vol. 76, no. 2, pp. 105–114, 2004.

[54] S. Singh, S. N. Sharma, and R. Prasad, "The effect of seeding and tillage methods on productivity of rice-wheat cropping system," *Soil and Tillage Research*, vol. 61, no. 3-4, pp. 125–131, 2001.

[55] D. L. Karlen, N. S. Eash, and P. W. Unger, "Soil and crop management effects on soil quality indicators," *American Journal of Alternative Agriculture*, vol. 7, no. 1-2, pp. 48–55, 1992.

[56] C. A. Campbell, V. O. Biederbeck, B. G. McConkey, D. Curtin, and R. P. Zentner, "Soil quality—effect of tillage and fallow frequency. Soil organic matter quality as influenced by tillage and fallow frequency in a silt loam in Southwestern Saskatchewan," *Soil Biology and Biochemistry*, vol. 31, no. 1, pp. 1–7, 1999.

[57] A. D. Halvorson, B. J. Wienhold, and A. L. Black, "Tillage, nitrogen, and cropping system effects on soil carbon sequestration," *Soil Science Society of America Journal*, vol. 66, no. 3, pp. 906–912, 2002.

[58] J. T. Spargo, M. A. Cavigelli, S. B. Mirsky, J. E. Maul, and J. J. Meisinger, "Mineralizable soil nitrogen and labile soil organic matter in diverse long-term cropping systems," *Nutrient Cycling in Agroecosystems*, vol. 90, no. 2, pp. 253–266, 2011.

[59] R. Lal, "Challenges and opportunities in soil organic matter research," *European Journal of Soil Science*, vol. 60, no. 2, pp. 158–169, 2009.

[60] A. Clark, *Managing Cover Crops Profitably*, Handbook Series Book 9, Sustainable Agriculture Network, 3rd edition, 2007.

[61] S. Millán, M. C. Sampedro, P. Gallejones et al., "Identification and quantification of glucosinolates in rapeseed using liquid chromatography-ion trap mass spectrometry," *Analytical and Bioanalytical Chemistry*, vol. 394, no. 6, pp. 1661–1669, 2009.

[62] S. Yasumoto, M. Matsuzaki, H. Hirokane, and K. Okada, "Glucosinolate content in rapeseed in relation to suppression of subsequent crop," *Plant Production Science*, vol. 13, no. 2, pp. 150–155, 2010.

[63] B. S. Choi, K. C. Hong, J. J. Nam et al., "Effect of rapeseed (*Brassica napus*) incorporated as green manure on weed growth in rice paddy, a pot experiment," *Korean Journal of Weed Science*, vol. 29, no. 1, pp. 39–45, 2009.

[64] E. Bernard, R. P. Larkin, S. Tavantzis et al., "Compost, rapeseed rotation, and biocontrol agents significantly impact soil microbial communities in organic and conventional potato production systems," *Applied Soil Ecology*, vol. 52, pp. 29–41, 2012.

4

Effect and Removal Mechanisms of 6 Different Washing Agents for Building Wastes Containing Chromium

Wang Xing-run,[1] **Zhang Yan-xia,**[1,2] **Wang Qi,**[1] **and Shu Jian-min**[1]

[1] State Key Laboratory of Environmental Criteria and Risk Assessment, Chinese Research Academy of Environmental Sciences, Beijing 100012, China
[2] College of Science, Northwest A&F University, Shaanxi 712100, China

Correspondence should be addressed to Wang Xing-run, xingrunwang@gmail.com

Academic Editors: W. K. Jo and R. A. Smith

With the building wastes contaminated by chromium in Haibei Chemical Plan in China as objects, we studied the contents of total Cr and Cr (VI) of different sizes, analyzed the effect of 6 different washing agents, discussed the removal mechanisms of 6 different washing agents for Cr in various forms, and finally selected applicable washing agent. As per the results, particle size had little impact on the contents of total Cr and Cr (VI); after one washing with water, the removal rate of total Cr and Cr (VI) was 75% and 78%, respectively, and after the second washing with 6 agents, the removal rate of citric acid was the highest, above 90% for total Cr and above 99% for hexavalent chromium; the pH of building wastes were reduced by citric acid, and under acid condition, hexavalent chromium was reduced to trivalent chromium spontaneously by organic acid, which led to better removal rate of acid soluble Cr and reducible Cr; due to the complexing action, citric acid had best removal rate for oxidizable trivalent chromium. In conclusion, citric acid is the most applicable second washing agent for building wastes.

1. Introduction

So far, there are 75 chromate enterprises in China. Due to the small scale, the backward techniques, and high environmental pollution of chromate enterprises in China, over 50 enterprises have been shut down. Not only the soil around the closed chromate enterprises had been severely contaminated, but there were serious environmental pollutions in the building wastes [1, 2]. In Minfeng Chemical Plant in Chongqing, the contaminated building wastes were about 30,000 m^3 and the content of hexavalent chromium was 2374 mg/kg in building wastes; in Haibei Chemical Plant in Qinghai Province, there were at least 7000 m^3 building wastes from sintering workshop and leaching workshop, and the content of hexavalent chromium in concrete of internal foundation was up to 6278 mg/kg. Due to the different composition of building wastes and soil, and the different forms of chromium occurrence in building wastes and soil the applicable treatment technologies for soil are not suitable for building wastes. As the progress of chromium contaminated

soil remediation all over all China, the treatment of building wastes is of great urgency.

In abroad, studies on contaminated soil remediation have been started for some time, and large amount of works have been done in Europe, USA and Japan [3–5]. Chemical washing can be used to separate and isolate hazardous substances or turn hazardous substances harmless [6], and besides, the technique is suitable for wide use based on its advantages, such as low energy consumption, low equipment investment, wide application scope and quick effect. The key point of chemical washing is to select and develop washing agents [7]. The commonly used washing agents includes water, acid [8], saline solution, chelating agent [9], surfactant [10], and so forth. The most economic and environmental protection washing agent is water, and one chromium-plating company in USA, United Chromium (Corvallis, OR), was using water to washing Cr (VI) in project site, which reduced the concentration of chromium from 1923 mg/kg to 65 mg/kg [11]. EDTA is also one of the commonly used washing agent and it can generate stable chelating agent by

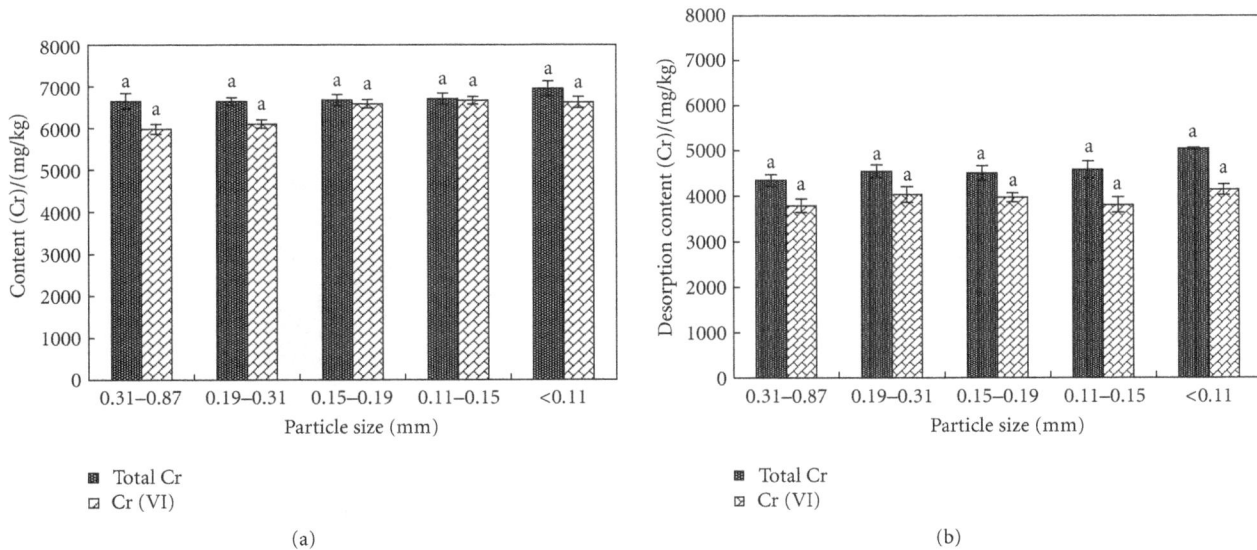

(a)

(b)

FIGURE 1: Contents and desorption contents of total Cr and Cr (VI) in 5 samples of different sizes.

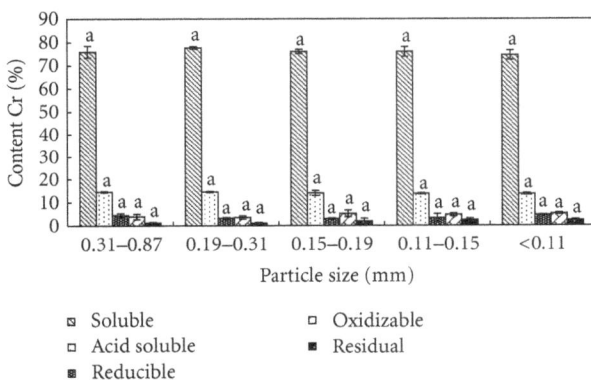

FIGURE 2: Different forms of Cr in 5 samples of different sizes.

reacting with most metals [12]. As indicated by Tampouris [13], the removal rate of Zn and Cd in contaminated soil by HCl + CaCl$_2$ was 78% and 70%, respectively. Lee et al. [14] found that the removal rate of As in sand at river bottom was 95% by citric acid washing. Bhattacharya et al. [15] adopted oxalates to remove the Cr in soil from lumber yard and 98% Cr was removed. However, there are few studies on the treatment of building wastes contaminated by chromium.

In this study, it adopted deionized water, EDTA, citric acid, oxalic acid, HCl, and acetic acid as washing agents to compare the removal effect of Cr in building wastes contaminated by Cr and select applicable agent and analyzed the removal mechanism of those 6 washing agents with the hope of providing technical supports for the disposal of building wastes contaminated by chromium.

2. Materials and Methods

2.1. Chemicals and Reagents.
The concrete paved inside the chromate production workshop in Beihai Chromate Plant was severely contaminated by Cr; the study took the concrete as object and sampled at 5 different locations; the collected concretes were dried in the air naturally and samples of 5 different sizes were fabricated by the crushing with 20, 50, 80, 100, and 150 mesh screens; the particle sizes were 0.31–0.87 mm, 0.19–0.31 mm, 0.15–0.19 mm, 0.11–0.15 mm and <0.11 mm; samples of the same size were mixed uniformly and stored properly for later use. Samples of 5 different sizes were mixed at 1 : 1 mass ratio and their physicochemical properties are shown in Table 1. Washing agent: deionized water, 0.05 mol/L and 0.1 mol/L EDTA-Na$_2$, 0.1 mol/L and 0.5 mol/L citric acid, 0.1 mol/L and 0.5 mol/L oxalic acid, 0.1 mol/L, 0.5 mol/L and 1 mol/L HCl, and 0.1 mol/L and 0.5 mol/L acetic acid.

2.2. Testing Methods.
Accurately weight 5.00 g sample and place it in a 250 mL conical flash, add 50 mL washing agent in the flash with solid-to-liquid ratio of 1 : 10 (i.e., 1 g : 10 mL), wash 40 min with electromagnetic stirring, and filter the mixture to collect eluent and solid sample. The washed sample shall be dried and properly stored for future use.

2.3. Sample Analysis.
Alkaline digestion [16] was applied to dissolve the Cr (VI) in solid sample, and atomic absorption spectrophotometry was used to test the concentration of Cr (VI) in dissolution. Microwave-assisted acid digestion [17] was used to dissolve the total Cr in solid sample, and atomic absorption spectrophotometry was used to test the concentration of total Cr in dissolution. Coprecipitation was used to dissolve the Cr (VI) in eluent and atomic absorption spectrophotometry was used to test the concentration of Cr (VI). Modified BCR sequential extraction was adopted to extract the soluble, acid soluble, reducible, oxidizable, and residual Cr in solid sample while atomic absorption spectrophotometry was used to test the concentration of Cr in extract.

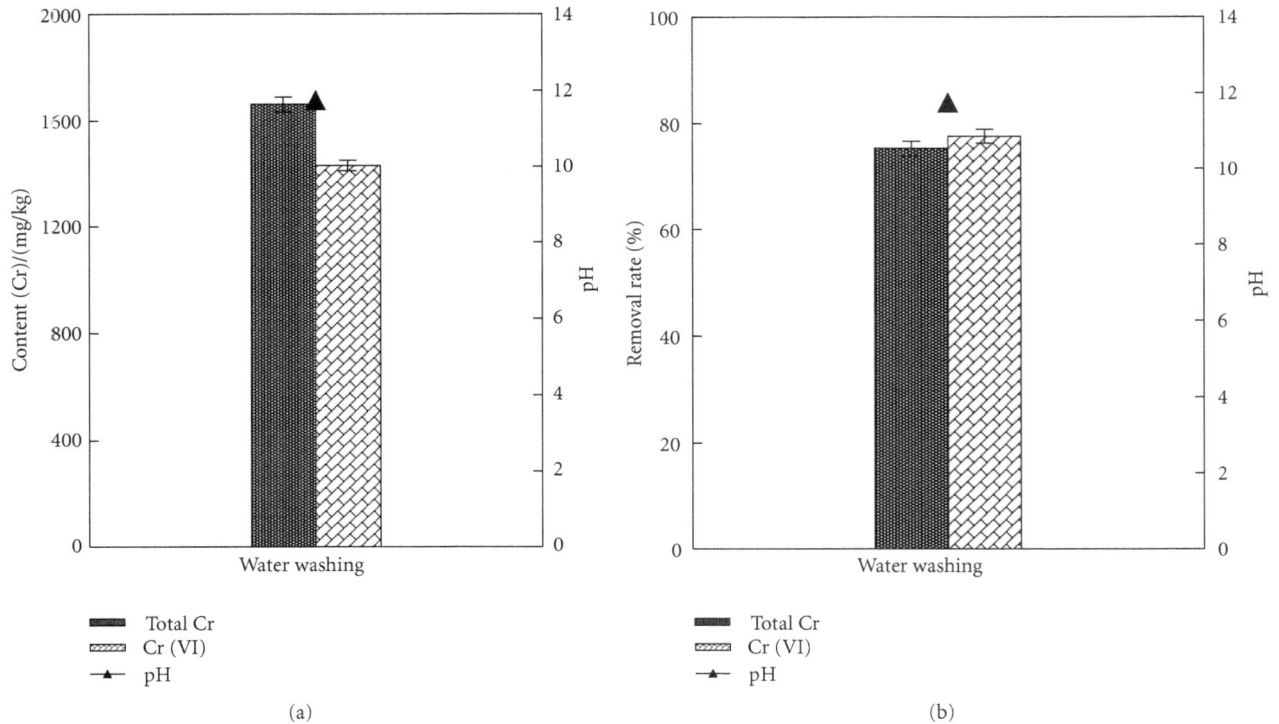

(a)   (b)

FIGURE 3: Content of total Cr and Cr (VI) and the pH in solid sample after water washing.

TABLE 1: Physicochemical properties of the sample.

| pH | w (total Cr)/(mg/kg) | w [Cr (VI)]/(mg/kg) | w (Cr)/(mg/kg) | | | | |
			Soluble	Acid soluble	Reducible	Oxidizable	Residual
11.91 ± 0.01	6714.67 ± 101.59	6387.96 ± 67.33	5164.80 ± 89.39	912.05 ± 5.71	250.02 ± 8.05	305.60 ± 8.45	126.06 ± 3.92

3. Results and Discussion

3.1. Impact of Particle Sizes

3.1.1. Impact on Cr Content in Different Sizes. The contents of total Cr and Cr (VI) in 5 samples of different sizes were measured, as shown in Figure 1(a). After being washed 40 min in deionized water, the desorption contents of total Cr and Cr (VI) in 5 samples were measured, as shown in Figure 1(b).

As indicated by Figure 1(a), the content of total Cr in sample was around 6600 mg/kg and the content of Cr (VI) was around 6300 mg/kg. Figure 1(a) also showed that there was no distinct difference on the contents of total Cr and Cr (VI) in different samples. As indicated by Figure 1(b), there was no distinct difference on the desorption contents of total Cr and Cr (VI) in 5 samples. Thus, particle size had no distinct impact either on the contents of total Cr and Cr (VI) in samples or on the desorption contents of total Cr and Cr (VI). Therefore, in the actual disposal of building wastes contaminated by Cr, it is not necessary to consider the impact of particle size and from the aspect of crushing cost, it shall select more economic crushing equipments. In following experiments, the study adopted 20 mesh screen after crushing.

3.1.2. Impact on Cr Forms in Different Sizes. The contents of Cr in different forms in samples of different sizes were tested, as shown in Figure 2. As indicated by Figure 2, for Cr in different forms in sample of same size, the major form of Cr was soluble, about 75% of the total, and the percentages of acid soluble, reducible, oxidizable, and residual Cr were 15%, 4%, 4% and 2%, respectively. For Cr of same form in different samples, there was no distinct difference on their contents, indicating that particle size had no distinct impact on the contents of Cr in different forms in samples.

3.2. Comparison of Washing Effects by Different Agents

3.2.1. Effect of Water Washing. The content of soluble Cr was high in sample, about 75%. Thus, it adopted washing by deionized water firstly and then tested the contents of total Cr and Cr (VI) in solid sample, as shown by Figures 3(a) and 3(b). The result showed that the remained content of total Cr and Cr (VI) after water washing was 1662.25 mg/kg and 1431.40 mg/kg while the removal rate of total Cr and Cr (VI) was 75.24% and 77.59%, respectively.

As shown by Table 2, the pH of sample was 11.74 after water washing, which was strongly alkaline. Cr (III) was mainly in the form of positive ions, such as Cr^{3+}, $Cr(OH)^{2+}$,

TABLE 2: Physicochemical properties of the sample after water washing.

pH	w (total Cr)/(mg/kg)	w [Cr (VI)]/(mg/kg)	w (Cr)/(mg/kg)				
			Soluble	Acid soluble	Reducible	Oxidizable	Residual
11.74 ± 0.04	1662.25 ± 65.45	1431.40 ± 41.29	1072.85 ± 54.65	347.25 ± 11.26	187.24 ± 4.56	156.68 ± 8.69	34.01 ± 5.93

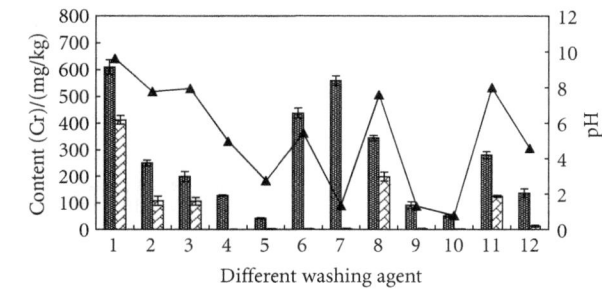

(a)

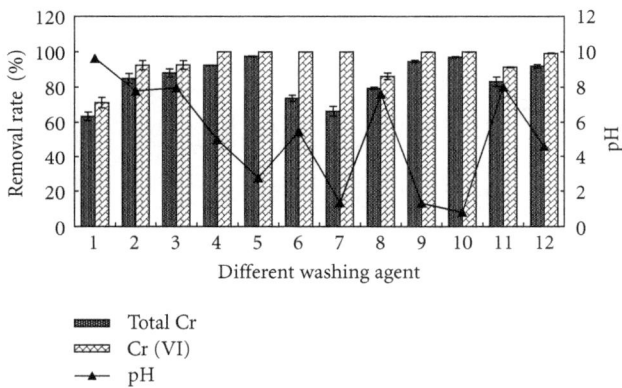

(b)

FIGURE 4: Comparison of total Cr, Cr (VI), and pH in solid sample after being washed by different agents. Washing agent: 1—water; 2—0.05 mol/L EDTA-Na$_2$; 3—0.1 mol/L EDTA-Na$_2$; 4—0.1 mol/L citric acid; 5—0.5 mol/L citric acid; 6—0.1 mol/L oxalic acid; 7—0.5 mol/L oxalic acid; 8—0.1 mol/L HCl; 9—0.5 mol/L HCl; 10—1 mol/L HCl; 11—0.1 mol/L acetic acid; 12—0.5 mol/L acetic acid.

and Cr(OH)$^{2+}$, and under alkaline conditions, it was hard for the hydroxides to dissolve in water. Cr (VI) was mainly in the form of negative ions, such as CrO$_4^{2-}$, HCr$_2$O$_7^{2-}$, HCrO$_4^{3-}$, and Cr$_2$O$_7^{2-}$, and it was easy for sodium, potassium, and ammonium salt to dissolve in water. Thus, it was mainly to remove Cr (VI) in water washing. Although the removal rate reached 75% after water washing, the content of Cr in sample was still high, and it was still quite harmful and needed further disposal by second washing.

3.2.2. Effect of Second Washing by Different Agent. After water washing, second washing was performed with different agents. The contents of total Cr and Cr (VI) after washing were tested, as shown in Figure 4. The result showed that

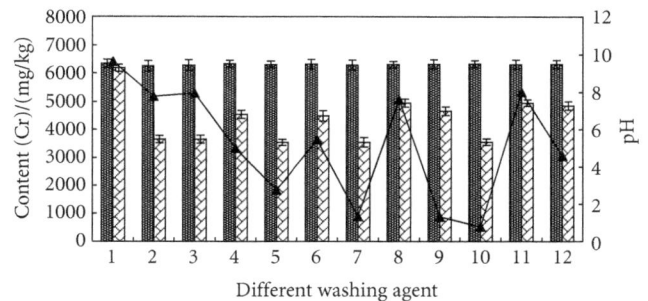

FIGURE 5: Total Cr and Cr (VI) content variation before and after washing. Washing agent: 1—water; 2—0.05 mol/L EDTA-Na$_2$; 3—0.1 mol/L EDTA-Na$_2$; 4—0.1 mol/L citric acid; 5—0.5 mol/L citric acid; 6—0.1 mol/L oxalic acid; 7—0.5 mol/L oxalic acid; 8—0.1 mol/L HCl; 9—0.5 mol/L HCl; 10—1 mol/L HCl; 11—0.1 mol/L acetic acid; 12—0.5 mol/L acetic acid.

there were distinct differences on the remained contents and the removal rates of total Cr and Cr (VI). The removal effects of total Cr by citric acid, concentrated HCl, and concentrated acetic acid were better, and the removal rates were all above 90%. The contents of remained total Cr were 41.5–136.67 mg/kg. The removal effects of Cr (VI) by citric acid, oxalic acid, concentrated HCl, and concentrated acetic acid were better, and the removal rates were above 99%. The contents of remained Cr (VI) were 0.96–12.66 mg/kg. Through comparison, we can see that the removal effect of citric acid was best.

3.2.3. Total Cr and Cr (VI) Content Variation. Test the contents of total Cr and Cr (VI) in solid sample and eluent after second washing, and calculate the total contents of Cr and Cr (VI) by adding the two values, as shown in Figure 5.

As indicated in Figure 5, the contents of total Cr remained the same, about 6600 mg/kg, and the contents of Cr (VI) was reduced, indicating that Cr (VI) was reduced into Cr (III). The pH of citric acid, oxalic acid, HCl, and acetic acid eluent was <6 and under acidic condition, the oxidation-reduction potential was larger than 0 and Cr (VI) was reduced into Cr (III), which reacted spontaneously, as shown in Figure 6. Thus, citric acid, oxalic acid, HCl, and acetic acid can remove Cr (VI) by both dissolution and reduction.

3.3. Removal of Soluble Cr. Figure 7 shows the remained contents and removal rates of soluble Cr in sample after

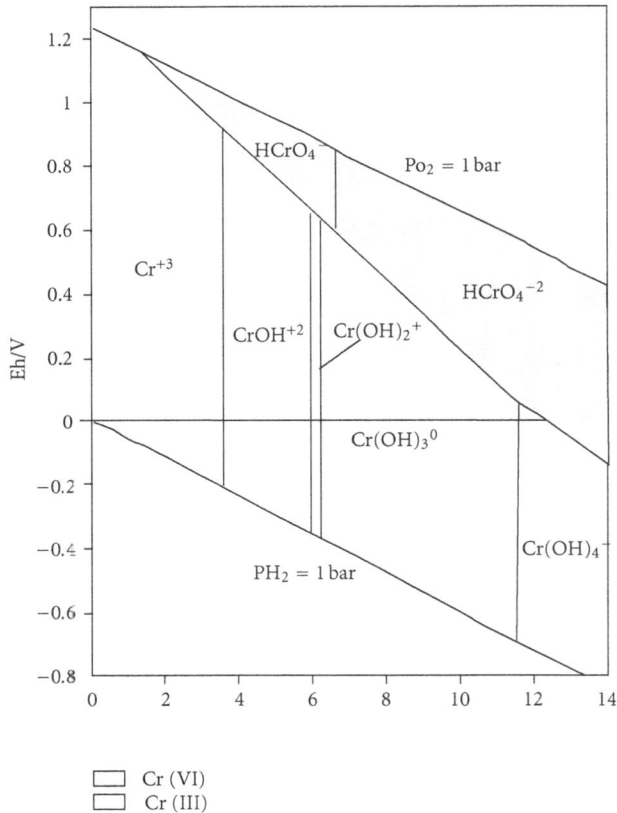

FIGURE 6: Eh-pH of Cr ion forms (25°C).

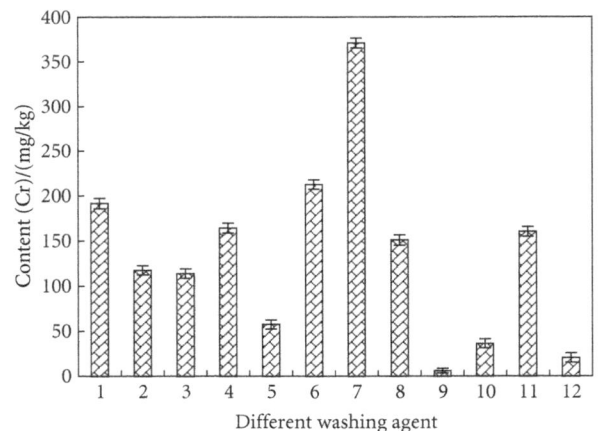

(a)

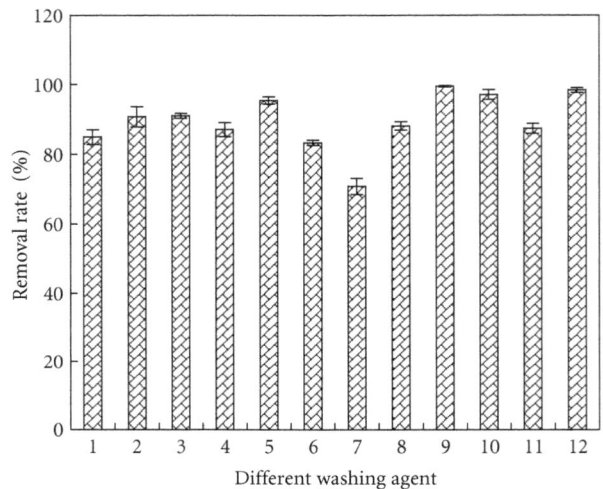

(b)

FIGURE 7: Removal effect of soluble Cr by different washing agents. Washing agent: 1—water; 2—0.05 mol/L EDTA-Na$_2$; 3—0.1 mol/L EDTA-Na$_2$; 4—0.1 mol/L citric acid; 5—0.5 mol/L citric acid; 6—0.1 mol/L oxalic acid; 7—0.5 mol/L oxalic acid; 8—0.1 mol/L HCl; 9—0.5 mol/L HCl; 10—1 mol/L HCl; 11—0.1 mol/L acetic acid; 12—0.5 mol/L acetic acid.

being washed by different agents. As indicated by the results, the second washing of deionized water reduced the content of soluble Cr from 1072.85 mg/kg to 193.34 mg/kg, and the removal rate was 83%. The best removal agents for soluble Cr are concentrated citric acid, concentrated hydrochloric acid, and concentrated acetic acid. With above washing agents, the remained contents of soluble Cr in sample were below 50 mg/kg, and the removal rates were all above 95%.

In sample, Cr (VI) is mainly in soluble form and the reason that concentrated citric acid, concentrated hydrochloric acid, and concentrated acetic acid can obtain better removal effect is that they cannot only dissolve the soluble Cr contained by sample in water, but reduce it into Cr (III) of other forms, which decreases the amount of soluble Cr.

3.4. Removal of Acid Soluble Cr. Figure 8 shows the remained contents and removal rates of acid soluble Cr in sample after being washed by different agents. The content of acid soluble Cr was reduced from 347.25 mg/kg to 227.41 mg/kg by deionized water and the removal rate was 34.5%. The removal rate of EDTA was 93.36–95.35%, that of citric acid was 94.86–99.00%, that of oxalic acid was 83.76–87.70%, that of HCl was 71.75–98.72%, and that of acetic acid was 82.56–96.67%. As indicated by Figure 4, the addition of above agents can reduce the pH of sample, and under acidic conditions, the removal rate of acid soluble Cr was better by all the previous agents.

3.5. Removal of Reducible Cr. Figure 9 shows the remained contents and removal rates of reducible Cr in sample after being washed by different agents. The content of reducible Cr was reduced from 187.24 mg/kg to 176.17 mg/kg by deionized water and the removal rate was 5.91%. The removal effects of citric acid and concentrated HCl were better with removal rate of 93.33% and 95.71%, respectively. After being washed by citric acid and concentrated HCl, the ramained content of reducible Cr in sample was 12.49 mg/kg and 8.04 mg/kg.

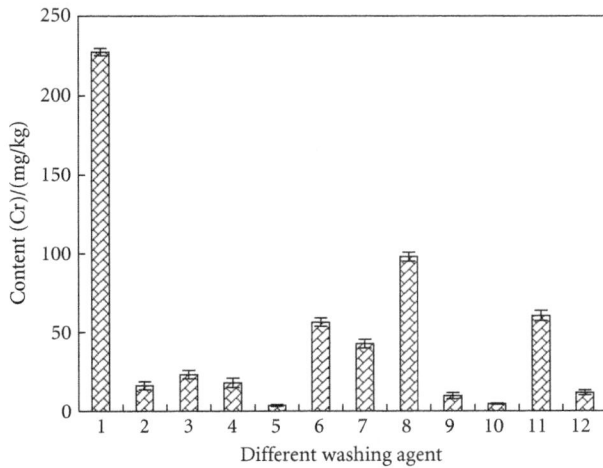

(a)

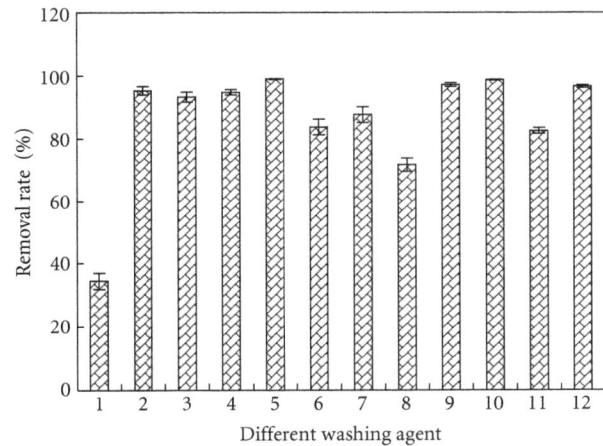

(b)

FIGURE 8: Removal effect of acid soluble Cr by different washing agents. Washing agent: 1—water; 2—0.05 mol/L EDTA-Na₂; 3—0.1 mol/L EDTA-Na₂; 4—0.1 mol/L citric acid; 5—0.5 mol/L citric acid; 6—0.1 mol/L oxalic acid; 7—0.5 mol/L oxalic acid; 8—0.1 mol/L HCl; 9—0.5 mol/L HCl; 10—1 mol/L HCl; 11—0.1 mol/L acetic acid; 12—0.5 mol/L acetic acid.

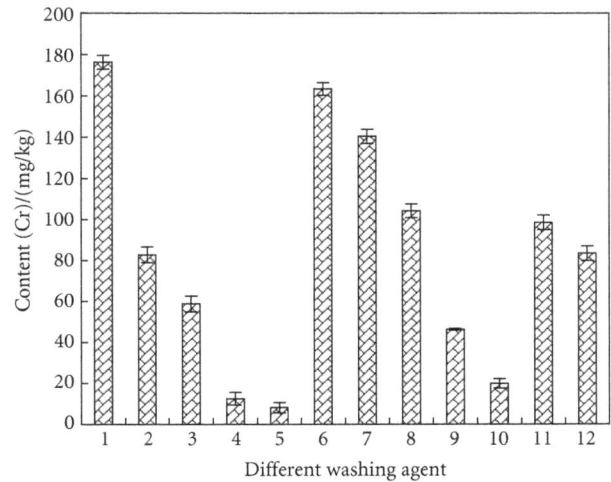

(a)

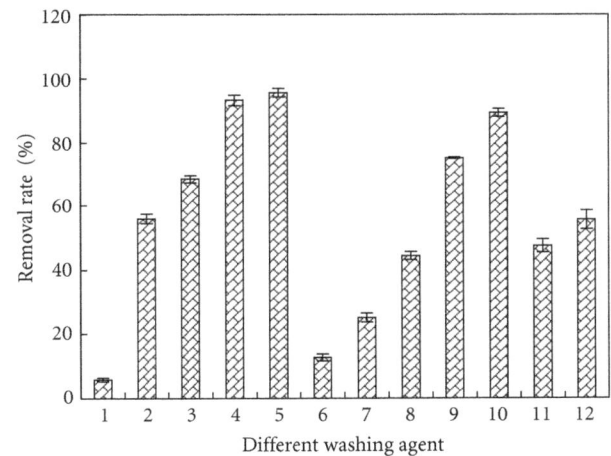

(b)

FIGURE 9: Removal effect of reducible Cr by different washing agents. Washing agent: 1—water; 2—0.05 mol/L EDTA-Na₂; 3—0.1 mol/L EDTA-Na₂; 4—0.1 mol/L citric acid; 5—0.5 mol/L citric acid; 6—0.1 mol/L oxalic acid; 7—0.5 mol/L oxalic acid; 8—0.1 mol/L HCl; 9—0.5 mol/L HCl; 10—1 mol/L HCl; 11—0.1 mol/L acetic acid; 12—0.5 mol/L acetic acid.

Reducible Cr was mainly made up of Cr (VI). Citric acid and concentrated HCl can reduce the pH of sample and under acidic condition, the oxidation-reduction potential was larger than 0 and Cr (VI) was reduced into Cr (III), which reacted spontaneously, which led to the better removal effect of reducible Cr. The add of citric improved the content of organic acid in samples, which assisted Cr (VI) being reduced into Cr (III).

3.6. Removal of Oxidizable Cr.

Figure 10 shows the remained contents and removal rates of oxidizable Cr in sample after being washed by different agents. The disposal effect of citric acid was the best, reducing the content of oxidizable Cr from 156.68 mg/kg to 10 mg/kg, and the removal rate was over 95%. Oxidizable Cr was mainly made up of Cr (III). Due to complexing action [18], EDTA, citric acid, oxalic acid, and acetic acid can better remove oxidizable Cr, especially citric acid, and the removal rates were all above 60%.

3.7. Removal of Residual Cr.

Figure 11 shows the remained contents and removal rates of residual Cr in sample after being washed by different agents. As indicated by the results,

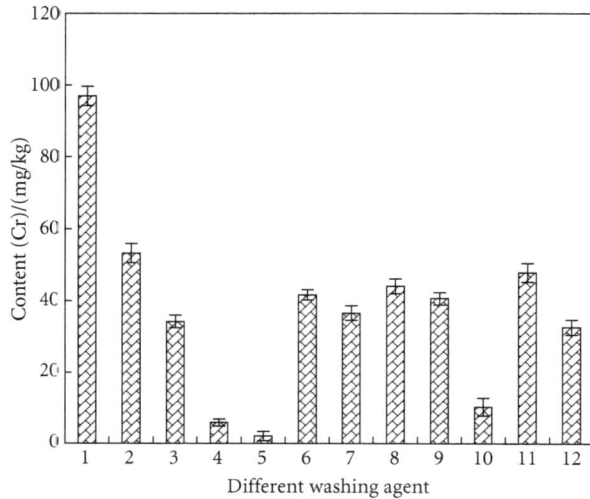

Oxidizable

(a)

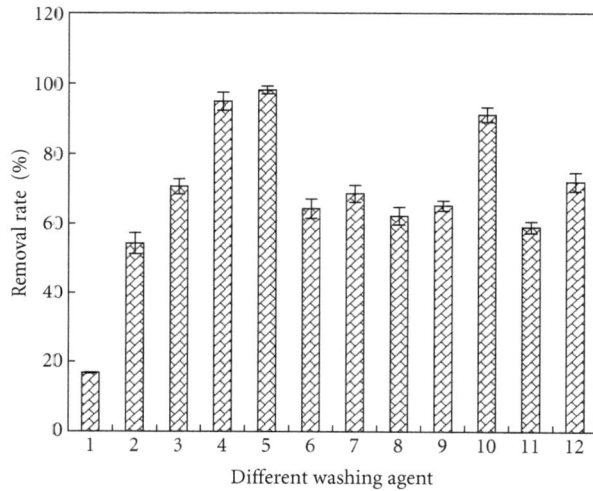

Residual

(a)

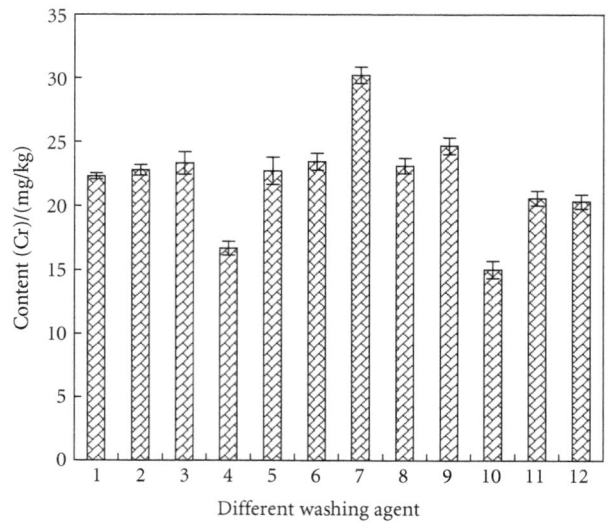

Oxidizable

(b)

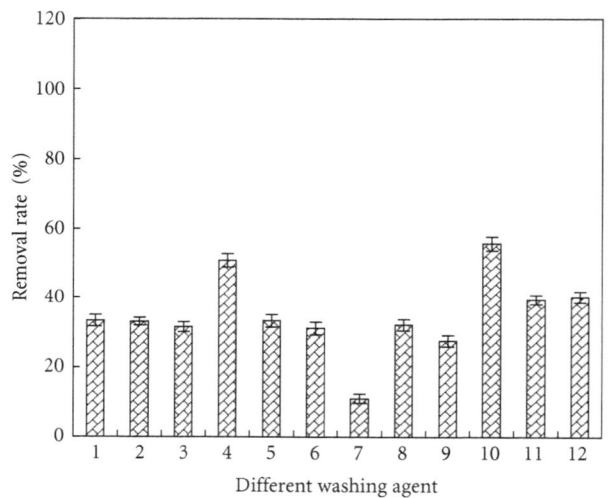

Residual

(b)

FIGURE 10: Removal effect of oxidizable Cr by different washing agents. Washing agent: 1—water; 2—0.05 mol/L EDTA-Na$_2$; 3—0.1 mol/L EDTA-Na$_2$; 4—0.1 mol/L citric acid; 5—0.5 mol/L citric acid; 6—0.1 mol/L oxalic acid; 7—0.5 mol/L oxalic acid; 8—0.1 mol/L HCl; 9—0.5 mol/L HCl; 10—1 mol/L HCl; 11—0.1 mol/L acetic acid; 12—0.5 mol/L acetic acid.

FIGURE 11: Removal effect of residual Cr by different washing agents. Washing agent: 1—water; 2—0.05 mol/L EDTA-Na$_2$; 3—0.1 mol/L EDTA-Na$_2$; 4—0.1 mol/L citric acid; 5—0.5 mol/L citric acid; 6—0.1 mol/L oxalic acid; 7—0.5 mol/L oxalic acid; 8—0.1 mol/L HCl; 9—0.5 mol/L HCl; 10—1 mol/L HCl; 11—0.1 mol/L acetic acid; 12—0.5 mol/L acetic acid.

the removal effects of all agents were not ideal and after disposal, the remained content of residual Cr was about 20–30 mg/kg. The removal rate was 30–50%.

4. Conclusions

(1) Particle size had no distinct impact on the contents of total Cr and Cr (VI) in samples. Therefore, in the

actual disposal, it shall select more economic crushing equipments regardless of the impact of particle size.

(2) Although the removal rate of total Cr and Cr (VI) was about 75% after water washing, the content of Cr in sample was still high, and it was still quite harmful and needed further disposal by second washing.

(3) Soluble Cr and reducible Cr were mainly made up of Cr (VI) and the reason that citric acid can obtain

better removal effect is that they not only can dissolve the soluble Cr, but can reduce it into Cr (III) of other forms due to the redox.

(4) Oxidizable Cr was mainly made up of Cr (III). Due to complexing action, EDTA, citric acid, oxalic acid, and acetic acid can better remove oxidizable Cr (III), especially citric acid.

(5) After water washing, citric acid was used for it can better remove soluble Cr, acid soluble Cr, reducible Cr, and oxidizable Cr, and the removal rate of Cr (VI) was over 99% after disposal. The remained content of Cr (VI) in sample was lower than 10 mg/kg.

Acknowledgment

This paper is supported by National High Technology Research and Development Program of China (863 Program), no. 2009AA063101-2.

References

[1] X. Liu, X. R. Wang, and Z. Q. Zhang, "Potential influences of pH and organic matter on the occurrence form s of chromium in chromium-contaminated soils," *Chinese Journal of Environmental Engineering*, vol. 4, pp. 1436–1440, 2010.

[2] X. R. Wang, X. Liu, X. H. Yan, and Q. Wang, "Selection of washing agents for remediation of chromium slag-contaminated soil," *Research of Environmental Sciences*, vol. 23, pp. 1405–1409, 2010.

[3] G. Dermont, M. Bergeron, and G. Mercier, "Soil washing for metal removal: a review of physical/chemical technologies and field applications," *Journal of Hazardous Materials*, vol. 152, no. 1, pp. 1–31, 2008.

[4] F. Khan, T. Husain, and R. Hejazi, "An overview and analysis of site remediation technologies," *Journal of Environmental Management*, vol. 71, no. 2, pp. 95–122, 2004.

[5] D. K. Elgh, Z. Arwidsson, and A. Camdzija, "Laboratory and pilot scale soil washing of PAH and arsenic from a wood preservation site: changes in concentration and toxicity," *Journal of Hazardous Materials*, vol. 172, no. 2-3, pp. 1033–1040, 2009.

[6] R. A. Griffiths, "Soil-washing technology and practice," *Journal of Hazardous Materials*, vol. 40, no. 2, pp. 175–189, 1995.

[7] H. B. Li, P. J. Li, and T. H. Sun, "Remediation of sediment contaminated by Cd and Pb in Zhangshi irrigation area using a washing technology," *Journal of Agro-Environment Science*, vol. 24, pp. 328–332, 2005.

[8] T. Makino, K. Sugahara, and Y. Sakurai, "Remediation of cadmium contamination in paddy soils by washing with chemicals: selection of washing chemicals," *Environmental Pollution*, vol. 144, no. 1, pp. 2–10, 2006.

[9] S. Kuo, M. S. Lai, and C. W. Lin, "Influence of solution acidity and CaCl$_2$ concentration on the removal of heavy metals from metal-contaminated rice soils," *Environmental Pollution*, vol. 144, no. 3, pp. 918–925, 2006.

[10] C. Mulligan and S. Wang, "Environmental applications for biosurfactants," *Environmental Pollution*, vol. 133, no. 2, pp. 183–198, 2005.

[11] J. X. Zhou and Z. Liu, "Recent developments in the remediation of chromium-contaminated soil," *Techniques and Equipment for Environmental Pollution Control*, vol. 1, pp. 51–55, 2000.

[12] Q. R. Zeng, S. Sauvé, and H. E. Allen, "Recycling EDTA solutions used to remediate metal-polluted soils," *Environmental Pollution*, vol. 133, no. 2, pp. 225–231, 2005.

[13] S. P. Tampouris, "Removal of contaminant metals from fine grained soils, using agglomeration, chloride solutions and pile leaching techniques," *Journal of Hazardous Materials*, vol. 84, no. 2-3, pp. 297–319, 2001.

[14] M. Lee, I. S. Paik, and W. H. Do, "Soil washing of as-contaminated stream sediments in the vicinity of an abandoned mine in Korea," *Environmental Geochemistry and Health*, vol. 29, no. 4, pp. 319–329, 2007.

[15] P. Bhattacharya, G. Jacks, and S. Nordqvist, "Metal contamination at a wood preservation site: characterisation and experimental studies on remediation," *Science of the Total Environment*, vol. 290, no. 1–3, pp. 165–180, 2002.

[16] US EPA, *Method 3060A Alkaline Digestion for Hexavalent Chromium*, United States Environment Protection Agency, Washington, DC, USA, 1996, http://www.epa.gov/osw/hazard/testmethods/sw846/pdfs/3060a.pdf.

[17] GB50853-2007, "Identification standards for hazardous wastes-Identification for extraction toxicity," Tech. Rep. GB 5085.3-2007, 2007.

[18] Z. H. Ding and X. Hu, "Application of chelants in remediation of heavy metals-contaminated soil," *Ecology and Environmental Sciences*, vol. 18, pp. 777–782, 2009.

5

Nitrate-Nitrogen Leaching and Modeling in Intensive Agriculture Farmland in China

Ligang Xu,[1] Hailin Niu,[2] Jin Xu,[3] and Xiaolong Wang[1]

[1] State Key Laboratory of Lake Science and Environment, Nanjing Institute of Geography and Limnology,
  Chinese Academy of Sciences, Nanjing, China
[2] College of Ecological and Environmental Engineering, Qinghai University, Xining, China
[3] Department of Environmental Engineering, Nanjing Institute of Technology, Nanjing, China

Correspondence should be addressed to Ligang Xu; lgxu@niglas.ac.cn

Academic Editors: F. Darve, C. García, P. F. Hudak, and A. Moldes

Protecting water resources from nitrate-nitrogen (NO_3-N) contamination is an important public health concern and a major national environmental issue in China. Loss of NO_3-N in soils due to leaching is not only one of the most important problems in agriculture farming, but is also the main factor causing nitrogen pollution in aquatic environments. Three typical intensive agriculture farmlands in Jiangyin City in China are selected as a case study for NO_3-N leaching and modeling in the soil profile. In this study, the transport and fate of NO_3-N within the soil profile and nitrate leaching to drains were analyzed by comparing field data with the simulation results of the LEACHM model. Comparisons between measured and simulated data indicated that the NO_3-N concentrations in the soil and nitrate leaching to drains are controlled by the fertilizer practice, the initial conditions and the rainfall depth and distribution. Moreover, the study reveals that the LEACHM model gives a fair description of the NO_3-N dynamics in the soil and subsurface drainage at the field scale. It can also be concluded that the model after calibration is a useful tool to optimize as a function of the combination "climate-crop-soil-bottom boundary condition" the nitrogen application strategy resulting for the environment in an acceptable level of nitrate leaching. The findings in this paper help to demonstrate the distribution and migration of nitrogen in intensive agriculture farmlands, as well as to explore the mechanism of groundwater contamination resulting from agricultural activities.

1. Introduction

Nitrogen levels in surface water and groundwater of agricultural lands have increased by 50% over the past two decades as a result of increases in the use of fertilizers and manure [1, 2]. As the nation with the largest agricultural production, China consumed 23 million tons of fertilizer N in 2000, accounting for about 28% of total world N consumption [3–5]. In China, the current nitrogen use had led to considerable nitrate-nitrogen (NO_3-N) losses through leaching. Nitrogen losses through leaching vary across a field due to differences in soil physical properties and N status of soil. The NO_3-N, a water-soluble nutrient, is transported to shallow groundwater as a leachate that consequently contaminates it. NO_3-N levels have been of worldwide concern due to the deteriorating quality of groundwater and surface water for the past four decades. As the NO_3-N exceeded the safe drinking water standards in drinking water, it always experiences some health problems. The major problems linked to NO_3-N contamination are methemoglobinemia (blue baby syndrome) infants and human birth defects [6]. The concern about the health and environmental effects of NO_3-N contaminated surface and groundwater has made it imperative to estimate NO_3-N losses from cropland and to evaluate the impact of crop production practices on NO_3-N leaching.

Numerical models are useful tools to predict the risk of NO_3-N$'$ contamination to surface water and groundwater. These models need to be calibrated and validated for the conditions under which they will be used. A properly validated model provides a fast and cost-effective way of estimating NO_3-N leaching under different agricultural management practices. Thus, the farmer can more accurately determine the amount of fertilizer to use on a crop to manage yield yet avoid over fertilization, while political decision makers can identify agricultural best practices. The number of nonpoint source agricultural models used to predict NO_3-N leaching through the root zone and into the underlying unsaturated soil zone has grown rapidly over the last two decades. The creation, calibration, and validation of water quality computer models impact agricultural practices leading to greater awareness and potential control of environmental impacts. These include nitrogen and carbon cycling in soil water and plant (NCSWAP) by Clemente et al. [7], pesticide root zonemodel (PRZM) by Cameira et al. [8], groundwater loading effects of agricultural management system (GLEAMS) by Leonard et al. [9], SLIM by Addiscott [10], nitrate leaching and economic analysis package (NLEAP) by Prasad [11], SOILSOILN by Bergstrom and Jarvis [12], GRAzing SIMulation Model (GRASIM) by Sarmah et al. [13], HYDRUS by Simunek et al. [14], RZWQM (Root ZoneWater Quality Model) by Hu et al. [15], and leaching estimation and chemistry model (LEACHM) by Hutson and Wagenet [16]. Evaluation of these models has also received increasing attention over the last decade [17–21]. If these models are appropriately validated with respect to their simulative capability under various conditions, the models will significantly improve the quantitative understanding of N cycling processes, which can be valuable tools in designing environmentally compatible and economically suitable agricultural systems [5].

Owing to intensive cropping of the orchard and vegetables combined with excessive use of fertilizer in China, the nitrogen pollution has become increasingly serious. However, it is not clear on the quantitive risk and the effects of NO_3-N leaching in intensive agricultural cropped soil in China. The objectives of this paper deal with (1) to explore the transport and fate characteristics of NO_3-N within the soil profile and the leaching loss from continuously typical intensive agriculture farmland, (2) to estimate the LEACHM model against the data in terms of its ability to simulate the process of NO_3-N leaching loss in field conditions, and (3) to determine the overall NO_3-N leached from the irrigation and natural rainfall and the leaching potential of nitrates under current traditional irrigation methods and suggest the best possible irrigation methods that can reduce nitrogen leaching.

2. Materials and Methods

2.1. Site Description.
The experimental site is located at the Qinshuihe catchment of Jiangyin cities through which the Yangtze River flows and has 200 km distance from Shanghai and Nanjing city of China. The average annual precipitation of the site is 1205.5 mm of which about 700 mm occurs during the crop growing period. Jiangyin city's GDP is one of the top three country cities in China. The utilization of agriculture is dominated by intensive cropped type. The risk and environmental effect on intensive agriculture cropped soil, especially on excessive nitrogen fertilization application, are not clear. Three kinds of typical intensive cropped soil were selected in Jiangyin city as the case study for NO_3-N

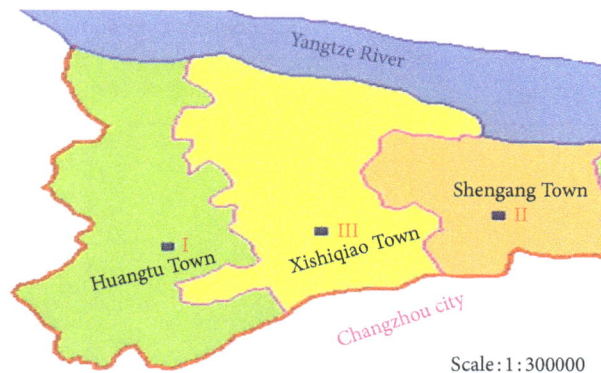

FIGURE 1: The distribution map of typical agriculture areas in Jiangyin City (I is grape orchard, II is vegetable base, and III is conventional planted farmland).

leaching and modeling study. The first one is a typical grape orchard located in the Huangtu town. The second one is a typical vegetable base located in the Shengang town. The last one is conventional cropped soil in Xishiqiao town. The map for the study area is illustrated in the Figure 1. The soil is sampled in three kinds of typical cropped soil, and the basic soil physical properties and background values for soil are given in Table 1.

2.2. Microclimate Monitoring and Crop Management.
A microclimate monitoring system by the Decagon Device Company was set up in the field to monitor an integrated climate variables (leaf wetness, precipitation, relative humidity, solar radiation, temperature, wind direction, and wind speed). A comprehensive questionnaire was delivered to the local farmers in three kinds of typical intensively cropped farmland on the following: the cropping patterns, crop varieties, seasonal crop inputs, the amount and date of fertilization, irrigation, medicine, and so forth.

2.3. Soil Moisture Monitoring and NO_3-N Sampling.
A field experiment is designed to monitor NO_3-N leaching losses from nitrogen fertilized and manured intensive cropped farmland. The EC-5 soil moisture made by the Decagon Device has been used to monitor the soil moisture in three different soil profiles. In each monitor station, the soil moisture sensors were set up at the depth of 20 cm, 40 cm, 60 cm, 80 cm, and 100 cm, respectively. Hourly data were recorded automatically with 24-hour cycle of each day at the three field stations. These data were used to calibrate soil hydraulic parameters and validate the model. The soil for three kinds of the cropped soil profile (1.0 m thick) was sampled at 0.2, 0.4, 0.6, 0.8, and 1.0 m depths for NO_3-N analyzing using an automated Cd reduction method (USEPA, 1979).

2.4. Groundwater Observation Well Monitoring in Field.
Three groundwater observation monitoring wells were built on three representative cropped farmland. The groundwater observation well is made of PVC with a protective casing on it to prevent the influx of garbage and insects (Figure 2).

TABLE 1: Selected physical properties and background values of soil samples.

Soil type	Depth (cm)	Bulk density (g·cm^{-3})	Soil porosity (%)	Organic matter content (%)	Soil texture (%)		
					Clay	Silt	Sand
	0–20	1.16	55.56	2.48	22	46	2
	20–40	1.42	47.26	1.92	24	44	4
Silt loam (Huangtu town)	40–60	1.49	44.79	0.58	24	44	4
	60–80	1.43	44.75	0.45	24	44	4
	80–100	1.40	44.74	0.33	18	50	4
	0–20	1.28	51.22	1.11	41	52	6
	20–40	1.30	51.88	2.11	43	49	6
Silt loam (Shengang town)	40–60	1.47	45.33	1.46	39	57	3
	60–80	1.49	44.79	0.36	26	67	5
	80–100	1.57	42.10	0.23	26	67	5
	0–20	1.08	58.20	2.88	48	50	3
	20–40	1.37	48.91	2.43	50	48	3
Silt loam (Xishiqiao town)	40–60	1.41	47.59	0.78	48	44	8
	60–80	1.41	47.49	1.50	46	43	10
	80–100	1.42	47.09	1.66	44	42	14

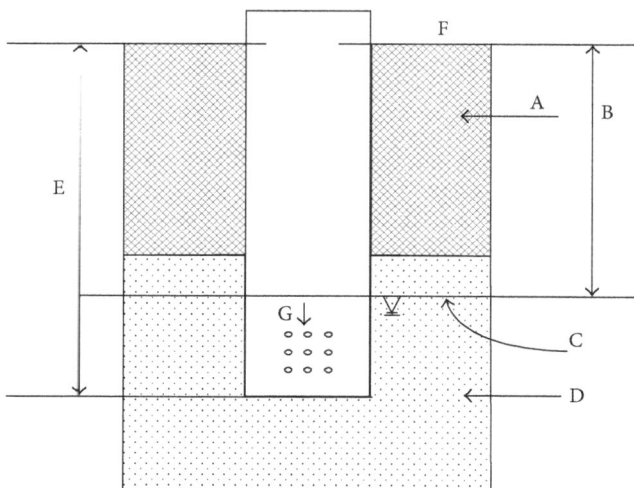

FIGURE 2: Schematic diagram of groundwater monitoring well (A: soil layer; B: depth of groundwater; C: groundwater level; D: sand layer; E: depth of groundwater monitoring well; F: soil surface; G: infall).

The groundwater sampling was usually performed once a week, but additional observation was conducted when there was a rainstorm. The water was promptly sent to the laboratory for analysis. Meanwhile, water level and farming conditions (fertilization, irrigation, and crop stage), the growing and physiological character, the yield of plants during their growing time, and crop harvest time were noted when they were happening. NO$_3$-N in percolation water was analyzed with a continuous-flow nitrogen analyzer (SKALAR, San Plus System, Netherlands).

2.5. Model Selection and Calibration Description. Modeling of water flow movement and NO$_3$-N transport vertically in the soil profile was conducted using LEACHM model (Hutson and Wagenet, 1997). This model had been used with varying degrees of success, primarily for determining the magnitude of nitrate leached below the plant root zone. This software package is a one-dimensional model for simulating the transient movement of water and multiple solutes in variably saturated media, with extensive capabilities, such as options to simulate crop root water uptake. It is easy to set flexible boundary conditions, time step, convergence conditions, and the output format for LEACHM, which greatly improves computational efficiency and simulation precision of the model. The LEACHM is selected for this study because it has subroutines to calculate water flow, NO$_3$-N leaching, evapotranspiration, rate constant adjustments for temperature and water content, and uptake. The previous studies in China have tested that LEACHM gives a fair description of the NO$_3$-N dynamics in the soil as other models (NCSWAP, GRASIM, HYDRUS,...). Currently, LEACHM has been widely used in the research of soil water, salt, and nitrogen transport. It would aid our understanding of the nitrogen migration and cycle. A detailed description of LEACHM can be found in Huston (1996, 2005, and 2009) and others [22, 23].

The soil profile was divided into 10 increments in the vertical direction with a uniform thickness of 0.1 m. The upper boundary condition was set as a fixed flux, while the bottom boundary was defined as a fixed water table depth condition. Each increment required the following data: bulk density, particle size distribution, initial N concentrations, initial soil water content, water retention parameters, saturated hydraulic conductivity, and dispersivity. Values for bulk density and saturated hydraulic conductivity were obtained from laboratory tests. The soil physical properties of each drainage class are presented in Table 1, and other model parameter values used in the simulations are presented in Table 2. The empirical constants of the Campbell retentively function were obtained by log transformation of (1).

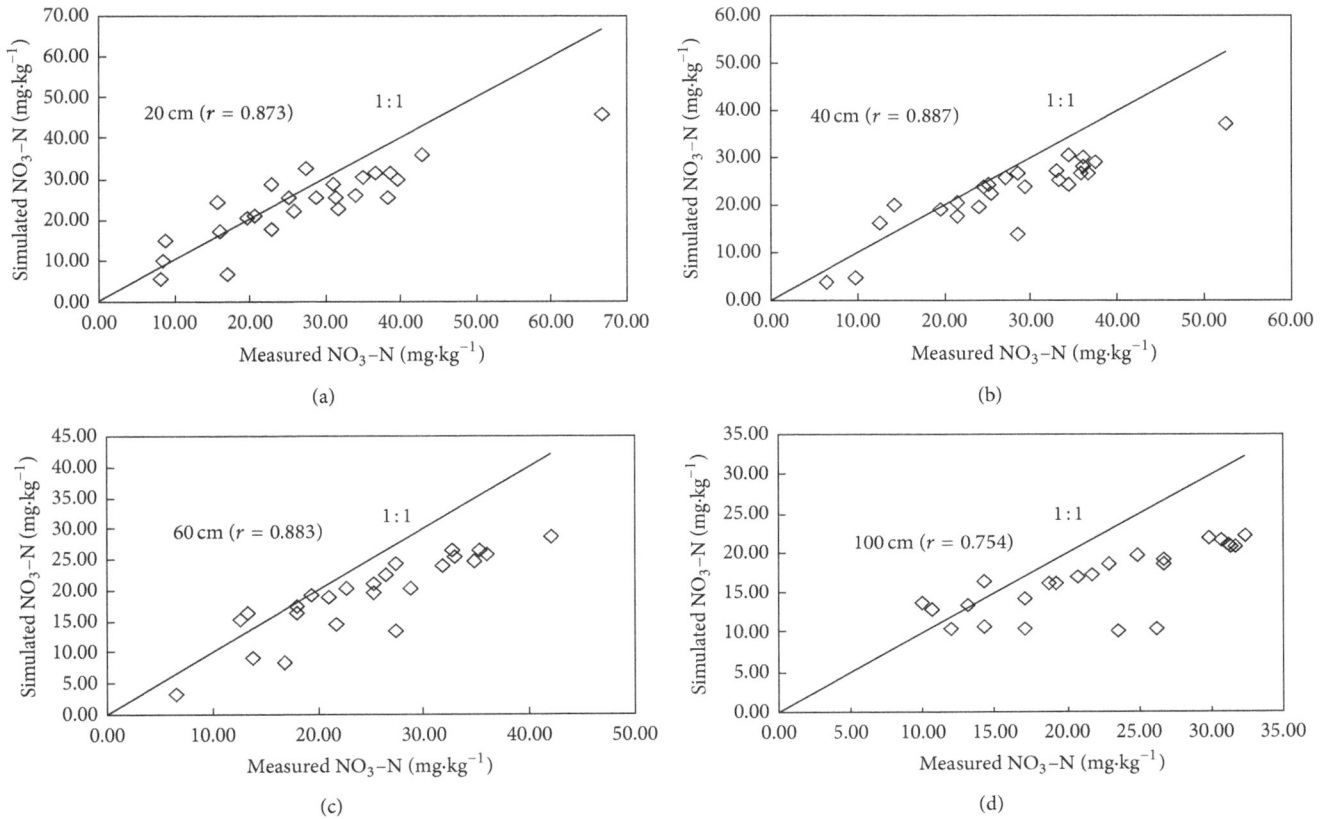

FIGURE 3: Comparison between simulated values and measured values of NO_3-N with different soil layer in intensive grape orchard.

TABLE 2: Input parameter values used in the LEACHM model during calibration.

Parameter	Input values
Partition coefficient, NH_4-N (m^3/kg)	0.6×10^{-3}
Partition coefficient, NO_3-N (m^3/kg)	0.55
Denitrification half saturation constant (mg/L)	10
Litter mineralization rate constant (per day)	0.01
Humus mineralization rate constant	7×10^{-5}
Q10 factor	2.0
C : N ratio for biomass and humus	10.0
Maximum NO_3^-/NH_4^+ ratio in solution to control	8.0
Nitrification rate	100
Molecular diffusion coefficient (mm^2/d)	140
Saturated hydraulic conductivity, (mm/d)	25
Water potential, kPa	1.3
Air-entry value, kPa	11.8
B parameters, kPa	

The only parameter calibrated for nitrogen transport was the dispersion coefficient by comparing the measured data and simulated data. The output time step was set up as 0.01 day.

In order to obtain a quantitative assessment of simulation results, correlation coefficient (r) was adopted to evaluate numerical simulation precision:

$$r = \frac{\sum_{i=1}^{N} \left(M_i - \overline{M}\right)\left(E_i - \overline{E}\right)}{\sqrt{\sum_{i=1}^{N} \left(M_i - \overline{M}\right)^2 \left(E_i - \overline{E}\right)^2}}, \quad (1)$$

where, M_i and E_i are, respectively, the ith measured values and simulated values; N is the observation frequency. The value range of correlation coefficient (r) is $[-1, +1]$, with a correlation coefficient of $+1$ indicating that the two variables have a perfect, upward-sloping ($+$) linear relationship and a correlation coefficient of -1 showing that the two variables have a perfect, downward-sloping ($+$) linear relationship. A correlation coefficient of 0 stands for nonlinear relationship between the variables.

3. Results and Discussion

3.1. NO_3-N Transport and Fate in Soil Profile. Based on the field monitoring, the LEACHM model was tested to simulate the transport and fate of NO_3-N in soil profile from May 21 to December 8 in 2010. Figures 3, 4, and 5 demonstrate the comparison between simulated values and measured values of NO_3-N at different depths in three different cropped

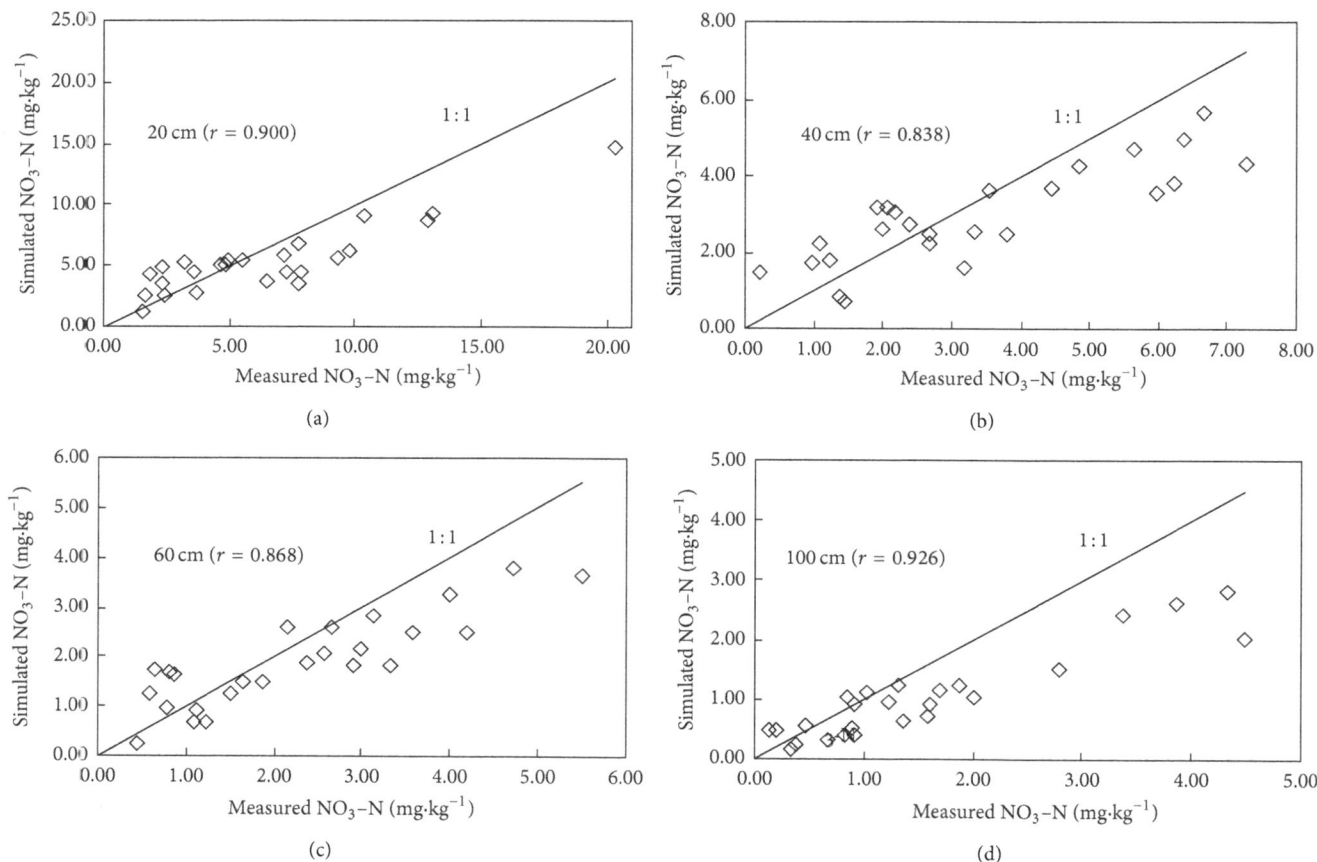

FIGURE 4: Comparison between simulated values and measured values of NO_3-N with different soil layer in intensive vegetable base.

lands, respectively. The LEACHM model was found to be able to successfully simulate the concentration of NO_3-N at different depths in soil. The simulations for all three kinds of typical cropped soil provided satisfactory results even through simulated values deviated from measured values in the initial stage. The average correlation analysis for three typical farmlands shows that 20 cm, 40 cm, 60 cm, and 100 cm soil layer between the simulated and measured values are 0.967, 0.942, 0.9243, and 0.893. Correlation coefficient of the soil reaches a significant level. It indicates that the model simulation of NO_3-N migration performs well. The simulated results reflect the regime of vertical migration of NO_3-N in soil under normal growing conditions. This discrepancy can be explained by the accumulation of water on the sand layer during the initial stage of the experiment leading to a lag in water drainage and thus leaching loss.

Overall, the results from this study show that the LEACHM model has the potential to predict the fate of N added to soil in relation to NO_3-N leaching loss below the 100 cm depth using parameters derived from previous experiments. Further field-testing using data from various soils, crops, management, and weather conditions is needed to evaluate the model's application to different field conditions.

3.2. NO_3-N Leaching from Irrigation and Natural Rainfall Condition.
Soil water percolation rate and NO_3-N leaching

loss rate are shown in Figures 6, 7, and 8, and it can be seen that the simulation results of NO_3-N for all three different planted farmland exhibit the same trend under different rainfall intensities. The increase of rainfall intensity has a clear positive correlation with leaching loss, such that an increase in rainfall intensity results in a corresponding increasing rate of leaching loss. The measured drainage water is lower than the simulated values in the initial stage of all three sets of experiments. This is due to the effect of the formation of a thin saturated layer at the sand area, which results in a short-term accumulation and lag of water drainage. This explained the lower measured values. After the initial stage, the simulated values and measured values match well and become stable. The simulated results show that the LEACHM model performs well and gives ideal simulations of water drainage in all three sets of experiments under different rainfall intensities.

Table 3 shows the comparison of soil water seepage and soil water storage changes in three typical cropped farmlands. In the conventional cropped farmland, the amount of soil water leakage within 0–100 cm is high as 1177.6 mm. It is significantly higher than the intensive grapes orchard (971.8 mm) and vegetable bases (963.8 mm). It indicates that the soil water leakage in the unsaturated soil under conventional cropped farmland increased significantly than the intensive cropped farmland although they have similar

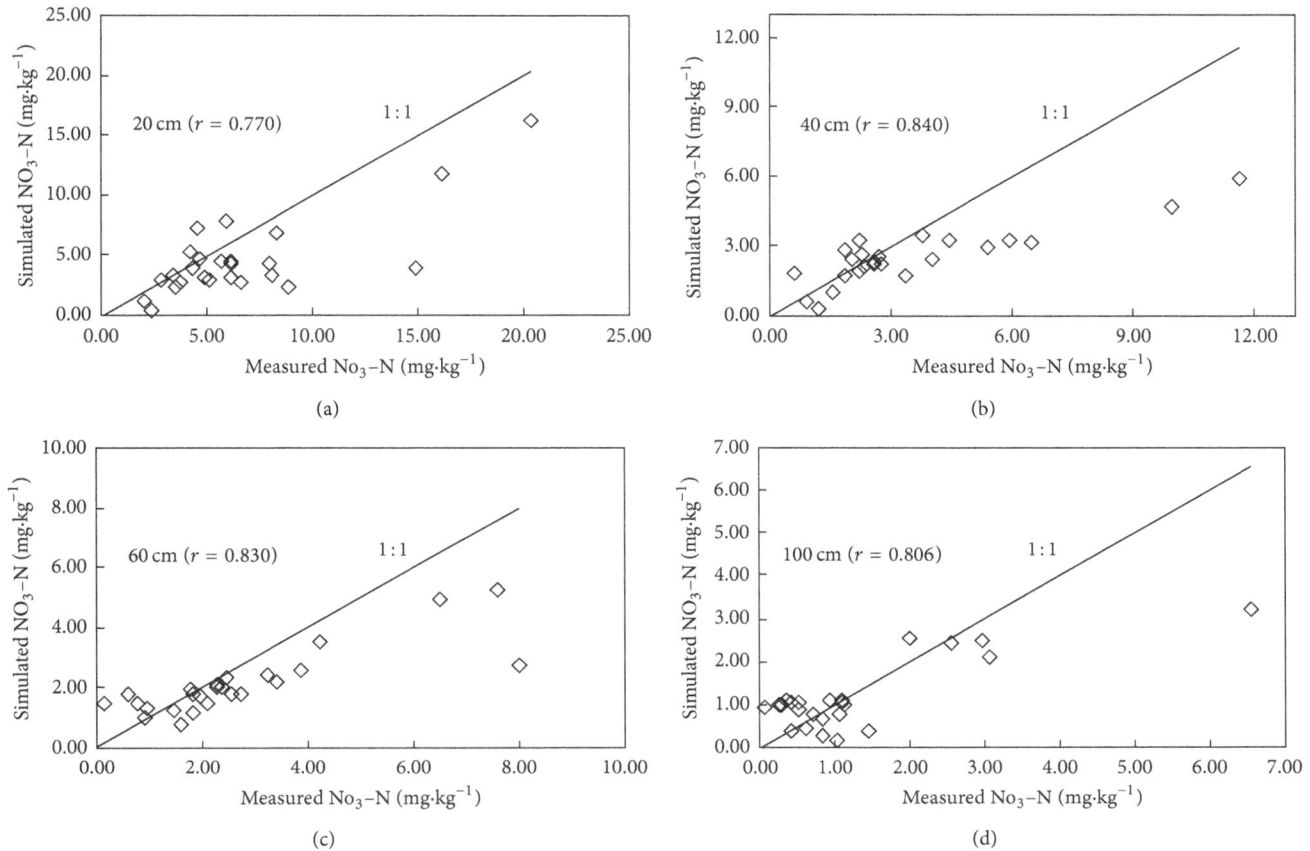

FIGURE 5: Comparison between simulated values and measured values of NO_3-N with different soil layer in conventional planted farmland.

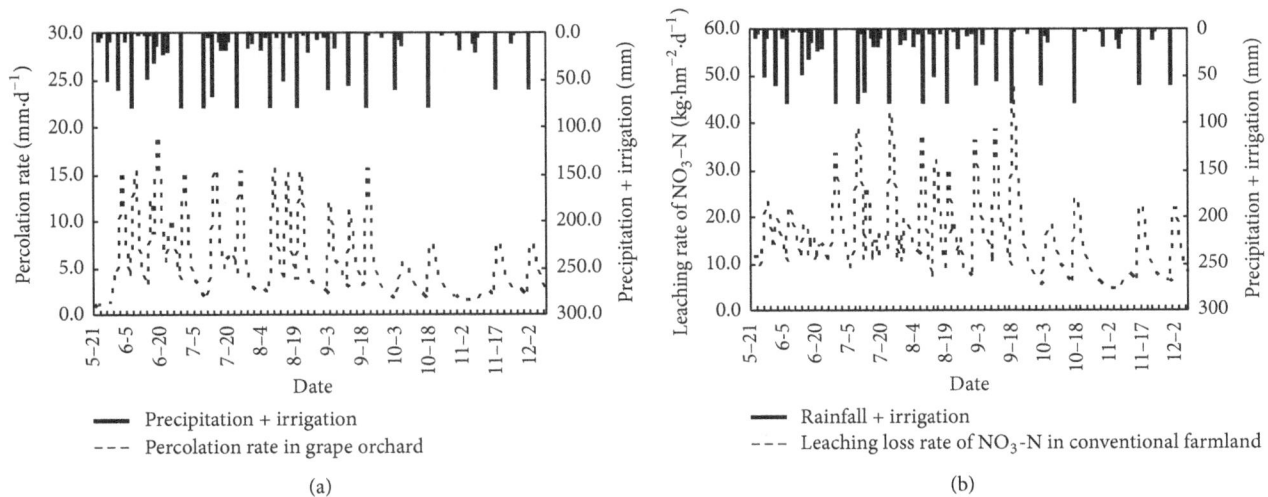

FIGURE 6: Variation curves of soil water percolation rate and NO_3-N leaching loss rate with date in intensive grape orchard.

precipitation and similar irrigation case. It is because of the conventional cropped farmland lack of scientific management and control of irrigation under normal growing conditions. That results in wasting of irrigation water to a certain extent. However, the intensive cropped grape orchard and vegetable bases used a scientific and rational irrigation

model to improve the utilization of irrigation water according to crop water demand characteristics. It reduced the amount of soil water leak effectively at different growth stages.

Table 3 also shows that the accumulated leachate of NO_3-N within 0–100 cm soil profile in grape orchard is 277.1 kg/hm^2, significantly higher than the vegetable bases

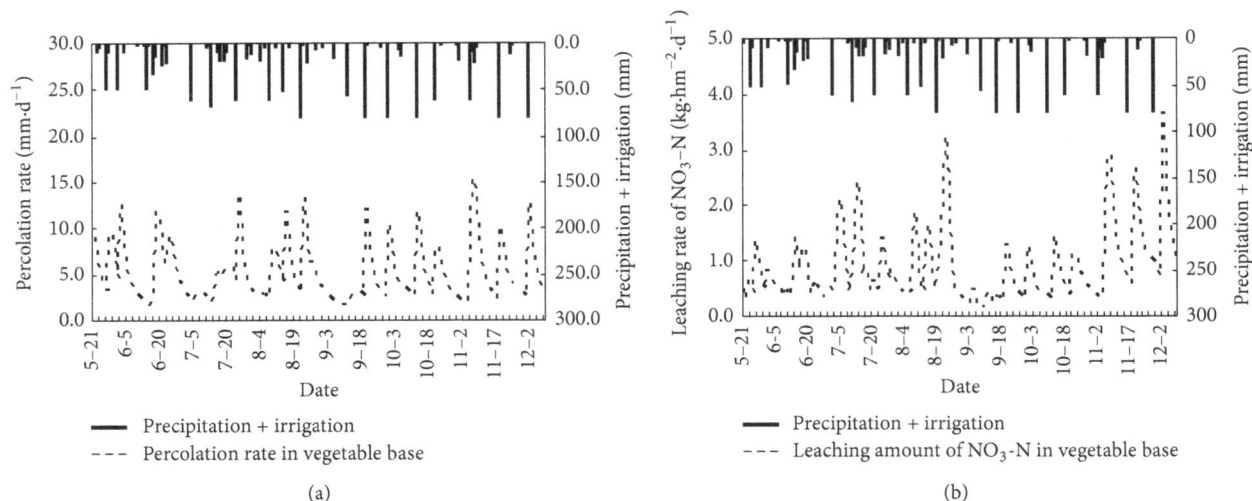

FIGURE 7: Variation curves of soil water percolation rate and NO_3-N leaching loss rate with date in intensive vegetable base.

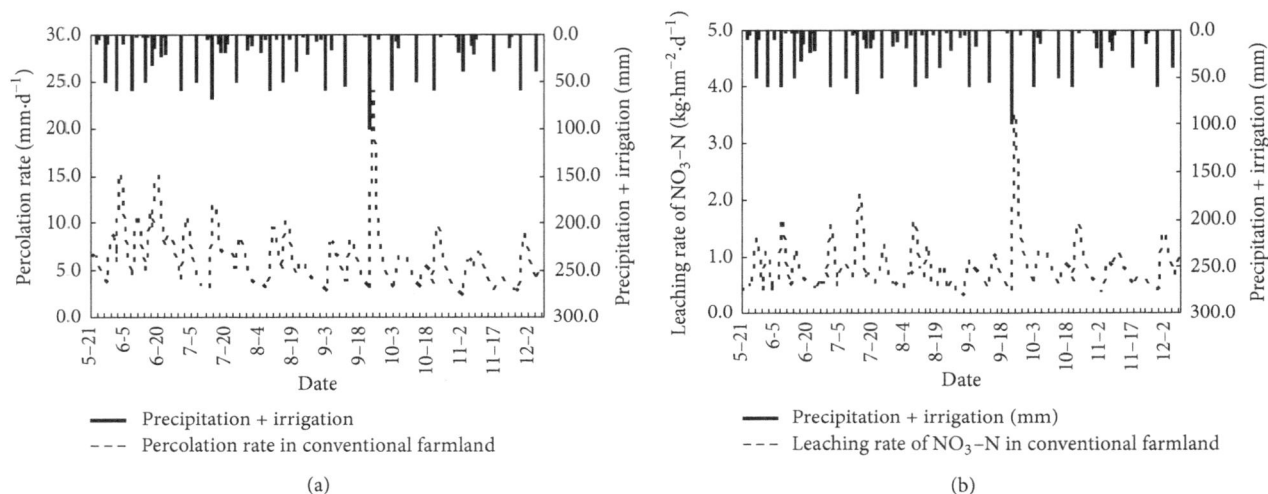

FIGURE 8: Variation curves of soil water percolation rate and NO_3-N leaching loss rate with date in conventional farmland.

and conventional cropped farmland areas, which are 91.3 kg/hm² and 15.2 kg/hm², respectively. It is mainly due to intensive application of organic fertilizer as basal fertilizer. Furthermore, there is some intermittent application of nitrogen fertilizers during the grape growing season. It resulted in long-term excess nitrogen accumulation in soil profile in the heavy rainfall. The strong leaching was caused by a large number of soil nitrate accumulation as rainstorm or irrigation happened. In contrast, intensive vegetable farmland deals with scientific methods to control water and fertilizer management of nitrogen fertilizer, effectively reducing the strength of the soil nitrate leaching; in conventional growing areas, due to the traditional model of low-intensity farming fertilization, soil nitrate accumulation is lower, which leads to a smaller amount of nitrate leaching.

According to the nitrogen migration characteristics and cumulative leaching loss amount, it can be concluded that the nitrogen leaching mainly takes place at the first stage

of rainfall. In addition, differences in precipitation patterns have a significant influence on the amount of total nitrogen leaching out of the soil column, and thus rainfall intensity was as critical as the total amount of precipitation during the experiment. The results showed that the NO_3-N leaching loss was affected by rainfall intensity and rainfall volume, and the NO_3-N moved relatively easily with water in the soil profile.

3.3. NO_3-N Leaching and Groundwater Quality. The groundwater quality monitoring results indicate that the variation of NO_3-N concentration fluctuated greatly in intensive grape orchard. Average concentration of NO_3-N was up to 15.97 mg/L, of which the peak value of nitrate nitrogen reached 22.95 mg/L (Figure 9). The leakage of groundwater samples exceeded the NO_3-N standard over a ratio of 100%, mainly due to high-frequency irrigation, fertilization in grape orchard. Moreover, long-term excessive fertilization also led to high nitrogen accumulation level in soil, and

TABLE 3: Simulated soil water leaching and soil water storage variation under different planting systems.

Planted type	Precipitation + irrigation (mm)	Soil water leaching (mm)	Soil water storage (mm)
Grape orchard	1631.8	971.8	6.5
Vegetable base	1521.8	963.8	−46.4
Conventional farmland	1581.8	1177.6	−17.2
Planted type	Fertilizer usage (kg·hm^{-2})	Leaching loss for NO_3-N (kg·hm^{-2})	Leaching rate for NO_3-N (%)
Grape orchard	695.5	277.1	39.8
Vegetable base	381.4	91.3	23.9
Conventional farmland	178.6	15.2	8.5

Noted: + means that the soil water storage increases; − means that the soil water storage decreases.

TABLE 4: Statistical comparison of groundwater concentration of NO_3-N in three different cropped farmlands.

Planted type	Average (mg/L)	Range (mg/L)	Coefficient of variation/(%)	Rate of exceed (%)
Grape orchard	15.97	9.10~22.95	24.48	100
Vegetable base	4.02	2.17~6.74	27.91	44
Conventional farmland	3.54	1.16~5.28	35.02	20

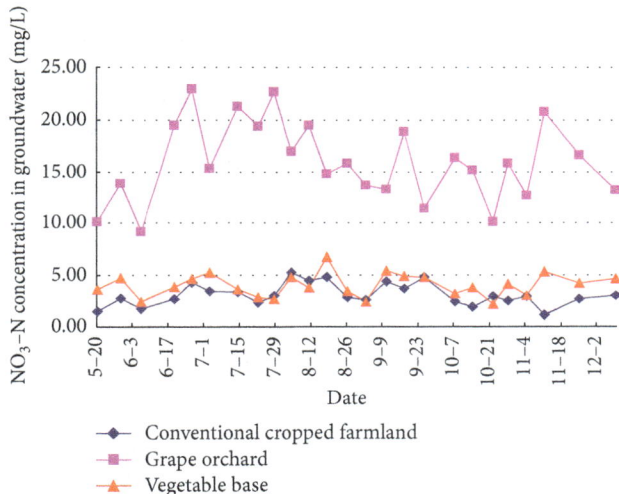

FIGURE 9: Change of NO_3-N concentrations in shallow groundwater of three kinds of typical cropped soil.

excessive irrigation increased the leaching degree of soil nitrogen. In the intensive vegetable farmland, the average groundwater concentration of NO_3-N was 4.02 mg/L, while the peak concentration of NO_3-N was high as 6.74 mg/L. It exceeded the standard over a ratio of 44%, indicating that during the cucumber-cabbage crop rotation period, there was little change in nitrate content of water leakage, and as a result there was minor agricultural shallow groundwater nitrogen pollution. Compared with intensive grape orchard and vegetable bases, the average groundwater concentration of NO_3-N in conventional planted farmland was 3.54 mg/L. It exceeded the standard over a ratio of 20%, which was due to lower fertilization rates in conventional planted farmland. Figure 9 shows that the groundwater concentration of NO_3-N in grape orchard was significantly higher than the level of intensive vegetable bases and conventional farmland, further indicating that excessive cropped farmland with intensive

fertilization and irrigation on local farmland had resulted in shallow water environmental pollution hazards. Excessive irrigation and fertilization aggravated nitrogen leaching loss in intensive grape orchard, which induced NO_3-N pollution in farmland groundwater environment. It proved that NO_3-N in intensive planting areas had large and higher varied amplitude, and excessive irrigation fertilization had caused environment pollution of groundwater.

Table 4 demonstrates the statistical comparison of groundwater concentration of NO_3-N in three cropped farmlands. Coefficient of variation in the conventional cultivation of farmland is high as 35.02%, while the coefficient of variation for NO_3-N in intensive grape orchard and intensive vegetable bases is 24.48% and 27.91%. It is mainly due to disorder fertilization which is strong conventional cropped farmland. Average concentration of NO_3-N in intensive vegetable base and conventional planted farmland is 4.02 mg/L and 3.54 mg/L, respectively, which are significantly lower than those in intensive grape orchard. According to the actual situation and management reality of the agricultural intensive planting areas, lots of controlling measures, such as optimizing irrigation and fertilizers, improving fertilization method to efficiently use fertilizer-nitrogen and reduce nitrogen leaching loss, and changing the traditional land use pattern, should be applied. The results in this paper also suggest increasing the efficient use of fertilizer-nitrogen, reducing nitrogen leaching loss, and putting forward feasible measurements for the management and control of nonpoint source pollution. It also provided the scientific basis for constituting the best management practices of the watershed. Research results were expected to provide scientific basis for optimizing field management practices, reducing farmland nitrogen leaching loss and controlling the agricultural nonpoint sources pollution.

3.4. Sensitivity Analysis. The model sensitivity analysis is helpful not only in reducing the complexity of the model and the workload of the data analysis and processing but

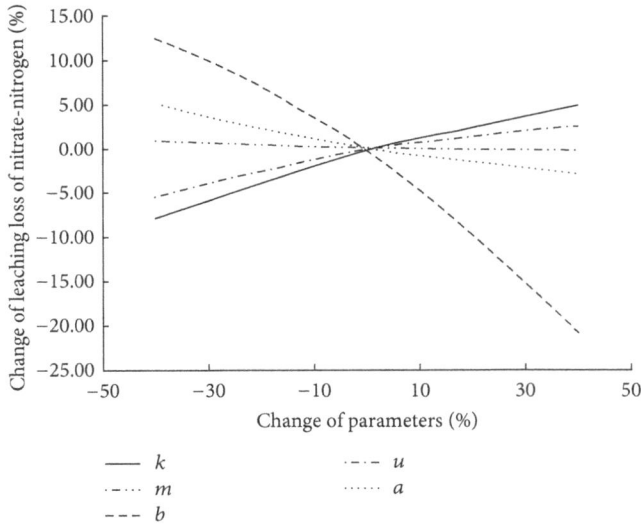

FIGURE 10: Sensitivity analysis for mainly parameters which affected the leaching loss of NO_3-N in soil.

also in improving greatly the accuracy of the model. The main factors which affect the LEACHM model for the water and nitrogen modeling are focused on the precipitation, soil texture, fertilizer types, and land use types. The accuracy of precipitation depends on the hydrological monitoring instrument. Saturated hydraulic conductivity (K) of soil water has obvious effects on the vertical migration of NO_3-N. Water characteristic curve reflects the relationship between soil water suction and soil moisture. It also has a direct impact on NO_3-N in the soil migration. Fertilization of different types and antimicrobial nitrification rate constants is different, and it will also affect the NO_3-N vertical migration. Furthermore, dispersion can penetrate the soil solute some time earlier than some time later than the penetration, which reflects the soil large and small gaps in the distribution. Therefore, in this paper hydraulic conductivity K, water retention curve parameters a ($a < 0$), b, and nitrification kinetic constants μ, m are selected to do sensitivity analysis in LEACHM model.

The sensitivity parameters based on the method of Lenhart [24] are classified as four types: (1) $0.00 \leq |I| < 0.05$, the low sensitivity parameters; (2) $0.05 \leq |I| < 0.2$, the middle sensitivity parameters; (3) $0.2 \leq |I| < 1.0$, high-sensitivity parameters; (4) $|I| \geq 1.0$, super high sensitivity parameters. Five parameters sensitivity analysis results based on this method were shown in Figure 10. The calculated sensitivity index of the two parameters (a, b) of the water retention curve results were $Ia = -0.096$, $Ib = -0.42$, the sensitivity of hydraulic conductivity K index $Ik = 0.16$, nitrification kinetic constants sensitivity index $Iu = 0.094$, and dispersion sensitivity index: $Im = -0.012$. It can be concluded that water retention curve parameters b ($Ib = -0.42$) are high sensitivity parameters, a ($Ia = -0.096$) is a sensitivity parameter, which affected the leaching loss of NO_3-N in soil profile. Both of these two parameters showed a negative correlation between the amount of leaching loss and the change of water retention curve parameters. Hydraulic conductivity is

the middle sensitivity parameters, and soil NO_3-N leakage was a positive correlation when this parameter increased in the model, the NO_3-N leakage with soil profile increased correspondingly. The nitrification kinetic constant is the sensitivity coefficient, and the NO_3-N leakage with soil profile had a positive correlation between the amounts of leakage and this parameter. Dispersion is low sensitivity parameters, which was negatively correlated with NO_3-N leakage within soil profile (Figure 10).

4. Conclusions

(1) The LEACHM model proves to be capable of simulating NO_3-N leaching loss in intensive agriculture cropped soil. Simulated results indicate that leaching amount of NO_3-N within 0–100 cm soil profile is 277.1 kg/hm^2 in intensive grape orchard during simulation period, but the leaching amount of NO_3-N within 0–100 cm soil profile in intensive vegetable farmland and in conventional planting farmland is only 91.3 kg/hm^2 and 15.2 kg/hm^2, respectively. The leaching amount and the leaching loss rate of NO_3-N in intensive cropped farmland are significantly higher than that of conventional planting areas. It shows that excessive fertilization increases the leaching risk of soil nitrate nitrogen and poses a potential threat to ecological environment in intensive grape orchard.

(2) Monitoring results of groundwater quality showed that groundwater concentration of NO_3-N in three typical planting areas is very large. The average concentration of NO_3-N in grape orchard is the highest as 15.97 mg/L among three typical planted farmlands. It further suggests that NO_3-N in intensive planting areas has large varied amplitude and higher over standard rate, and excessive irrigation and fertilization have caused environmental pollution of groundwater.

(3) Sensitivity analysis demonstrates that water retention curve parameters (a) and nitrification kinetic constants μ are highly sensitive parameters, followed by the hydraulic conductivity (K), and the sensitivity is the smallest dispersion (m).

(4) The work presented in this paper is also believed to be useful in formulating management strategies for intensive cropped catchment to reduce diffusive pollution from agricultural activities.

Acknowledgments

The authors thank Dr. Hongkun Jiang and Rui Cui for help with laboratory work. This research was jointly supported by National Basic Research Program of China (973 Program, No. 2012CB417005), National Natural Science Foundation of China (41101465, 41271034), Special Key Projects for the Alliances Support between the National Academy of Sciences and Chinese Academy of Sciences and Education Department of Jiangsu Province (11KJB610003).

References

[1] C. Stopes, E. I. Lord, L. Philipps, and L. Woodward, "Nitrate leaching from organic farms and conventional farms following best practice," *Soil Use and Management*, vol. 18, pp. 256–263, 2002.

[2] F. Guo, T. Tao, D. Cheng, and R. Gao, "Research on the influence of irrigation amount on dynamic changes and accumulation of soil nitric nitrogen under film hole irrigation," *Advanced Materials Research*, vol. 518–523, pp. 5292–5296, 2012.

[3] X. Wang, J. G. Zhu, R. Gao, H. Yasukazu, and K. Feng, "Nitrogen cycling and losses under rice-wheat rotations with coated urea and urea in the Taihu Lake region," *Pedosphere*, vol. 17, no. 1, pp. 62–69, 2007.

[4] D. Wang, Z. Xu, J. Zhao, Y. Wang, and Z. Yu, "Excessive nitrogen application decreases grain yield and increases nitrogen loss in a wheat-soil system," *Acta Agriculturae Scandinavica Section B: Soil and Plant Science*, vol. 61, no. 8, pp. 681–692, 2011.

[5] X. Wang, G. Huang, L. Yu, and Q. Huang, "Coupled simulation on soil-water-nitrogen transport and transformation and crop growth," *Transactions of the Chinese Society of Agricultural Engineering*, vol. 27, no. 3, pp. 19–25, 2011.

[6] S. Guo, J. Wu, T. Dang et al., "Impacts of fertilizer practices on environmental risk of nitrate in semiarid farmlands in the Loess Plateau of China," *Plant and Soil*, vol. 330, no. 1, pp. 1–13, 2010.

[7] R. S. Clemente, R. D. Jong, H. N. Hayhoe, W. D. Reynolds, and M. Hares, "Testing and comparison of three unsaturated soil water flow models," *Agricultural Water Management*, vol. 25, no. 2, pp. 135–152, 1994.

[8] M. R. Cameira, R. M. Fernando, L. Ahuja, and L. Pereira, "Simulating the fate of water in field soil-crop environment," *Journal of Hydrology*, vol. 315, no. 1–4, pp. 1–24, 2005.

[9] R. A. Leonard, W. G. Knisel, and D. A. Still, "GLEAMS–groundwater loading effects of agriculture mangement systems," *Transactions of the American Society of Agricultural Engineers*, vol. 30, no. 5, pp. 1403–1418, 1987.

[10] T. M. Addiscott, "Measuring and modelling nitrogen leaching: parallel problems," *Plant and Soil*, vol. 181, no. 1, pp. 1–6, 1996.

[11] R. Prasad, "Sustainable agriculture and fertilizer use," *Current Science*, vol. 77, no. 1, pp. 38–43, 1999.

[12] L. Bergstrom and N. J. Jarvis, "Prediction of nitrate leaching losses from arable land under different fertilization intensities using the SOIL-SOILN models," *Soil Use & Management*, vol. 7, no. 2, pp. 79–85, 1991.

[13] A. K. Sarmah, M. E. Close, L. Pang, R. Lee, and S. R. Green, "Field study of pesticide leaching in a Himatangi sand (Manawatu) and a Kiripaka bouldery clay loam (Northland). 2. Simulation using LEACHM, HYDRUS-1D, GLEAMS, and SPASMO models," *Australian Journal of Soil Research*, vol. 43, no. 4, pp. 471–489, 2005.

[14] J. Simunek, M. T. van Genuchten, and M. Sejna, *The HYDRUS-1D Software Package for Simulating the One-Dimensional Movement of Water, Heat, and Multiple Solutes in Variably-Saturated Media (Version 3. 0)*, Department of Environmental Sciences, University of California Riverside, Riverside, Calif, USA, 2005.

[15] C. Hu, S. A. Saseendran, T. R. Green, L. Ma, X. Li, and L. R. Ahuja, "Evaluating nitrogen and water management in a double-cropping system using RZWQM," *Vadose Zone Journal*, vol. 5, no. 1, pp. 493–505, 2006.

[16] J. L. Hutson and R. J. Wagenet, "Simulating nitrogen dynamics in soils using a deterministic model," *Soil Use & Management*, vol. 7, no. 2, pp. 74–78, 1991.

[17] B. R. Khakural and P. C. Robert, "Soil nitrate leaching potential indices: using a simulation model as a screening system," *Journal of Environmental Quality*, vol. 22, no. 4, pp. 839–845, 1993.

[18] X. Wang, S. Yang, C. M. Mannaerts, Y. Gao, and J. Guo, "Spatially explicit estimation of soil denitrification rates and land use effects in the riparian buffer zone of the large Guanting reservoir," *Geoderma*, vol. 150, no. 3-4, pp. 240–252, 2009.

[19] H. Y. Hu, J. Jin, D. J. Cheng, and R. Gao, "Influence of plant space on dynamic changes of soil nitric nitrogen under film hole irrigation," *Procedia Engineering*, vol. 28, pp. 246–251, 2012.

[20] L. Zhao, G. Yang, T. Zhao et al., "Soil nutrient dynamics and loss risks in a chicken-forage mulberry-medicinal plant intercropping system," *Acta Ecologica Sinica*, vol. 32, no. 12, pp. 3737–3744, 2012.

[21] USEPA, "Methods for chemical analysis of water and wastes," Tech. Rep. EPA 600/4-79-020, USEPA, Cincinnati, Ohio, USA, 1979.

[22] C. Ramos and E. A. Carbonell, "Nitrate leaching and soil moisture prediction with the LEACHM model," *Fertilizer Research*, vol. 27, no. 2-3, pp. 171–180, 1991.

[23] L. Xu, Q. Zhang, and L. Huang, "Nitrogen leaching in a typical agricultural extensively cropped catchment, China: experiments and modelling," *Water and Environment Journal*, vol. 24, no. 2, pp. 97–106, 2010.

[24] H.-J. Lenhart, "Effects of river nutrient load reduction on the eutrophication of the North Sea, simulated with the ecosystem model ERSEM," *Senckenbergiana Maritima*, vol. 31, no. 2, pp. 299–311, 2001.

6

Dual Inoculation with Mycorrhizal and Saprotrophic Fungi Applicable in Sustainable Cultivation Improves the Yield and Nutritive Value of Onion

Jana Albrechtova,[1,2] **Ales Latr,**[3] **Ludovit Nedorost,**[4]
Robert Pokluda,[4] **Katalin Posta,**[5] **and Miroslav Vosatka**[2,3]

[1] *Department of Experimental Plant Biology, Faculty of Science, Charles University in Prague, 12844 Vinicna 5, Czech Republic*
[2] *Institute of Botany of the Academy of Sciences of the Czech Republic, 25243 Pruhonice, Czech Republic*
[3] *Symbiom Ltd., Sazava 170, 56301 Lanskroun, Czech Republic*
[4] *Department of Vegetable Sciences and Floriculture, Mendel University in Brno, Valticka 337, 69144 Lednice, Czech Republic*
[5] *Microbiology and Environmental Toxicology Group, Plant Protection Institute, Szent István University, 2100 Gödöllő, Hungary*

Correspondence should be addressed to Jana Albrechtova, albrecht@natur.cuni.cz

Academic Editors: O. K. Douro Kpindou, A. Ferrante, R. L. Jarret, and A. Roldán Garrigós

The aim of this paper was to test the use of dual microbial inoculation with mycorrhizal and saprotrophic fungi in onion cultivation to enhance yield while maintaining or improving the nutritional quality of onion bulbs. Treatments were two-factorial: (1) arbuscular mycorrhizal fungi (AMF): the mix corresponding to fungal part of commercial product Symbivit (*Glomus etunicatum, G. microaggregatum, G. intraradices, G. claroideum, G. mosseae,* and *G. geosporum*) (M1) or the single-fungus inoculum of *G. intraradices* BEG140 (M2) and (2) bark chips preinoculated with saprotrophic fungi (mix of *Gymnopilus* sp., *Agrocybe praecox,* and *Marasmius androsaceus*) (S). The growth response of onion was the highest for the M1 mix treatment, reaching nearly 100% increase in bulb fresh weight. The effectiveness of dual inoculation was proved by more than 50% increase. We observed a strong correlation ($r = 0.83$) between the growth response of onion bulbs and AM colonization. All inoculation treatments but the single-fungus one enhanced significantly the total antioxidant capacity of bulb biomass, was the highest values being found for M1, S + M1, and S + M2. We observed some induced enhancement of the contents of mineral elements in bulb tissue (Mg and K contents for the M2 and M2, S, and S + M2 treatments, resp.).

1. Introduction

Soil organisms play a crucial role in the functioning of soil agricultural ecosystems. The functions performed by the soil biota have major direct and indirect effects on soil quality, crop growth and quality, its disease resistance, and thus on the sustainability of crop production systems [1]. Sustainable agriculture centres its focus on developing new comprehensive farming practices including management of soil microorganisms that are safe and environmentally friendly fostering the development of multidisciplinary studies [2].

Among soil microorganisms, arbuscular mycorrhizal fungi (AMF) are regarded as essential components of sustainable soil-plant systems. Since the "first green revolution," which saw the intensification of agriculture relying on high-dosage fertilization, however, less attention has been given to beneficial soil microorganisms in general and to AM fungi in particular [3, 4]. Mycorrhiza has numerous benefits for sustainable crop production. It can function as an ecological biofertilizer, a biocontrol agent against soil-borne pathogens, a bioprotectant against toxic stresses, or a soil-improver acting as a soil antierosion factor [5, 6]. Moreover, new functions of mycorrhiza in crop production are currently

Dual Inoculation with Mycorrhizal and Saprotrophic Fungi Applicable in Sustainable Cultivation Improves the Yield and
Nutritive Value of Onion

51

being explored. For example, it has been reported that AM fungi are useful for phytoremediation of contaminated soils, for example, by organophosphorus pesticides, concentration of which is high worldwide [7]. AM fungi show promising potential for reducing organophosphorus pesticide residues in plant tissues as it was shown in recent study on green onion [7].

Native AMF populations are often reduced by soil management practices of conventional agriculture such as tillage, high-dosage use of systemic fungicides, and soluble phosphate fertilizers. These practices select against conditions favourable for the survival and development of AMF. The AMF biodiversity is often reduced in high-input systems compared to low-input ones [8, 9].

Synergistic inoculations bring benefits to plant production, so they are already sought after. This, for example, applies to dual inoculations consisting of AMF and rhizobacteria, which promote plant-growth together [10]. Dual inoculations involving saprotrophic fungi can exhibit a beneficial effect on the growth of mycorrhizal plants. Members of the saprobic genus *Trichoderma* have emerged as an especially promising group of microbial inoculants when *Trichoderma* genotypes supporting plant growth were found [11, 12].

Onion production in high-input cropping systems relies on high dosage of fertilizers to achieve high yields [13] since the yields are higher by almost 50% compared to organic systems, as has been documented in the Netherlands [14]. Phosphorus availability, which often determines the plant root mycorrhizal colonization and response to mycorrhiza, is usually lower in organic systems compared to conventional ones. This has been confirmed in an extensive study of onion fields in which the average phosphorus concentration (P_w) in organic soils was up to 27% lower than in conventional soils [14]. A pioneering study on the performance of mycorrhizal onion under field conditions has shown that smaller growth benefits from mycorrhiza are observed under high phosphorus-levels than in phosphorus-deficient soils [15]. New strategies are, therefore, sought to improve phosphorus uptake and use by onion plants. Researchers focus, for example, on breeding towards improving the root system and its architecture and enhancing the responsiveness to mycorrhiza, which aids water and nutrient uptake in plants [14].

Onion contains high amounts of a variety of antioxidants, mainly of flavonoid character (quercetin, luteolin, kaempferol, etc.), of which quercetin glycosides represent the highest portion [16]. Onion is, therefore, considered a fundamental vegetable that has been valued for its medicinal qualities since ancient times. Modern research has revealed that onion possesses antibiotic, anticarcinogenic, anti-inflammatory, and antioxidative properties [17]. Biotests suggest that a diet including onion may be beneficial for the elderly as a means of improving antioxidant status [18]. Onion extracts have been reported to be effective in treating cardiovascular diseases thanks to their hypocholesterolemic, hypolipidemic, antihypertensive, antidiabetic, antithrombotic, and antihyperhomocysteinemia effects [19].

Recent studies document that mycorrhiza can enhance the nutritive value of crop plants, for example, increase the content of antioxidants in artichoke [20]. Leaf antioxidant content can also be increased by adding an organic fertilizer instead of a conventional nitrogen source, which simultaneously promotes soil biota activity and mycorrhizal colonization, as shown for highbush blueberries [21].

The aim of the present study was to evaluate a possible sustainable, ecological way of producing onion using synergistic microbial treatments with mycorrhizal and saprotrophic fungi while increasing yield and maintaining or improving the nutritional quality of onion bulbs.

2. Materials and Methods

2.1. Experiment Site, Plant, and Substrate Materials. Pot experiment was located outdoors in field conditions at the Horticultural Faculty in Lednice (Location: 48°47′54.502″N; 16°48′0.39″E, Czech Republic) in 2009. Seeds of *Allium cepa* L. (Alliaceae) cv. "ALICE" (SEMO a.s.) were sown on 24th February, and seedlings were planted into experimental 10 litre pots on 23rd April. For both the sowing and the pot experiment, we used a sterile substrate mixture composed of zeolite : peat : bark chips: bentonite (3 : 3 : 2 : 1; v/v/v/v). Substrate sterilization was done by gamma radiation (min· 25 kGy, company Artim s.r.o., Prague, CZ, http://www.artim.cz/). Basic chemical and physical properties of the substrate were 75.76% of dry matter, 14 647 mg·kg^{-1} of K, 533 mg·kg^{-1} of P, 1 159 mg·kg^{-1} of N, 0.015 mg·kg^{-1} of S, 1 250 mg·kg^{-1} of Mg, 2 536 mg·kg^{-1} of Ca, and pH 4.47. Before seedlings were planted, the substrate was fertilized, based on a chemical analysis, with 660 mg of P_2O_5 and amended with 5.4 g of $CaCO_3$ per pot to achieve pH = 6.4. The electric conductance of the substrate was 0.18 mS·m^{-1}.

2.2. Experimental Design with Microbial Treatments. Microbial treatments consisted of combination of two factors: (1) an AM fungal treatment (AMF)—either a mixed *Glomus* sp. inoculum (M1) or a single-isolate (M2); (2) bark chips preinoculated with saprotrophic fungi (S). The treatment M1 was a mixture of AM fungi corresponding to the composition of a fungal part of a commercial product Symbivit (Symbiom s.r.o., http://www.symbiom.cz/): *Glomus intraradices* BEG140, *G. mosseae* BEG95, *G. etunicatum* BEG92, *G. claroideum* BEG96, *G. microaggregatum* BEG56, and *G. geosporum* BEG199, only without the bioadditives that are used in the standard commercial product. The treatment M2 was a single-fungus inoculum of *G. intraradices* BEG140. The treatment S represented bark chips preinoculated with saprotrophic fungi. Saprotrophic fungi effective in wood decomposition (*Gymnopilus* sp. isolate IZO24, *Agrocybe praecox* isolate AER1, and *Marasmius androsaceus* isolate MAN1) were inoculated on nonsterile pine bark chips (chip size 7–15 mm, TerraSan, http://www.terrasan.de/online/index.php) in a mixture (1 : 3.3, v/v) and were left to grow for 3 months at 25°C. The AER-1 strain was derived from a fruit body in 2005 and identified based on morphological characteristics and ITS rDNA sequence similarity to AM905094 (derived from an *A. praecox* sporocarp) and AY194531 (derived from an

A. praecox sporocarp, voucher MSC 378486). The final treatments tested were Ctrl (control, nontreated plants), M1, M2, S and S + M1, S + M2 combinations of two mycorrhizal inoculants with saprotrophic amendment. Each of 6 treatments involved 7 pot replicates, each pot contained 3 seedlings.

Fungal inocula were supplied by the company Symbiom s.r.o. The mycorrhizal inoculum (M1 or M2) was added as a mixture of the cultivation substrate, colonized roots, and mycelium fragments (grown as single AMF cultures on maize as the host plant for 5 months in zeolite) in a dose of 120 g into each planting hole approx. 3 cm below each seedling. Bark chips preinoculated with saprotrophic-fungi (S) were added to the bottom of cultivation pots in the amount of 500 mL mixed with 2.5 L of cultivation substrate per 10 L pot (in S treatments thus replacing part of the cultivation substrate by uninoculated bark chips). Plants were regularly fertilized with the leaf fertilizer Wuxal super (containing 98 g/L N, 98 g/L P_2O_5, 73 g/L K_2O and trace elements B, Fe, Cu, Mn, Mo, Zn in physiological concentrations; product of AgroBio s.r.o. Opava, CZ, http://www.agrobio.cz/intro/) starting 8 weeks after planting (the fertilization dates: 08/06, 20/06, 02/07, 13/07, 22/07, 19/08).

2.3. Harvests, Plant Analyses, Mycorrhizal Parameters.
The plants were harvested, measured, and sampled for plant analyses after 4 months of cultivation at 21st August 2009. The dry mass values of bulbs, shoots, and roots were recorded after drying to constant weight at 105°C in a drying oven Sterimat 574.2 (BMT, Czech Republic) for at least 24 h. Roots of all plants were sampled to determine AM colonization. All onion bulbs were sampled for analysis of antioxidant capacity and contents of mineral elements (a 10 g sample for each), and mixed 10 g-samples per treatment were prepared for the vitamin C content analysis.

2.4. Antioxidant Capacity Determination FRAP Assay.
Onion bulb samples (10 g) were homogenized in a blender with 30 mL ethyl alcohol (50% concentration), homogenate was added with ethyl alcohol (50% concentration) to amount of 50 mL, filtered, centrifuged (3800 rt/min for 10 minutes), and the supernatant was used for the measurement. Total antioxidant capacity (TAC) was determined using the ferric-reducing antioxidant power (FRAP) assay developed by Benzie and Strain [22]. In the FRAP assay, reductants ("antioxidants") present in the extract reduce Fe(III)-tripyridyltriazine (TPTZ) complex to its intensely blue ferrous form with an absorption maximum at 593 nm. The working FRAP reagent was prepared fresh on the day of the analysis by mixing acetate buffer, 10 mM TPTZ solution, and 20 mM ferric chloride solutions in the ratio of 10 : 1 : 1 (v/v/v). The mixture was incubated at 37°C. The absorbance was monitored for 4 min in a temperature-controlled cuvette held at 37°C using a JENWAY 6100 spectrophotometer (AIR, UK).

The final total antioxidant capacity was expressed in mg equivalent of Trolox in 100 g of fresh biomass (mg Trolox·100 g^{-1}). For quantification, a calibration curve of Trolox was prepared with dilutions from 50 mg/L to 700 mg/L.

2.5. Contents of Mineral Elements.
Contents of Na, K, Ca, and Mg were determined by the method of capillary isotachophoresis using the IONOSEP 2003 device (RECMAN, CZ) following method described by Blatny et al. [23]. Onion bulb samples (10 g) were homogenized and diluted with distilled water (1 : 40) and then analysed. The head of electrolyte in the analysis was 5 mL mM H_2SO_4 + 7 mM-18-crown-6 + 0.1% HPMC[1](hydroxypropyl methylcellulose). The terminating electrolyte was 10 mM BTP[1] (bis-tris propane). The drive current was 100 μA at the beginning and 50 μA at the end. The amount of each mineral element was expressed as mg per kg of onion bulb fresh weight.

2.6. Vitamin C—Ascorbic Acid (AA) Content.
The concentration of vitamin C (ascorbic acid) was determined only in compound 10 g samples combined from bulbs harvested in each treatment by HPLC according to Arya et al. [24] with slight modification. Onion bulb samples were homogenized in a blender with 75 mL of 0.1 M oxalic acid. The homogenate was topped up with oxalic acid to the volume of 100 mL, filtered, and centrifuged (3800 rt/min for 10 minutes), and the supernatant was used for measurement. The analyses were performed by RP-HPLC in a LCO-101 column placed in an Ecom thermostat (t = 30°C), mobile phase TBAH (tetrabutylamonium hydroxide) : 0.1 M oxalic acid : water in the ratio of 10 : 20 : 70 (v/v/v), flow 0.5 mL/min at 254 nm using a UV-VIS detector. The amount of AA was expressed as mg·100 g^{-1} of fresh weight.

2.7. Mycorrhizal Colonization.
Mycorrhizal root colonization was evaluated in root samples taken from root systems of experimental plants (3 samples corresponding to 3 plants per pot). Samples were stained with 0.05% trypan blue in lactoglycerol [25] and quantified by the modified grid-line intersect method [26] using an ocular grid at a 100x magnification.

2.8. Statistical Analysis.
We analysed the results of our experiments using Statistica 9.0 software (StafSoft Inc. 1984–2009). We tested the data for normal distribution and homogeneity of variance by Bartlett's test. The effects of experimental factors were evaluated by the analysis of variance (ANOVA), and comparisons between means were carried out using Tukey HSD test at the significance level of $P < 0.05$. Data on root colonization were arcsine/logarithmically transformed in order to meet the requirements of ANOVA prior to the statistical analysis. Measure of variability of a mean value throughout the text is expressed by ±SD. Linear correlation was evaluated with Pearson coefficient r, $P < 0.05$.

3. Results

Onion growth measured as the bulb fresh biomass was significantly enhanced by three experimental inoculation treatments (Figure 1(a)).

Bulb fresh weight was the highest for the M1 mix treatment, reaching nearly a 100% increase in bulb fresh biomass comparing to control. The effectiveness of dual inoculation

Dual Inoculation with Mycorrhizal and Saprotrophic Fungi Applicable in Sustainable Cultivation Improves the Yield and Nutritive Value of Onion

53

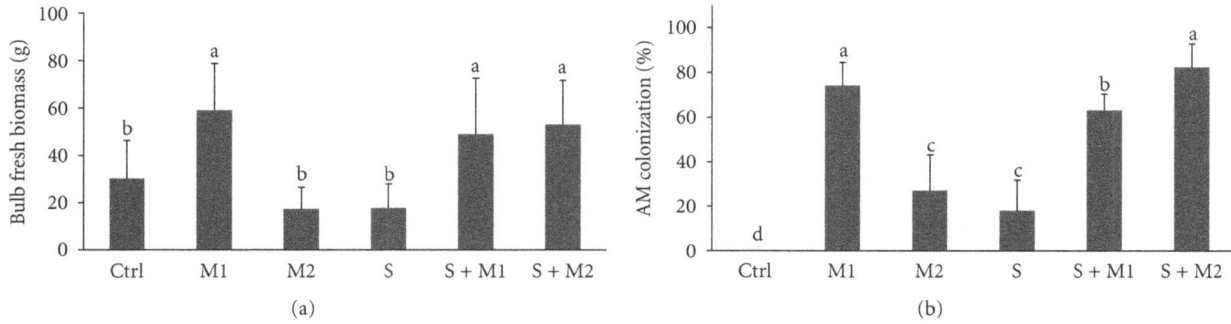

FIGURE 1: Response of onion to inoculations: (a) bulb fresh biomass, (b) AM colonization. Treatments: Ctrl—Control, M1—the mix of *Glomus* sp. (*G. intraradices* BEG140, *G. mosseae* BEG95, *G. etunicatum* BEG92, *G. claroideum* BEG96, *G. microaggregatum* BEG56, *G. geosporum* BEG199), M2—*G. intraradices* BEG140, S—saprophytic fungi preinoculated bark chips (*Gymnopilus* sp. IZO24, *Agrocybe praecox* AER1, *Marasmius androsaceus* MAN1), S + M1, S + M2. Means ± SD, columns marked with the same letters are not significantly different at the level $P < 0.05$, Tukey HSD Test, $n = 19$.

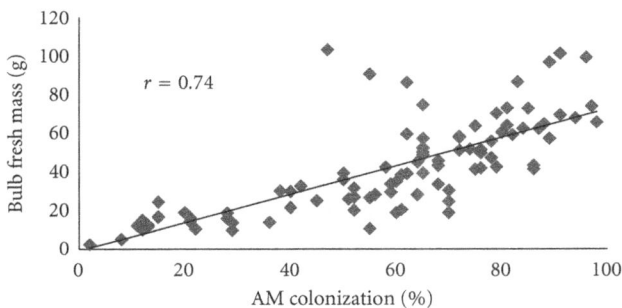

FIGURE 2: Relationship between root AM colonization (%) and bulb fresh weight (g) for all plants in inoculated treatments (M1, M2, S, S + M1, S + M2) expressed as linear correlation, r—Pearson correlation coefficient, $P < 0.05$, $n = 92$.

in both treatments (S + M1, S + M2) was proved by a more than 50% significant increase comparing to control. The lowest bulb yield was in the control and in the treatment inoculated with the saprotrophic fungus or the *G. intraradices* M2 inoculum.

Roots of onion plants of all inoculated treatments were colonized with the exception of the control, noninoculated onion plants (Ctrl). As expected, the highest rate of colonization was found for the treatments that involved mycorrhizal inoculation (M1 74.3 ± 10%, M2 27.2 ± 16.2%, S + M1 63.3 ± 7.3% S + M2 82.6 ± 10.4%) (Figure 1(b)). The observed AM colonization of plants inoculated with saprobes only (S) was rather unexpected, but AM colonization caused by airborne mycorrhizal propagules could not been fully prevented since the experiment was kept outdoor. Mycorrhizal colonization enhanced the yield response, and a strong correlation ($r = 0.83$, $P < 0.05$) was observed between AM colonization and the onion bulb growth (Figure 2).

Regarding the nutritional value of plants under different inoculation treatments, the values were either positively affected by the inoculation treatment or were not changed significantly. All inoculation treatments, but the single-fungus, (M1, S, S + M1, S + M2) enhanced significantly the

total antioxidant capacity of bulb mass (Figure 3(b)). The highest values were found for the mix M1 (27.5 ± 3.98 mg Trolox.100 g^{-1}) and both dual inoculations (S + M1 and S + M2 were 23.2 ± 2.51 and 19.3 ± 3.48 mg Trolox·100 g^{-1}, resp.). The *G. intraradices* inoculum itself did not enhance TAC but did cause a TAC increase when combined with saprotrophic fungi. The contents of nutritionally important elements in bulb tissue were enhanced in response to the microbial treatments with only some significant values. The Mg content (Figure 3(a)) was significantly increased in bulbs under the M2 treatment (127.6 ± 12.8 mg·kg^{-1}) compared to the control plants (Ctrl 73.1 ± 12.4 mg·kg^{-1}). In the case of K content (Figure 3(b)), significantly increased values were found in treatment combinations involving of the AMF mix inoculum and in saprotrophic treatments (M2 1411 ± 84 mg·kg^{-1}, S 1769 ± 22 mg·kg^{-1}, S + M2 1879 ± 25 mg·kg^{-1}), compared to the control plants (Ctrl 1411 ± 84 mg·kg^{-1}). The contents of Ca and Na were not found to be affected significantly by the inoculation treatment and reached on average 46.5 ± 13.5 and 18.7 ± 5.9 mg·kg^{-1}, respectively. Values of Ca content for individual treatments were Ctrl 35 ± 19.0 mg·kg^{-1}, M1 61 ± 7.7 mg·kg^{-1}, M2 35 ± 14.5 mg·kg^{-1}, S 55 ± 4.8 mg·kg^{-1}, S + M1 48 ± 5.7 mg·kg^{-1}, S + M2 44 ± 5.1 mg·kg^{-1}, and the values of Na content for individual treatments were Ctrl 19 ± 2.5 mg·kg^{-1}, M1 20 ± 3.4 mg·kg^{-1}, M2 16 ± 8.1 mg·kg^{-1}, S 22 ± 4.8 mg·kg^{-1}, S + M1 20 ± 3.8 mg·kg^{-1}, and S + M2 15 ± 10.6 mg·kg^{-1}. The content of vitamin C in onion was not found to be affected by the inoculation treatments, and its values ranged from 11 to 20 mg·kg^{-1} with the average value of 15.1 ± 4 mg·g^{-1}.

4. Discussion

Allium species and *A. cepa* in particular are regarded as highly AMF responsive plants [27]. Some of the reported increases are much higher than that observed in our study, which was only twice as high for the treatment with the M1 mix inoculum compared to control, uninoculated plants. For *A. cepa*, Hayman and Mosse [28] recorded an up to 18-fold increase in the weight of mycorrhizal plants compared

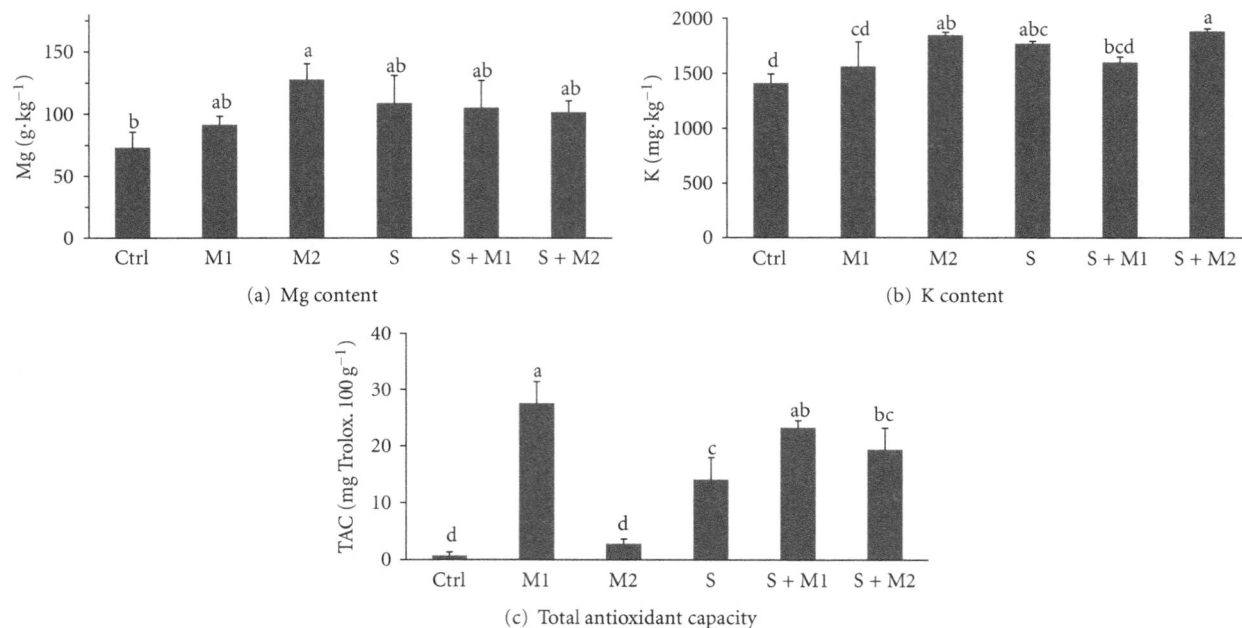

(a) Mg content

(b) K content

(c) Total antioxidant capacity

FIGURE 3: Contents of mineral elements Mg (a), K (b) and total antioxidant capacity determined by FRAP method (c) in onion bulb mass. Treatments: Ctrl—Control, M1—the mix of *Glomus* sp. (*G. intraradices* BEG140, *G. mosseae* BEG95, *G. etunicatum* BEG92, *G. claroideum* BEG96, *G. microaggregatum* BEG56, and *G. geosporum* BEG199), M2—*G. intraradices* BEG140, S—saprophytic fungi preinoculated bark chips (*Gymnopilus* sp. IZO24, *Agrocybe praecox* AER1, *Marasmius androsaceus* MAN1), S + M1, S + M2. Means ± SD, columns marked with the same letters are not significantly different at the level $P < 0.05$, Tukey HSD Test, $n = 7$.

to nonmycorrhizal ones. Onion has a coarse root system without root hairs [29]. Such plants are often obligate mycorrhizal crops that are unable to complete their life cycle in the absence of AMF because of insufficient P uptake and hence insufficient growth [30, 31]. Mycorrhiza helps plants with such a shallow sparse root system to increase phosphorus uptake. AM fungi are, to different degree, capable of promoting phosphorus availability by acidifying the soil and, consequently, exploiting the phosphorus in nutrient patches and by facilitating the growth and development of host plants [32]. The observed high mycorrhizal responsiveness to mixed inoculum has been reported previously [33, 34], and it is thus often used in commercial products.

A similarly strong correlation ($r = 0.70$) between arbuscular colonization (%) and onion yield (tons/ha) as in our case was found in an extensive survey of onion under conventional management conducted in the Netherlands [14], but the correlation was found to be much lower ($r = 0.47$) for onion grown in a field under organic management.

AMF directly or indirectly affects the soil ecosystem in different ways through multiple interactions with other soil organisms. Although they are not saprotrophs, AMF can enhance the rate of decomposition of organic material [35], indirectly influencing decomposition through interactions with other soil microorganisms [36]. Various interactions between AMF and saprotrophic fungi have been reported. *Trichoderma* in particular is now in the centre of attention because of its antagonistic effects against root pathogens and a possible synergistic interaction with AMF that helps promote plant growth. This synergy in the interaction appears

to be very genotype-specific, however. Tiunov and Scheu [37] showed in their study that *G. mosseae* BEG12 suppressed the abundance of *Trichoderma harzianum* Rifai and *Exophiala* sp. while promoting the development and abundance of *Ramichloridium schulzeri* (Sacc.) de Hoog. *Trichoderma* genotypes exhibiting synergistic effects with AMF on growth of mycorrhizal plants have already been identified. Camprubi et al. [11] reported synergistic effects between *G. intraradices* and *T. aureoviride* on the growth of *Citrus reshni* in organic substrates. Tchameni et al. [12] observed positive effects of a mixture of two different mycorrhizal fungi (*Gigaspora margarita* and *Acaulospora tuberculata*) with the PR11 strain of *Trichoderma asperellum* promoting the growth of cacao trees. This mixture also induced resistance of cacao tree against the soil-borne pathogen *Phytophthora megakarya*.

The saprotrophic fungi used in the present study were selected because they are efficient saprotrophic decomposers of wood and litter. To our knowledge, this is the first time that such a combination of saprotrophic basidiomycete fungi, preinoculated on bark chips, has been used together with AMF inoculation to improve vegetable plants. It has been reported that *A. praecox* and other basidiomycetous decomposers *in vitro* under intermediate nitrogen supply promoted carbon mineralization and induced high levels of ligninase activity in *A. praecox* cultures grown on wood of black spruce (*Picea mariana* (Mill.) Britton, Sterns & Poggenb.) [38]. This property of saprotrophic basidiomycetes, which helps plants retrieve nutrients from organic matter supplied in the form of woody material, thus may be utilized under conditions of low-input production schemes.

Dual Inoculation with Mycorrhizal and Saprotrophic Fungi Applicable in Sustainable Cultivation Improves the Yield and
Nutritive Value of Onion

55

The source of organic matter that is used as fertilizer in crop production may affect the achieved mycorrhizal responsiveness remarkably. Organic matter amendment in the form of ground leaves was found to lower the dry weight of shoots of AM cucumber plants compared to non-AM plants [39]. In the form of dried stems, it led to an increase in biomass of AM plants compared to non-AM plants of the tropical crop *Desmodium ovalifolium* L. [40]. Wheat bran amendment resulted in lowered biomass of non-AM tomato plants, but the biomass increased remarkably in plants treated either with a single inoculum containing AMF and the saprotrophic fungus *Clonostachys rosea* or a dual inoculation, which increased plant growth synergistically [41].

Plant genotype selection for reduced functioning of AM symbiosis under agricultural intensification was postulated a long time ago [42]. A recent genetic analysis of *Allium* species suggests that modern onion breeding does not select against the response to AMF [31], as has been suggested before for other cultivated species, for example, wheat [43]. In the present study, we therefore did not focus on genotypic specificity of host-plant microbe combinations. We instead focused on *A. cepa* cv. "ALICE", a genotype that is commonly used in high-input systems in the Czech Republic. Our results suggest promising applicability of the tested cultivation technology in commercial field production and small-scale cultivation alike.

During the last decades, improvements of nutritional quality brought about by microbial treatments, mycorrhiza in particular, have received increasing attention. Nell et al. [34], for example, reported positive effects of the same mixed inoculum as the one used in this study (M1) on the production of secondary metabolites (sesquiterpenic acids) by *Valeriana* sp. Recent studies also show that AM plants can contain higher amounts of nutritionally important elements. AMF inoculation, for example, significantly increased the total dry weight, leaf P, K, Ca, Mg, Fe, Cu, and Mn contents in trifoliate orange (*Poncirus trifoliata* L. Rafin.) seedlings [44], or Cu and Fe contents in lettuce [45]. Increases in K, Ca, Mg, and Fe content were also found to be an effect of dual inoculation with AMF and saprobic fungi [46]. Researchers have recently recorded positive effects of AM on the content of different groups of compounds with antioxidative properties in harvested parts of crop plants, for example, organosulfur compounds in bunching onion [47], anthocyanins, carotenoids, and, to a lesser extent, phenolics in lettuce [45]. Arbuscular mycorrhizal fungi may support the production of organosulfur compounds under field conditions. In our study, we measured only the total antioxidant capacity, not specific compounds or chemical species, so we could not determine which species the observed increase should be ascribed to.

AMF inoculation might become very important for sustainable agronomical management, especially in cases when the efficiency of native inocula is poor [48]. Positive effects of AM inoculation with nonnative species on the growth and mycorrhizal colonization of onion plants were observed in five previous field experiments [49]. In last decades, the mycorrhizal industry has been developing an entire range of mycorrhizal products although the concern remains about inoculum quality [50]. An ongoing debate remains whether introduced AMF can survive in a form in which mycorrhizal symbionts are efficient in natural field conditions. Sequencing of fungal ITS has already provided several lines of evidence about the persistence of inoculated AMF after 2 years suggesting [20]. Another concern in application of AMF inoculants is about lowering AMF diversity in target ecosystems caused by outcompeting indigenous fungi by introduced nonnative fungal strains. Applications of tuned AM inocula derived from isolated native AMF strains from a particular ecosystem could overcome this problem and have been already the strategy of some commercial companies [5, 6].

5. Conclusions

Our results support the conclusion that there is a synergistic effect between dual microbial inoculation containing both AMF and saprotrophic fungi supplied together with organic matter. This ecological way of cultivation can lead to improvement of the parameters of onion plants as well as their nutritional value in sustainable production. Proper tuning of responsive genotypes of host plants, AMF, and saprotrophic fungi can bring not only a biomass increase but can also lead to improvement of nutritional quality. We believe that such synergistic dual fungal inoculations involving mycorrhizal and saprotrophic fungi together with organic matter supply have high potential for sustainable, environment-friendly production systems not only of onion but of crops in general.

Abbreviations

AMF: Arbuscular mycorrhizal fungi
AM: Arbuscular mycorrhiza
TAC: Total antioxidant capacity.

Acknowledgments

Support of the COST project Mykotech OC09057 in the framework of the COST Action 870 and Eurostars Project Microfruit E!4366 funded by the Ministry of Education, Youth and Sports of the Czech Republic is acknowledged. The technical help and contribution of Klara Prochazkova is acknowledged. The authors declare that they do not have a conflict of interests with any identities mentioned in the present paper. The company Symbiom Ltd. with its scientific division led by Dr. Ales Latr participated in the research study as the principal investigator of the Microfruit project and coinvestigator of the Mykotech project, which was led by Charles University in Prague.

References

[1] J. Roger-Estrade, C. Anger, M. Bertrand, and G. Richard, "Tillage and soil ecology: partners for sustainable agriculture," *Soil and Tillage Research*, vol. 111, no. 1, pp. 33–40, 2010.

[2] E. Lichtfouse, R. Habib, J. M. Meynard, and F. Papy, "Agronomy for sustainable development," *Agronomie*, vol. 24, no. 8, p. 445, 2004.

[3] R. Lal, "Soils and sustainable agriculture. A review," *Agronomy for Sustainable Development*, vol. 28, no. 1, pp. 57–64, 2008.

[4] S. Gianinazzi, A. Gollotte, M. N. Binet, D. van Tuinen, D. Redecker, and D. Wipf, "Agroecology: the key role of arbuscular mycorrhizas in ecosystem services," *Mycorrhiza*, vol. 20, no. 8, pp. 519–530, 2010.

[5] S. Gianinazzi and M. Vosátka, "Inoculum of arbuscular mycorrhizal fungi for production systems: science meets business," *Canadian Journal of Botany*, vol. 82, no. 8, pp. 1264–1271, 2004.

[6] M. Vosatka and J. Albrechtova, "Theoretical aspects an practical uses of mycorrhizal technology in floriculture and horticulture," in *Floriculture, Ornamental and Plant Biotechnology. Advances and Topical Issues*, J. A. T. da Silva, Ed., vol. 5, pp. 466–479, Global Science Books Ltd, 2008.

[7] F. Y. Wang, R. J. Tong, Z. Y. Shi, X. F. Xu, and X. H. He, "Inoculations with Arbuscular mycorrhizal fungi increase vegetable yields and decrease phoxim concentrations in carrot and green onion and their soils," *PLoS ONE*, vol. 6, no. 2, e16949, 2011.

[8] F. Oehl, E. Sieverding, K. Ineichen, P. Mäder, T. Boller, and A. Wiemken, "Impact of land use intensity on the species diversity of arbuscular mycorrhizal fungi in agroecosystems of Central Europe," *Applied and Environmental Microbiology*, vol. 69, no. 5, pp. 2816–2824, 2003.

[9] F. Oehl, E. Sieverding, P. Mäder et al., "Impact of long-term conventional and organic farming on the diversity of arbuscular mycorrhizal fungi," *Oecologia*, vol. 138, no. 4, pp. 574–583, 2004.

[10] M. Gryndler, M. Vosátka, H. Hršelová, V. Catská, I. Chvátalová, and J. Jansa, "Effect of dual inoculation with arbuscular mycorrhizal fungi and bacteria on growth and mineral nutrition of strawberry," *Journal of Plant Nutrition*, vol. 25, no. 6, pp. 1341–1358, 2002.

[11] A. Camprubi, C. Calvet, and V. Estaun, "Growth enhancement of *Citrus reshni* after inoculation with *Glomus intraradices* and *Trichoderma aureoviride* and associated effects on microbial populations and enzyme activity in potting mixes," *Plant and Soil*, vol. 173, no. 2, pp. 233–238, 1995.

[12] S. N. Tchameni, M. E. L. Ngonkeu, B. A. D. Begoude et al., "Effect of *Trichoderma asperellum* and arbuscular mycorrhizal fungi on cacao growth and resistance against black pod disease," *Crop Protection*, vol. 30, no. 10, pp. 1321–1327, 2011.

[13] A. D. Bosch-Serra and L. Currah, "Agronomy of onions," in *Allium Crop Science: Recent Advances*, H. D. Rabinowitch and L. Currah, Eds., pp. 187–232, CAB International, Wallingford, UK, 2002.

[14] G. A. Galván, I. Parádi, K. Burger et al., "Molecular diversity of arbuscular mycorrhizal fungi in onion roots from organic and conventional farming systems in the Netherlands," *Mycorrhiza*, vol. 19, no. 5, pp. 317–328, 2009.

[15] J. M. Phillips and D. S. Hayman, "Improved procedures for clearing roots and staining parasitic and vesicular-arbuscular mycorrhizal fungi for rapid assessment of infection," *Transactions of the British Mycological Society*, vol. 55, pp. 158–161, 1970.

[16] K. H. Miean and S. Mohamed, "Flavonoid (myricetin, quercetin, kaempferol, luteolin, and apigenin) content of edible tropical plants," *Journal of Agricultural and Food Chemistry*, vol. 49, no. 6, pp. 3106–3112, 2001.

[17] M. Corzo-Martínez, N. Corzo, and M. Villamiel, "Biological properties of onions and garlic," *Trends in Food Science and Technology*, vol. 18, no. 12, pp. 609–625, 2007.

[18] J. Park, J. Kim, and M. K. Kim, "Onion flesh and onion peel enhance antioxidant status in aged rats," *Journal of Nutritional Science and Vitaminology*, vol. 53, no. 1, pp. 21–29, 2007.

[19] M. A. Vazquez-Prieto and R. M. Miatello, "Organosulfur compounds and cardiovascular disease," *Molecular Aspects of Medicine*, vol. 31, no. 6, pp. 540–545, 2010.

[20] N. Ceccarelli, M. Curadi, L. Martelloni, C. Sbrana, P. Picciarelli, and M. Giovannetti, "Mycorrhizal colonization impacts on phenolic content and antioxidant properties of artichoke leaves and flower heads two years after field transplant," *Plant and Soil*, vol. 335, no. 1, pp. 311–323, 2010.

[21] R. Montalba, C. Arriagada, M. Alvear, and G. E. Zúñiga, "Effects of conventional and organic nitrogen fertilizers on soil microbial activity, mycorrhizal colonization, leaf antioxidant content, and Fusarium wilt in highbush blueberry (*Vaccinium corymbosum* L.)," *Scientia Horticulturae*, vol. 125, no. 4, pp. 775–778, 2010.

[22] I. F. F. Benzie and J. J. Strain, "The ferric reducing ability of plasma (FRAP) as a measure of 'antioxidant power': the FRAP assay," *Analytical Biochemistry*, vol. 239, no. 1, pp. 70–76, 1996.

[23] P. Blatny, F. Kvasnicka, R. Loucka, and H. Safarova, "Determination of ammonium, calcium, magnesium, and potassium in silage by capillary isotachophoresis," *Journal of Agricultural and Food Chemistry*, vol. 45, no. 9, pp. 3554–3558, 1997.

[24] S. P. Arya, M. Mahajan, and P. Jain, "Non-spectrophotometric methods for the determination of Vitamin C," *Analytica Chimica Acta*, vol. 417, no. 1, pp. 1–14, 2000.

[25] R. E. Koske and J. N. Gemma, "A modified procedure for staining roots to detect VA mycorrhizas," *Mycological Research*, vol. 92, pp. 486–505, 1989.

[26] M. Giovannetti and B. Mosse, "Evaluation of techniques for measuring vesicular-arbuscular mzcorrhiyal infection in roots," *New Phytologist*, vol. 84, no. 3, pp. 489–500, 1980.

[27] C. Plenchette, J. A. Fortin, and V. Furlan, "Growth responses of several plant species to mycorrhizae in a soil of moderate P-fertility—I. Mycorrhizal dependency under field conditions," *Plant and Soil*, vol. 70, no. 2, pp. 199–209, 1983.

[28] D. S. Hayman and B. Mosse, "Plant growth responses to vesicular-arbuscular mycorrhiza—I. Growth of endogone-inoculated plants in phosphate-deficient soils," *New Phytologist*, vol. 70, no. 1, pp. 19–27, 1971.

[29] C. A. M. Portas, "Development of root systems during the growth of some vegetable crops," *Plant and Soil*, vol. 39, no. 3, pp. 507–518, 1973.

[30] G. Charron, V. Furlan, M. Bernier-Cardou, and G. Doyon, "Response of onion plants to arbuscular mycorrhizae 1. Effects of inoculation method and phosphorus fertilization on biomass and bulb firmness," *Mycorrhiza*, vol. 11, no. 4, pp. 187–197, 2001.

[31] G. A. Galván, T. W. Kuyper, K. Burger et al., "Genetic analysis of the interaction between *Allium* species and arbuscular mycorrhizal fungi," *Theoretical and Applied Genetics*, vol. 122, no. 5, pp. 947–960, 2011.

[32] Z. Shi, F. Wang, C. Zhang, and Z. Yang, "Exploitation of phosphorus patches with different phosphorus enrichment by three arbuscular mycorrhizal fungi," *Journal of Plant Nutrition*, vol. 34, no. 8, pp. 1096–1106, 2011.

[33] C. Tu, F. L. Booker, D. M. Watson et al., "Mycorrhizal mediation of plant N acquisition and residue decomposition: impact of mineral N inputs," *Global Change Biology*, vol. 12, no. 5, pp. 793–803, 2006.

[34] M. Nell, C. Wawrosch, S. Steinkellner et al., "Root colonization by symbiotic arbuscular mycorrhizal fungi increases

Dual Inoculation with Mycorrhizal and Saprotrophic Fungi Applicable in Sustainable Cultivation Improves the Yield and Nutritive Value of Onion

57

sesquiterpenic acid concentrations in *Valeriana officinalis* L," *Planta Medica*, vol. 76, no. 4, pp. 393–398, 2010.

[35] A. Hodge, C. D. Campbell, and A. H. Fitter, "An arbuscular mycorrhizal fungus accelerates decomposition and acquires nitrogen directly from organic material," *Nature*, vol. 413, no. 6853, pp. 297–299, 2001.

[36] J. Leigh, A. H. Fitter, and A. Hodge, "Growth and symbiotic effectiveness of an arbuscular mycorrhizal fungus in organic matter in competition with soil bacteria," *FEMS Microbiology Ecology*, vol. 76, no. 3, pp. 428–438, 2011.

[37] A. V. Tiunov and S. Scheu, "Arbuscular mycorrhiza and Collembola interact in affecting community composition of saprotrophic microfungi," *Oecologia*, vol. 142, no. 4, pp. 636–642, 2005.

[38] S. D. Allison, D. S. LeBauer, M. R. Ofrecio, R. Reyes, A. M. Ta, and T. M. Tran, "Low levels of nitrogen addition stimulate decomposition by boreal forest fungi," *Soil Biology and Biochemistry*, vol. 41, no. 2, pp. 293–302, 2009.

[39] A. Albertsen, S. Ravnskov, H. Green, D. F. Jensen, and J. Larsen, "Interactions between the external mycelium of the mycorrhizal fungus *Glomus intraradices* and other soil microorganisms as affected by organic matter," *Soil Biology and Biochemistry*, vol. 38, no. 5, pp. 1008–1014, 2006.

[40] C. L. Boddington and J. C. Dodd, "The effect of agricultural practices on the development of indigenous arbuscular mycorrhizal fungi. II. Studies in experimental microcosms," *Plant and Soil*, vol. 218, no. 1-2, pp. 145–157, 2000.

[41] S. Ravnskov, B. Jensen, I. M. B. Knudsen et al., "Soil inoculation with the biocontrol agent *Clonostachys rosea* and the mycorrhizal fungus *Glomus intraradices* results in mutual inhibition, plant growth promotion and alteration of soil microbial communities," *Soil Biology and Biochemistry*, vol. 38, no. 12, pp. 3453–3462, 2006.

[42] N. C. Johnson, "Can fertilization of soil select less mutualistic mycorrhizae?" *Ecological Applications*, vol. 3, no. 4, pp. 749–757, 1993.

[43] R. J. H. Sawers, C. Gutjahr, and U. Paszkowski, "Cereal mycorrhiza: an ancient symbiosis in modern agriculture," *Trends in Plant Science*, vol. 13, no. 2, pp. 93–97, 2008.

[44] Q. S. Wu and Y. N. Zou, "The effect of dual application of arbuscular mycorrhizal fungi and polyamines upon growth and nutrient uptake on trifoliate orange (*Poncirus trifoliata*) seedlings," *Notulae Botanicae Horti Agrobotanici Cluj-Napoca*, vol. 37, no. 2, pp. 95–98, 2009.

[45] M. Baslam, I. Garmendia, and N. Goicoechea, "Arbuscular mycorrhizal fungi (AMF) improved growth and nutritional quality of greenhouse-grown Lettuce," *Journal of Agricultural and Food Chemistry*, vol. 59, no. 10, pp. 5504–5515, 2011.

[46] C. Arriagada, I. Sampedro, I. Garcia-Romera, and J. Ocampo, "Improvement of growth of *Eucalyptus globulus* and soil biological parameters by amendment with sewage sludge and inoculation with arbuscular mycorrhizal and saprobe fungi," *Science of the Total Environment*, vol. 407, no. 17, pp. 4799–4806, 2009.

[47] H. Perner, S. Rohn, G. Driemel et al., "Effect of nitrogen species supply and mycorrhizal colonization on organosulfur and phenolic compounds in onions," *Journal of Agricultural and Food Chemistry*, vol. 56, no. 10, pp. 3538–3545, 2008.

[48] E. Pellegrino, S. Bedini, L. Avio, E. Bonari, and M. Giovannetti, "Field inoculation effectiveness of native and exotic arbuscular mycorrhizal fungi in a Mediterranean agricultural soil," *Soil Biology and Biochemistry*, vol. 43, no. 2, pp. 367–376, 2011.

[49] M. Vosatka, "Influence of inoculation with arbuscular mycorrhizal fungi on the growth and mycorrhizal infection of transplanted onion," *Agriculture, Ecosystems and Environment*, vol. 53, no. 2, pp. 151–159, 1995.

[50] M. Vosatka, J. Albrechtova, and R. Patten, "The international market development for mycorrhizal technology," in *Mycorrhiza*, A. Varma, Ed., chapter 21, pp. 438–419, Springer, 2008.

7

Computed Tomography to Estimate the Representative Elementary Area for Soil Porosity Measurements

Jaqueline Aparecida Ribaski Borges,[1] Luiz Fernando Pires,[1] and André Belmont Pereira[2]

[1] Laboratory of Soil Physics and Environmental Sciences, Department of Physics, State University of Ponta Grossa (UEPG), Avenue Carlos Cavalcanti 4748, 84030-900 Ponta Grossa, PR, Brazil
[2] Department of Soil Science, State University of Ponta Grossa (UEPG), Avenue Carlos Cavalcanti 4748, 84030-900 Ponta Grossa, PR, Brazil

Correspondence should be addressed to Jaqueline Aparecida Ribaski Borges, jaqueribaski@gmail.com

Academic Editors: F. Knollmann and H. Mori

Computed tomography (CT) is a technique that provides images of different solid and porous materials. CT could be an ideal tool to study representative sizes of soil samples because of the noninvasive characteristic of this technique. The scrutiny of such representative elementary sizes (RESs) has been the target of attention of many researchers related to soil physics field owing to the strong relationship between physical properties and size of the soil sample. In the current work, data from gamma-ray CT were used to assess RES in measurements of soil porosity (ϕ). For statistical analysis, a study on the full width at a half maximum (FWHM) of the adjustment of distribution of ϕ at different areas (1.2 to 1162.8 mm^2) selected inside of tomographic images was proposed herein. The results obtained point out that samples with a section area corresponding to at least 882.1 mm^2 were the ones that provided representative values of ϕ for the studied Brazilian tropical soil.

1. Introduction

Computed tomography (CT) is proven an efficient technique that can be largely used in studies related to soil structure [1–3]. It has been seen as an important tool to be adopted by new generation's tomographs designed exclusively for research carried out with porous materials [4, 5]. The success of the aforementioned technique is ascribed to a method that is noninvasive to determine physical properties in a cross-section of a material. Another advantage of such technique is that CT also provides 2D and 3D images with micro- and millimetric resolutions and allows qualitative and quantitative analyses [6].

Among several practical applications [7–9], CT is also an excellent technique employed to assess representative sizes of soil samples, as well as to scrutinize soil physical properties. This is because it is possible to select volumes, areas, or lengths of different sizes in the inside of tomographic images, depending on the generation of the equipment [10, 11].

The concept of representative elementary size (RES) was first introduced to the continuum mechanics by Jacob Bear

in 1972 as a tool to be employed to describe flow in porous media. The approach deals with the definition of a minimal size or physical point of a sample necessary for representing its characteristics of interest. In other words, it refers to as the size at which a measured parameter turns out to be independent of the size of the sample [12].

The analysis applied to RES is commonly made by selecting consecutive sizes around a central point in the image of the sample. It is reported in the literature that adjacent selections within the same image and centered in different points can also be utilized [11, 13]. The representative size is then defined as that one corresponding to the domain transition of the microscopic effects (region I) to the domain of a porous media (region II) (Figure 1).

The main concern with the use of samples with representative sizes is due to the relationship between soil physical properties and size of soil samples [14, 15]. However, such sizes are normally investigated for properties of a particular interest in homogeneous media, such as spherical glass beads and sands [10]. Moreover, representative elementary volume (REV) in particular became a parameter that demonstrates

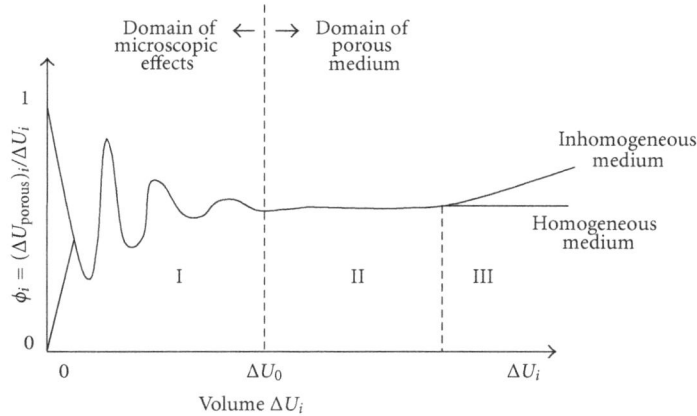

FIGURE 1: Representative elementary volume (REV, ΔU_0) for the porosity (ϕ). ΔU_i represents any volume in the porous media. REA is defined from the concept of REV and shows similar behavior. Source: adapted from Bear [12].

the quality of the measurements made via third-generation CTs [13].

The available research in the literature for nonhomogeneous media with porosity and/or particles size varying in space is still scarce. In these cases, a difficulty for determining the elementary size is observed under experimental conditions whenever slight fluctuations occur for the analyzed physical property. In order to overcome such a problem, statistical tools have been applied to delimitate the maximum allowable variation in each case, setting up therefore, a reliability level for RES [15].

Porosity (ϕ) is an important index that reveals the structural quality of a soil and represents the volume of a soil not occupied by solid particles, including all porous spaces occupied by water and air. Porosity is fundamentally linked to the root growth and movement of air, water, and solutes in the soil. For instance, a well-structured soil generally possesses intraaggregate (textural) and interaggregate (structural) pores, being the macroporosity an index that expresses the structural quality of the soil. Soils with a good quantity of macropores (10%) will favor the gas exchanges and the development of the root system for crop production [16].

Faced with the scarcity of published manuscripts that deals with representative sizes for nonhomogeneous samples in conjunction with the agricultural relevance of representative measurements of soil porosity, the current contribution aims to make use of computed tomography to determine the representative elementary area for measurements of ϕ.

2. Materials and Methods

2.1. Soil Sampling.
Eighteen soil samples were collected from an experimental field belonging to the University of São Paulo-(ESALQ/USP)-located at Piracicaba, SP, Brazil (22°42′S and 47°38′W, 580 m above sea level). Samples were collected in triplicate, being selected 6 collection points along a transection of 200 m long. The volumes of the soil clod sample varied from 50 to 100 cm^3. The sampling was made at the surface layer (0–15 cm) inside of a small trench. Shortly

before the opening of a trench the crop above the soil surface was removed.

The clay soil (43% clay, 24% sand, and 33% silt) was classified as an Eutric Nitosol [17]. It presents a soil particle density (ρ_p) of 2.65 g cm^{-3}. As to its chemical characteristics, the soil possesses 20.2 g dm^{-3} of organic matter; pH 5.3 (in CaCl$_2$), and 29.0, 20.0, and 4.3 mol m^{-3} of Ca, Mg, and K [18].

2.2. Gamma-Ray Computed Tomography.
CT utilized is a first-generation scanner with the source and detector fixed and with rotation and translation movement of the sample. The scanner was built by engineers from EMBRAPA/CNPDIA (São Carlos, Brazil) and presents the following modules: (1) ^{241}Am (59.54 keV, 3.7 GBq) gamma ray source mounted in a Pb shield castle; (2) NaI(Tl) scintillation crystal detector (7.62 × 7.62 cm) coupled to a photomultiplier tube; (3) electronic modules (preamplifier, high-voltage supplier, single channel analyzer, counter and timer); (4) step motor (rotational and translational movements); (5) Pb collimators; (6) software for CT data acquision [19]. The counter, with an RS-232 interface, makes communication with an IBM PC that controls the step motor.

Lead collimators with 1 and 4.5 mm were placed in front of the source and detector to collimate the beam. The matrixes of tomographic units (TUs) data obtained were of 80 × 80 for all tomographs. The resolution obtained for the clod samples was of 1.1 × 1.1 mm^2. A 2D section image for each clod was obtained at the center of the sample.

TU is proportional to the linear attenuation coefficient, μ (cm^{-1}), and the reference media for the TU is the air, showing the lowest attenuation indices. For the soil system, TU corresponds to the contribution of the mineral particles, organic matter, water, and air, generating different values of μ for each crossed sample path by the radiation beam [20].

The relationship between TU and different physical properties of the soil, such as soil bulk density (ρ_s) and its volumetric water content (θ) is given by (1) [3, 21, 22]:

$$TU = \alpha(\mu_{ms}\rho_s + \mu_{mw}\rho_w\theta), \tag{1}$$

TABLE 1: Areas (mm^2) adopted for the REA definition.

Area	Size (mm^2)	Area	Size (mm^2)	Area	Size (mm^2)	Area	Size (mm^2)
01	1.2	05	98.0	09	349.7	13	756.3
02	10.9	06	146.4	10	436.8	14	882.1
03	30.3	07	204.5	11	533.6	15	1017.6
04	59.3	08	272.3	12	640.1	16	1162.8

where ρ_w (g cm^{-3}) is the water density, α is the angular coefficient of the calibration straight line of the tomographic system, μ_{ms} and μ_{mw} (cm^2 g^{-1}) are the mass attenuation coefficients of soil and water, respectively.

It is possible to calculate ϕ for each TU data of the dry soil sample scanned by using a combination of the equation used to determine ϕ by conventional methods [23] and (1) as follows:

$$\phi = \left(1 - \frac{\rho_s}{\rho_p}\right) = \left(1 - \frac{\mathrm{TU}}{\alpha\mu_{ms}\rho_p}\right). \tag{2}$$

For the calibration of the scanner, samples of the following homogeneous materials were used: acrylic, ethanol, water, nylon, and glycerin. 2D section images were obtained at the center of samples used for calibration. A thorough description of the calibration process of the first-generation scanner can be found in Crestana et al. and Pires et al. [24, 25].

2.3. Attenuation Coefficient.

In order to evaluate the soil linear attenuation coefficient (μ_s), samples were air dried and sieved with a 2.0 mm mesh. After sieving, the soil was transferred to an acrylic box with the dimensions of 4.9 × 5.1 × 5.5 cm. The intensities of monoenergetic photons were obtained at three different positions of the acrylic box filled with soil. Five replicates were taken for each position. The same procedure was performed to assess the linear attenuation coefficient of water (μ_w).

The mass attenuation coefficients of soil and water were calculated dividing the experimental μ by the ρ of the samples ($\mu_m = \mu/\rho$). For the case of water, the density was considered as $\rho_w = 1$ g cm^{-3}.

2.4. Representative Elementary Area.

Matrixes of TU data obtained via CT (80 × 80) were firstly converted into density and porosity matrixes by means of (1) and (2). The images were reconstructed using the software Microvis [26]. Darker regions in the images correspond to the lowest values of density and, consequently, to the lowest values of TU. In the current work, the darker regions represent larger values of density.

For the REA evaluation, the largest possible rectangular area at the center of the sample was delimitated with no interference of the edges in a tomographic image. The edges were avoided since the interface sample-air might generate artifacts that can affect the analysis of soil physical properties by CT [19, 27].

The reference points at each vertex of a rectangular area were selected in the tomographic image by means of Microvis

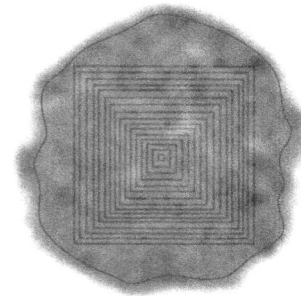

FIGURE 2: Schematic drawn of the area construction on the tomographic images. The area next to edge corresponds to the free area (FA). Darker regions represent higher soil bulk density values.

software and identified and demarked in the matrix of TU afterwards. Then, consecutive concentric quadrangular areas (Figure 2) were selected without extrapolating the maximum area previously chosen. The initial area was obtained from a 1 : 1 square matrix (1.1 mm × 1.1 mm, Table 1). The number of delimited areas for each sample image varied in compliance with its size and differences in its shape. Plus an area with an irregular shape containing almost the entire tomographic image was also selected to refer to the free area (FA).

The linear steps of the tomographic system were of 1.0 to 1.1 mm. The discrepancies in the linear steps are due to different dimensions of the samples of the analyzed soil.

The ϕ was determined for each one of the quadrangular areas, as well as for the FA. Soil porosity obtained via CT corresponds to the mean value of such a physical property once CT allows for its analysis point by point (from "pixel" to "pixel").

The identification of REA was established taking into account the criterion used by Vandenbygaart and Protz [15]. Below are listed the analyses made and the criteria adopted to define REA as a function of ϕ:

(i) determination of ϕ values frequency within each area, for the FA, the computed frequency corresponds to the interval of porosity present in the selected FA for each sample;

(ii) elaboration of graphs that show the frequency of ϕ (%), such a procedure was performed for each one of the areas;

(iii) determination of the full width at a half maximum (FWHM) of each distribution by means of Gaussian adjustments, which were obtained from the fifth area of each sample (matrix 9 × 9), in the present

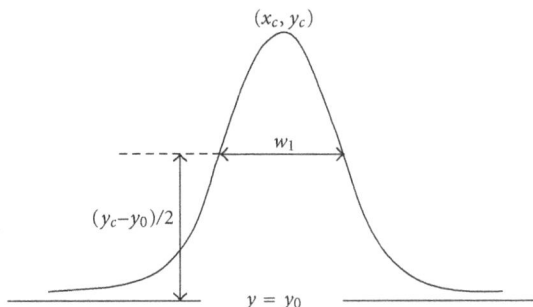

FIGURE 3: Schematic representation of the parameters used to calculate the full width at a half maximum (FWHM) for a normal distribution. Adapted from Origin User Guide [28].

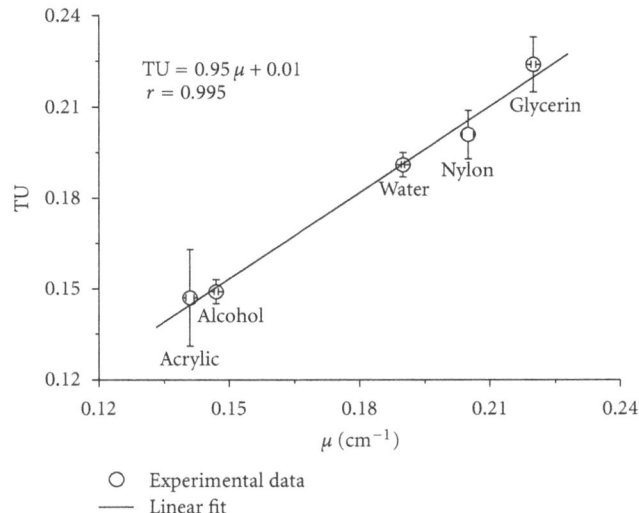

FIGURE 4: Experimental relationship between the tomographic units (TU) of the images obtained by the tomograph and linear attenuation coefficients (μ) for homogeneous substances used during the calibration of the tomograph. The vertical error bars represent the standard deviation of the TU values in the matrix of selected data. The horizontal error bars depict the mean standard deviation ($n = 3$).

study, the parameter FWHM was used to describe the distribution of ϕ of the samples in each area, the adjustment equation is

$$y_c = y_0 + \frac{A}{w\sqrt{\pi/2}}e^{-2(x-x_c)^2/w^2},\qquad(3)$$

where y_c is the maximum height of the adjustment curve; y_0 is the basis of the curve; A is the area below the curve; w is the parameter given by

$$w = \frac{w_1}{\sqrt{\ln(4)}},\qquad(4)$$

where w_1 is the FWHM of the curve; x is any position in the abscissas axis and x_c the central position of the adjustment curve in the same axis (Figure 3);

(iv) determination of relative deviations between the value of FWHM corresponding to the distribution of ϕ for the last rectangular area and each one of its previous areas;

(v) REA for distribution of ϕ is reached whenever three consecutive areas did not present deviations higher than 10%;

(vi) graphs for the values of FWHM for each area of the samples (including FA) were made, for the samples that reached REA, corresponding areas were demarked.

3. Results and Discussion

The coefficient of correlation (r) obtained during the calibration of the CT system was of 0.995 (Figure 4). Such a good correlation between the experimental data is of a great importance in studies conducted to acquire representative measurements of soil physical properties through CT [25].

The μ_{ms} and μ_{mw} values were 0.3339 ± 0.0029 and 0.2001 ± 0.0004 cm^2 g^{-1}, respectively. Such values are coherent when compared to the experimental and theoretical outcomes obtained by other researchers for water and soil with the same texture of that one studied herein [29, 30].

The ϕ values calculated by the CT technique were compared to the values obtained by the paraffin-sealed method (PSM) for the same soil (Figure 5). The mean values of ϕ obtained by means of PSM and CT were of 37 and 36%, respectively. A part from that, a correlation study between methods was performed ($r = 0.75$) and a relatively strong positive correlation was observed between the variables in study (Figure 5(a)). The Bland-Altman analysis [31] was also performed to reveal the relationship between the differences and the magnitude of measurements (Figure 5(b)). A good agreement was found since the mean difference is close to zero and not statistically significant (test statistic t, 0.05). From the plot, it is also possible to observe that 1/18 (5.5%) of the points are beyond the limits of agreement (±2sd lines).

In Figure 6, we can visualize tomographic images of some studied samples. Besides being capable of conducting quantitative studie on soil physical properties via CT images, we can also have a qualitative idea of its spatial variability. For instance, S 01 (Figure 6) does not show great discrepancies on its structure, being more homogeneous (see grey scale) in relation to the other samples.

By elaborating graphs for frequency distribution of ϕ in each selected area of images, we observed that such images presented a Gaussian distribution of ϕ from the fifth selected area. Such a fact was explained in compliance with the central theory of the limit, which assumes that each randomized variable with no particular distribution approaches the normal in so far as the sample size increases [32].

Figures 7 and 8 show the distributions obtained experimentally for S 01 and S 10. The sample S 01 (Figure 7) reveals that the superior limit of the distribution of ϕ remains

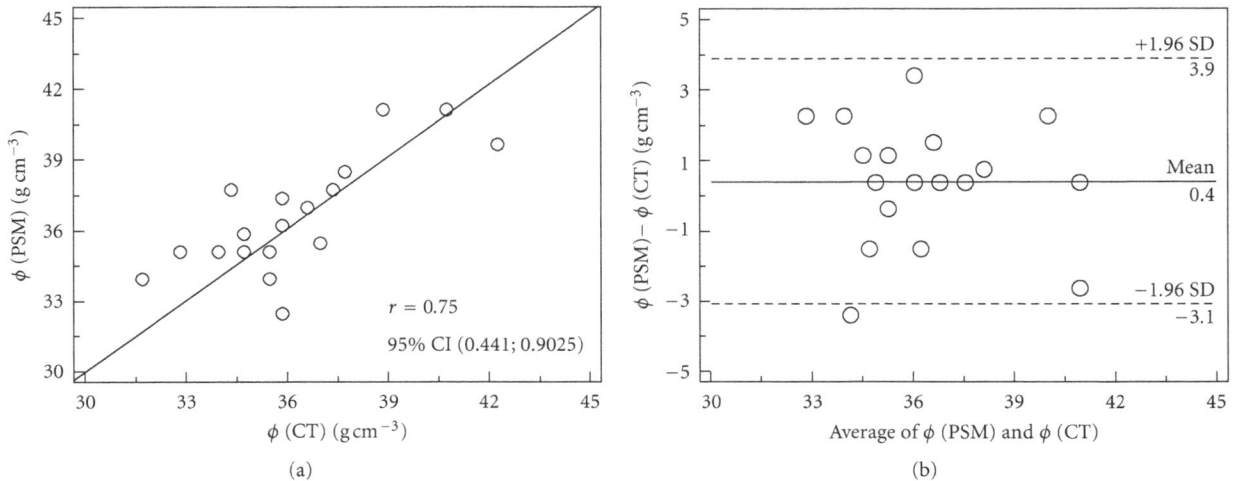

FIGURE 5: (a) Soil porosity (ϕ) measured by the computed tomography (CT) and paraffin-sealed methods (PSMs), with line of equality; (b) Bland-Altman plot.

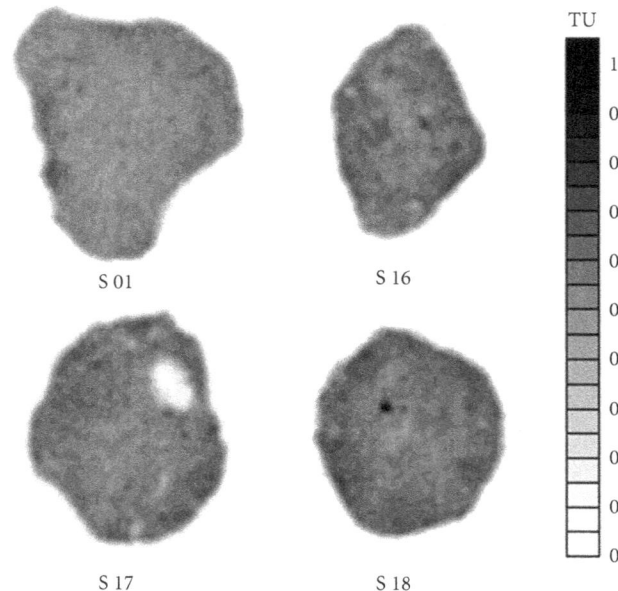

FIGURE 6: Tomographic images referent to the samples 01, 16, 17, and 18. The images are in a grey scale of tomographic units (TU), where the darker regions indicate the highest value of TU.

roughly the same, whereas the inferior limit turns from 40% (5th area) to 24% (FA) of ϕ. On the other hand, for S 10 (Figure 8), such an inferior limit varies from 37% to 17% of ϕ. The superior limit of the distribution of ϕ for S 10 also shows a great variation, from 54% to 62%. However, such variations at the extremities of the curves have a despicable influence on FWHM of the distribution, since this variable is directly influenced by the distribution of the central values. Although FWHM comes to being normally used for resolution measurements in spectral analysis, we opted to use it in the current study aiming at generating a new parameter to predict REA for ϕ.

The coefficient of determination (r^2) obtained for each one of the distributions is quite high. Its value is equal or

greater than 0.90 in 16 out of 24 distributions presented—a fact that indicates a good agreement between the variables studied.

The central position of the distribution at the abscissas axis (x_c) demonstrated a slight variation within the same sample for consecutive areas. However, the variation is more pronounced when we compare FA for previous areas. This happens because the FA shows an interval of a large distance of the quadrangular areas, whilst they were selected consecutively. Moreover, FA includes practically the entire sample, which causes the heterogeneity of it to be greater than that one of small areas.

Figure 9 shows the graphs for FWHM from the 5th area selected in 18 samples and its respective REA for a deviation

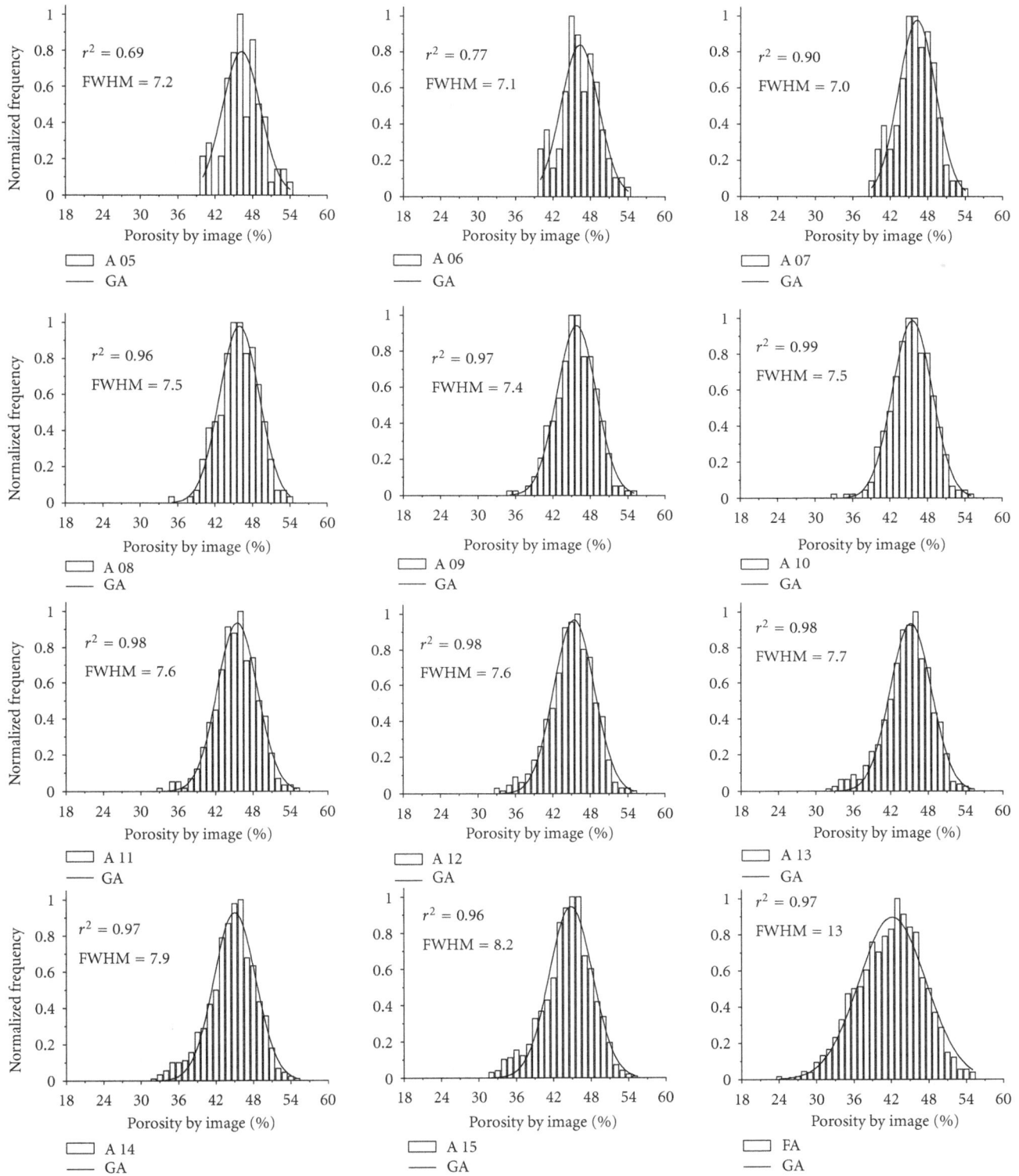

FIGURE 7: Normalized frequency of the porosity by image (%) from the 5th selected area in the sample 01 (S 01). A 05, A 06, ..., and A 15 correspond to the sequence of quadrangular areas selected in the image and free area (FA). GA represents the Gaussian adjustment, r^2 the coefficient of determination, and FWHM the full width at a half maximum of the curve.

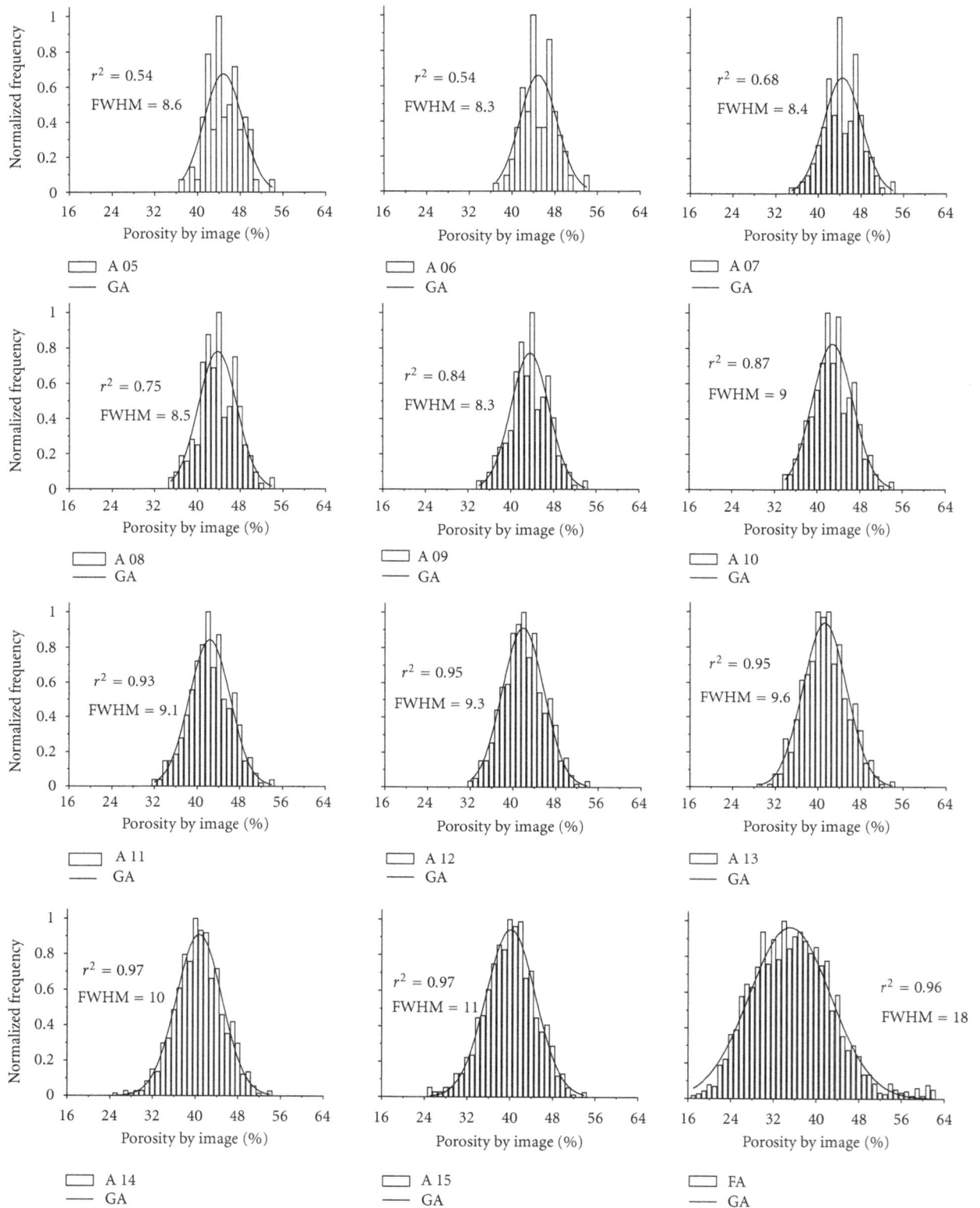

FIGURE 8: Normalized frequency of the porosity by image (%) from the 5th selected area in the sample 10 (S 10). A 05, A 06, ..., and A 15 correspond to the sequence of quadrangular areas selected in the image and free area (FA). GA represents the Gaussian adjustment, r^2 the coefficient of determination, and FWHM the full width at a half maximum of the curve.

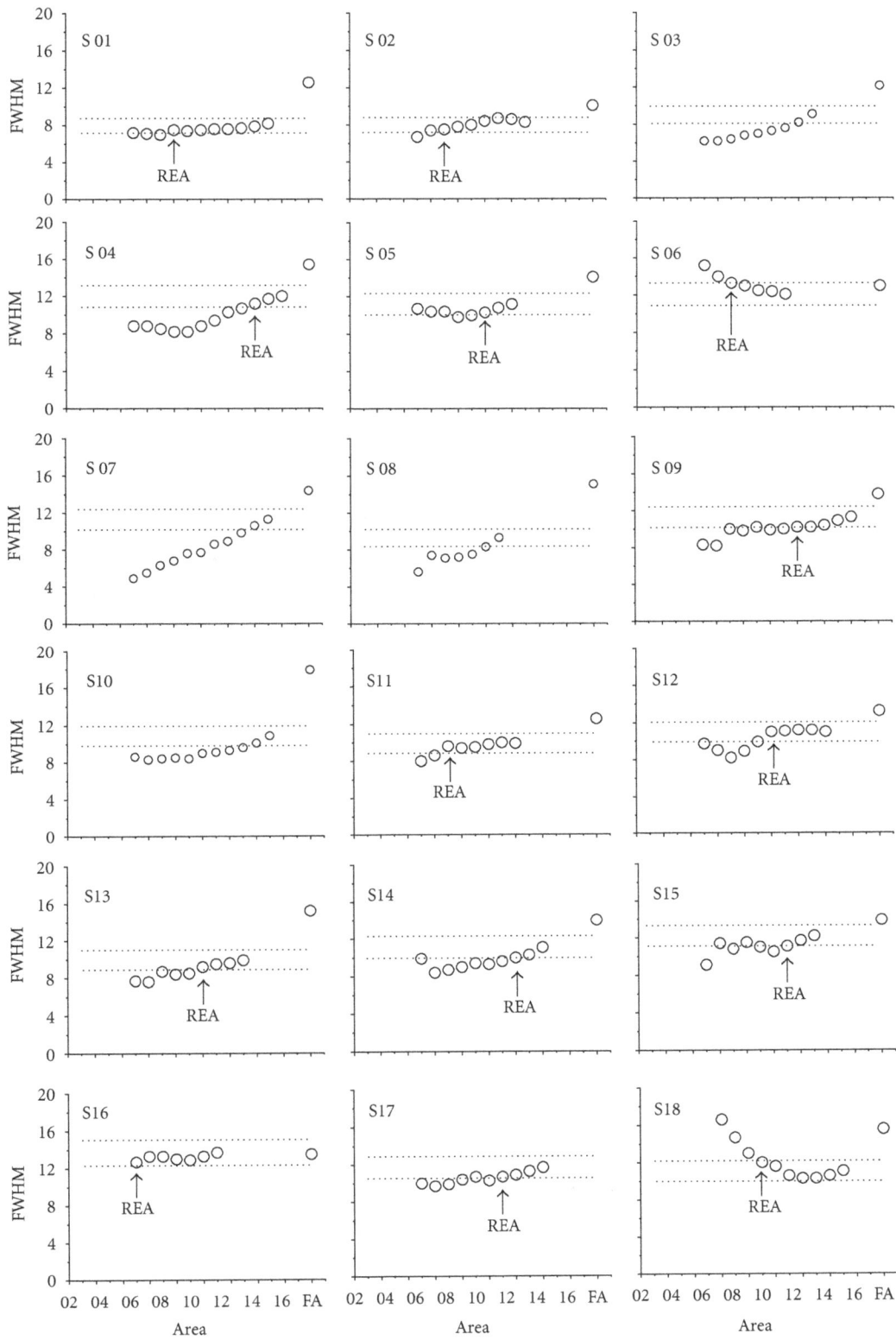

FIGURE 9: Graphs to depict values of FWHM from the 5th area selected in 18 samples (S 01, S 02, . . ., and S 18) and its respective representative elementary areas (REA) for a deviation of 10% when the REA was reached.

of 10%, when REA was actually reached. The curves obtained for four samples in Figure 6 might be better explained by analyzing the variability of ϕ for such samples.

Taking the criterion of variation adopted (item 2.4) into consideration, only 14 out of 18 samples reached the REA for ϕ. Such a fact might be explained by the spatial variability presented by this physical property [33]. For instance, in order to compare it to ρ_s, the intervals of variation coefficients (CV) of ϕ and ρ_s are 7–11% and 3–26%, respectively. The interval of CV for ρ_s is higher; however, its inferior limit is smaller than the inferior limit of ϕ.

The different values of REA found in this study might be ascribed to a nonhomogeneity of the samples. For instance, Pires et al. [34] analyzed the different values of TU obtained within the same sample, divided into 15 adjacent areas. The authors observed a significant variation among TU, a point that brought about variations on soil structure as a result of natural or artificial processes. The largest difference obtained between the areas of the same sample was of 111 TU, whereas the smallest one corresponded to 40 TU.

In the current work, we observed small stones and/or big voids (biopores) in some of the samples studied, while others are in its totality denser in relation to the others (e.g., S 14 and S 18). Such characteristics of each sample could be visualized and quantified in the tomographic images in such a nondestructive way, a procedure that might not be adopted by making use of traditional methods to measure ϕ.

Among the eighteen cases studied herein, only the S 16 did not show a crescent FWHM in relation to FA. However, the difference between FWHM of the distribution of ϕ of the last analyzed area and FA of this sample is just 0.2. This might be attributed to the hypothesis that such a sample probably presents a high value of ϕ in its central region (Figure 6). Thus, the variation of ϕ is already included in smaller areas, causing just an increase in frequency of the central values as the counting is made for FA. With regard to S 17, it was not possible to insert the point referent to FA. This can be explained due to the distribution of frequency, which presented a bimodal behavior for this area and, therefore, impaired the measurement of FWHM for FA by means of the criteria adopted for the other areas.

Following up the 10% variation criterion, as adopted by Vandenbygaart and Protz [15], the 14 samples that reached the REA were the ones that did it so up to the 14th area (Figure 10). In this case, samples with a section area of at least $882.1\,\mathrm{mm}^2$ give representative values of ϕ.

It is important to notice that the representative minimum size of an area may vary as a function of the physical property investigated and also according to the material utilized. For instance, by Al-Raoush and Papadopoulos [14] such a point was discussed and they reached the conclusion that a study carried out to look at the porosity of an analyzed media cannot be used as a basis for representative measurements of other parameters of interest. This is because other parameters require a size large enough to provide representative measurements. Thus, the usage of the same representative size for measurement of different properties is going to be dependent on the fact whether or not it includes REA in all parameters analyzed.

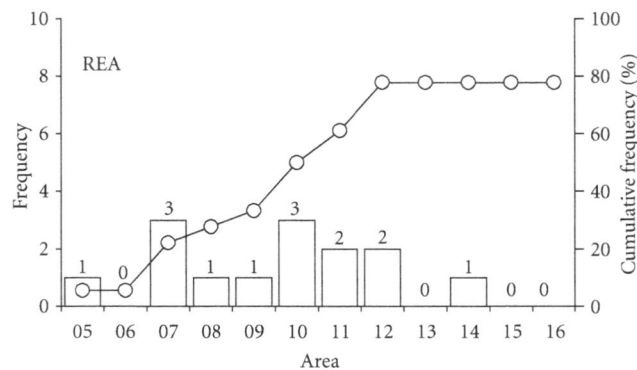

FIGURE 10: Graphs of full width at a half maximum (FWHM) values from the 5th selected area in 18 samples and its respective representative elementary areas (REA) for a 10% deviation, when REA was reached.

It is important to mention that although this work presents results for only one type of soil (clay soil); measurements for different soil types can be easily made. However, it is necessary to obtain the TU data matrix, after CT scanning, in which the different areas to determine RES will be selected. Another condition for the reproducibility of this study is the sample size. Samples larger than those used in this work can present artifacts due to the excessive attenuation of the ^{241}Am radiation beam. Low spatial resolution will also be obtained for large samples affecting the quality of RES analysis [13].

4. Conclusions

CT employed to assess REA made possible the distribution of soil porosity (ϕ) to be analyzed in both qualitative and quantitative terms. The study of FWHM of the adjustment curves of the distribution of ϕ for crescent selected areas in tomographic images showed to be satisfactory to determine REA.

FWHM expresses the distribution of ϕ prevailing in the investigated area, since the extreme values of such property (samples with small stones) cause insignificant implications in its value. However, when sample presents a region with stones and macropores, the distribution of ϕ might turn out to be bimodal and difficulties on the analysis of FWHM can be found.

Faced with the evaluation employed for REA, the results indicate that samples with sizes of at least $882.1\,\mathrm{mm}^2$ provided the representative values of ϕ for the investigated soil.

Acknowledgments

Many thanks are owed to the Brazilian Federal Funding Agencies: CNPq for the provision of the productivity fellowship in research and CNEN/CAPES for the Ph.D Scholarship, as well as to Dr. Osny O. S. Bacchi from the Laboratory of Soil Physics of the Center for Nuclear Energy in Agriculture, Piracicaba, Brazil, for the infrastructure used to obtain the tomographic images.

References

[1] L. F. Pires, R. C. J. Arthur, O. O. S. Bacchi, and K. Reichardt, "Application of γ-ray computed tomography to evaluate the radius of influence of soil solution extractors and tensiometers," *Nuclear Instruments and Methods in Physics Research. Section B*, vol. 259, no. 2, pp. 969–974, 2007.

[2] P. E. Cruvinel and F. A. Balogun, "Compton scattering tomography for agricultural measurements," *Engenharia Agricola*, vol. 26, no. 1, pp. 151–160, 2006.

[3] A. Pedrotti, E. A. Pauletto, S. Crestana, F. S. R. Holanda, P. E. Cruvinel, and C. M. P. Vaz, "Evaluation of bulk density of Albaqualf soil under different tillage systems using the volumetric ring and computerized tomography methods," *Soil and Tillage Research*, vol. 80, no. 1-2, pp. 115–123, 2005.

[4] A. Macedo, C. M. P. Vaz, J. M. Naime, P. E. Cruvinel, and S. Crestana, "X-ray microtomography to characterize the physical properties of soil and particulate systems," *Powder Technology*, vol. 101, no. 2, pp. 178–182, 1999.

[5] I. A. Taina, R. J. Heck, T. R. Elliot, and N. Scaiff, "Micromorphological and X-ray μCT study of Orthic Humic Gleysols under different management conditions," *Geoderma*, vol. 158, no. 3-4, pp. 110–119, 2010.

[6] L. F. Pires, J. A. R. Borges, O. O. S. Bacchi, and K. Reichardt, "Twenty-five years of computed tomography in soil physics: a literature review of the Brazilian contribution," *Soil and Tillage Research*, vol. 110, no. 2, pp. 197–210, 2010.

[7] R. Tippkötter, H. Eickhorst, H. Taubner, B. Gredner, and G. Rademaker, "Detection of soil water in macropores of undisturbed soil using microfocus X-ray tube computerized tomography (μCT)," *Soil and Tillage Research*, vol. 105, no. 1, pp. 12–20, 2009.

[8] T. R. Elliot, W. D. Reynolds, and R. J. Heck, "Use of existing pore models and X-ray computed tomography to predict saturated soil hydraulic conductivity," *Geoderma*, vol. 156, no. 3-4, pp. 133–142, 2010.

[9] S. Tomioka, T. Kozaki, H. Takamatsu et al., "Analysis of microstructural images of dry and water-saturated compacted bentonite samples observed with X-ray micro CT," *Applied Clay Science*, vol. 47, no. 1-2, pp. 65–71, 2010.

[10] M. R. Razavi, B. Muhunthan, and O. Hattamleh, "Representative elementary volume analysis of sands using X-ray computed tomography," *Geotechnical Testing Journal*, vol. 30, no. 3, pp. 212–219, 2007.

[11] P. Baveye, H. Rogasik, O. Wendroth, I. Onasch, and J. W. Crawford, "Effect of sampling volume on the measurement of soil physical properties: simulation with x-ray tomography data," *Measurement Science and Technology*, vol. 13, no. 5, pp. 775–784, 2002.

[12] J. Bear, *Dynamics of Fluids in Porous Media*, American Elsevier, New York, NY, USA, 1972.

[13] M. S. Costanza-Robinson, B. D. Estabrook, and D. F. Fouhey, "Representative elementary volume estimation for porosity, moisture saturation, and air-water interfacial areas in unsaturated porous media: data quality implications," *Water Resources Research*, vol. 47, no. 7, 2011.

[14] R. Al-Raoush and A. Papadopoulos, "Representative elementary volume analysis of porous media using X-ray computed tomography," *Powder Technology*, vol. 200, no. 1-2, pp. 69–77, 2010.

[15] A. J. Vandenbygaart and R. Protz, "The representative elementary area (REA) in studies of quantitative soil micromorphology," *Geoderma*, vol. 89, no. 3-4, pp. 333–346, 1999.

[16] D. Hillel, *Environmental Soil Physics*, Academic Press, San Diego, Calif, USA, 1998.

[17] Fao, "World reference base for soil resources," World Soil Resources number 84, FAO, ISRIC and ISSS, Rome, Italy, 1998.

[18] L. F. Pires, O. O. S. Bacchi, and K. Reichardt, "Gamma ray computed tomography to evaluate wetting/drying soil structure changes," *Nuclear Instruments and Methods in Physics Research, Section B*, vol. 229, no. 3-4, pp. 443–456, 2005.

[19] P. E. Cruvinel, R. Cesareo, S. Crestana, and S. Mascarenhas, "X- and τ-rays computerized minitomograph scanner for soil science," *IEEE Transactions on Instrumentation and Measurement*, vol. 39, no. 5, pp. 745–750, 1990.

[20] S. Crestana, P. E. Cruvinel, S. Mascarenhas, C. I. Biscegli, L. N. Martin, and L. A. Colnago, Eds., *Agricultural Instrumentation: Contributions on the Threshold of New Century*, EMBRAPA-SPI, Brasília, Brazil, 1996.

[21] L. F. Júnior, J. C. M. Oliveira, L. H. Bassoi et al., "Computed Tomography for the soil density evaluation of samples of semiarid Brazilian soils," *Revista Brasileira de Ciência do Solo*, vol. 26, pp. 835–842, 2002 (Portuguese).

[22] C. M. P. Vaz, S. Crestana, S. Mascarenhas, P. E. Cruvinel, K. Reichardt, and R. Stolf, "Using a computed tomography miniscanner for studying tillage induced soil compaction," *Soil Technology*, vol. 2, no. 3, pp. 313–321, 1989.

[23] R. E. Danielson and P. L. Sutherland, "Porosity," in *Methods of Soil Analysis. Part 1: Physical and Mineralogical Methods*, A. M. Klute, Ed., no. 9 Soil Science Society of America Book Series, pp. 443–461, American Society of Agronomy, Madison, Wis, USA, 1986.

[24] S. Crestana, P. E. Cruvinel, C. M. P. Vaz, R. Cesareo, S. Mascarenhas, and K. Reichardt, "Calibration and use of a computerized tomography in soil science," *Revista Brasileira de Ciência do Solo*, vol. 16, pp. 161–167, 1992.

[25] L. F. Pires, R. C. J. Arthur, O. O. S. Bacchi, and K. Reichardt, "Representative gamma-ray computed tomography calibration for applications in soil physics," *Brazilian Journal of Physics*, vol. 41, pp. 21–28, 2011.

[26] Microvis, *Program Manual for Reconstruction and Visualization of Tomographic Images*, Embrapa Instrumentação Agropecuária, São Carlos, Brazil, 2000.

[27] A. C. Kak and M. Slaney, *Principles of Computerized Tomographyc Imaging*, IEE Press, New York, NY, USA, 1999.

[28] Origin 8, *User Guide*, OriginLab Corporation, Northampton, UK, 1st edition, 2007.

[29] E. S. B. Ferraz and R. S. Mansell, "Determining water content and bulk density of soil by gamma ray attenuation methods," *Technical Bulletin*, no. 807, p. 51, IFAS, Fla, USA, 1979.

[30] J. H. Hubbell and S. M. Seltzer, "Tables of X-Ray mass attenuation coefficients and mass energy-absorption coefficients 1 keV to 20 MeV for elements Z=1 to 92 and 48 additional substances of dosimetric interest," NISTIR number 5632, Gaithersburg, US Department of Commerce, National Institute of Standards and Technology, Physics Laboratory, Ionizing Radiation Division, 1995.

[31] J. M. Bland and D. G. Altman, "Measuring agreement in method comparison studies," *Statistical Methods in Medical Research*, vol. 8, no. 2, pp. 135–160, 1999.

[32] D. Downing and J. Clark, *Applied Statistics*, Saraiva, São Paulo, Brazil, 2 edition, 2000.

[33] W. A. Jury and R. Horton, *Soil Physics*, Wiley, NJ, USA, 6th edition, 2004.

[34] L. F. Pires, O. O. S. Bacchi, K. Reichardt, and L. C. Timm, "Application of γ-ray computed tomography to analysis of soil structure before density evaluations," *Applied Radiation and Isotopes*, vol. 63, no. 4, pp. 505–511, 2005.

8

Lattice Boltzmann Method for Evaluating Hydraulic Conductivity of Finite Array of Spheres

Mário A. Camargo, Paulo C. Facin, and Luiz F. Pires

Laboratory of Soil Physics and Environmental Sciences, Department of Physics, State University of Ponta Grossa (UEPG), 84.030-900 Ponta Grossa, PR, Brazil

Correspondence should be addressed to Luiz F. Pires, lfpires@uepg.br

Academic Editor: Donald Tanaka

The hydraulic conductivity (K) represents an important hydrophysical parameter in a porous media. K direct measurements, usually demand a lot of work, are expensive and time consuming. Factors such as the media spatial variability, sample size, measurement method, and changes in the sample throughout the experiment directly affect K evaluations. One alternative to K measurement is computer simulation using the Lattice Boltzmann method (LBM), which can help to minimize problems such as changes in the sample structure during experimental measurements. This work presents K experimental and theoretical results (simulated) for three regular finite arrangements of spheres. Experimental measurements were carried out aiming at corroborating the LBM potential to predict K once the smallest relative deviation between experimental and simulated results was 1.4%.

1. Introduction

Hydraulic conductivity (K) is an important parameter in processes of fluid flow in porous media. K indicates how easily certain fluid is transported through porous media, and depends on the media properties as well as percolating fluid characteristics. Pore size distribution, type of pores, tortuosity, and connectivity are some of the factors related to the porous media. Regarding the percolating fluid, its viscosity (ν) is the main factor related to K measurements [1]. For example, increase in water temperature reduces its viscosity and potentially increases K.

K determinations are usually characterized by great variability due to factors such as media spatial variability, sample size, measurement method, changes in the sample throughout the experiment, and others [2]. Therefore, representative determination of K requires several measurements and samples. Direct K measurements are usually expensive and demand hard and thorough technical work [3].

Theoretical models and numeric simulations which enable K measurement from information about the porous media structure might be an interesting alternative to predict this physical parameter [4–7].

One theoretical tool that can be successfully used to predict K is the Lattice Boltzmann method (LBM) [8, 9]. The LBM is based on evolution of a relaxation equation for fluid particles distribution function, which is related to density and fluid macroscopic momentum. In the LBM equation, there is input data, which is the relaxation time that is related to the number of time steps so that the thermodynamic equilibrium is reached defining fluid viscosity [10]. The LBM can reproduce the macroscopic behavior of a fluid according to the Navier-Stokes (NS) equation. Relative easiness of computation implementation and numeric stability, in a great variety of flow conditions, makes the LBM ideal for treatment of fluid flow in porous media [11].

The development of LB models had important advances in recent decades; for example, today it is possible to simulate compressible, heat, and multiphase flows [10, 12]. However reliability of these applications has occurred through validations without regard to analytical results, which are usually associated with simplified cases of reality. There are few studies in which the LBM is validated with experimental results and the structure of porous media to be simulated is usually obtained indirectly by techniques (computed tomography or image reconstruction) that have their intrinsic sources of error [9, 11, 13].

LBM has been used for problems in porous media under several aspects [9, 11, 14, 15]. However, such studies have posed extra difficulties to the LBM such as image capture and reconstruction of the media as a representative porous media and results presented take into consideration possible deviations related to these difficulties.

In order to create higher possibility of comparison between results of LBM simulations and experimental results, this work suggests simulating the fluid flow through a layer of spheres. This is due to the fact that spheres have the simplest form to be digitally built and their symmetry enables the control of their superficial irregularity on the results, since by increasing their diameter such irregularities are less perceived by the flow.

We propose an experiment where the construction of the porous media is greatly facilitated—a finite array of spheres. The unique source of error is the roughness (discretized surface) of the sphere, which can be controlled with its diameter increase. So, we present in this work experimental and simulated results of K measurements in three porous media constituted of regular arrangements of spheres.

The main objective is to show that the LBM method can be used to evaluate K in the arrangements analyzed. This objective was achieved through comparisons between experimental and simulated K. The success of results proposed in this work points to the future use of this method in representative measurements of more complex porous media such as soil samples.

2. Materials and Methods

2.1. Experimental Methods. An experimental apparatus tested by Camargo et al. [8] was used for K experimental measurements (Figure 1(a)). The steps below were followed to determine K: (a) porous media saturation (acrylic box with a certain sphere arrangement) with glycerin, $C_3H_5(OH)_3$, (manufactured by Biotec, 99.5% purity); (b) H length measurement (hydraulic load) and L (porous media height); (c) percolated glycerin mass measurement to obtain its volume, where the glycerin density is known; (d) measurement of the necessary time interval for the glycerin to percolate; (e) use of Darcy's Law to calculate conductivity using

$$K = \left[\frac{V}{(At)}\right]\left[\frac{L}{(L+H)}\right],\tag{1}$$

V = volume of percolated glycerin (cm^3), A = cross sectional area of the box containing spheres (cm^2), and t = time interval for a given volume of glycerin to percolate (s) [16]; (f) glycerin viscosity measurement using its flow through a $D = 0.27\,cm$ diameter acrylic cylinder and the analytical expression for the cylinder conductivity $v = (D^2/32)(g/K_{cylinder})$: g = gravity acceleration ($cm\,s^{-2}$); (g) $K_{cylinder}$ was measured using the steps (a) to (e).

In both K measurement cases a sufficiently big H was guaranteed so that the glycerin would not form drops when leaving the spheres for the K measurement, or the cylinder

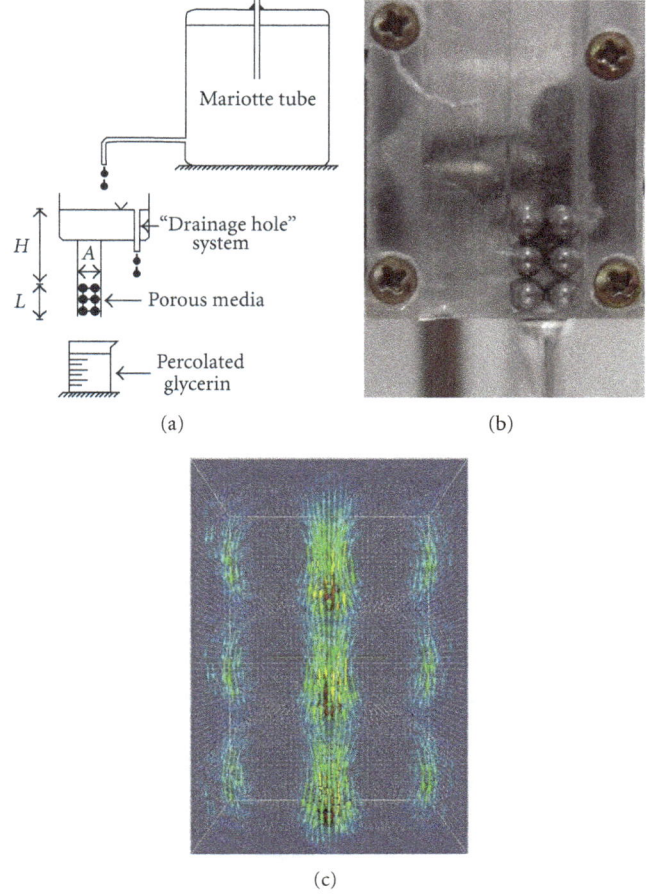

(a)   (b)

(c)

FIGURE 1: (a) Schematic drawing of the experimental apparatus used; (b) example of a regular arrangement of spheres. The box used to contain the spheres has the following measurements 0.635 cm × 0.635 cm × 4.7 cm (width × height × length) and the diameter of spheres used is 0.3175 cm; (c) velocities field for 3 layers of spheres.

for the viscosity measurement, which would delay the measurement time used in Darcy's equation.

Glycerin was used because it presents high v leading to a Reynolds number (Re) smaller than 1, where Darcy's Law is valid [17]. Once the glycerin v is highly susceptible to temperature variations, its measurement was carried out for each layer of spheres added during the experimental arrangements (Figure 1(b)).

2.2. The Lattice Boltzmann Method. In order to simulate K through the LBM a 3D media was built (Figures 2(a)–2(c)) similar to the real porous media being represented in a binary language 0 (solid) and 1 (porous) distributed along the vertices of a regular lattice. Once the porous media was built, the computer program that simulates (2) was used. The program returns the media's intrinsic permeability (k) as well as the pressure and flow velocity fields (Figure 1(c)).

The lattice used in the simulations was the cubic D3Q19. The 18 direction vectors of this lattice (Figure 2(d)) connect the sites one to another and also represent the possible velocity vectors, and there is still the null velocity (19 velocities).

Being a site in the lattice located by the vector $\vec{X}$ and having b_m close neighbors, the evolution equation for the particle of fluid distribution function $N_i(\vec{X}, T)$ is given by the Lattice Boltzmann Equation:

$$N_i\left(\vec{X} + \vec{c}_i, T + 1\right) = N_i\left(\vec{X}, T\right) + \Omega_i\left(\vec{X}, T\right), \qquad (2)$$

$\vec{X}$ = vector coordinates of the site in the lattice (lattice units), T = time step variable $(0, 1, 2, 3, \ldots)$. The duration of a time step is taken to be unity that represents the time interval for the particle of fluid travel between the closest neighbors, $N_i(\vec{X}, T)$ = number of fluid particles (direction i) located at site $\vec{X}$ at time T, $\Omega_i(\vec{X}, T)$ = collision operator that represents the collision of $N_i(\vec{X}, T)$ fluid particles with others at time T (see (4)), i = direction of one of the closest b_m neighbors $(0, 1, 2, 3, \ldots, 19)$, and $\vec{c}_i$ = velocity vector in direction i. This vector coincides with the lattice vectors, because in a unit time step, a particle travels from one site to adjacent one. The term $i = 0$ represents the b_r resting particles.

Variables $\vec{X}$ and T are given in the called lattice units and scale factors are necessaries for these variables assume length (h) and time (δ) dimensions, that is, it can be assumed that, for example, 5 units of the lattice is equivalent to 1 mm ($h = 1$ mm/5) or 5 time steps equivalent to 1 second ($\delta = 1$ s/5). Thus, the scale factor for the velocity vector becomes h/δ.

In this work we assume that variables without units will be represented as lattice units, that is, length and time variables have as unit the lattice spacing and the time step, respectively. So, velocity, viscosity, pressure, and other properties will be represented by lattice units.

The mesoscopic dynamics occurs in two steps: (1) propagation step represented by (2); (2) collision step represented by (3), which simulates the molecular collisions needed so that thermodynamic equilibrium occurs. This step is given by the action of collision operator $\Omega_i(\vec{X}, T)$ on the $N_i(\vec{X}, T)$:

$$N_i'\left(\vec{X}, T\right) = N_i\left(\vec{X}, T\right) + \Omega_i\left(\vec{X}, T\right), \qquad (3)$$

$N_i'(\vec{X}, T)$ = "collided" distribution function that will present a new value (number of fluid particles) at site $\vec{X}$, in i direction, and time T.

A simple and sufficient form of collision operator which recovers the Navier-Stokes macroscopic equation is known as BGK (variables $\vec{X}$ and T were omitted from here to reduce notations) operator [19]:

$$\Omega_i = \frac{N_i^{eq} - N_i}{\tau}, \qquad (4)$$

τ = relaxation time, which is a function of fluid viscosity, N_i^{eq} = equilibrium distribution (see (8)).

So, if $N_i < N_i^{eq}$, $\Omega_i > 0$ and the amount Ω_i will be added to N_i making N_i tend to N_i^{eq}.

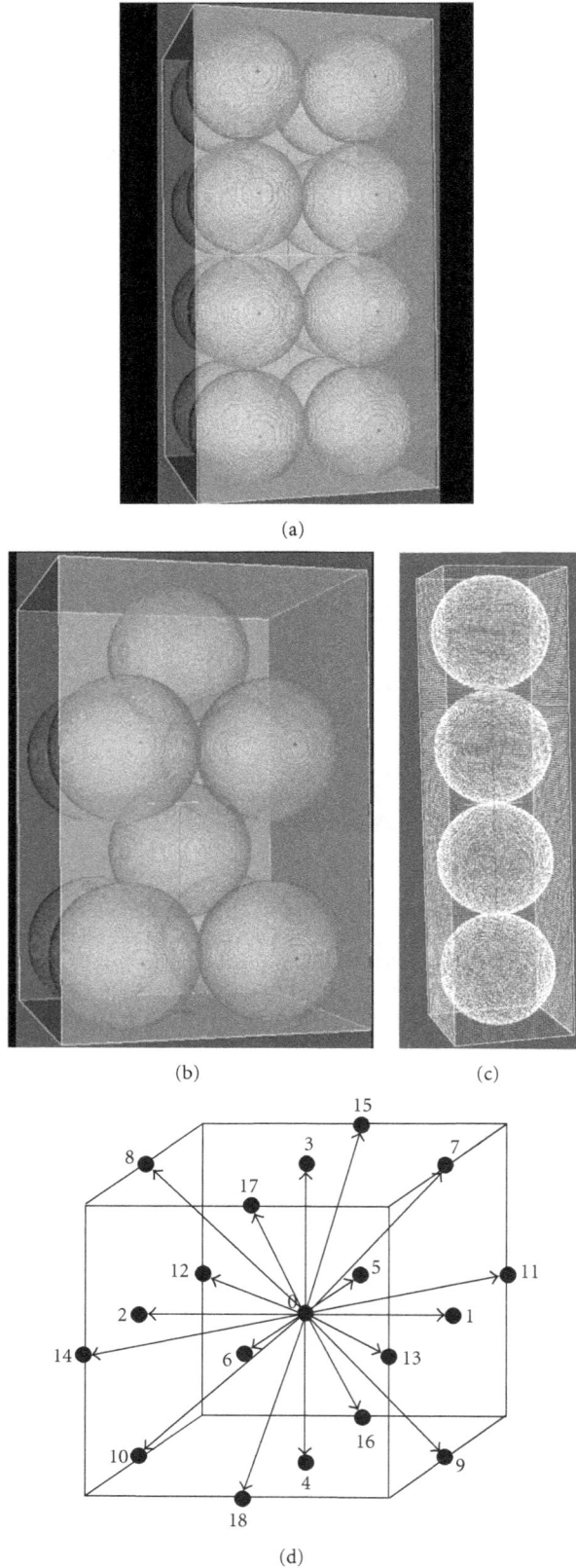

FIGURE 2: Examples of tridimensional porous media built for the computer simulation (arrangement with four solid layers—SL). (a) QQ; (b) QU; (c) OO and (d) direction vectors of the D3Q19 net site used in simulations [18].

FIGURE 3: Media built with one solid layer (SL) of four spheres (a) $D = 1$, (b) $D = 5$, (c) $D = 41$, and (d) $D = 121$. The flow occurs coming in (or out) the leaf.

The macroscopic particle density and the macroscopic momentum in a site are given by:

$$\sum_{i=0}^{b} N_i = \rho, \qquad \sum_{i=1}^{b_m} N_i \vec{c_i} = \rho \vec{u}, \qquad (5)$$

ρ = density at site $\vec{X}$ at time T.

Taking that into consideration, the collision operator conserves mass and momentum,

$$\sum_{i=0}^{b} \Omega_i = 0, \qquad \sum_{i=1}^{b_m} \Omega_i \vec{c_i} = 0, \qquad (6)$$

where $b = b_r + b_m$.

Particle distribution for the N_i^{eq} is usually obtained through the N_i^{eq} expansion, in power series at the macroscopic velocity ($\vec{u}$), being $O(u^2)$ sufficient so that Navier-Stokes

equation is recovered. For the low Mach number the pressure (p) is given by:

$$p = \frac{b_m c^2}{b D_e} \rho, \qquad (7)$$

where D_e is the Euclidian dimension of space in which the lattice is immerse and $c^2 = |\vec{c_i}|^2$. With this, the balance distribution form for moving particles is given by:

$$N_i^{eq} = \frac{\rho}{b} + \frac{\rho D_e}{b_m c^2} c_{i\alpha} u_\alpha + \frac{\rho D_e (D_e + 2)}{2 b_m c^4} c_{i\alpha} u_\alpha c_{i\beta} u_\beta - \frac{\rho D_e}{2 b_m c^2} u^2. \qquad (8)$$

In the main directions x, y, and z, the balance distribution must be doubled ($N_i^{eq} = 2N_i^{eq}$) so that viscosity is isotropic [20].

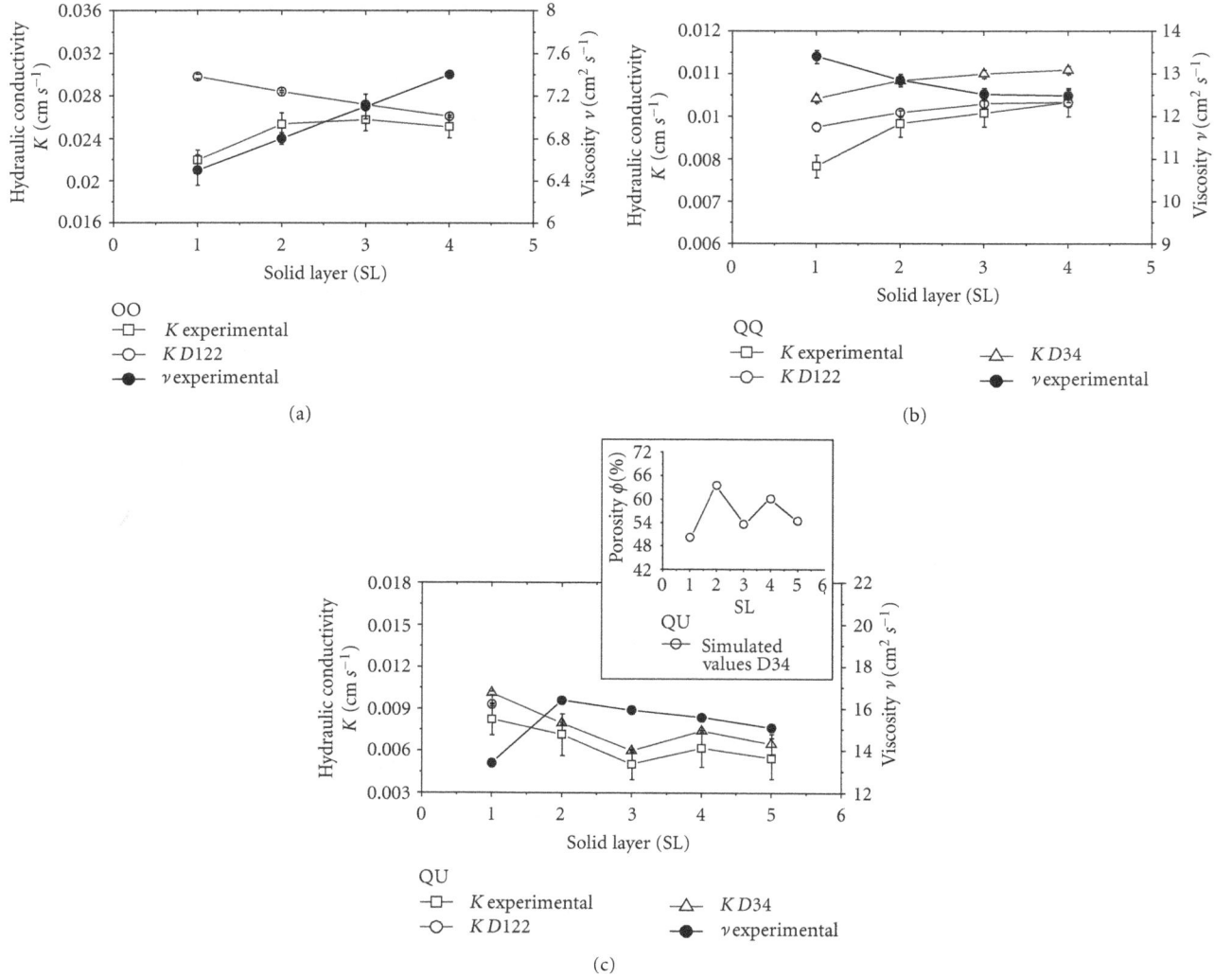

FIGURE 4: Experimental and simulated results of the hydraulic conductivity (K), glycerin experimental viscosity (ν), and experimental porosity (ϕ) for: (a) OO arrangement; (b) QQ arrangement, and (c) QU arrangement. $D34$ and $D122$ represent the spheres diameter (D) with 34 and 122 sites.

Resting particles have the following balance distribution:

$$N_o^{\text{eq}} = \frac{\rho}{b}b_r - \frac{\rho}{c^2}u^2. \tag{9}$$

For a macroscopic analysis of the dynamics proposed by (2), time δ and space scales h are usually used and the Knudsen variable $k_n = h/L = \delta/T_c$ is defined, where L and T_c are, respectively, the macroscopic characteristic length and time. With this, the Chapman-Enskog method [20] can be used, considering the equilibrium distribution disturbance, to show that (2) becomes the Navier-Stokes, given by (10), disregarding the $\mathrm{O}(k_n^2)$ contributions,

$$\partial_t\left(\rho u_\beta\right) + \partial_\alpha(p)\delta_{\alpha\beta} + \partial_\alpha\left(\rho u_\alpha u_\beta\right) = \nu\rho\partial_\alpha\left(\partial_\alpha u_\beta + \partial_\beta u_\alpha\right), \tag{10}$$

in which α and β are indexes which represent spatial coordinates x, y, or z; for these indexes Einstein's notation

is seen (sum over repeated indexes). Equation (10) is the β component of the Navier-Stokes equation with kinematic viscosity $\nu = \eta/\rho$ given by:

$$\nu = \frac{h^2 c^2}{\delta(D_e + 2)}\left[\tau - \frac{1}{2}\right]. \tag{11}$$

So when the hydrodynamic limit is imposed $L \gg h$, the macroscopic dynamic given by (10) is reproduced by (2).

The boundary conditions used in the simulation are periodic, that is, the fluid which leaves one end of the cavity is injected in the other end. The interaction between fluid and solid occurs so that there is no sliding, in this case, the "bounce back" condition was adopted, where the fluid which collides with the walls has its velocity inverted. The program returns permeability (k), which is calculated (12) with the Santos method [11], where in a stationary flow, the strength applied to the fluid is equal to the loss of momentum on the walls:

$$k = \nu\phi \frac{\langle mu_x \rangle}{\langle mu_x \rangle_{\text{lost}}}. \tag{12}$$

ϕ = porosity, $\langle mu_x \rangle$ = fluid average momentum in the porous media, $\langle mu_x \rangle_{\text{lost}}$ = fluid average momentum lost in the collisions with the porous media walls in a propagation step.

In order to obtain K from k, the relation below is used:

$$K = k\frac{g}{\nu}, \tag{13}$$

k = permeability, ν = kinematic viscosity (see (11)).

2.3. Computational Simulations. Different diameters (D) of spheres were investigated in the simulations (Figure 3), in order to better compare K theoretical results with the ones obtained experimentally (Figure 4).

The contour conditions used in simulations are periodic and make the fluid that leaves one end of the simulation dominium to enter the other end, as in an infinite array of spheres. Therefore, in the QQ arrangement simulation (Figure 2(a)) the k permeability of a single layer of spheres is the same as the permeability of any other number of layers. Despite that, the conductivity might be different, once viscosity is altered by the room temperature in the K experimental measurement of each layer.

3. Comparison between Experimental and Simulated Results

Larger D (Figure 4) provides better approximation between simulated and experimental results. This is due to the reduction in the sphere surface discretization. In the QQ arrangement (Figure 4(b)), porosity is kept constant by the arrangement symmetry, in the QU (Figure 4(c)) the same does not occur and it is necessary to simulate the flow on all sphere layers, which was carried out for $D = 34$ (scale factor $h = 0.3175$ mm/34). Concerning the OO arrangement (Figure 4(a)), the spheres have twice the diameter of the two previous cases. The K discrepancy from the first layer to the others remains and is a case to be further investigated.

Experimental hydraulic conductivities were plotted against the computed K based on the LBM simulations (Figure 5). The data obtained clearly shows that K based on computed simulations are in very good agreement (high positive correlation coefficients) with the laboratory measurements. Analyzing the correlation graphics (Figures 5(a) and 5(b)) most of the results are within the variability limits of the laboratory measurements as indicated by the experimental error bars.

In relation to discrepancies found in conductivity values from the first to the other layers, simulations were carried out with a buffer zone up to 100 sites length at the input and output of the flow to check the influence of the periodic contour condition, which does not exist in this experiment. However, no alteration was detected in the conductivities. The interfacial tension might have some effect in the flow

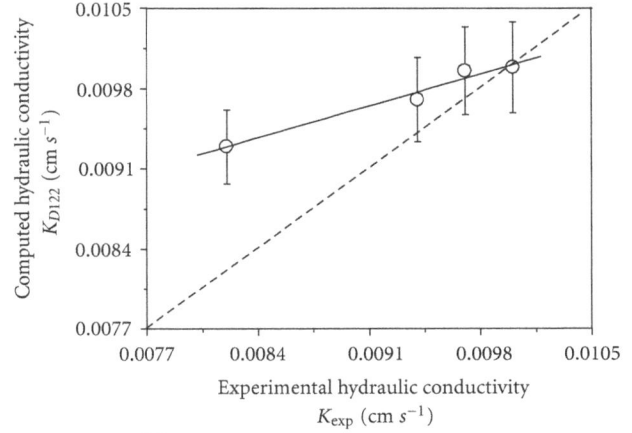

(a)

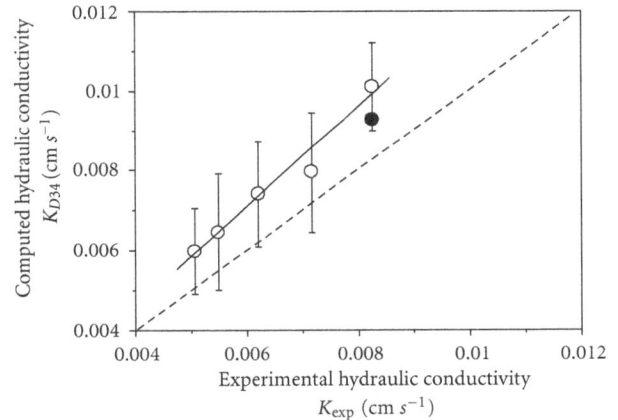

(b)

FIGURE 5: Experimental K measured against LBM computer simulated K for: (a) QQ arrangement and (b) QU arrangement.

output since in the cases in which the hydraulic load is not big enough there is some glycerin dropping. When the hydraulic load is big enough, a stream forms in the flow output, where there are two fluids, glycerin and air; the same does not happen in simulations where there is a monophasic flow. Although there are collision operator models to simulate biphasic flows [10], it is still complicated to control the huge difference between the viscosities of glycerin and air.

Due to some limitation of the PC RAM memory recognition by the software used ($\sim$2 Gbytes), it was only possible to simulate one layer of spheres with a maximum diameter $D = 122$ (the same first layer as the QQ case). However, the maximum and minimum deviation for $D = 34$ were 18.6% and 15.5% for QU. It is relevant to point out that the deviations obtained might be minimized with the use of clusters and parallel processing for simulations with larger diameter spheres, mainly in the QU case which represents a more complex media than QQ and OO.

4. Concluding Remarks

In this work, an experimental physical reality was pursued with flow in Reynolds low number and spherical symmetry to facilitate the media digital construction, avoiding image acquisition problems. This provided a close comparison between the experimental K measurement and the one simulated via LBM, with maximum and minimum deviation (using the experimental value as a reference) from 18.8% and 1.4% for QQ, with the maximum deviation happening only in the first layer. For the OO case, the deviations were 35.6% and 3.9% with the maximum deviation being observed in the first layer again.

In the soil science area, LBM can be used in association with the X-ray computed tomography (CT) utilized to acquire more real 3D soil pore structures. The selection of adequate image analysis procedures, for example, threshold, will allow to accurately reconstructing the pore system structure used to simulate K for heterogeneous and nonsymmetrical media such as soil. It is important to mention that no extra computational difficulty is included in this case.

The use of 3D soil images will make it possible to simulate the 3D fluid flow allowing the evaluation of important hydraulic soil properties such as k and K. As K direct measurements, usually demand a lot of work, are expensive and time consuming, LBM can be an interesting tool to its simulation. With LBM it will also be possible to access the computed flow velocity, which can be utilized for instance to better understand some important phenomena that occur in the soil such as water fingering.

Acknowledgments

The authors would like to thank the Brazilian agencies Conselho Nacional de Desenvolvimento Científico e Tecnológico (CNPq) for the research fellowship to L. F. Pires and Coordenação de Aperfeiçoamento de Pessoal de Nível Superior (CAPES) for the M.S. fellowship granted to M. A. Camargo.

References

[1] R. H. Brooks and A. T. Corey, *Hydraulic Properties of Porous Media*, Colorado State University, Fort Collins, Colo, USA., 1964.

[2] M. C. Falleiros, O. Portezan, J. C. M. Oliveira, O. O. S. Bacchi, and K. Reichardt, "Spatial and temporal variability of soil hydraulic conductivity in relation to soil water distribution, using an exponential model," *Soil and Tillage Research*, vol. 45, no. 3-4, pp. 279–285, 1998.

[3] M. Kutilek and D. R. Nielsen, *Soil Hydrology*, Catena, Cremlingen, Germany, 1994.

[4] A. R. Dexter and G. Richard, "The saturated hydraulic conductivity of soils with n-modal pore size distributions," *Geoderma*, vol. 154, no. 1-2, pp. 76–85, 2009.

[5] W. L. Quinton, T. Elliot, J. S. Price, F. Rezanezhad, and R. Heck, "Measuring physical and hydraulic properties of peat from X-ray tomography," *Geoderma*, vol. 153, no. 1-2, pp. 269–277, 2009.

[6] K. Nakano and T. Miyazaki, "Predicting the saturated hydraulic conductivity of compacted subsoils using the non-similar media concept," *Soil and Tillage Research*, vol. 84, pp. 145–153, 2005.

[7] M. G. Schaap and F. J. Leij, "Using neural networks to predict soil water retention and soil hydraulic conductivity," *Soil and Tillage Research*, vol. 47, no. 1-2, pp. 37–42, 1998.

[8] M. A. Camargo, P. C. Facin, and L. F. Pires, "Theoretical-experimental analyses of simple geometry saturated conductivities for a Newtonian fluid," *Brazilian Journal of Physics*, vol. 40, no. 4, pp. 393–397, 2010.

[9] X. Zhang, L. K. Deeks, A. Glyn Bengough, J. W. Crawford, and I. M. Young, "Determination of soil hydraulic conductivity with the lattice Boltzmann method and soil thin-section technique," *Journal of Hydrology*, vol. 306, no. 1–4, pp. 59–70, 2005.

[10] P. C. Facin, *Modelo de Boltzmann Baseado em Mediadores de Campo Para Fluidos Imiscíveis*, Tese de Doutorado, Programa de Pós-Graduação em Engenharia Mecânica, Universidade Federal de Santa Catarina, Florianópolis, Brazil, 2003.

[11] L. O. E. Santos, C. E. P. Ortiz, H. C. Gaspari, G. E. Haverroth, and P. C. Philippi, "Prediction of intrinsic permeabilities with lattice Boltzmann method," in *Proceedings of the 18th International Congress of Mechanical Engineering (COBEM '05)*, Ouro Preto, Brazil, 2005.

[12] P. C. Philippi, L. A. Hegele Jr., L. O. E. Dos Santos, and R. Surmas, "From the continuous to the lattice Boltzmann equation: the discretization problem and thermal models," *Physical Review E*, vol. 73, no. 5, Article ID 056702, 2006.

[13] M. E. Kutay, A. H. Aydilek, and E. Masad, "Laboratory validation of lattice Boltzmann method for modeling pore-scale flow in granular materials," *Computers and Geotechnics*, vol. 33, no. 8, pp. 381–395, 2006.

[14] F. J. Jiménez-Hornero, J. V. Giráldez, and A. Laguna, "Estimation of the role of obstacles in the downslope soil flow with a simple erosion model: the analytical solution and its approximation with the lattice Boltzmann model," *Catena*, vol. 57, no. 3, pp. 261–275, 2004.

[15] F. J. Jiménez-Hornero, E. Gutiérrez de Ravé, J. V. Giráldez, and A. M. Laguna, "The influence of the geometry of idealised porous media on the simulated flow velocity: a multifractal description," *Geoderma*, vol. 150, no. 1-2, pp. 196–201, 2009.

[16] D. Hillel, *Environmental Soil Physics*, Academic Press, San Diego, Calif, USA, 1998.

[17] J. Bear, *Dynamics of Fluids in Porous Media*, Dover, New York, NY, USA, 1988.

[18] S. De, K. Nagendra, and K. N. Lakshmisha, "Simulation of laminar flow in a three-dimensional lid-driven cavity by lattice Boltzmann method," *International Journal of Numerical Methods for Heat and Fluid Flow*, vol. 19, no. 6, pp. 790–815, 2009.

[19] Y. H. Qian, D. D'humieres, and P. Lallemand, "Lattice BGK models for Navier-Stokes equation," *Europhysics Letters*, vol. 17, no. 6, pp. 479–484, 1992.

[20] D. H. Rothman and S. Zaleski, *Lattice-Gas Cellular Automata: Simple Models of Complex Hydrodynamics*, Cambridge University Press, Cambridge, UK, 1997.

9

Solubility and Leaching Risks of Organic Carbon in Paddy Soils as Affected by Irrigation Managements

Junzeng Xu,[1,2] Shihong Yang,[1] Shizhang Peng,[1] Qi Wei,[1] and Xiaoli Gao[1,2]

[1] *State Key Laboratory of Hydrology-Water Resources and Hydraulic Engineering, Hohai University, Nanjing 210098, China*
[2] *College of Water Conservancy and Hydropower Engineering, Hohai University, Nanjing 210098, China*

Correspondence should be addressed to Shizhang Peng; szpeng@hhu.edu.cn

Academic Editors: C. García, P. F. Hudak, and M. D. Mingorance Alvarez

Influence of nonflooding controlled irrigation (NFI) on solubility and leaching risk of soil organic carbon (SOC) were investigated. Compared with flooding irrigation (FI) paddies, soil water extractable organic carbon (WEOC) and dissolved organic carbon (DOC) in NFI paddies increased in surface soil but decreased in deep soil. The DOC leaching loss in NFI field was $63.3 \, kg \, C \, ha^{-1}$, reduced by 46.4% than in the FI fields. It indicated that multi-wet-dry cycles in NFI paddies enhanced the decomposition of SOC in surface soils, and less carbon moved downward to deep soils due to less percolation. That also led to lower SOC in surface soils in NFI paddies than in FI paddies, which implied that more carbon was released into the atmosphere from the surface soil in NFI paddies. Change of solubility of SOC in NFI paddies might lead to potential change in soil fertility and sustainability, greenhouse gas emission, and bioavailability of trace metals or organic pollutants.

1. Introduction

Dissolved organic carbon (DOC) in soil, which is present in soil solution and interacts with colloids and clays, plays an important role in soil carbon cycling [1, 2]. It is highly related to the greenhouse gas (CO_2 and CH_4) emissions [3–5], nutrient availability [6, 7], as well as the mobilization, translocation, and toxicity of several inorganic and organic pollutants in soil [8–13]. The DOC losses from soil ecosystems, via leaching or surface runoff, account for numerous pollution problems of surface water and groundwater [14, 15]. Agricultural practices impact the timing and magnitude of DOC export from soils to rivers or ditches [16–18]. However, information on the effects of agricultural practices on soil DOC leaching is still limited, although it is a crucial component of the ecosystem carbon balance [19–22].

Rice is one of the most important crops in the Asian monsoon region [23]. The rice field ecosystem is commonly characterized by flooding conditions and high percolation rate. A great deal attention is paid to nitrogen and phosphorus with regard to leaching risks of nutrients in rice fields [24–28]. Leaching loss of DOC from paddy soil, which is relatively rich in organic matter, is always overlooked. With increasing water

scarcity, water saving irrigation techniques, such as nonflooding controlled irrigation (NFI), alternate dry-wet irrigation (AWDI), and the rice intensification (SRI) system, are applied widely [29–33]. Soil wetting and drying cycles influence a large number of biological and chemical processes [34–37]. Solubility of soil organic carbon (SOC) and its leaching risks will change when the rice field is exposed to nonflooding conditions under water-saving irrigation management.

In the present study, water extractable organic carbon (WEOC) contents in soils and DOC in soil solutions, as well as DOC leaching risks, were measured in rice paddies under different irrigation managements to reveal the influence of NFI on soil DOC dynamics.

2. Materials and Methods

2.1. Site Description and Experimental Design. The study was conducted in rice paddies at the Kunshan irrigation and drainage experiment station ($31°15'15''N$ $120°57'43''E$) in the Tai Lake region in China. The study area has a subtropical monsoon climate, with an average annual air temperature of 15.5°C and a mean annual precipitation of 1,097.1 mm. The paddy soil is Gleyic-Stagnic Anthrosols, developed from

TABLE 1: Limits for irrigation in different stages of rice for non-flooding controlled irrigation.

| Stages | Regreening | Tillering | | | Jointing and booting | | Heading and flowering | Milk maturity | Yellow maturity |
		Former	Middle	Later	Former	Later			
Upper limit	30 mm	θs_1	θs_1	θs_1	θs_2	θs_2	θs_3	θs_3	Drying
Lower limit	10 mm	$0.7\theta s_1$	$0.65\theta s_1$	$0.6\theta s_1$	$0.7\theta s_2$	$0.75\theta s_2$	$0.8\theta s_3$	$0.7\theta s_3$	
Monitored soil depth (cm)	—	0–20	0–20	0–20	0–30	0–30	0–40	0–40	—

θs_1, θs_2, and θs_3 are the saturated water content of the soil in different stages of rice.

alluvial deposits. The soil texture in the plowed layer (0–20 cm) is clay, with a total nitrogen content of 1.03 g kg^{-1}, total phosphorus content of 1.35 g kg^{-1}, total potassium content of 20.8 g kg^{-1}, and pH of 7.4 (soil : water = 1 : 2.5 by weight). SOC contents for soil depths of 0–10, 10–20, 20–40, and 40–60 cm are 13.8, 12.1, 11.4, and 10.3 g kg^{-1}; soil bulk densities are 1.28 g, 1.33^3, 1.36, and 1.35 g cm^{-3}, respectively. The saturated soil water contents (v/v) for the layers of 0–20, 0–30, and 0–40 cm are 52.4, 49.7, and 47.8%, respectively. The cropping system used is a rice-wheat rotation system. Winter wheat was harvested on 16-17 May before the experiment. The wheat straw was removed, whereas the root and about 10 cm stubble were buried by plowing. The variety of rice planted was Japonica Rice NJ46. The rice was transplanted with 13 cm × 25 cm hill spacing on 23 June, and harvested on 26 October in 2009.

Two irrigation treatments were used, namely, flooding irrigation (FI) and nonflooding controlled irrigation (NFI). A randomized complete block design and three replications were established in 6 plots (5 m × 7 m). The adjacent plots were separated by plastic membrane which was inserted into the ridges at a depth of 500 mm, to isolate the water within different plots and avoid hydraulic exchange between adjacent plots. In the FI rice fields, a depth of 3–5 cm standing water was always maintained after transplanting, except when drying in the later tillering and yellow maturity periods. In the NFI rice fields, standing water depth was kept between 5 and 25 mm during the first 7-8 days after transplanting (DAT) in regreening period; irrigation was applied only to keep soil saturated in other stages. Standing water was avoided in other stages, except during rain harvesting period and the pesticide or fertilizer application period. Table 1 presents the root zone soil water content criteria in different growth stages. The same fertilizer doses for each split were applied into each plot according to the local conventional fertilizer application method.

2.2. Field Measurements.
Irrigation water volumes were recorded by water meters installed on the pipes. Soil moisture in rice field was monitored with three replications using a time domain reflectometer (TDR, soil moisture, USA) and with 20 cm waveguides installed at 0–20, 20–40, and 40–60 cm depths. Water layer depth was monitored using a vertical ruler fixed in the field. Daily meteorological data, including precipitation volume, wind speed, temperature (maximum, minimum, and average), sunshine duration, and relative humidity, were recorded by an automatic weather station (ICT, Australia). Soil temperature and soil redox

potentials (Eh) at 5 cm depth were measured using mercury thermometers and oxidation-reduction potential meters, with three replications in situ. Rice was harvested on 26 October 2009, and yield was determined for each plot.

2.3. Soil Sampling and WEOC Contents Measurement.
Soil samples were collected during the rice season with a hand auger for soil WEOC measurement. The sampling was conducted at five locations for each treatment plot at five depths, 0–10, 10–20, 20–30, 30–40, and 40–60 cm. Then fresh samples of the same depth in each plot were homogenized by mixing and separated from debris and crop residues. Five grams of fresh soil samples were then extracted in distilled water (soil : water = 1 : 10 by weight) on a shaker for 60 min. DOC in the extract was determined using a TOC-1020 A analyzer (Elementar, High TOC II, Germany). Using the same method, soil samples were collected at pretransplanting and postharvesting periods in depths of 0–10, 10–20, 20–40, and 40–60 cm for SOC measurement. Ten grams of fresh soil samples were selected for SOC measurement using the potassium dichromate oxidation method, with 0.8 mol L^{-1} $K_2Cr_2O_4$-H_2SO_4 solution at 170–180°C (oil bath) [38]. Soil samples moisture contents were determined using an oven-dried method, and WEOC and SOC contents were calculated as milligram C per gram dry soil. Change in SOC storage was calculated based on the values obtained at pretransplant and at harvest.

2.4. Soil Solution Sampling and DOC Contents Measurement.
Ceramic suction cups (2 cm in inner diameter and 7 cm in length) with numerous pores (about 2 μm in diameter) were installed vertically at 7–14 cm, 27–34 cm, and 47–54 cm depths to collect soil solutions with three replications. To acquire a field-equilibrated status and eliminate the sorb of DOC by suction cups [39], the cups, which were cleaned by 0.1 molar HCl and deionised water, were installed firmly into the soil one year ago in June 2008 [28]. The clay suction cup was embedded in a polyvinyl chloride pipe, allowing the water to be pumped out. Soil solutions were collected and stored in 100 mL polytetrafluoroethylene bottles and then taken to the laboratory immediately. DOC contents in the soil solutions were determined using the TOC-1020 A analyzer.

2.5. Leaching Loss of DOC.
Seasonal DOC leaching losses were calculated based on DOC contents in 47–54 cm soil solutions and deep percolation (DP) rate. Daily DP was

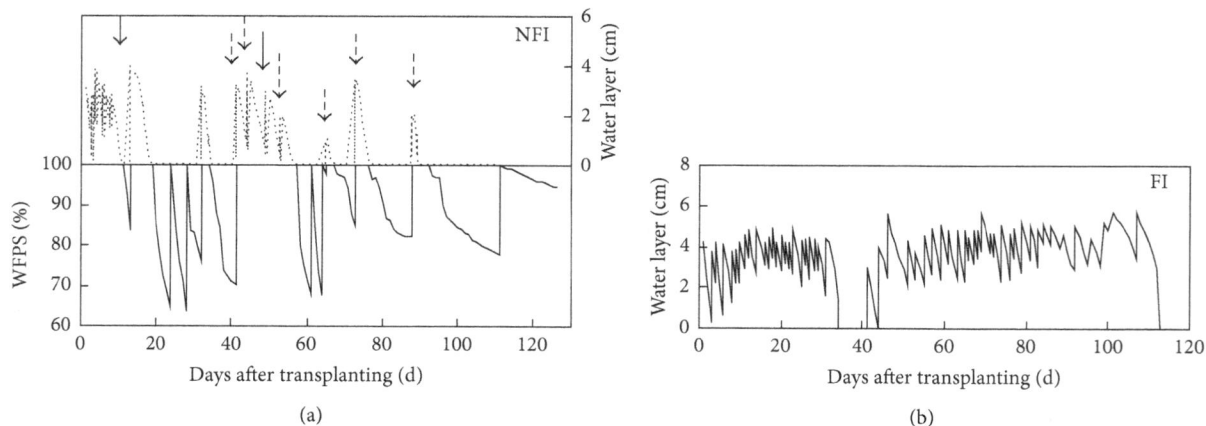

FIGURE 1: Water depth and soil moisture in FI and NFI rice fields (solid and dashed arrows denote irrigations for fertilizer and pesticides application in NFI rice fields).

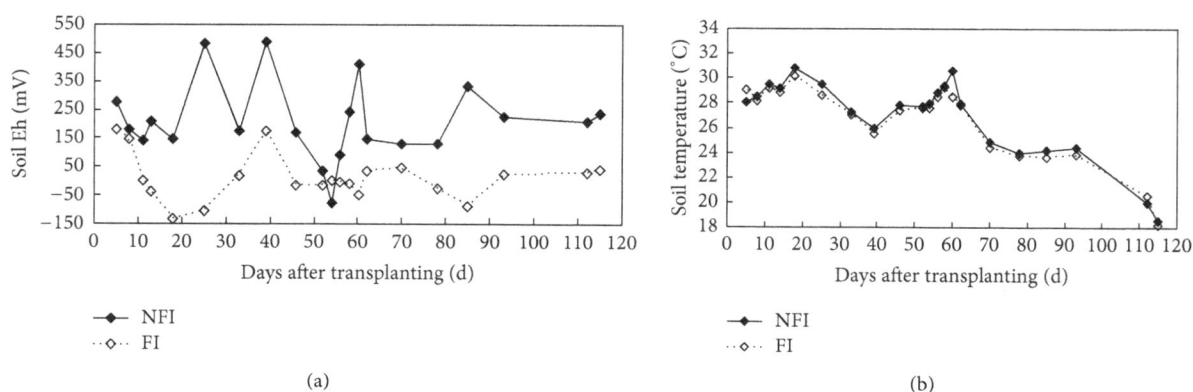

FIGURE 2: Seasonal variation of soil Eh and soil temperature at 5 cm depth in rice fields under different water managements.

calculated by following the water balance principle based on field measurements:

$$DP_t = W_{t-1} - W_t + I_t + P_t - D_t - ET_t, \quad (1)$$

where DP is the volume of the percolation water, W is the flooding depth or the soil water content in the root zone. I, P, and D are the water volumes of irrigation, precipitation, and drainage, respectively. ET is the evapotranspiration, which was calculated using the water balance principle based on measurement in bottom-sealed lysimeters (nonweighted, 40 cm in diameter and 60 cm in depth with 4 rice hills) with the same irrigation management as the plot.

3. Results

3.1. Water Regimes and Soil Characteristics. Eleven wet-dry cycles were observed in NFI fields, with more than 72 days of nonflooding (Figure 1). About two-thirds of the total growth season was nonflooding in the NFI fields, which was much longer than those reported in zero-drainage or alternate wetting and drying irrigation rice fields [40, 41]. Multi-wet-dry cycles led to huge change in soil properties in the NFI fields. Soil redox potentials (Eh) at 5 cm depth ranged from

−77.9 to +488.9 mV in the NFI fields, much higher than those in the FI fields (from −134.43 to +181.86 mV) (Figure 2). Drying in the NFI fields was always accompanied by a rapid increase in Eh values, whereas rewetting caused a sharp decrease. Soil Eh at 5 cm depth increased from a negative value to as high as +480 mv in the NFI fields (25 DAT and 39 DAT). In the FI paddies, midseason drainage also led to a significant Eh increase, from −104.1 to +131.1 mV (33–39 DAT). Soil temperatures at 5 cm depth were slightly higher in the NFI fields than those in the FI fields during most of the rice season (Figure 2).

3.2. Rice Yields and Water Consumption. Evapotranspiration and percolation were 404.6 and 368.8 mm in the NFI fields, which were reduced by 111.7 and 276.8 mm compared with the FI treatment (Table 2). Irrigation volumes in the NFI and FI fields were 233.3 and 635.9 mm, whereas water consumption volumes were 773.4 and 1,161.9 mm, respectively. Irrigation volumes and water consumption in the NFI fields were reduced by 63.3 and 33.4%, compared with the FI fields. Rice yield for the NFI treatment was 10,335.8 kg ha^{-1}; it was the same as the yield for FI treatment (9,889.7 kg ha^{-1}). Water use efficiency greatly increased in the NFI paddies due to the

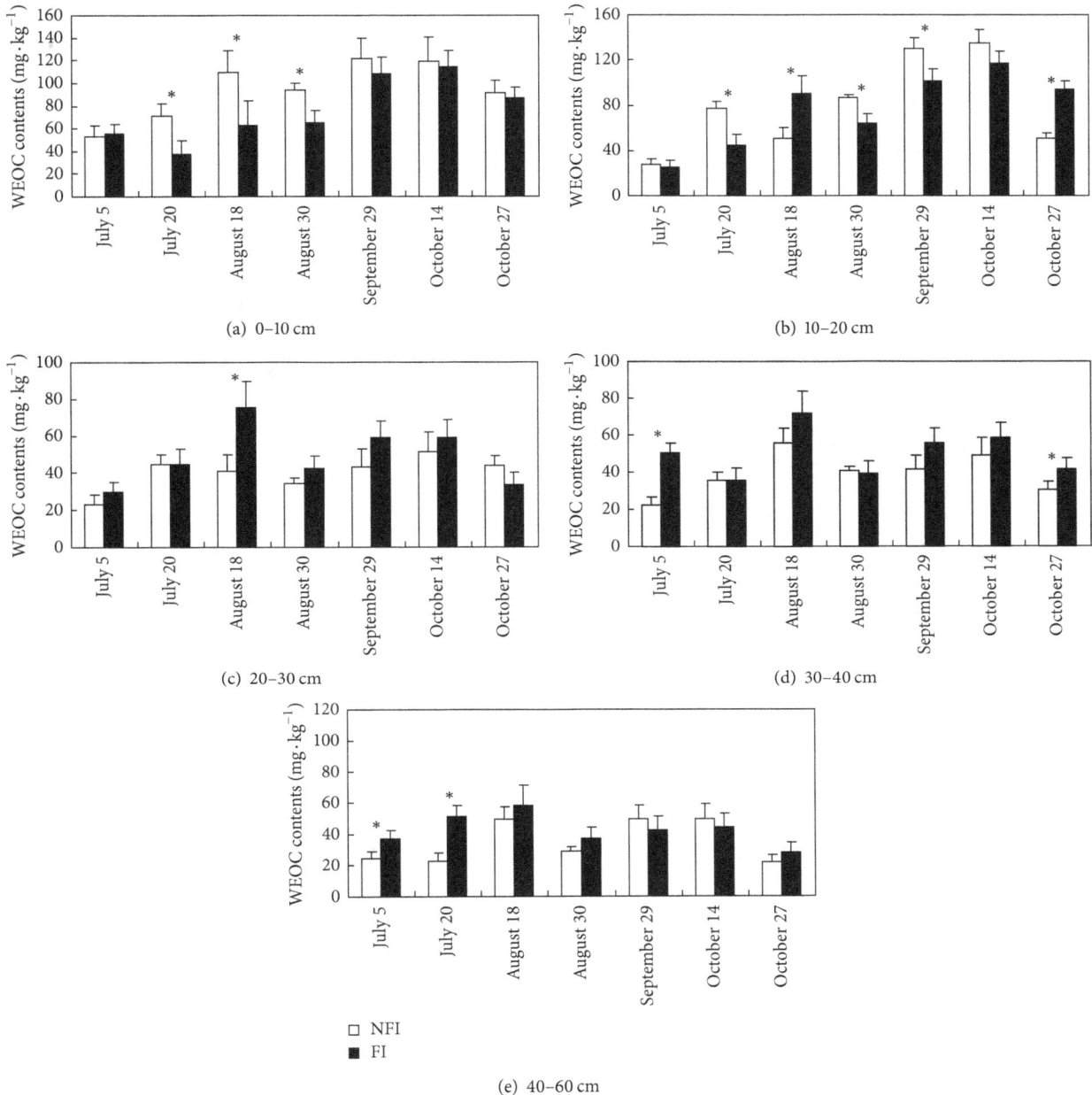

(a) 0–10 cm

(b) 10–20 cm

(c) 20–30 cm

(d) 30–40 cm

(e) 40–60 cm

FIGURE 3: Water extractable organic carbon (WEOC) contents in paddy soils under different irrigation managements (*indicates difference between NFI and FI that is significant at $P < 0.05$).

large reduction in water consumption and irrigation volume. It indicated that NFI is able to get the same yield as FI with a lower cost irrigation and higher water use efficiency than FI.

3.3. Soil WEOC Contents.

Soil WEOC contents in both FI and NFI fields at different stages decreased with the increase of soil depth (Figure 3). Soil WEOC contents were more variable in the top 0–20 cm soils than in the 30–40 cm and 40–60 cm soils. Soil WEOC contents were always high in the middle stage, when the crop growth and agronomic activities were intensive. Soil WEOC contents at 0–10, 10–20, 20–30, 30–40, and 40–60 cm depths varied in the range of 52.4–121.0,

27.4–134.5, 23.5–51.3, 22.2–55.8, and 22.2–49.7 mg C kg^{-1} in the NFI fields, whereas those in the FI fields varied in the range of 37.7–114.0, 25.0–116.5, 29.6–75.6, 35.3–71.7, and 28.4–58.0 mg C kg^{-1}. But the WEOC contents in the NFI and FI fields were lower than the results obtained by Zhan et al. (2010) [42] in paddy soils (0.44–0.83 g C kg^{-1}) in Hubei China.

WEOC contents in surface soils at 0–10 and 10–20 cm depths in the NFI fields were mostly significantly higher than those in the FI fields. However, WEOC contents in the NFI fields were frequently lower than in FI fields in deep soils at 20–30, 30–40, and 40–60 cm depths, but only a few results

TABLE 2: Rice yields and water consumption under different irrigation managements.

Treatment	Yield Kg ha^{-1}	Irrigation mm	Evapotranspiration mm	Water consumption mm	Deep seepage mm
NFI	10335.8[a]	233.3[a]	404.6[a]	773.4[a]	368.8[a]
FI	9889.7[a]	635.9[b]	516.3[b]	1161.9[b]	634.7[b]

Different letters in each column represent significant difference between the treatments at $P = 0.05$ by t-test.

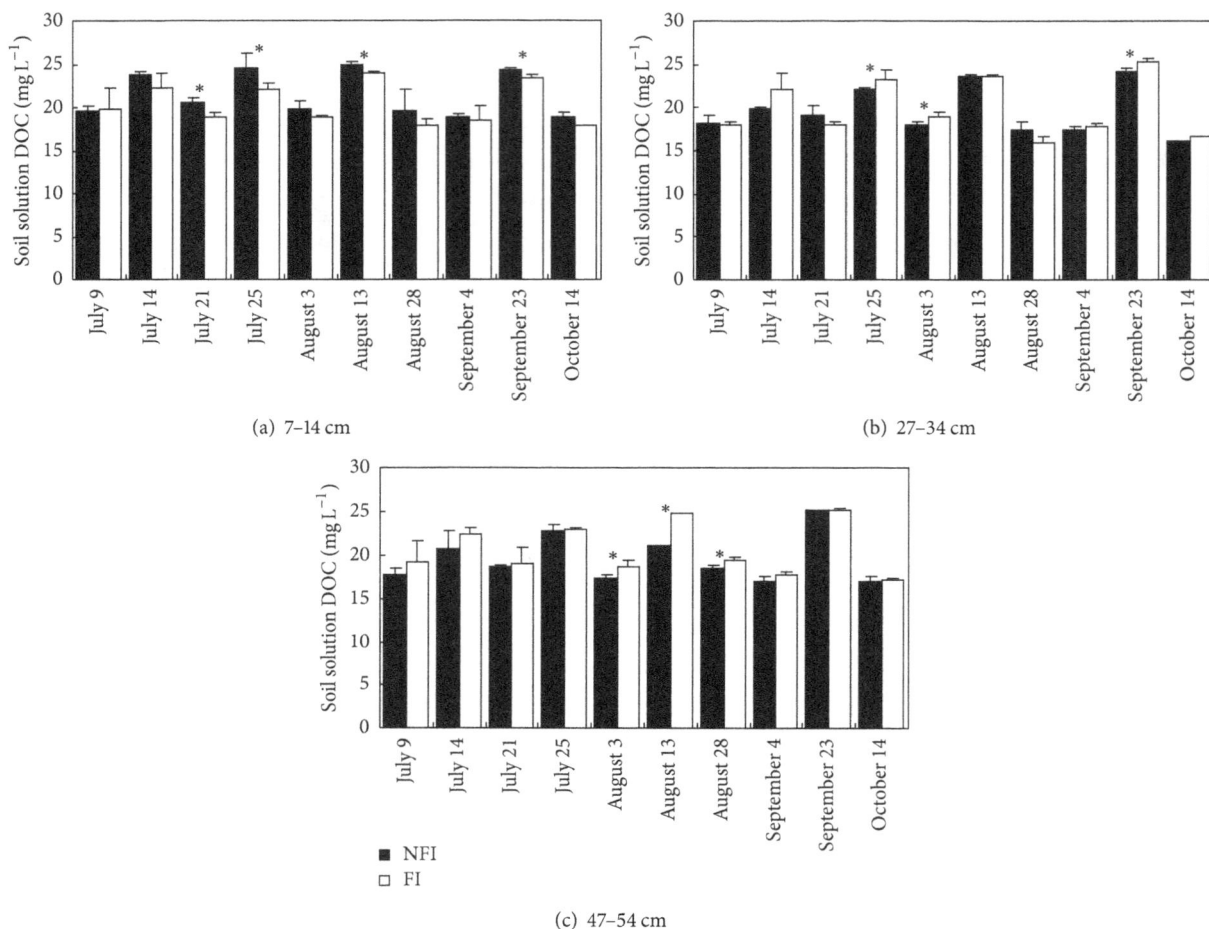

(a) 7–14 cm

(b) 27–34 cm

(c) 47–54 cm

FIGURE 4: Dissolved organic carbon (DOC) concentrations in paddy soil solutions at different depths under different irrigation managements (*indicates difference between NFI and FI that is significant at $P < 0.05$).

are significantly lower. Soil WEOC contents at 0–10 cm and 10–20 cm depths in the NFI fields increased by an average of 18.5 mg C kg^{-1} (24%) and 2.7 mg C kg^{-1} (4%) compared with those in the FI fields. Soil WEOC contents at 20–30, 30–40, and 40–60 cm depths in the NFI fields decreased by 8.7 mg C kg^{-1} (18%), 11.0 mg C kg^{-1} (22%), and 7.3 mg C kg^{-1} (17%). Therefore, the long duration of nonflooding aerobic condition and multi wet-dry cycles in NFI enhanced the soil organic decomposition and mineralization at 0–20 cm depth.

3.4. DOC Concentrations in Soil Solutions. DOC concentrations of soil solutions were slightly higher in surface soils than in deep soils (Figure 4). Compared with FI soils, DOC concentrations in soil solutions at 7–14 cm depth in NFI paddies

were slightly higher, which increased by 0.81–2.49 mg L^{-1}. The DOC concentrations in soil solutions further confirm that long duration of nonflooding aerobic condition and wet-dry cycles in NFI enhanced the soil organic decomposition and mineralization in surface soils. DOC concentrations in NFI soil solutions at 47–54 cm depth decreased slightly by 0.05–3.61 mg L^{-1} compared with those in FI paddies, because the decreased percolation led to less carbon moving downward to deep soils. However, the DOC concentrations in NFI soil solutions at 27–34 cm depth were highly variable, sometimes higher than those in FI and sometimes lower. But generally there is no significant difference between DOC concentrations in NFI and FI fields, with only a few number of differences are significant between the two treatments.

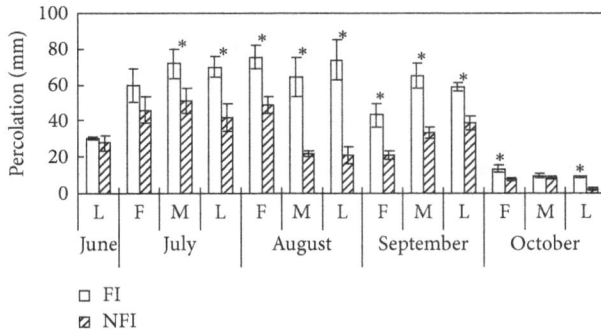

FIGURE 5: Percolation volumes from paddy soils under different irrigation managements (*indicates difference between NFI and FI that is significant at $P < 0.05$).

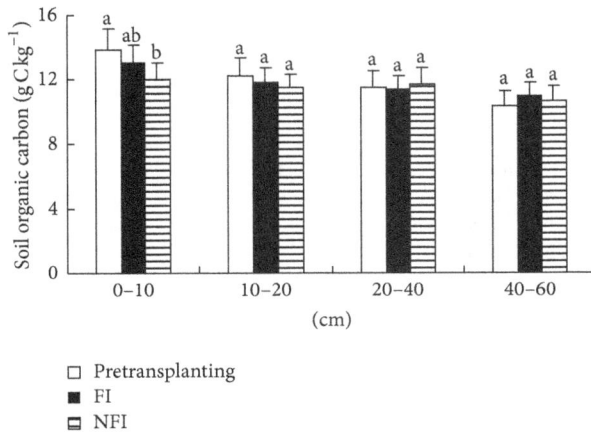

FIGURE 6: Soil organic carbon (SOC) contents in paddy soils under different irrigation management (pretransplanting means the SOC contents in soil before transplanting. Difference between the columns with the same letter in the same soil depth is not significant at $P < 0.05$).

3.5. *Leaching Risks of DOC.* Ten-day deep seepage ranged from 2.3 mm to 46.2 mm in NFI fields, which was significantly lower than the corresponding values (from 7.2 mm to 72.6 mm) in FI fields except in later June and middle October (Figure 5). Seasonal percolation in NFI paddies was 368.8 mm, which was reduced by 42.9% compared with the FI treatment. Seasonal leaching loss of DOC was 63.3 kg C ha^{-1} from NFI soils, which was reduced by 46.4% compared with those from FI fields (118.1 kg C ha^{-1}). Although DOC concentration in surface soil (0–20 cm) and soil solutions (7–14 cm) was increased in NFI fields, the reduced percolation in the NFI fields led to lower risk of DOC leaching loss than in FI fields. Several studies focused on DOC losses from forest soil [21, 43]. However, few studies on DOC leaching losses in rice paddies have been reported. In the current study, the seasonal leaching losses of DOC from FI paddies fell in the range reported by Katoh et al. (2004) [44] in typical rice fields in Japan (85 to 170 kg C ha^{-1}). The seasonal DOC leaching losses of FI paddies were less than the lower limit of 85 kg C ha^{-1}.

3.6. *Soil Organic Carbon.* SOC contents at harvest were reduced at 0–10 and 10–20 cm depth for both treatments but increased at 20–40 and 40–60 cm depth in the NFI and FI fields, respectively (Figure 6), but the reduction is insignificant in the short-term experiment except for 0–10 cm soil in NFI paddies. The SOC in different soil depth was calculated based on the SOC content (in Figure 6) and soil bulk density. The SOC storage in NFI paddies was reduced by 231 and 96 g m^{-2} at 0–10 cm and 10–20 cm depths but increased by 59 and 83 g m^{-2} at 20–40 and 40–60 cm depths. SOC storage in FI paddies was reduced by 118, 66, and 52 g m^{-2} at 0–10, 10–20, and 20–40 cm, and increased by 161 g m^{-2} at 40–60 cm. The obtained values confirm that long duration of nonflooding aerobic condition and wet-dry cycles enhanced the soil organic decomposition and mineralization in paddy soils, leading to increased SOC loss [45, 46]. Compared with FI paddies, less carbon was accumulated in deep soils (40–60 cm), and more SOC was lost in surface soils (0–20 cm) of NFI paddies. SOC content of NFI paddies indicate that more carbon was released into the atmosphere from surface soil than in FI paddies. A study regarding the greenhouse gas emission from NFI paddies [47] reported that seasonal CH$_4$ emission from NFI paddies was 1.17–1.35 g m^{-2}, which was much lower than that (6.62–7.20 g m^{-2}) from FI paddies. Thus, we can deduce that more CO$_2$ was released from NFI paddies than FI paddies because the aerobic condition favored carbon decomposition [45].

4. Discussions

4.1. *Solubility and Mobility of Soil Organic Carbon.* The long duration of nonflooding aerobic condition and multi-wet-dry cycles in NFI enhanced the solubility of organic carbon in 0–20 cm soil (Figures 3 and 4). Several studies confirmed the effect of wet-dry cycles, oven drying, or air drying with incubation experiments by measuring the soil carbon mineralization rate, soil respiration, soil microbial biomass carbon [35, 37, 48–51], or soil DOC contents [52–54]. Higher WEOC contents in deep soils of 20–30, 30–40, and 40–60 cm in the FI fields indicate that more DOC was transferred from the topsoils to the deep soils due to high percolation rate (Table 2). The relationship between soil WEOC distribution in the soil profile and water flows was indicated by Mertens et al. (2007) [55] and Junod et al. (2009) [56] on arable soils. But DOC contents in deep soil solutions in NFI paddies were always lower than those in FI paddies (Figure 4). It indicated that downward moving of DOC was determined more by deep seepage volume than the DOC contents in surface soils.

4.2. *Potential Environment Impacts.* High WEOC and DOC contents in NFI soils are the consequence of the high microbial oxidative breakdown of soil organic matter and turnover of microbial biomass [2]. The high microbial activity in NFI soil will be accompanied with high soil respiration [57, 58], which led to greenhouse gas (CO$_2$ and CH$_4$) emission. The lower SOC contents in NFI surface soil also confirmed it. Compared with FI paddies, less carbon was accumulated in deep soils (40–60 cm) and more SOC was lost in surface soils

(0–20 cm) of NFI paddies (Figure 6), although the differences were mostly insignificant for one-year experiment in current research. If the NFI is applied to rice paddies in long-term, the effect on soil carbon pool and soil carbon output will be accumulated and get even significant. The reduced SOC content in NFI surface soils indicates that more carbon was released into the atmosphere from surface soil than in FI paddies. A study regarding the greenhouse gas emission from NFI paddies [47] reported that seasonal CH_4 emission from NFI paddies was 1.17–$1.35\,\mathrm{g\,m^{-2}}$, which was much lower than that (6.62–$7.20\,\mathrm{g\,m^{-2}}$) from FI paddies. Thus, we can deduce that more CO_2 was released from NFI paddies than FI paddies because the aerobic condition favored carbon decomposition [45]. The reduced percolation in the NFI fields also led to lower DOC leaching loss than in FI fields that is helpful to reduce the risk of groundwater pollution. In addition, solubility of SOC concentration (especially DOC content) is also an important factor for the translocation of trace metals [11–13] and organic compound pollutants [8–10]. Thus, soil respiration rate, SOC fractions, and translocation of heavy metals and organic compounds should be studied to help illustrate the ecoenvironment effect of water saving irrigation on rice paddies.

4.3. Soil Fertility and Sustainability. Generally, flooding condition in rice paddies frequently results in high SOC contents compared with the upland's seasonal soil carbon accumulation, or results in long-term SOC continuous accumulation [59–61]. As a result of enhanced decomposition and mineralization of SOC in NFI surface soil, SOC in surface NFI soil decreased. Long-term application of NFI in rice fields might lead to more release of carbon from surface soil and consequently lead to degradation in the soil fertility and sustainability. Thus, future studies should look into the combinations of water and carbon (residue or biochar) management practices to enhance soil carbon storage and soil sustainability in NFI rice paddies.

5. Conclusions

WEOC contents in soils and soil solutions, soil organic carbons, and DOC leaching risks were observed in rice paddies under different irrigation managements. The results indicated that long duration of nonflooding aerobic condition and wet-dry cycles in NFI enhanced the soil organic decomposition and mineralization and consequently led to high solubility of SOC in surface soil. WEOC contents in soils and DOC in soil solution increased in NFI paddies in the surface soil layer but decreased in the deep soil layer. Less carbon moved downward to deep soils due to the decrease in percolation. The leaching losses of DOC in NFI fields were reduced by 46.4% compared with those from FI fields. SOC in surface soil was decreased in NFI paddies, indicating that more carbon was released into the atmosphere from the surface soil than in FI. The influence of irrigation management on soil organic carbon dynamics, soil respiration, and net CO_2 exchange are important problems that should be discussed in future studies. Moreover, the influence on soil carbon fraction,

which strongly related to changes in translocation of heavy metals and organic compounds, must also be considered.

Authors' Contribution

Junzeng Xu performed the data analysis and wrote the paper; Qi Wei and Xiaoli Gao contributed to the field experiment and laboratory assay; Shizhang Peng is the corresponding author, conceived, and designed the experiments; Shihong Yang contributed significantly to data analysis and paper preparation.

Acknowledgments

The research was financially supported by the National Natural Science Foundation of China (no. 51179051 and no. 51209066), the Fundamental Research Funds for the Central Universities (Program no. 2012B07514), and the Qinglan Project of Jiangsu Province.

References

[1] K. Kalbitz, S. Solinger, J.-H. Park, B. Michalzik, and E. Matzner, "Controls on the dynamics dissolved organic matter in soils: a review," *Soil Science*, vol. 165, no. 4, pp. 277–304, 2000.

[2] B. Michalzik, K. Kalbitz, J.-H. Park, S. Solinger, and E. Matzner, "Fluxes and concentrations of dissolved organic carbon and nitrogen—a synthesis for temperate forests," *Biogeochemistry*, vol. 52, no. 2, pp. 173–205, 2001.

[3] J. Iqbal, R. Hu, M. Feng, S. Lin, S. Malghani, and I. M. Ali, "Microbial biomass, and dissolved organic carbon and nitrogen strongly affect soil respiration in different land uses: a case study at Three Gorges Reservoir Area, South China," *Agriculture, Ecosystems and Environment*, vol. 137, no. 3-4, pp. 294–307, 2010.

[4] Y. Lu, R. Wassmann, H.-U. Neue, and C. Huang, "Dynamics of dissolved organic carbon and methane emissions in a flooded rice soil," *Soil Science Society of America Journal*, vol. 64, no. 6, pp. 2011–2017, 2000.

[5] M. Zhan, C. Cao, J. Wang et al., "Dynamics of methane emission, active soil organic carbon and their relationships in wetland integrated rice-duck systems in Southern China," *Nutrient Cycling in Agroecosystems*, vol. 89, no. 1, pp. 1–13, 2010.

[6] T. Filep and M. Rékási, "Factors controlling dissolved organic carbon (DOC), dissolved organic nitrogen (DON) and DOC/DON ratio in arable soils based on a dataset from Hungary," *Geoderma*, vol. 162, no. 3-4, pp. 312–318, 2011.

[7] M. L. A. Silveira, "Dissolved organic carbon and bioavailability of N and P as indicators of soil quality," *Scientia Agricola*, vol. 62, pp. 502–508, 2005.

[8] H. V. Mott, "Association of hydrophobic organic contaminants with soluble organic matter: evaluation of the database of Kdoc values," *Advances in Environmental Research*, vol. 6, no. 4, pp. 577–593, 2002.

[9] A. J. Sweetman, M. D. Valle, K. Prevedouros, and K. C. Jones, "The role of soil organic carbon in the global cycling of persistent organic pollutants (POPs): interpreting and modelling field data," *Chemosphere*, vol. 60, no. 7, pp. 959–972, 2005.

[10] E. González-Pradas, M. Fernández-Pérez, F. Flores-Céspedes et al., "Effects of dissolved organic carbon on sorption of 3,4-dichloroaniline and 4-bromoaniline in a calcareous soil," *Chemosphere*, vol. 59, no. 5, pp. 721–728, 2005.

[11] Y. Yang, L. Liang, and D. Y. Wang, "Effect of dissolved organic matter on adsorption and desorption of mercury by soils," *Journal of Environmental Sciences*, vol. 20, no. 9, pp. 1097–1102, 2008.

[12] P. N. Williams, H. Zhang, W. Davison et al., "Organic matter-solid phase interactions are critical for predicting arsenic release and plant uptake in Bangladesh paddy soils," *Environmental Science and Technology*, vol. 45, no. 14, pp. 6080–6087, 2011.

[13] L. H. Wu, L. Zhu, A. Ikuko et al., "Effects of arganic amendments on Cd, Zn and Cu bioavailability in soil with repeated phytoremediation by sedum plumbizincicola," *International Journal of Phytoremediation, Vol*, vol. 14, no. 10, pp. 1024–1038, 2012.

[14] S. Tao and B. Lin, "Water soluble organic carbon and its measurement in soil and sediment," *Water Research*, vol. 34, no. 5, pp. 1751–1755, 2000.

[15] M. D. Ruark, B. A. Linquist, J. Six et al., "Seasonal losses of dissolved organic carbon and total dissolved solids from rice production systems in northern California," *Journal of Environmental Quality*, vol. 39, no. 1, pp. 304–313, 2010.

[16] P. J. Hernes, R. G. M. Spencer, R. Y. Dyda, B. A. Pellerin, P. A. M. Bachand, and B. A. Bergamaschi, "The role of hydrologic regimes on dissolved organic carbon composition in an agricultural watershed," *Geochimica et Cosmochimica Acta*, vol. 72, no. 21, pp. 5266–5277, 2008.

[17] H. F. Wilson and M. A. Xenopolous, "Effects of agricultural land use on the composition of fluvial dissolved organic carbon," *Nature Geoscience*, vol. 2, pp. 37–41, 2009.

[18] L. Wang, C. Song, Y. Song, Y. Guo, X. Wang, and X. Sun, "Effects of reclamation of natural wetlands to a rice paddy on dissolved carbon dynamics in the Sanjiang Plain, Northeastern China," *Ecological Engineering*, vol. 36, no. 10, pp. 1417–1423, 2010.

[19] I. A. Janssens, A. Freibauer, P. Ciais et al., "Europe's terrestrial biosphere absorbs 7 to 12% of European anthropogenic CO_2 emissions," *Science*, vol. 300, no. 5625, pp. 1538–1542, 2003.

[20] J. Siemens and I. A. Janssens, "The european carbon budget: a gap," *Science*, vol. 302, no. 5651, p. 1681, 2003.

[21] B. Gielen, J. Neirynck, S. Luyssaert, and I. A. Janssens, "The importance of dissolved organic carbon fluxes for the carbon balance of a temperate Scots pine forest," *Agricultural and Forest Meteorology*, vol. 151, no. 3, pp. 270–278, 2011.

[22] R. Kindler, J. Siemens, K. Kaiser et al., "Dissolved carbon leaching from soil is a crucial component of the net ecosystem carbon balance," *Global Change Biology*, vol. 17, no. 2, pp. 1167–1185, 2011.

[23] K. Kyuma, *Paddy Soil Science*, Kyoto University Press, Kyoto, Japan, 2004.

[24] B. K. Pathak, F. Kazama, and I. Toshiaki, "Monitoring of nitrogen leaching from a tropical paddy in Thailand," *Journal of Scientific Research and Development*, vol. 6, pp. 1–11, 2004.

[25] X. J. Xie, W. Ran, Q. R. Shen, C. Y. Yang, J. J. Yang, and Z. H. Cao, "Field studies on 32P movement and P leaching from flooded paddy soils in the region of Taihu Lake, China," *Environmental Geochemistry and Health*, vol. 26, no. 2, pp. 237–243, 2004.

[26] K. S. Yoon, J. K. Choi, J. G. Son, and J. Y. Cho, "Concentration profile of nitrogen and phosphorus in leachate of a paddy plot during the rice cultivation period in southern Korea," *Communications in Soil Science and Plant Analysis*, vol. 37, no. 13-14, pp. 957–972, 2006.

[27] X. Zhao, Y.-X. Xie, Z.-Q. Xiong, X.-Y. Yan, G.-X. Xing, and Z.-L. Zhu, "Nitrogen fate and environmental consequence in paddy soil under rice-wheat rotation in the Taihu lake region, China," *Plant and Soil*, vol. 319, no. 1-2, pp. 225–234, 2009.

[28] S.-Z. Peng, S.-H. Yang, J.-Z. Xu, Y.-F. Luo, and H.-J. Hou, "Nitrogen and phosphorus leaching losses from paddy fields with different water and nitrogen managements," *Paddy and Water Environment*, vol. 9, no. 3, pp. 333–342, 2011.

[29] Z. Mao, Water efficient irrigation and environmentally sustainable irrigated rice production in China, International Commission on Irrigation and Drainage, http://www.icid.org/wat_mao.pdf, 2001.

[30] W. A. Stoop, N. Uphoff, and A. Kassam, "A review of agricultural research issues raised by the system of rice intensification (SRI) from Madagascar: opportunities for improving farming systems for resource-poor farmers," *Agricultural Systems*, vol. 71, no. 3, pp. 249–274, 2002.

[31] D. F. Tabbal, B. A. M. Bouman, S. I. Bhuiyan, E. B. Sibayan, and M. A. Sattar, "On-farm strategies for reducing water input in irrigated rice; case studies in the Philippines," *Agricultural Water Management*, vol. 56, no. 2, pp. 93–112, 2002.

[32] P. Belder, B. A. M. Bouman, R. Cabangon et al., "Effect of water-saving irrigation on rice yield and water use in typical lowland conditions in Asia," *Agricultural Water Management*, vol. 65, no. 3, pp. 193–210, 2004.

[33] B. A. M. Bouman, R. M. Lampayan, and T. P. Tuong, *Water Management in Irrigated Rice: Coping With Water Scarcity*, International Rice Research Institute, Los Baños, Philippines, 2007.

[34] E. A. Davidson, "Sources of nitric oxide and nitrous oxide following wetting of dry soil," *Soil Science Society of America Journal*, vol. 56, no. 1, pp. 95–102, 1992.

[35] A. Prechtel, C. Alewell, B. Michalzik, and E. Matzner, "Different effect of drying on the fluxes of dissolved organic carbon and nitrogen from a Norway spruce forest floor," *Journal of Plant Nutrition and Soil Science*, vol. 163, no. 5, pp. 517–521, 2000.

[36] Z. Mao, "Water saving irrigation for rice and its effect on environment," *Engineering Science*, vol. 4, no. 7, pp. 8–16, 2002.

[37] S.-R. Xiang, A. Doyle, P. A. Holden, and J. P. Schimel, "Drying and rewetting effects on C and N mineralization and microbial activity in surface and subsurface California grassland soils," *Soil Biology and Biochemistry*, vol. 40, no. 9, pp. 2281–2289, 2008.

[38] S. T. Bao, *Soil and Agro-Chemistry Analysis*, Agricultural Press, Beijing, China, 2000.

[39] G. Guggenberger and W. Zech, "Sorption of dissolved organic carbon by ceramic P80 suction cups," *Journal of Plant Nutrition and Soil Science*, vol. 155, pp. 151–155, 1992.

[40] R. J. Cabangon, T. P. Tuong, E. G. Castillo et al., "Effect of irrigation method and N-fertilizer management on rice yield, water productivity and nutrient in typical lowland rice conditions in China," *Paddy and Water Environment*, vol. 2, no. 4, pp. 195–206, 2004.

[41] H. Li, X. Liang, Y. Chen, G. Tian, and Z. Zhang, "Ammonia volatilization from urea in rice fields with zero-drainage water management," *Agricultural Water Management*, vol. 95, no. 8, pp. 887–894, 2008.

[42] M. Zhan, C.-G. Cao, Y. Jiang, J.-P. Wang, L.-X. Yue, and M.-L. Cai, "Dynamics of active organic carbon in a paddy soil under different rice farming modes," *Chinese Journal of Applied Ecology*, vol. 21, no. 8, pp. 2010–2016, 2010.

[43] K. Fujii, M. Uemura, C. Hayakawa et al., "Fluxes of dissolved organic carbon in two tropical forest ecosystems of East Kalimantan, Indonesia," *Geoderma*, vol. 152, no. 1-2, pp. 127–136, 2009.

[44] M. Katoh, J. Murase, M. Hayashi, K. Matsuya, and M. Kimura, "Nutrient leaching from the plow layer by water percolation and accumulation in the subsoil in an irrigated paddy field," *Soil Science and Plant Nutrition*, vol. 50, no. 5, pp. 721–729, 2004.

[45] C. Yang, L. Yang, and Z. Ouyang, "Organic carbon and its fractions in paddy soil as affected by different nutrient and water regimes," *Geoderma*, vol. 124, no. 1-2, pp. 133–142, 2005.

[46] R. J. Buresh and S. M. Haefele, *Changes in Paddy Soils Under Transition to Water-Saving and Diversified Cropping Systems*, World Congress of Soil Science, Brisbane, Australia, 2010.

[47] H. J. Hou, S. Z. Peng, J. Z. Xu, S. H. Yang, and Z. Mao, "Seasonal variations of CH_4 and N_2O emissions in response to water management of paddy fields located in Southeast China," *Chemosphere*, vol. 89, no. 7, pp. 884–892, 2012.

[48] N. Fierer and J. P. Schimel, "Effects of drying-rewetting frequency on soil carbon and nitrogen transformations," *Soil Biology and Biochemistry*, vol. 34, no. 6, pp. 777–787, 2002.

[49] M. M. Mikha, C. W. Rice, and G. A. Milliken, "Carbon and nitrogen mineralization as affected by drying and wetting cycles," *Soil Biology and Biochemistry*, vol. 37, no. 2, pp. 339–347, 2005.

[50] D. L. Jones and V. B. Willett, "Experimental evaluation of methods to quantify dissolved organic nitrogen (DON) and dissolved organic carbon (DOC) in soil," *Soil Biology and Biochemistry*, vol. 38, no. 5, pp. 991–999, 2006.

[51] S.-H. Yao, B. Zhang, and F. Hu, "Soil biophysical controls over rice straw decomposition and sequestration in soil: the effects of drying intensity and frequency of drying and wetting cycles," *Soil Biology and Biochemistry*, vol. 43, no. 3, pp. 590–599, 2011.

[52] K. Kaiser, M. Kaupenjohann, and W. Zech, "Sorption of dissolved organic carbon in soils: effects of soil sample storage, soil-to-solution ratio, and temperature," *Geoderma*, vol. 99, no. 3-4, pp. 317–328, 2001.

[53] S. Klitzke, *Mobilization Mechanisms of Soluble and Dispersible Heavy Metals and Metalloids in Soils*, Technischen Universität, Berlin, Germany, 2007.

[54] G. F. Koopmans and J. E. Groenenberg, "Effects of soil oven-drying on concentrations and speciation of trace metals and dissolved organic matter in soil solution extracts of sandy soils," *Geoderma*, vol. 161, no. 3-4, pp. 147–158, 2011.

[55] J. Mertens, J. Vanderborght, R. Kasteel et al., "Dissolved organic carbon fluxes under bare soil," *Journal of Environmental Quality*, vol. 36, no. 2, pp. 597–606, 2007.

[56] J. Junod, E. Zagal, M. Sandoval, R. Barra, G. Vidal, and M. Villarroel, "Effect of irrigation levels on dissolved organic carbon soil distribution and the depth mobility of chlorpyrifos," *Chilean Journal of Agricultural Research*, vol. 69, no. 3, pp. 435–444, 2009.

[57] V. A. Orchard and F. J. Cook, "Relationship between soil respiration and soil moisture," *Soil Biology and Biochemistry*, vol. 15, no. 4, pp. 447–453, 1983.

[58] R. Mancinelli, S. Marinari, V. Di Felice, M. C. Savin, and E. Campiglia, "Soil property, CO_2 emission and aridity index as agroecological indicators to assess the mineralization of cover crop green manure in a Mediterranean environment," *Ecological Indicator*, vol. 34, pp. 31–40, 2013.

[59] W. Zhang, Y.-Q. Yu, W.-J. Sun, and Y. Huang, "Simulation of soil organic carbon dynamics in Chinese rice paddies from 1980 to 2000," *Pedosphere*, vol. 17, no. 1, pp. 1–10, 2007.

[60] S. Nishimura, S. Yonemura, T. Sawamoto et al., "Effect of land use change from paddy rice cultivation to upland crop cultivation on soil carbon budget of a cropland in Japan," *Agriculture, Ecosystems and Environment*, vol. 125, no. 1–4, pp. 9–20, 2008.

[61] M. F. Pampolino, E. V. Laureles, H. C. Gines, and R. J. Buresh, "Soil carbon and nitrogen changes in long-term continuous lowland rice cropping," *Soil Science Society of America Journal*, vol. 72, no. 3, pp. 798–807, 2008.

10

Log-Cubic Method for Generation of Soil Particle Size Distribution Curve

Songhao Shang[1,2]

[1] State Key Laboratory of Hydroscience and Engineering, Tsinghua University, Beijing 100084, China
[2] Department of Hydraulic Engineering, Tsinghua University, Beijing 100084, China

Correspondence should be addressed to Songhao Shang; shangsh@tsinghua.edu.cn

Academic Editors: F. Darve, B.-L. Ma, and A. Moldes

Particle size distribution (PSD) is a fundamental physical property of soils. Traditionally, the PSD curve was generated by hand from limited data of particle size analysis, which is subjective and may lead to significant uncertainty in the freehand PSD curve and graphically estimated cumulative particle percentages. To overcome these problems, a log-cubic method was proposed for the generation of PSD curve based on a monotone piecewise cubic interpolation method. The log-cubic method and commonly used log-linear and log-spline methods were evaluated by the leave-one-out cross-validation method for 394 soil samples extracted from UNSODA database. Mean error and root mean square error of the cross-validation show that the log-cubic method outperforms two other methods. What is more important, PSD curve generated by the log-cubic method meets essential requirements of a PSD curve, that is, passing through all measured data and being both smooth and monotone. The proposed log-cubic method provides an objective and reliable way to generate a PSD curve from limited soil particle analysis data. This method and the generated PSD curve can be used in the conversion of different soil texture schemes, assessment of grading pattern, and estimation of soil hydraulic parameters and erodibility factor.

1. Introduction

Particle size distribution (PSD) is a fundamental physical property of soils, which can be described by the PSD curve of cumulative particle percentage versus logarithm of particle size. The PSD curve provides detailed information about the soil, such as grading pattern and the sand, silt, and clay fractions to determine the soil textural classes [1]. It is also useful for the conversion of different soil texture schemes [2]. What is more, these textural fractions are more readily available from particle size analysis or existing soil database, so they are usually taken as main inputs to estimate other soil properties difficult to obtain, such as hydraulic properties [3–5] and soil erodibility factor [6, 7].

In the practice of particle size analysis, only limited data of cumulative particle percentage versus particle size are available. Traditionally, these limited data were plotted on semilogarithmic coordinates, and then these points were smoothly connected by hand to generate a smooth and monotone PSD curve. After the generation of a PSD curve, cumulative percentage at unmeasured size and characteristic particle size corresponding to specified cumulative percentage can be estimated from the curve graphically. However, the previous processes are subjective, which may lead to significant uncertainty in the freehand PSD curve and graphically estimated cumulative particle percentage and characteristic particle size [8]. To overcome the subjectivity of freehand PSD curve, regression and interpolation methods and similarity procedure had been used to estimate cumulative particle percentages at unmeasured particle sizes.

Regression method was used to fit the PSD curve with various empirical formulae [9, 10], which had been evaluated with measured data from different part of the world [11–13]. These empirical PSD curves can represent the trend of cumulative percentage varying with particle size and can be used in the estimation of soil hydraulic properties [4, 14].

However, the fitted empirical curves may not be flexible enough to depict PSD of diverse soil types. Besides, they usually do not pass through measured data, which is not in accord with the essential requirement of a PSD curve.

The similarity procedure to estimate cumulative percentage at specified unmeasured size of a soil sample is based on the similarity of PSD between soil under consideration and an external reference data set [15], on condition that data corresponding to the specified particle size were available from the reference data set. Therefore, this procedure is not suitable for the generation of a continuous PSD curve. Besides, because a large external reference data set is required to find soils with similar PSDs, this procedure was not often used due to the lack of appropriate reference data set.

Interpolation method was also used to approximate the PSD with a function passing through measured data, which is similar to artificially plotted PSD curve. Main methods for the interpolation of PSD curves include the log-linear interpolation [15, 16] and the cubic spline [8]. The log-linear interpolation curve can ensure the monotonicity of the PSD curve, but it is not smooth. The cubic spline is smooth, but it is monotone only in specified conditions of measured data [17]. In some cases, cubic spline may produce impractical results, which can be overcome by modifying impractical results with regression analysis results or dividing the whole range of particle size into two segments and constructing a spline for each segment [8]. However, these modifications may be only applicable to specific conditions. It is still necessary to find a simple and reliable method to generate a PSD curve from limited data.

The main purpose of this study was to propose a log-cubic method to generate the PSD curve from limited soil particle analysis data, which is based on a monotone piecewise cubic interpolation method [17]. This method was evaluated with the leave-one-out cross-validation method for 394 soil samples extracted from UNSODA database [18].

2. Materials and Methods

2.1. The Log-Cubic Method. Generally, cumulative particle percentages are available for limited sizes from soil particle size analysis. Suppose that n points, (d_i, P_i), $i = 1, 2, \ldots, n$, were available, where P_i is the cumulative particle percentage corresponding to particle size d_i. Since a PSD curve is both smooth and monotone, it can be approximated with the monotone piecewise cubic interpolation method [17].

Considering that the particle size covers several orders of magnitude and the PSD curve is plotted on semilogarithmic coordinates, the logarithm of particle size ($\ln d$) was used in the interpolation. Similar to the log-linear method, the proposed method was named as the log-cubic method. The log-cubic interpolation function, $P_c(d)$, is composed of $n - 1$ cubic polynomial segments defined in particle size intervals

$[d_i, d_{i+1}]$, $i = 1, 2, \ldots, n-1$, which can be described by (1) and (2):

$$P_c(d) = P_{ci}(d), \quad d_i \le d \le d_{i+1}, \ i = 1, 2, \ldots, n-1, \quad (1)$$

$$P_{ci}(d) = P_{i+1} \frac{3hs^2 - 2s^3}{h^3} + P_i \frac{h^3 - 3hs^2 + 2s^3}{h^3}$$
$$+ f_{i+1} \frac{s^2(s-h)}{h^2} + f_i \frac{s(s-h)^2}{h^2}, \quad i = 1, 2, \ldots, n-1, \quad (2)$$

where $P_{ci}(d)$ is the segment of $P_c(d)$ for particle size interval of $[d_i, d_{i+1}]$, $h = \ln d_{i+1} - \ln d_i$, $s = \ln d - \ln d_i$, and f_i and f_{i+1} denote the slope of $P_c(d)$ at knots d_i and d_{i+1}, respectively. The slope at a knot can be estimated from the lengths and the first divided differences of two adjacent intervals [17, 19]. This interpolation method has been used in several fields of soil and agricultural sciences [20–22].

The log-cubic interpolation function defined in (1) passes through all measured points and is both smooth and monotone, which meets essential requirements of a PSD curve.

For comparison, commonly used log-linear [15, 16] and log-spline [8] methods were also used. The log-linear interpolation function for PSD can be described by (3):

$$P_l(d) = P_{li}(d), \quad d_i \le d \le d_{i+1}, \ i = 1, 2, \ldots, n-1,$$
$$P_{li}(d) = P_i \frac{h-s}{h} + P_{i+1} \frac{s}{h}, \quad i = 1, 2, \ldots, n-1, \quad (3)$$

where $P_{li}(d)$ is the segment of log-linear interpolation function, $P_l(d)$, for $d_i \le d \le d_{i+1}$.

The log-spline method for PSD is based on the cubic spline interpolation of P versus $\ln d$, which can be described by (4):

$$P_s(d) = P_{si}(d), \quad d_i \le d \le d_{i+1}, \ i = 1, 2, \ldots, n-1,$$
$$P_{si}(d) = M_i \frac{(h-s)^3}{6h} + M_{i+1} \frac{s^3}{6h} + \left(\frac{P_i}{h} - \frac{M_i h}{6} \right)(h-s)$$
$$+ \left(\frac{P_{i+1}}{h} - \frac{M_{i+1} h}{6} \right)s, \quad i = 1, 2, \ldots, n-1, \quad (4)$$

where $P_{si}(d)$ is the segment of log-spline interpolation function, $P_s(d)$, for $d_i \le d \le d_{i+1}$; M_i is second derivative at knot x_i, $i = 1, 2, \ldots, n$, which can be determined from the continuous condition of the first derivative for the cubic spline [19].

The interpolation procedure was accomplished by the "interp1" function of the Matlab package [19], as described by (5):

$$Pc = \text{interp1}(\ln d, P, \text{lndc}, \text{"method"}), \quad (5)$$

where $\ln d$ and P are n-dimensional arrays of the logarithm of measured particle size and cumulative percentage, respectively; lndc is an array representing a desired classification of the logarithm of particle size; Pc is the interpolated cumulative percentage corresponding to lndc; and "method" specifies interpolation methods, where "linear," "spline," and "cubic" refer to piecewise linear interpolation, cubic spline interpolation, and monotone piecewise cubic interpolation, respectively.

2.2. Evaluation of the Log-Cubic, Log-Linear, and Log-Spline Methods. A leave-one-out cross-validation method was performed to assess the performance of the log-cubic, log-linear, and log-spline methods, using particle size data extracted from UNSODA database [18].

In the cross-validation procedure, data of one particle size were left out and interpolated with remaining data. Considering the boundary effect of interpolation, data of the first two and last two particle sizes were not left out in the validation process. Interpolated values were compared with omitted measured values to calculate the mean error (ME) and root mean square error (RMSE) with (6), which were then used to assess the performance of interpolation methods:

$$\mathrm{ME} = \frac{1}{n-4} \sum_{i=3}^{n-2} \left(P_i - P_{i,i}\right),$$

$$\mathrm{RMSE} = \left[\frac{1}{n-4} \sum_{i=3}^{n-2} (P_i - P_{i,i})^2\right]^{1/2},$$

(6)

where n is the number of particle size grade and $P_{i,i}$ is the interpolated values of cumulative particle percentage.

Considering data requirement for interpolation and cross-validation, soils with five or more particle size grades $n \geq 5$ were considered. As a result, 394 soil samples were available from the UNSODA database [18] with the grade of particle size from 5 to 16, among which 94.2% has the grade number from 6 to 9.

Among these samples, particle size data of 353 samples covers the range from $2\,\mu m$ to $2000\,\mu m$, which can be used to determine the clay ($<2\,\mu m$), silt ($2\sim50\,\mu m$), and sand ($50\sim 2000\,\mu m$) fractions directly or through interpolation. The particle size distribution of these 353 soil samples covers a wide range of soil textures (Figure 1).

2.3. Generation of the PSD Curve. Once the interpolation method for the PSD curve was chosen, the PSD curve can be generated automatically from measured data, (d_i, P_i), $i = 1, 2, \ldots, n$, with the following procedure.

(1) Determination of particle sizes for interpolation. Each particle size interval, $[d_i, d_{i+1}]$, $i = 1, 2, \ldots, n - 1$, is divided into m subintervals with the endpoints defined in (7):

$$d_{i,j} = \exp\left[\ln d_i + \frac{j\left(\ln d_{i+1} - \ln d_i\right)}{m}\right], \quad j = 1, 2, \ldots, m,$$

(7)

Consequently, $m(n - 1) + 1$ points are used for interpolation. To guarantee the smoothness of the PSD curve, m should not be less than 50, and $m = 100$ was used in this study.

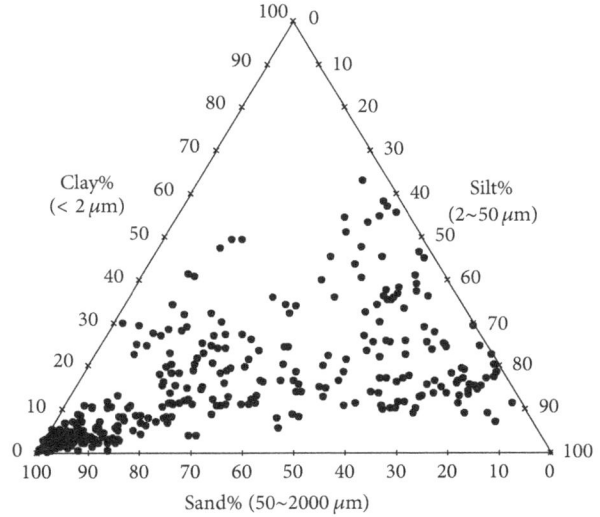

FIGURE 1: Textural composition of 353 soils samples used in the evaluation of interpolation methods.

(2) Interpolation of cumulative particle percentages ($P_{i,j}$) corresponding to $d_{i,j}$ from measured data using an appropriate interpolation method.

(3) Plot points for measured particle size analysis data, (d_i, P_i), $i = 1, 2, \ldots, n$, in a semilogarithmic coordinates.

(4) Connect interpolated points, $(d_{i,j}, P_{i,j})$, $i = 1, 2, \ldots, n-1, j = 1, 2, \ldots, m$, with line segments in succession in the same semilogarithmic coordinates.

3. Results and Discussion

3.1. Evaluation of Interpolation Methods. The monotonicity and the smoothness are essential requirements of PSD curves. From the principles of three interpolation methods, the log-linear method can guarantee the monotonicity of interpolated PSD curve, the log-spline method can guarantee the smoothness of interpolated PSD curve, while the log-cubic method can guarantee both the monotonicity and the smoothness of interpolated PSD curve. Therefore, the log-cubic method is essentially superior to log-linear and log-spline methods.

The log-linear, log-spline, and log-cubic methods were evaluated by the leave-one-out cross-validation method for 394 soil samples extracted from UNSODA database [18]. Average values of ME of these three methods are −0.4%, −0.6%, and −0.6%, respectively, which are all very close to the perfect value of 0. Average values of RMSE are 8.6%, 6.8%, and 6.3%, respectively, which indicates that the log-cubic method is superior to the other two methods in the average sense.

ME and RMSE deciles of these three methods are shown in Figure 2. The results show that most of the ME deciles of the log-cubic method are more close to the perfect value of 0 than those of log-linear and log-spline methods. As far

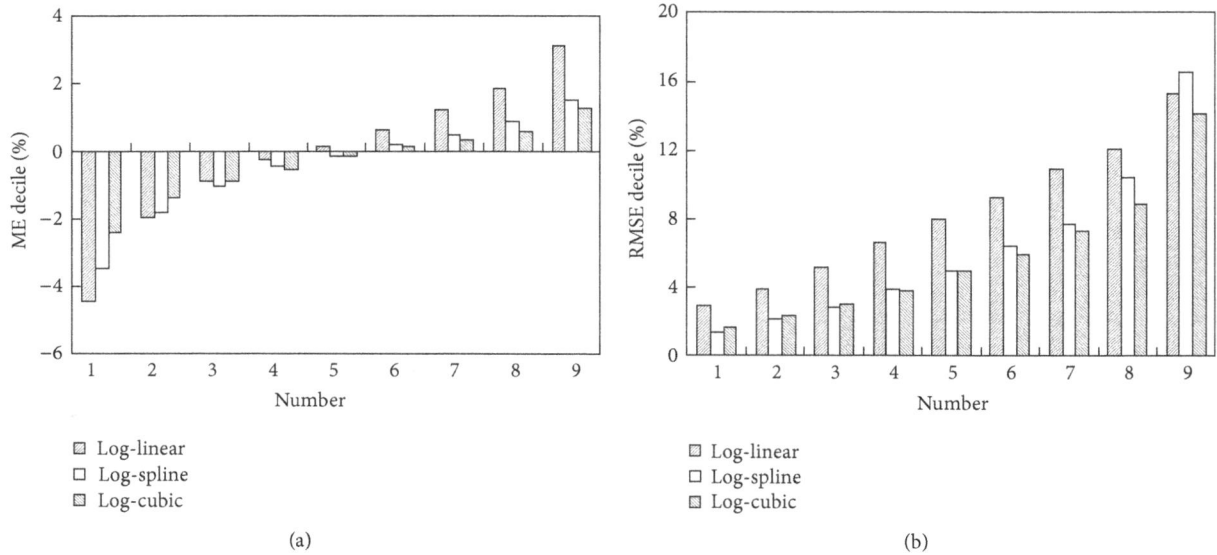

FIGURE 2: ME (a) and RMSE (b) deciles of the log-linear, log-spline, and log-cubic methods.

as ME is concerned, the log-cubic method outperforms the log-linear and log-spline methods for 75% and 50% of soil samples, respectively. On the other hand, RMSE deciles of the log-cubic method are all less than those of the log-linear method and less than or slightly greater than those of the log-spline method. As far as RMSE is concerned, the log-cubic method outperforms the log-linear and log-spline methods for 95% and 84% soil samples, respectively. Therefore, the log-cubic method is superior to the log-linear and log-spline methods for interpolating PSDs.

Nemes et al. (1999) [15] used four procedures to estimate cumulative percentage at unmeasured particle size, including log-linear and spline interpolation methods, Gompertz curve regression method, and similarity method. Their results of cross-validation indicate that the similarity method is the most effective, which yielded the lowest RMSE ranging from 2% to 11% for different distances between particle size limits. RMSE value of the present log-cubic method is 6.3%, which corresponds to lower or medium RMSE values of the similarity method at smaller distances between particle size limits [15]. Moreover, the log-cubic method does not require a large external reference data set and can be used to estimate cumulative particle percentages at any size in the PSD range. The computation of the log-cubic method is much simpler than that of the similarity method. Therefore, the log-cubic method is appropriate to be used in the estimation of cumulative percentage at unmeasured particle size.

3.2. Comparison of Generated PSD Curves.

Using the log-linear, log-spline, and log-cubic methods, cumulative particle percentages at unmeasured sizes can be estimated from measured data and used for the generation of PSD curve. Figure 3 illustrates two PSD curves for two soil samples generated from 7 measured data. All generated PSD curves pass through measured data, which is an essential requirement of interpolation methods. For the log-linear method,

the generated PSD curves are monotone but not smooth. For the log-spline method, the generated PSD curves are smooth but not monotone since an interpolated cubic spline is monotone only in specified conditions of measured data [17]. Impractical fluctuations in these PSD curves show that the log-spline method itself is not appropriate to generate a PSD curve, unless it is modified in some way [8]. While for the log-cubic method, the generated PSD curves are both smooth and monotone, which meet the essential requirements of a PSD curve. PSD curves of other soils generated with the log-cubic method all follow these essential requirements. These characters are the same as those of traditional freehand PSD curve, but the present result is objective and independent on persons involved.

Compared with other methods to generate a PSD curve, the log-cubic method is superior to the freehand method in its objectivity, superior to the log-linear method in the smoothness and accuracy of the interpolated PSD curve, and superior to the log-spline method in the monotonicity of the interpolated PSD curve. Therefore, the proposed log-cubic method provides an objective and reliable way to generate a PSD curve from limited soil particle analysis data with satisfactory precision.

3.3. Application of the Log-Cubic Method.

The log-cubic method and the generated PSD curves can be used to estimate cumulative particle percentages at unmeasured sizes and to estimate characteristic particle sizes corresponding to specified cumulative particle percentages. These results can be further used in the conversion of different soil texture schemes, assessment of grading pattern, and estimation of soil hydraulic parameters and erodibility factor.

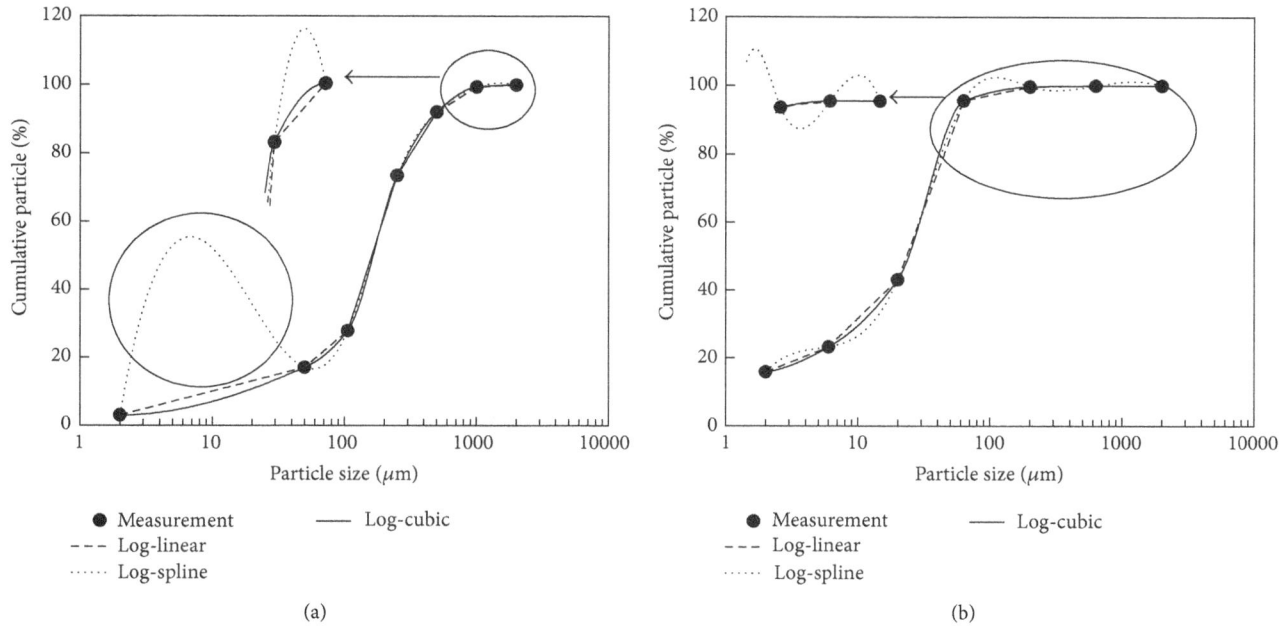

FIGURE 3: Examples of PSD curves generated from measured data (dots) using the log-linear (dashed lines), log-spline (dot lines), and log-cubic (solid lines) methods for soils nos. 1010 (a) and 1490 (b). Circles indicate particle size ranges in which PSD curves generated with the log-spline method are not monotonic.

3.3.1. Estimation of Cumulative Particle Percentage at Unmeasured Size. One direct use of the log-cubic method and the generated PSD curve is to estimate cumulative particle percentages at unmeasured sizes, which is necessary in the conversion of different soil texture schemes.

There are various classification schemes of soil particle size in different countries and different fields [1], such as schemes of the International Society of Soil Science (ISSS), U.S. Department of Agriculture (USDA). Incompatibility of different schemes may cause considerable confusions and inconveniences, and it is necessary to convert different classification schemes into the desired one [2, 15]. This conversion is essential in achieving compatibility among different classification schemes [15]. Besides, it is also useful in estimating soil hydraulic parameters and erodibility factor [8] from available pedotransfer functions when different classification schemes are used. For example, in the ROSETTA program for estimating soil hydraulic parameters with pedotransfer functions [3], soil hydraulic parameters were estimated from soil textural fractions based on the USDA scheme and other physical properties. When the ROSETTA program is used for soils with soil textural fractions of other classification schemes, conversion of the available scheme to the USDA scheme is required.

The log-cubic method can be used in this conversion process, which was illustrated later with two soil samples shown in Figure 3.

For soil No. 1010 shown in Figure 3(a), clay ($<2\,\mu$m), silt (2~$50\,\mu$m), and sand (50~$2000\,\mu$m) fractions of the USDA scheme can be determined directly from measured data, which are 3%, 14%, and 83%, respectively. The soil texture is classified as loamy sand by the USDA scheme. However, the

textural fractions are not available for the ISSS scheme, where the limit between silt and sand is $20\,\mu$m. The cumulative particle percentage at $20\,\mu$m can be estimated with the log-cubic method, which is 10.7%. Consequently, clay ($<2\,\mu$m), silt (2~$20\,\mu$m), and sand (20~$2000\,\mu$m) fractions of the ISSS scheme can be estimated to be 3%, 7.7%, and 89.3%, respectively, and the soil texture is classified as sand.

On the contrary, soil textural fractions of the ISSS scheme can be determined directly from measured data for soil no. 1490 (Figure 3(b)), which are 15.8%, 27.3%, and 56.9% for clay, silt, and sand particles, respectively. The soil texture is classified as sandy clay loam. The cumulative particle percentage at $50\,\mu$m can be estimated with the log-cubic method, which is 90.0%. Then clay, silt, and sand fractions of the USDA scheme can be estimated to be 15.8%, 74.2%, and 10.0%, respectively, and the soil texture is classified as silt loam in the USDA scheme.

3.3.2. Estimation of Characteristic Particle Sizes Corresponding to Specified Cumulative Particle Percentages. Characteristic particle sizes are sizes corresponding to specified cumulative particle percentages, such as d_{10}, d_{30}, d_{50}, and d_{60} corresponding to cumulative particle percentages of 10, 30, 50, and 60, respectively. These characteristic particle sizes are essential data in the assessment of soil grading pattern with the uniformity index defined as d_{60}/d_{10} and in the estimation of soil hydraulic properties from empirical or semiempirical formulae [23, 24]. However, these particle sizes cannot be obtained directly from particle size analysis results. They were usually determined graphically from the freehand PSD curve, which is subjective and error-prone. This problem can be solved by inverse interpolation of the PSD curve.

In the process to generate a PSD curve, dense points of particle size versus cumulative particle percentage were available using the log-cubic method, which are $(d_{i,j}, P_{i,j})$, $i = 1, 2, \ldots, n - 1$, $j = 1, 2, \ldots, m$. These points can be used in the inverse interpolation to estimate characteristic particle sizes similar to (5), as described in (8):

$$dc = \exp\left(\text{interp1}\left(P, \ln d, \text{Pc}, \text{"method"}\right)\right), \qquad (8)$$

where P and $\ln d$ refer to $P_{i,j}$ and $\ln d_{i,j}$, $i = 1, 2, \ldots, n - 1$, $j = 1, 2, \ldots, m$, respectively; Pc is the desired cumulative particle percentages, such as 10, 30, 50, and 60; dc is the interpolated characteristic particle sizes corresponding to Pc.

For example, d_{10}, d_{30}, d_{50}, and d_{60} of soil no. 1010 shown in Figure 3(a) are estimated using the log-cubic interpolation method, which are 17.8 μm, 114.0 μm, 164.5 μm, and 192.0 μm, respectively. Then the uniformity index can be estimated to be 10.8, indicating a well-graded soil.

Two points need to be stressed about the previously mentioned inverse interpolation process. First, the desired cumulative particle percentage should be in the range of measured values, since the extrapolation results may be impractical. For example, d_{10} for the soil no. 1490 (Figure 3(b)) cannot be estimated satisfactorily using the previously mentioned inverse interpolation method or other similar methods. Second, values of cumulative particle percentage should be distinct, which is required by interpolation methods. For example, values of cumulative particle percentage for soil no. 1490 (Figure 3(b)) are kept constant (100) in the particle size range from 630 μm to 2000 μm, which should be omitted in the inverse interpolation. Using data from 2 μm to 630 μm, d_{30}, d_{50}, and d_{60} of this soil are estimated to be 10.3 μm, 24.2 μm, and 29.2 μm, respectively.

Considering the satisfactory precision of the log-cubic method, this method provides an objective and reliable way for the conversion of different soil texture schemes and the estimation of characteristic particle sizes.

4. Conclusions

To overcome the subjectivity and uncertainty of the freehand PSD curve generated from limited data of particle size analysis, the log-cubic method was proposed for the interpolation of cumulative particle percentages at unmeasured sizes to generate the PSD curve automatically. The generated PSD curve passes through all measured data and is both smooth and monotone, which meets essential requirements of a PSD curve. Results of the leave-one-out cross-validation using 394 soil samples show that the interpolation precision of the log-cubic method is satisfactory compared with other methods.

The proposed log-cubic method provides an objective and reliable way to generate a PSD curve from limited soil particle analysis data. This method and the generated PSD curve can be used to estimate cumulative particle percentages at unmeasured sizes and to estimate characteristic particle

sizes corresponding to specified cumulative particle percentages, which can be further used in the conversion of different soil texture schemes, assessment of grading pattern, and estimation of soil hydraulic parameters and erodibility factor.

Acknowledgments

This work was supported by the National Natural Science Foundation of China (Grant no. 51279077) and the National Key Technology R&D Program of China (Grant no. 2011BAD25B05).

References

[1] W. Jury and R. Horton, *Soil Physics*, John Wiely & Sons, Hoboken, NJ, USA, 6th edition, 2004.

[2] S. S. Rousseva, "Data transformations between soil texture schemes," *European Journal of Soil Science*, vol. 48, no. 4, pp. 749–758, 1997.

[3] M. G. Schaap, F. J. Leij, and M. T. van Genuchten, "Rosetta: a computer program for estimating soil hydraulic parameters with hierarchical pedotransfer functions," *Journal of Hydrology*, vol. 251, no. 3-4, pp. 163–176, 2001.

[4] A. Nemes and W. J. Rawls, "Evaluation of different representations of the particle-size distribution to predict soil water retention," *Geoderma*, vol. 132, no. 1-2, pp. 47–58, 2006.

[5] K. H. Liao, S. H. Xu, J. C. Wu, S. H. Ji, and Q. Lin, "Assessing soil water retention characteristics and their spatial variability using pedotransfer functions," *Pedosphere*, vol. 21, no. 4, pp. 413–422, 2011.

[6] K. G. Renard, G. R. Foster, G. A. Weesies, D. K. McCool, and D. C. Yoder, *Predicting Soil Erosion By WAter: A Guide To conservAtion plAnning With the Revised universAl Soil Loss equAtion (RUSLE)*, US Department of Agriculture, Washington, DC, USA, 1997.

[7] K. L. Zhang, A. P. Shu, X. L. Xu, Q. K. Yang, and B. Yu, "Soil erodibility and its estimation for agricultural soils in China," *Journal of Arid Environment*, vol. 72, no. 6, pp. 1002–1011, 2008.

[8] Y. M. Cai, K. L. Zhang, and S. C. Li, "Study on the conversionof different soils texture," *Acta Pedologica Sinica*, vol. 40, pp. 511–517, 2003.

[9] M. D. Fredlund, D. G. Fredlund, and G. W. Wilson, "An equation to represent grain-size distribution," *Canadian Geotechnical Journal*, vol. 37, no. 4, pp. 817–827, 2000.

[10] G. D. Buchan, K. S. Grewal, and A. B. Robson, "Improved models of particle-size distribution: an illustration of model comparison techniques," *Soil Science Society of America Journal*, vol. 57, no. 4, pp. 901–908, 1993.

[11] S. I. Hwanga, K. P. Leeb, D. S. Leeb, and S. E. Powers, "Models for estimating soil particle-size distributions," *Soil Science Society of America Journal*, vol. 66, no. 4, pp. 1143–1150, 2002.

[12] J. L. Liu, S. H. Xu, H. Liu, and F. Guo, "Application of parametric models to description of particle size distribution in loamy soils," *Acta Pedologica Sinica*, vol. 41, pp. 375–379, 2004.

[13] V. Bagarello, G. Provenzano, and A. Sgroi, "Fitting particle size distribution models to data from Burundian soils for the BEST procedure and other purposes," *Biosystems Engineering*, vol. 104, no. 3, pp. 435–441, 2009.

[14] S. Assouline, D. Tessier, and A. Bruand, "A conceptual model of the soil water retention curve," *Water Resources Research*, vol. 34, no. 2, pp. 223–231, 1998.

[15] A. Nemes, J. H. M. Wösten, A. Lilly, and J. H. Oude Voshaar, "Evaluation of different procedures to interpolate particle-size distributions to achieve compatibility within soil databases," *Geoderma*, vol. 90, no. 3-4, pp. 187–202, 1999.

[16] O. Tietje and V. Hennings, "Accuracy of the saturated hydraulic conductivity prediction by pedo-transfer functions compared to the variability within FAO textural classes," *Geoderma*, vol. 69, no. 1-2, pp. 71–84, 1996.

[17] F. N. Fritsch and R. E. Carlson, "Monotone piecewise cubic interpolation," *SIAM Journal on Numerical Analysis*, vol. 17, no. 2, pp. 238–246, 1980.

[18] F. J. Leij, W. J. Alves, M. T. van Genuchten, and J. R. Williams, *The UNSODA Unsaturated Soil Hydraulic Database User's Manual Version 1. 0*, National Risk Management Research Laboratory, Cincinnati, Ohio, USA, 1996.

[19] C. Moler, *Numerical Computing with MATLAB*, The Math-Works, Natick, Mass, USA, 2004.

[20] S. C. Iden and W. Durner, "Free-form estimation of the unsaturated soil hydraulic properties by inverse modeling using global optimization," *Water Resources Research*, vol. 43, no. 7, Article ID W07451, 2007.

[21] A. Peters and W. Durner, "Simplified evaporation method for determining soil hydraulic properties," *Journal of Hydrology*, vol. 356, no. 1-2, pp. 147–162, 2008.

[22] S. H. Shang, "Temporal downscaling of crop coefficient and crop water requirement from growing stage to substage scales," *The Scientific World Journal*, vol. 2012, Article ID 105487, pp. 1–6, 2012.

[23] R. P. Chapuis, "Predicting the saturated hydraulic conductivity of sand and gravel using effective diameter and void ratio," *Canadian Geotechnical Journal*, vol. 41, no. 5, pp. 787–795, 2004.

[24] J. X. Song, X. H. Chen, C. Cheng, D. M. Wang, S. Lackey, and Z. X. Xu, "Feasibility of grain-size analysis methods for determination of vertical hydraulic conductivity of streambeds," *Journal of Hydrology*, vol. 375, no. 3-4, pp. 428–437, 2009.

11

Updating Categorical Soil Maps Using Limited Survey Data by Bayesian Markov Chain Cosimulation

Weidong Li,[1] **Chuanrong Zhang,**[1] **Dipak K. Dey,**[2] **and Michael R. Willig**[3]

[1] *Department of Geography and Center for Environmental Sciences & Engineering, University of Connecticut, Storrs, CT 06269, USA*
[2] *Department of Statistics, University of Connecticut, Storrs, CT 06269, USA*
[3] *Center for Environmental Sciences & Engineering and Department of Ecology & Evolutionary Biology, University of Connecticut, Storrs, CT 06269, USA*

Correspondence should be addressed to Weidong Li; weidongwoody@gmail.com

Academic Editors: J. C. Domec and J. J. Wang

Updating categorical soil maps is necessary for providing current, higher-quality soil data to agricultural and environmental management but may not require a costly thorough field survey because latest legacy maps may only need limited corrections. This study suggests a Markov chain random field (MCRF) sequential cosimulation (Co-MCSS) method for updating categorical soil maps using limited survey data provided that qualified legacy maps are available. A case study using synthetic data demonstrates that Co-MCSS can appreciably improve simulation accuracy of soil types with both contributions from a legacy map and limited sample data. The method indicates the following characteristics: (1) if a soil type indicates no change in an update survey or it has been reclassified into another type that similarly evinces no change, it will be simply reproduced in the updated map; (2) if a soil type has changes in some places, it will be simulated with uncertainty quantified by occurrence probability maps; (3) if a soil type has no change in an area but evinces changes in other distant areas, it still can be captured in the area with unobvious uncertainty. We concluded that Co-MCSS might be a practical method for updating categorical soil maps with limited survey data.

1. Introduction

Soil is an important natural resource and is also an essential component of ecosystems. The spatial distribution of different soils represents a special kind of natural landscapes (called soilscape). Soils are traditionally classified into a number of types and delineated as categorical maps based on multiple attributes observed at sample profiles, tacit knowledge of experienced surveyors, remotely sensed landscape features, and a specific classification system. Categorical soil maps are widely used in ecological and agricultural studies and provide crucial information for natural resource and environmental management. Because existing soil maps may be of low quality or too outdated to reflect current soil distributions, map update is necessary for providing current, more accurate, or more detailed information to meet the requirements of applications. For example, most soil series maps in United States (e.g., the USDA Soil Survey Geographic Database) were made on the basis of field surveys carried out in the 1950s, and

they may not have been effectively updated to reflect recent soil changes. However, large-scale detailed soil survey is too costly to be carried out frequently for generating new high-quality maps. If an existing soil map is of sufficient quality and appropriately scaled, updating may not require a new full-coverage soil survey for a revised soil map because the types of soils at most places in the legacy map may not have changed. Consequently, we may be able to update a legacy soil map with only limited new survey data on soil distribution. When qualified legacy soil maps are available, we may only need to address areas where the previously determined soil types have a large possibility of type change due to some reasons (e.g., internal or environmental changes, incorrect mapping, or taxonomy change), identified by careful map examination with ancillary information. Changes can be found through a limited soil update survey or simply map examination by experts. Other reasons of using legacy soil maps and survey data together to create current categorical soil maps include that: (1) historical field survey data were not well kept or were kept without accurate coordinates and (2) legacy soil maps were based on drawings of experienced soil surveyors during field surveys, but most observed soil profiles were not sampled for laboratory analysis or recorded into a database. In general, we may incorporate information from a legacy soil map into the current soil map based on limited survey data if the legacy soil map contains valuable information that cannot be replaced by a limited survey.

A variety of quantitative modeling methods have been used or developed to predict spatially explicit soil categorical characteristics. These methods may have their own merits in different contexts. One group of methods is soil-landscape models, which use environmental soil-forming factors to predict soil patterns over unvisited areas. These methods include multinomial logistic regressions (MLRS), classification and regression tree analysis, and fuzzy methods; see applications in predictive categorical soil mapping [1–9]. This group of methods generally does not incorporate spatial autocorrelations. The other group considers spatial statistical models, mainly including indicator geostatistics, maximum entropy models, and Markov chain random fields (MCRF); see related studies [10–12] in mapping categorical soil variables. These methods are based on spatial autocorrelations of categories, but legacy data and remotely sensed landscape data may also be incorporated as auxiliary information. Other spatial statistical methods that were suggested for mapping categorical variables may also be used or adapted for mapping soil categories [13, 14]. In addition, some qualitative methods such as the rule-based method [15] and the pure remote sensing method [16] were introduced recently for mapping soil types, but only for special soil types such as peat lands or gypsic soils.

Recently, Markov chains were extended into a new spatial statistical approach, that is, the MCRF approach, for simulating categorical spatial variables [17]. This approach uses transiograms [18] to measure class spatial auto- and cross-correlations and uses MCRF models (usually simplified models) to estimate the local conditional probability distribution of a categorical spatial variable at an unobserved location. MCRFs may be regarded as an extension of Markov mesh random fields [19] toward conditional simulation on sample data or as a special kind of causal Markov random fields in accordance with the Bayesian inference principle. MCRF-based sequential simulation algorithms can be used to generate simulated realizations in single sweeps, similar to other geostatistical sequential simulation algorithms. This approach may incorporate various interclass relationships, thus effectively reducing the uncertainty associated with prediction and generating more accurate simulated realizations that strictly obey class neighboring relationships [20]. Nonetheless, currently implemented MCRF algorithms do not incorporate auxiliary or legacy data by cosimulations, thus requiring further extensions.

It is easy to understand that legacy soil data, whether they are map data or observed point data, contain valuable information that is relevant to present soil patterns. Legacy soil maps also contain the tacit knowledge of experienced surveyors, who were intensively trained for soil survey but may not be available at the time of soil map updating [21].

Therefore, proper use of legacy soil data may appreciably improve the prediction of soil spatial distributions. In fact, the use of legacy soil data in digital soil modeling has become a commonplace [22]. If densely distributed survey data are not available, a legacy soil map available at a similar scale may be used as auxiliary data to create the current soil map with limited survey data.

In this study, we assume that the legacy soil maps from the last update or made from last extensive soil surveys need limited corrections related to natural or anthropogenic soil changes or other reasons. Consequently, update is only necessary in altered areas or erroneously mapped locations. As such, we assume that the legacy soil maps are mainly outdated rather than being of low quality, and that update is necessary for a variety of reasons. This is reasonable because (1) many high-quality soil maps were made by extensive soil surveys, usually commissioned by government agencies, and (2) many soil types only change slowly as a result of natural processes, except for some special soil groups (e.g., hydric soils). Such an assumption may be applicable to many situations in the United States, where detailed large-scale categorical soil maps exist for each county in many states. To incorporate legacy soil maps through cosimulations for categorical soil map creation with limited survey data, the MCRF sequential simulation (MCSS) algorithm proposed in [20] was extended into a MCRF sequential cosimulation (Co-MCSS) algorithm and its workability was demonstrated by a case study on synthetic data in this study. The main objective is to suggest a suitable cost-efficient method for updating legacy categorical soil maps that only requires limited new survey data, mainly in the changed areas. It should be noted that although limited map changes in categorical soil map update may be carried out using a conventional hand-delineating method, a spatial statistical method would be appreciated due to many reasons, such as efficiency, objectivity in soil type boundary determination, and availability of uncertainty information associated with the updating.

2. Methods

2.1. Markov Chain Random Fields. The chief obstacle to extending one-dimensional Markov chains to multidimensional causal random field models such as Markov mesh models [19] is the lack of a natural ordering for a multidimensional grid and hence the lack of a natural notion of causality in the spatial data. As a result, an artificial ordering for spatial data must be assumed, which often yields directional artifacts in simulated images [23, 24]. The MCRF theory solved this problem and other related issues that hindered conditional Markov chain simulations on sparse sample data. The initial ideas of MCRFs aimed to correct the flaws of a two-dimensional Markov chain model for subsurface characterization [17, 24]. The ideas were generalized into a theoretical framework for a new geostatistical approach for simulating categorical fields [17]. Wide applications of this approach lie within further extensions of MCRF models and the development of simulation algorithms that can effectively deal with data clustering (or redundancy), ancillary information, and multiple-point statistics.

A MCRF refers to a random field defined by a single spatial Markov chain that moves or jumps in space and decides its state at any uninformed (i.e., unobserved and unvisited in a simulation process) location by interactions with its nearest neighbors in different directions and its last stay (i.e., visited) location [17]. The interactions within a neighborhood are performed through a sequential Bayesian updating process [25]. Therefore, a MCRF is a spatial Markov chain with local Bayesian updating. Here, a "state" means a category (or class) for a categorical spatial variable. For a MCRF $Z(\mathbf{u})$, if we assume that i_1 to i_m are the states of the nearest neighbors in different directions around an uninformed location $\mathbf{u}_0$ plus the state of the last visited location of the spatial Markov chain, the local conditional probability distribution of $Z(\mathbf{u})$ at the current uninformed location $\mathbf{u}_0$ can be denoted as $p[i_0(\mathbf{u}_0) \mid i_1(\mathbf{u}_1), \ldots, i_m(\mathbf{u}_m)]$, where i_0 refers to the state of $z(\mathbf{u}_0)$ being estimated. Emphasizing the single-chain nature of a MCRF and the last visited location, this local conditional probability distribution can be factorized as

$$
\begin{aligned}
p\left[i_0\left(\mathbf{u}_0\right) \mid i_1\left(\mathbf{u}_1\right), \ldots, i_m\left(\mathbf{u}_m\right)\right] \\
= \frac{1}{A} p\left[i_m\left(\mathbf{u}_m\right) \mid i_0\left(\mathbf{u}_0\right), \ldots, i_{m-1}\left(\mathbf{u}_{m-1}\right)\right] \\
\cdots p\left[i_2\left(\mathbf{u}_2\right) \mid i_0\left(\mathbf{u}_0\right), i_1\left(\mathbf{u}_1\right)\right] p\left[i_0\left(\mathbf{u}_0\right) \mid i_1\left(\mathbf{u}_1\right)\right],
\end{aligned}
\tag{1}
$$

where $A = p[i_1(\mathbf{u}_1), \ldots, i_m(\mathbf{u}_m)]/p[i_1(\mathbf{u}_1)]$ is a normalizing constant and $\mathbf{u}_1$ indicates the last visited location or the location that the spatial Markov chain goes through to the current location $\mathbf{u}_0$ [17]. This explicit full general solution of MCRFs is essentially a multiple-point geostatistical model, composed of a series of two- to $m + 1$-point statistics (or cliques) involving the current uninformed location $\mathbf{u}_0$. These two- and multiple-point statistics are also functions of directional lag distances because these points are usually not immediately adjacent in a space of sample data. They may be estimated from training images but the computation is much complex. Note that the local joint probability distribution of $Z(\mathbf{u})$; that is, $p[i_0(\mathbf{u}_0), i_1(\mathbf{u}_1), \ldots, i_m(\mathbf{u}_m)]$ can be similarly factorized.

If we consider (1) in the Bayesian inference formulation, $p[i_0(\mathbf{u}_0) \mid i_1(\mathbf{u}_1), \ldots, i_m(\mathbf{u}_m)]$ is the posterior probability distribution, $p[i_0(\mathbf{u}_0) \mid i_1(\mathbf{u}_1)]$ (i.e., a transition probability, or a transiogram if regarded as a function of the lag distance) is the prior probability distribution, and the other part of the right-hand side excluding the constant is the likelihood component. The prior probability indicates the single Markov chain nature of a MCRF. The likelihood component is composed of multiple terms (one for each nearest neighbor), which update the prior probability using nearest neighbors in different directions by a manner of recursion as follows:

$$
\begin{aligned}
\text{posterior}_1 &= \text{prior} \\
\text{posterior}_2 &\propto L_2 \times \text{posterior}_1 \\
&\cdots \\
\text{posterior}_m &\propto L_m \times \text{posterior}_{m-1} \propto L_m \times \cdots \times L_2 \times \text{prior},
\end{aligned}
\tag{2}
$$

where L_k refers to the likelihood term for the kth nearest neighbor, that is, $p[i_k(\mathbf{u}_k) \mid i_0(\mathbf{u}_0), \ldots, i_{k-1}(\mathbf{u}_{k-1})]$. Thus, when no nearest neighbor other than the last visited location is available, we get a posterior probability equal to the prior probability (the likelihood term L_1 is 1). But when there are nearest neighbors other than the last visited location available, update begins on each datum in turn, and in each time of update the posterior of last update serves as the new prior. Therefore, a MCRF model can be explained from the viewpoint of Bayesian inference. The generation of a MCRF may be regarded as a dynamic Bayesian inference process. Because the above Bayesian updating process is conducted simultaneously within a neighborhood rather than an iterative updating algorithm, it can be simply written as [25]

$$
\begin{aligned}
\text{posterior} &\propto \text{likelihood}\left[i_m\left(\mathbf{u}_m\right)\right] \\
&\times \cdots \times \text{likelihood}\left[i_2\left(\mathbf{u}_2\right)\right] \times \text{prior}.
\end{aligned}
\tag{3}
$$

This sequential Bayesian updating process on nearest neighbors starts from nearest neighbor $i_2(\mathbf{u}_2)$ and ends at nearest neighbor $i_m(\mathbf{u}_m)$ in a Markov-type neighborhood around the uninformed location $\mathbf{u}_0$ being estimated (see Figure 1(a) as an example). This updating process may not need to follow a fixed sequence of nearest neighbors because earlier considered nearest neighbors within the neighborhood become the conditioning data of later updates, and all updates are conditioned on the datum $i_0(\mathbf{u}_0)$ being estimated. Such a spatial estimation method is different from existing spatial estimation methods such as kriging and conventional Markov random field models.

If the spatial Markov chain is stationary and its last visited location is far away from the current uninformed location, the influence of the last visited location may be ignored (i.e., the transition probabilities from the last visited location to the current location decay to corresponding marginal probabilities). Thus, the local conditional probability distribution $p[i_0(\mathbf{u}_0) \mid i_1(\mathbf{u}_1), \ldots, i_m(\mathbf{u}_m)]$ can be factorized differently as

$$
\begin{aligned}
p\left[i_0\left(\mathbf{u}_0\right) \mid i_1\left(\mathbf{u}_1\right), \ldots, i_m\left(\mathbf{u}_m\right)\right] \\
= \frac{1}{A} p\left[i_m\left(\mathbf{u}_m\right) \mid i_0\left(\mathbf{u}_0\right), \ldots, i_{m-1}\left(\mathbf{u}_{m-1}\right)\right] \\
\cdots p\left[i_1\left(\mathbf{u}_1\right) \mid i_0\left(\mathbf{u}_0\right)\right] p\left[i_0\left(\mathbf{u}_0\right)\right],
\end{aligned}
\tag{4}
$$

where $A = p[i_1(\mathbf{u}_1), \ldots, i_m(\mathbf{u}_m)]$ is a normalizing constant and $\mathbf{u}_1$ is not the last visited location but just a nearest neighbor. Equation (4) is a special case of (1). If we consider this equation in the Bayesian inference formulation, $p[i_0(\mathbf{u}_0) \mid i_1(\mathbf{u}_1), \ldots, i_m(\mathbf{u}_m)]$ is still the posterior, $p[i_0(\mathbf{u}_0)]$ (i.e., a marginal probability) becomes the prior, and the other part of the right-hand side excluding the constant is the likelihood component. For this special case, the sequential Bayesian updating process on nearest neighbors starts from nearest neighbor $i_1(\mathbf{u}_1)$ and ends at nearest neighbor $i_m(\mathbf{u}_m)$ in a Markov-type neighborhood around the location $\mathbf{u}_0$ being estimated (see Figure 1(b) as an example).

Because (1) involves complex multiple-point statistics that are difficult to estimate from sparse sample data, simplification is necessary. If we invoke the conditional independence

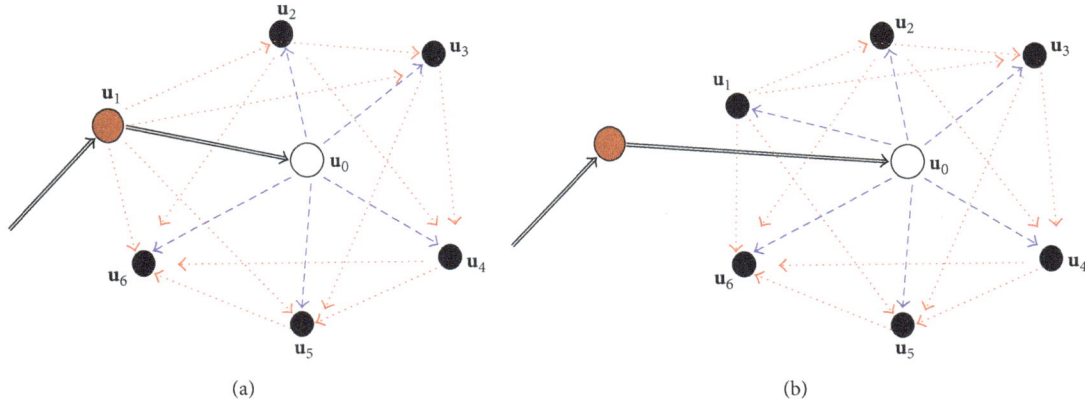

$$(a) \qquad\qquad\qquad (b)$$

FIGURE 1: Neighborhood structures with six nearest neighbors and the sequential Bayesian updating process in basic Markov chain random fields: (a) assuming $\mathbf{u}_1$ to be the last visited location; (b) assuming the last visited location is far away (outside the neighborhood). Data interactions across the uninformed location $\mathbf{u}_0$ being estimated are ignored according to the Markov property.

assumption, a simplified general solution for MCRFs can be obtained from (1) as follows:

$$p\left[i_0\left(\mathbf{u}_0\right) \mid i_1\left(\mathbf{u}_1\right),\ldots,i_m\left(\mathbf{u}_m\right)\right]$$

$$= \frac{p_{i_1 i_0}\left(\mathbf{h}_{10}\right)\prod_{g=2}^{m}p_{i_0 i_g}\left(\mathbf{h}_{0g}\right)}{\sum_{f_0=1}^{n}\left[p_{i_1 f_0}\left(\mathbf{h}_{10}\right)\prod_{g=2}^{m}p_{f_0 i_g}\left(\mathbf{h}_{0g}\right)\right]}, \qquad (5)$$

where $p_{i_0 i_g}\left(\mathbf{h}_{0g}\right)$ represents a transiogram (i.e., transition probability function) from class i_0 at location $\mathbf{u}_0$ to class i_g at location $\mathbf{u}_g$ with the lag distance $\mathbf{h}_{0g}$; $i_1\left(\mathbf{u}_1\right)$ represents the nearest neighbor from or across which the spatial Markov chain moves to the current location $\mathbf{u}_0$; m represents the number of nearest neighbors plus the last visited location; i and f all represent states (i.e., classes) in the state space $S = (1,\ldots,n)$ of the categorical field under study. This simplified general solution is still in accordance with the Bayesian inference formulation. Because this simplified solution involves only two-point statistics—transiograms, which can be estimated from sample data, it is computable directly using sample data. In addition, because class proportions are not involved in this solution, no assumption is required concerning their stationarity. This simplified solution did not consider the data clustering issue and left it to model extension and specific algorithm design. Data clustering apparently impacts the contributions of nearest neighbors to the local conditional probability distribution at the current location being estimated, thus accounting for this effect is preferable when it is possible. For example, one may consider applying a set of power parameters to the transition probability terms based on the neighborhood configuration, but the computation load will inevitably largely increase.

If the spatial Markov chain is stationary and its last visited location is far from the current location $\mathbf{u}_0$ (i.e., outside the neighborhood), we have $p_{i_1 i_0}(\mathbf{h}_{10}) \approx p_{i_0}$ due to $\mathrm{Lim}_{\mathbf{h}_{10}\to\infty}p_{i_1 i_0}(\mathbf{h}_{10}) = p_{i_0}$, which is a basic property of one-dimensional first-order stationary Markov chains. Thus, if we

still assume that there are m nearest neighbors, (5) may be rewritten as

$$p\left[i_0\left(\mathbf{u}_0\right) \mid i_1\left(\mathbf{u}_1\right),\ldots,i_m\left(\mathbf{u}_m\right)\right]$$

$$= \frac{p_{i_0}\prod_{g=1}^{m}p_{i_0 i_g}\left(\mathbf{h}_{0g}\right)}{\sum_{f_0=1}^{n}\left[p_{f_0}\prod_{g=1}^{m}p_{f_0 i_g}\left(\mathbf{h}_{0g}\right)\right]}, \qquad (6)$$

where p_{i_0} refers to the marginal probability of class i_0, which is approximately equal to the mean value of the class proportion for omni- and bi-directional transition probabilities or a large study area. Equation (6) also can be obtained by simplifying (4) based on the conditional independence assumption or by transforming (5) using the relationship of $p_{i_0 i_1}(\mathbf{h}_{01})p_{i_0}/p_{i_1} = p_{i_1 i_0}(\mathbf{h}_{10})$. However, for spatial data, this relationship only holds in stationary situations and does not hold for nonstationary situations and unidirectional transiograms. Therefore, (6) is a special stationary case of (5) and is included in (5).

2.2. MCRF Cosimulation Model. To incorporate auxiliary variables, we need to expand (5) into a Co-MCRF model. The contributions of auxiliary variables may be incorporated by using the formulation of addition (to some extent similar to cokriging), that is, by including one contribution term for each auxiliary variable. Such a formulation must be renormalized or allocate weights to its contribution terms to ensure the total probability of occurrences of all states (i.e., classes) at location $\mathbf{u}_0$ sums to unity. Alternatively, the formulation of multiplication can be used to incorporate auxiliary variables. In this scenario, we regard the data of auxiliary variables as nearest neighbors of the uninformed location $\mathbf{u}_0$ in different variable spaces. Here, we use the multiplication formulation to construct the Co-MCRF model. We consider only the colocated cosimulation case because it is what we need for revising categorical soil maps while the involved auxiliary variables, for example, the legacy categorical soil map, provide exhaustive data. The colocated Co-MCRF model with k auxiliary variables can be written as

$$p\left[i_0\left(\mathbf{u}_0\right) \mid i_1\left(\mathbf{u}_1\right), \ldots, i_m\left(\mathbf{u}_m\right); r_0^{(1)}\left(\mathbf{u}_0\right); \ldots; r_0^{(k)}\left(\mathbf{u}_0\right)\right]$$

$$= \frac{p_{i_1 i_0}\left(\mathbf{h}_{10}\right) \prod_{g=2}^{m} p_{i_0 i_g}\left(\mathbf{h}_{0g}\right) \prod_{l=1}^{k} b_{i_0 r_0^{(l)}}}{\sum_{f_0=1}^{n}\left[p_{i_1 f_0}\left(\mathbf{h}_{10}\right) \prod_{g=2}^{m} p_{f_0 i_g}\left(\mathbf{h}_{0g}\right) \prod_{l=1}^{k} b_{f_0 r_0^{(l)}}\right]},$$

$$(7)$$

where $r_0^{(k)}$ represents the state of the kth auxiliary variable at the colocation $\mathbf{u}_0$. The cross-transiograms from the primary variable to auxiliary variables reduce to cross transition probabilities $b_{i_0 r_0}$ due to the colocation property. We may call this kind of cross-transition probabilities (and transiograms) between classes of two different categorical fields *cross-field transition probabilities* (and *transiograms*). The cross-field transition probabilities, however, have to be estimated separately. In this equation, we do not deal with cross-correlations between auxiliary variables and practically consider them to be independent of each other.

In this study, we consider only one auxiliary variable in the form of a legacy soil map. Hence, (7) further reduces to

$$p\left[i_0\left(\mathbf{u}_0\right) \mid i_1\left(\mathbf{u}_1\right), \ldots, i_m\left(\mathbf{u}_m\right); r_0\left(\mathbf{u}_0\right)\right]$$

$$= \frac{b_{i_0 r_0} p_{i_1 i_0}\left(\mathbf{h}_{10}\right) \prod_{g=2}^{m} p_{i_0 i_g}\left(\mathbf{h}_{0g}\right)}{\sum_{f_0=1}^{n}\left[b_{f_0 r_0} p_{i_1 f_0}\left(\mathbf{h}_{10}\right) \prod_{g=2}^{m} p_{f_0 i_g}\left(\mathbf{h}_{0g}\right)\right]}. \quad (8)$$

If an auxiliary variable has no correlation with the primary variable, the cross-field transition probabilities will equal the corresponding class mean proportions of the auxiliary variable, and the corresponding cross-field transition probability terms in (8) will be canceled from the numerator and denominator.

2.3. MCRF Sequential Cosimulation Algorithm. The conditional independence assumption was assumed for nearest neighbors in different directions to derive the simplified general solution of MCRFs. Such an assumption is practical, often used in nonlinear probability models [26]. However, the conditional independence of adjacent neighbors in cardinal directions for a rectangular lattice is a property of Pickard random fields, a kind of unilateral Markov models [27–29]. For the situation of the four (or less) nearest neighbors found in cardinal directions, the conditional independence property of Pickard random fields may be applied to the sparse data situation [17, 24]. This supports the neighborhood choice of using four nearest neighbors in four cardinal directions or quadrants in MCRF algorithm design to reduce data clustering effects [20].

In fact, it is also unnecessary and difficult to consider many nearest neighbors in different directions in applications. Nearest neighbors outside correlation ranges can be eliminated from consideration. The influence of remotely located data on the current uninformed location is typically screened by closer data within a certain angle. In addition, the conditional independence assumption apparently does

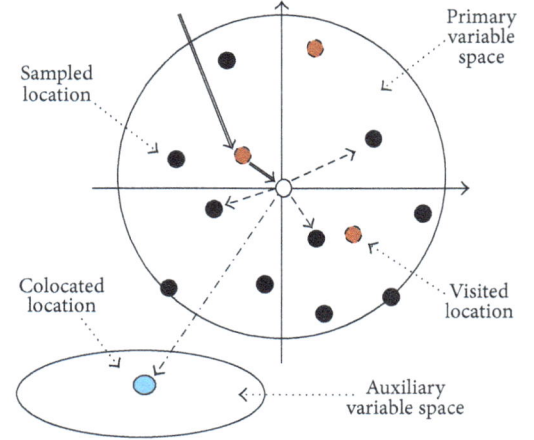

FIGURE 2: Illustration of the Markov chain random field colocated cosimulation model with quadrant search and one auxiliary variable for random-path sequential simulation. Double arrows represent the moving directions of the spatial Markov chain. Dashed arrows represent the interactions of the spatial Markov chain with nearest neighbors and auxiliary data.

not hold for clustered sample data. Therefore, it is proper for MCRF-based Markov chain models to consider only the nearest neighbors in several cardinal directions within a search range to both approximately meet the conditional independence assumption and increase the computation efficiency.

The four nearest neighbors in four cardinal directions can be regarded as conditionally independent given the state of the surrounded central location in a sparse data space [17]. Consequently, the neighborhood choice for the Co-MCRF model needs only to use the four nearest neighbors in four cardinal directions, allowing (8) to be further simplified to

$$p\left[i_0\left(\mathbf{u}_0\right) \mid i_1\left(\mathbf{u}_1\right), \ldots, i_4\left(\mathbf{u}_4\right); r_0\left(\mathbf{u}_0\right)\right]$$

$$= \frac{b_{i_0 r_0} p_{i_1 i_0}\left(\mathbf{h}_{10}\right) \prod_{g=2}^{4} p_{i_0 i_g}\left(\mathbf{h}_{0g}\right)}{\sum_{f_0=1}^{n}\left[b_{f_0 r_0} p_{i_1 f_0}\left(\mathbf{h}_{10}\right) \prod_{g=2}^{4} p_{f_0 i_g}\left(\mathbf{h}_{0g}\right)\right]}. \quad (9)$$

Here, we assume that the last visited location of the spatial Markov chain is always within the four nearest neighbors; if it is not so, we assume that the spatial Markov chain comes through one of them (Figure 2). Such a simplified Co-MCRF model provides the MCRF approach the capability of dealing with large data sets.

A tolerance angle is required because nearest neighbors in a neighborhood may not be located exactly along cardinal directions. To cover the whole space of a search area, sectors can be substituted for cardinal directions, and we can seek one nearest neighbor from each sector to represent the neighborhood (Figure 2). If we consider four cardinal directions, the sectors representing cardinal directions are quadrants. There may be no data to occur in some quadrants within a search range at the boundary strips or at the beginning of a simulation when sample data are very sparse. Consequently, the size of a neighborhood may be less than four. Equation (9) can always be adapted to the situation. In case no data can be found in the whole search area, we assume the spatial Markov chain comes from a location outside the search range. By choosing a suitable search radius based on the density of sample data, this situation rarely occurs.

The MCSS algorithm was developed based on the above quadrant search method and was effective in simulating multinomial classes in two horizontal dimensions [20]. The colocated Co-MCSS algorithm used in this paper is an extension of the random-path MCSS algorithm; therefore, their computation processes are similar.

2.4. Transiogram Modeling and Cross-Field Transition Probability Matrix. To perform simulations using Co-MCSS, transiogram models are needed to provide transition probability values at any needed lag distances. The transiogram was formally established in recent years to meet the needs of related Markov chain models [18]. The initial purpose of proposing such a spatial correlation measure was to provide a practical way to estimate multistep transition probabilities from sparse point sample data [30]. Later, it was found that the transiogram could also be an excellent independent spatial measure to characterize the spatial variability of categorical spatial variables [18]. This spatial measure is related to some pioneer studies [31–35], which used or explored transition probability curves in some special conditions. There are different ways to get continuous transiogram models [36]. One is using nonparametric methods such as linear interpolation to interpolate experimental transiograms into continuous models. The second is using parametric methods (i.e., mathematical models) to fit experimental transiograms. Because the latter is relatively tedious and the sample data for soil map updating are usually sufficient for estimating reliable experimental transiograms, the first approach was chosen in this study.

For a colocated cosimulation conditioned on one auxiliary variable, one cross-field transition probability matrix (CTPM) is sufficient. Transition probabilities in a CTPM can be estimated by counting point-to-point frequencies of different class pairs from the sample data of the primary variable to the colocated data of the auxiliary variable using the following equation:

$$b_{ik} = \frac{f_{ik}}{\sum_{j=1}^{n} f_{ij}}, \qquad (10)$$

where f_{ik} represents the frequency of transitions from class i of the primary variable to class k of the auxiliary variable and n is the number of classes of the auxiliary variable.

3. Case Study for Method Testing

3.1. Data, Parameters, and Outputs. The major purpose of this case study was to test the method proposed in this paper, rather than a real application. Because a real field soil survey was unavailable to us, synthetic data extracted from a piece of a real soil series map ($9 \, km^2$ area) [20] was used in this case study. However, the spatial pattern and spatial relationships among the soil series can mimic some real-world situation,

thus still providing an effective test to the proposed spatial statistical method.

The area was discretized into a 175×128 grid of 22,400 pixels, with a square pixel area of $400 \, m^2$. The soil map has seven soil types. Here, the exact soil series names are not our concern. For convenience, we denote them as S1, S2, S3, S4, S5, S6, and S7. This soil series map (Figure 3(a)) served as the legacy soil map for this study. The soil survey for delineating the legacy soil map was mainly done in the 1950s [37]. After five decades, such a soil map is likely outdated and would be improved by revision. We assumed that the legacy soil map from USDA was made with high-quality data at the mapping time, but that is now inaccurate. We further assumed that only a few of small areas in the legacy soil map were subject to soil type changes. For testing the suggested soil map update method, we designed the following soil series changes in the study area: S5 is joined to S3; S1 is joined to S7; part of S6 became S7 at the bottom middle east; and part of S7 became S6 at the top-right corner. As a result, we have five new soil series: SU2 (i.e., S2), SU3 (i.e., S3 + S5), SU4 (i.e., S4), SU6 (i.e., S6 + part of S7), and SU7 (i.e., S7 + S1 + part of S6). Soil series of S2, S3, and S4 were assumed to have no changes confirmed in the updated survey. The resulting new soil series distribution map (Figure 3(b)) was used as the reference soil map for checking simulated results.

Because we assumed only a few of small areas were subject to soil type changes, our limited field survey was also confined to these small areas. Thus, the survey data are insufficient and also biased for estimating the parameters (e.g., transiogram models) used in the cosimulation. Our suggestion is to use pseudosample data, that is, sample data directly extracted from unchanged areas in the legacy soil map. Therefore, we sampled a sparse data set of 646 points (Figure 3(c)) from the reference soil map, which cover both the changed and unchanged areas. These samples are randomly distributed, not purposively arranged with respect to soil type changes. Using this data set, we examined simulated results for other points to see how well our suggested method predicted soil type characteristics, both those that were unchanged and changed compared to the legacy map. The rationalities behind the sample data are that (1) for areas where soil types have changed, a field survey or visual observation through remote sensing is necessary to identify the changes on the map, and both methods may produce survey sample data for the update; and (2) for areas where soil types did not change, no matter how the judgment is made (from a field survey, remote sensing, or expertise), pseudosample data may be simply extracted from the legacy soil map. Pseudosample data extraction from a legacy map or from the combination of a legacy map and remotely sensed imagery can be carried out through human-computer interactions. Thus, it is not difficult to obtain sufficient sample data with a limited soil survey (i.e., a small set of real soil survey data).

Experimental transiograms were estimated from the sample data to generate transiogram models for conditional simulations. Two subsets of omnidirectional transiogram models interpolated from the experimental transiograms are provided in Figure 4 and show that cross-transiogram models

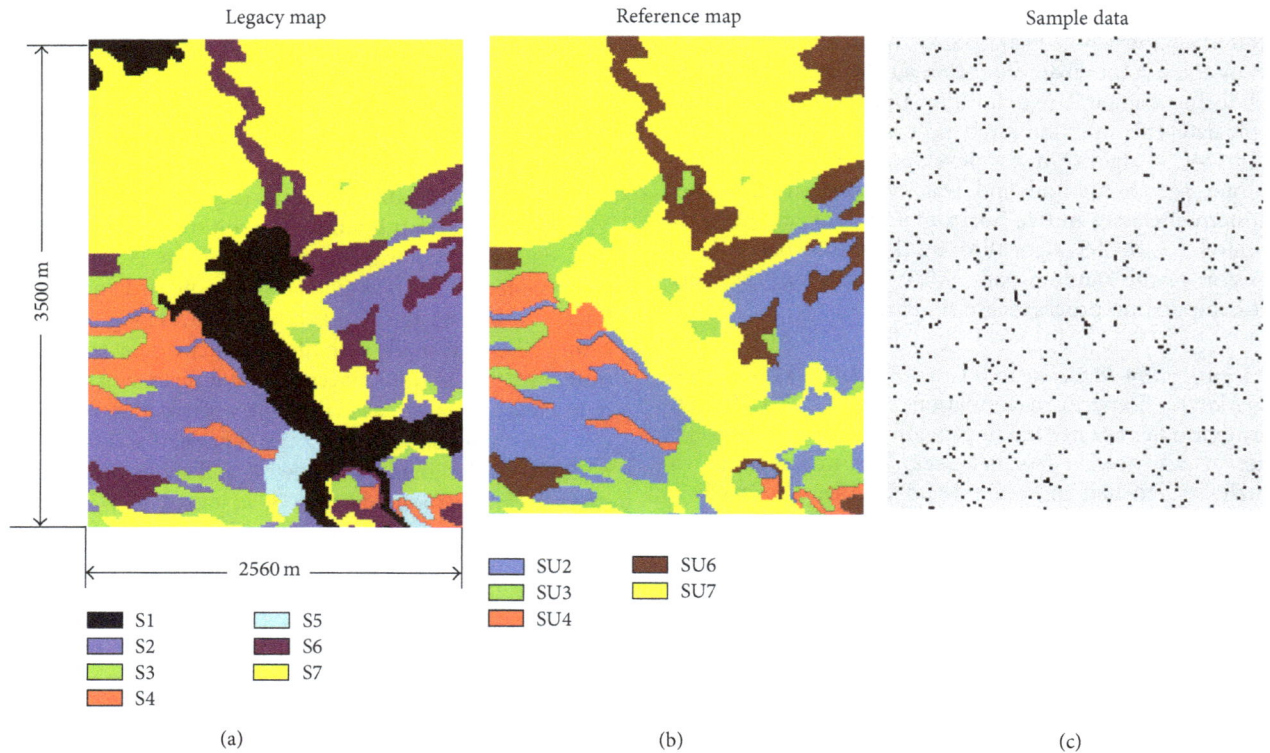

FIGURE 3: The data for categorical soil map update by Markov chain cosimulation: (a) the legacy soil map; (b) the reference soil map, representing the current distribution of soil series; (c) the sample data set (646 points), including field survey data and pseudosample data directly extracted from the unchanged areas in the legacy soil map. Previous soil series: S1, S2, S3, S4, S5, S6, and S7. Updated soil series: SU2, SU3, SU4, SU6, and SU7. SU2 = S2, SU3 = S3 + S5, SU4 = S4, SU6 = S6 + part of S7, and SU7 = S7 + S1 + part of S6.

have very different sills, related to their tail class proportions. Anisotropy was not considered because no identifiable anisotropic direction can characterize all soil types in the whole area while partial anisotropy is difficult to account for. The CTPM from the sample data set to the legacy soil map data is provided in Table 1. The numbers of columns and rows in the CTPM can be different and the classes in columns and rows need not have the same physical meanings, as they represent two different categorical variables, respectively. But for each row the transition probability values still sum to unity. Such a CTPM was used to express the cross-correlations between sample data and the legacy map. The sample data set has five soil types while the legacy soil map has seven soil types; thus, they have five types in common. These five soil types show strong cross-field autocorrelations, and two of them have no changes (i.e., cross-field transition probabilities are 1.0).

The search radius chosen is 30 pixels (i.e., 600 m). One hundred realizations were generated for the cosimulation conditioned on both the sample data and the legacy soil map using Co-MCSS, and occurrence probability maps were estimated from those realizations. The optimal prediction map was obtained from maximum occurrence probabilities. For the purpose of comparison, the same was done without conditioning on the legacy soil map using MCSS. The PCC (percentage of correctly classified locations) values were estimated for the optimal prediction map and realization

maps against the reference soil map (sample data being excluded) to verify the simulation accuracies.

3.2. Results of Cosimulation. The updated categorical soil maps include the optimal prediction map, a series of simulated realization maps, and occurrence probability maps. But the most important should be the optimal prediction map generated from maximum occurrence probabilities that reflect the best predictions for a chosen method and available data. The optimal prediction map of the soil series and the corresponding maximum occurrence probability map (Figure 5) were estimated from simulated realizations generated by Co-MCSS, conditioned on both the sample data and the legacy soil map. The maximum occurrence probability map reflects the uncertainty of the optimal prediction map against the conditioning data. Comparing with the legacy soil map and the reference map (Figure 3) shows that the unchanged S2 and S4 were exactly reproduced as SU2 and SU4, respectively, and that the S3, which was merged with a minor soil series (S5) without other changes, was also exactly reproduced as SU3 in the optimal prediction map (Figure 5(a)). However, the S6 and S7, which changed into each other in some areas, were only approximately captured (as SU6 and SU7, resp.) with apparent uncertainty (see shallow gray areas in Figure 5(b)). The uncertainty mainly occurred at the boundary zones between these two soil series.

TABLE 1: Cross-field transition probability matrix from sample data (5 soil series) to colocated data in the legacy soil map (7 soil series).

Data	Soil series[†]	Legacy soil map						
		S1	S2	S3	S4	S5	S6	S7
	SU2	.0000	1.0000	.0000	.0000	.0000	.0000	.0000
	SU3	.0000	.0000	.9011	.0000	.0989	.0000	.0000
Sample data	SU4	.0000	.0000	.0000	1.0000	.0000	.0000	.0000
	SU6	.0000	.0000	.0000	.0000	.0000	.8143	.1857
	SU7	.2169	.0000	.0000	.0000	.0000	.0271	.7560

[†]S1 is a soil series in the legacy soil map. SU2 is a soil series in the updated soil map.

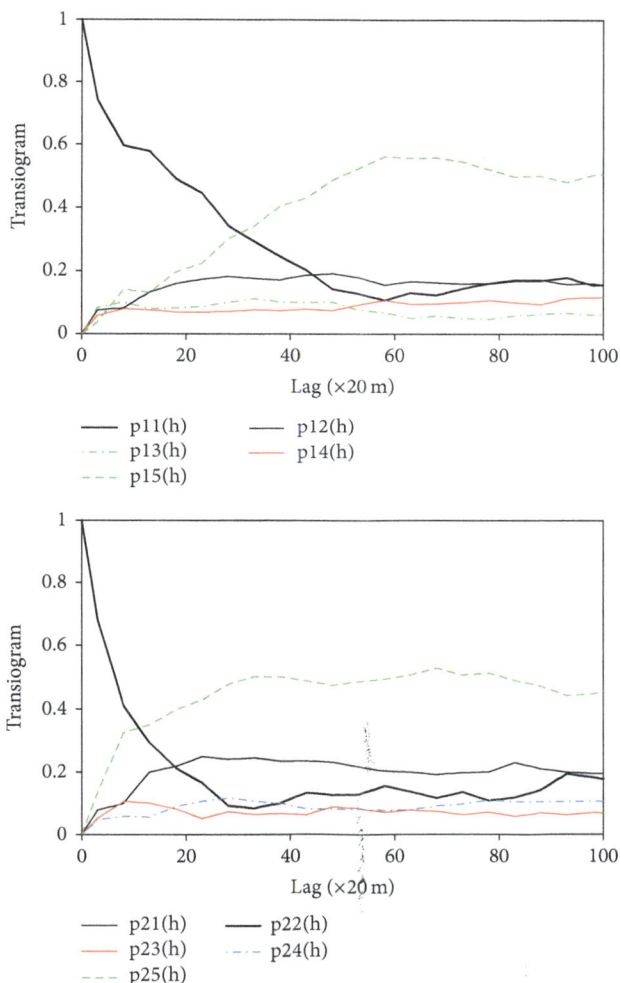

FIGURE 4: Two subsets of transiogram models interpolated from experimental transiograms estimated from the sample data. The numbers in transiogram labels (1 to 5) refer to the five updated soil series (i.e., SU2, SU3, SU4, SU6, and SU7), respectively.

Those areas of these two soil series that are located far away from each other were also well reproduced. Although soil type changes were confirmed by sample data only in two small areas (i.e., the top-left corner and the bottom-middle east) for S6 and S7, such changes caused the uncertainty of these two soil types in other areas in the updated map. This is reasonable because if a soil series is found to have changed at some places,

it is quite possible that it may also have changed at other places, even if the changes at other places were not verified by the new survey. The changed areas of S6 and S7 (i.e., the top-left corner and the bottom-middle east) were well captured in the optimal prediction map (Figures 3(a) and 3(b)). The merging of S1 into S7 only increased the total area of SU7 and did not affect its uncertainty caused by the transformation in some areas between S6 and S7.

Similar to hand-delineated maps, optimal prediction maps of categorical spatial variables normally also have an omission effect: minor classes are underrepresented because of their lower occurrence probabilities at most unsampled locations and major classes are consequently overrepresented [20, 38]. This situation also occurred on the predicted soil maps using the MLR method [3]. Because of the contribution of the legacy soil map and the lack of apparent minor classes, this effect is visually not obvious in the optimal prediction map by Co-MCSS compared with the simulated realization maps (Figure 6), which normally do not have such an effect.

The simulated realization maps (Figure 6) and occurrence probability maps of single soil series (Figure 7) further verify the judgments based on the results provided in Figure 5. Between those two realizations (Figure 6), soil series SU2, SU3, and SU4 do not show differences in pattern details, but some differences for SU6 and SU7 do exist. In Figure 7, occurrence probability maps of SU2, SU3, and SU4 are simply binary maps (i.e., 0 and 1), meaning that they are simply inherited from the legacy map because no changes other than taxonomic adjustments were confirmed by sample data. This does not mean that these soil series in the updated soil map are correct. They just made no changes against the legacy map. Uncertainties in the occurrence probability maps of SU6 and SU7 are clear but mainly appear along the boundaries between them.

3.3. Comparison with MCSS. To verify the improvement and advantages of Co-MCSS over MCSS, which cannot incorporate auxiliary information, we also used the MCSS method to conduct a simulation conditioned on the same sample data. Comparing optimal prediction and maximum occurrence probability maps (Figure 8) generated by MCSS to those generated by Co-MCSS (see Figure 5) clearly indicates distinct differences: (1) unlike MCSS, Co-MCSS can capture pattern details, particularly linear features; and (2) MCSS generated much more uncertainty than did Co-MCSS. Thus, the contribution of the legacy soil map to the accuracy of the simulated results by Co-MCSS is huge due to the assumed

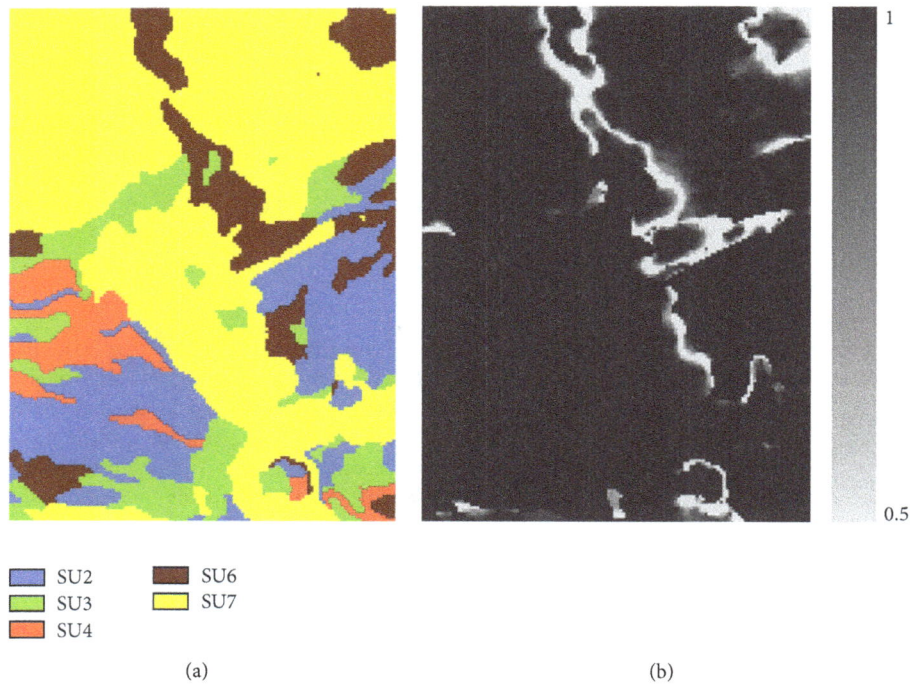

SU2		SU6
SU3		SU7
SU4		

(a)   (b)

FIGURE 5: The optimal prediction map (a) and the maximum occurrence probability map (b) of updated soil series conditioned on sample data and the legacy soil map using the Co-MCSS method.

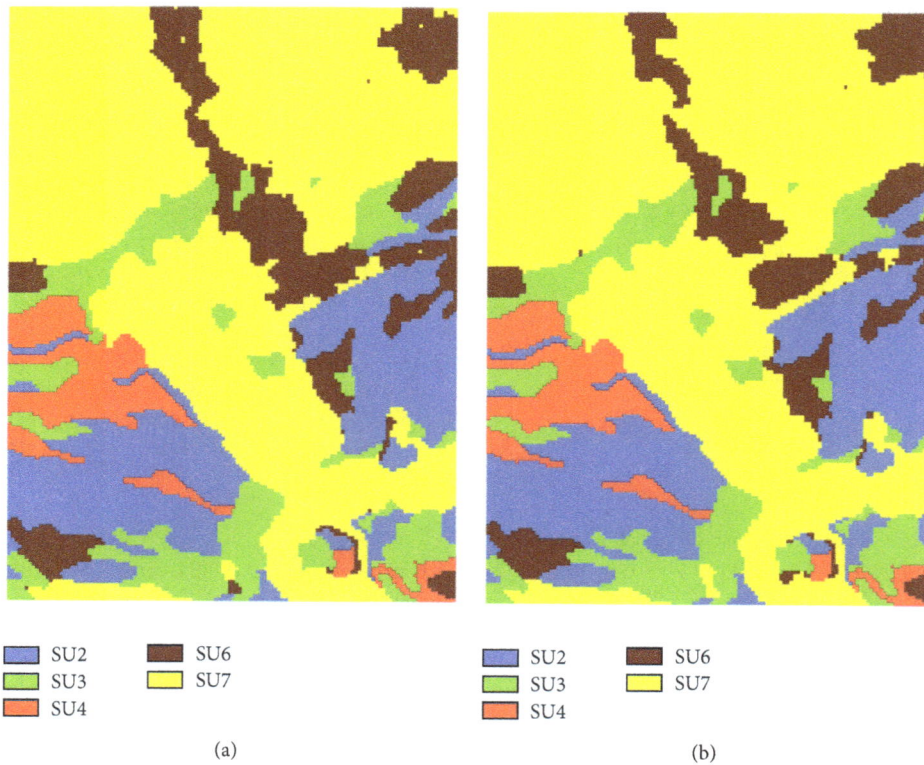

SU2		SU6		SU2		SU6
SU3		SU7		SU3		SU7
SU4				SU4		

(a)   (b)

FIGURE 6: Two simulated realization maps of updated soil series conditioned on sample data and the legacy soil map using the Co-MCSS method.

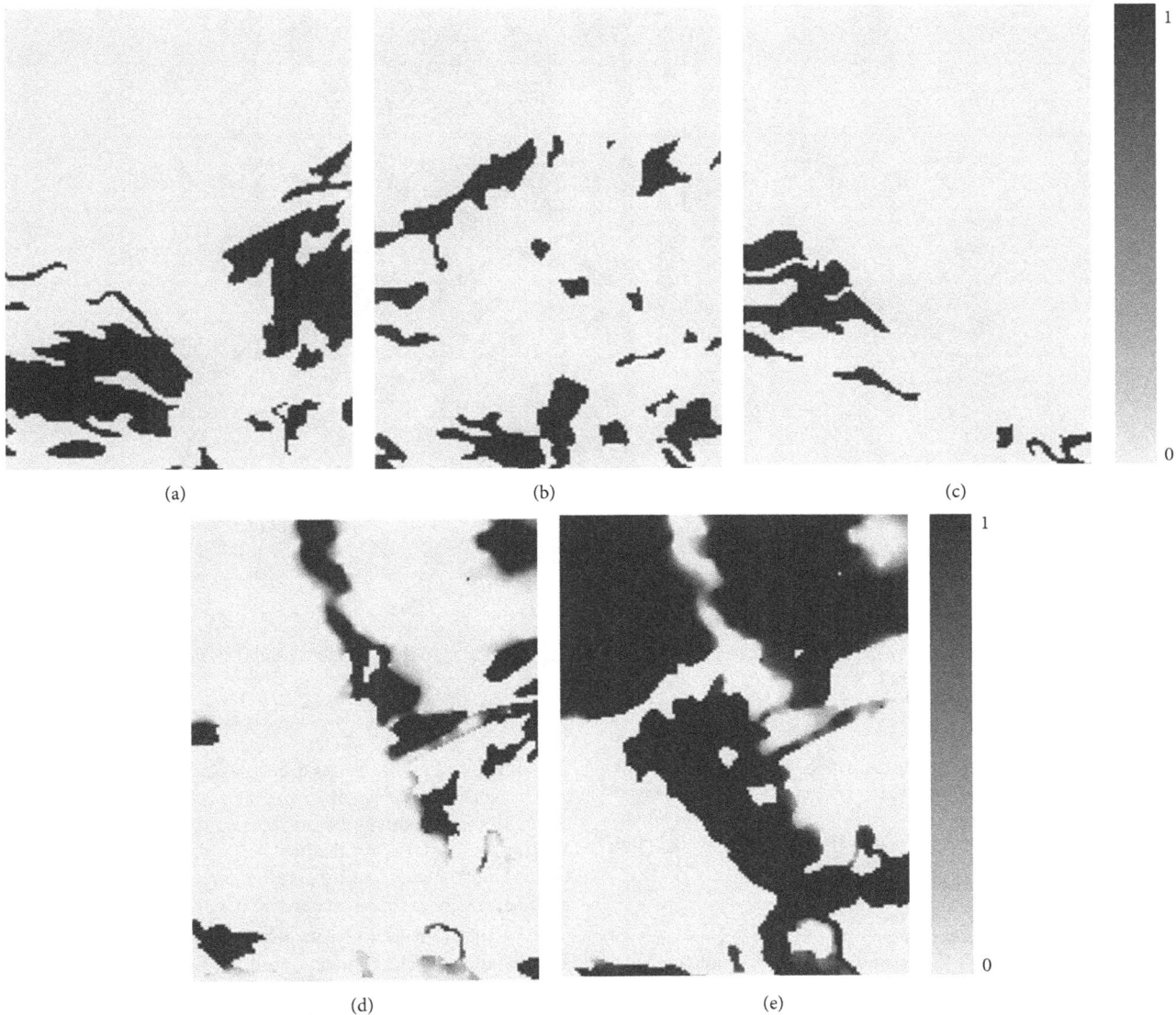

FIGURE 7: Occurrence probability maps of updated single soil series conditioned on the sample data and the legacy soil map using the Co-MCSS method. (a) SU2; (b) SU3; (c) SU4; (d) SU6; and (e) SU7.

good quality of the legacy map. The simulated realization maps and occurrence probability maps of single soil series by MCSS are omitted (one may see similar simulations in [20]) but corroborated the same conclusions.

The PCC value represents the accuracy of a classified map compared to reference data. Using the reference map modified from the legacy soil map (Figure 3(c)), we calculated the PCC values of the optimal prediction maps and the averaged PCC values of the simulated realization maps generated by Co-MCSS and MCSS. The results (Table 2) show that the optimal prediction map and the simulated realization maps by Co-MCSS have a substantive improvement in simulation accuracy over those by MCSS. For the optimal prediction map, the improvement is about 16% in absolute values and about 19% in relative values, whereas for the realization maps, the improvement is about 18% in absolute values and about 23% in relative values. These accuracy improvements are

attributed to the legacy soil map as auxiliary information. The accuracies of the optimal prediction map and the realization maps by Co-MCSS are generally above 97%. Such a high accuracy should be related to the relatively small soil changes. If the soil series in the study area have large changes since last survey or if the legacy soil map is of lower quality with many errors confirmed, the legacy map cannot contribute so much in improving simulation accuracy. Of course, unidentified errors in the legacy map will be brought into the updated soil map invisibly.

Sample data directly extracted from the unchanged areas of the legacy soil map are not real survey data for map updating. They were used for fairly estimating the transiogram models and the cross-field transition probability parameters and also for conditioning the simulations. This study does not show that the conditioning of the extracted pseudosample data for unchanged soil series (including merged unchanged

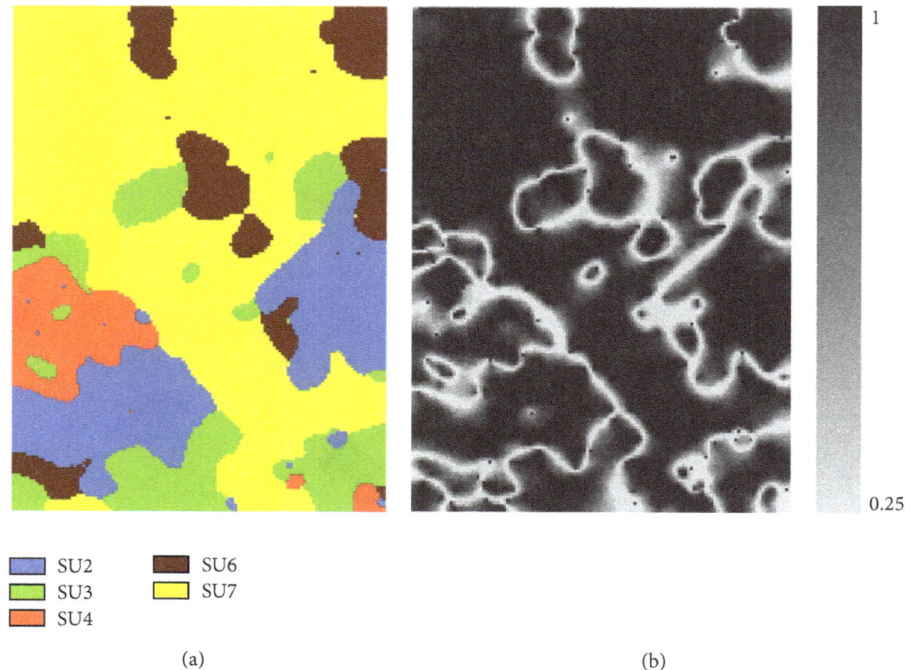

SU2 SU6
SU3 SU7
SU4

(a)   (b)

FIGURE 8: The optimal prediction map (a) and the maximum occurrence probability map (b) of updated soil series conditioned only on the sample data using the MCSS method.

TABLE 2: Percentages of correctly classified locations (PCCs) of optimal prediction maps and simulated realizations (averaged from 100 realizations) generated by Co-MCSS and MCSS. PCCs (%) are estimated relative to the reference soil map with sample data being excluded.

Item	Accuracy	
	Optimal prediction map	Realization maps
MCSS	82.50	79.32
Co-MCSS	98.25	97.23
Absolute improvement[†]	15.75	17.91
Relative improvement[‡]	19.09	22.58

[†]Absolute improvement = PCC of Co-MCSS − PCC of MCSS. [‡]Relative improvement = absolute improvement/PCC of MCSS × 100.

soil series) in simulations is necessary, as these unchanged soil series are simply reproduced from the legacy soil map. But if a soil type change is confirmed at a place by a survey sample datum, pseudosample data should not be extracted nearby unless they are surely correct because pseudosample data confirm the unchanged status of soil series at their locations.

4. Conclusions

Updating categorical soil maps is necessary for many reasons, such as being outdated or of low quality. We assumed that the most recent legacy soil maps may need only limited corrections due to modest natural and anthropogenic soil changes occurring during the intervening time period. As a result,

updates to the legacy maps are necessary in only the changed and mistakenly mapped areas. In essence, we assume that the legacy soil maps were outdated but of good quality. Such a situation may be applicable to the soil map update of the United States, where quite detailed large-scale categorical soil maps exist for each county in most states.

We introduced the random-path Co-MCSS algorithm, which extended the random-path MCSS algorithm, for revising categorical soil maps and applied it to a case study of synthetic data that involved the revision of a legacy soil series map using limited survey data. Simulated results show that (1) Co-MCSS can greatly improve simulation accuracy of soil types via the contribution of legacy soil maps, and (2) the accuracy of the optimally predicted soil map by Co-MCSS is better than that by MCSS, at least in a situation characterized by the use of limited survey sample data. Co-MCSS demonstrated the following merits: (1) if a soil type has no changes confirmed in an update survey or if it is decided to be reclassified into another type that is deemed to have no change, it will be simply reproduced in updated soil maps; (2) if a soil type has changes in some areas (e.g., an update survey confirmed the changes or map examination found previous wrong classification), it will be simulated with uncertainty; (3) if a soil type has no change in an area but has changes in other distant areas, it still can be captured with little changes in the area. The occurrence probability maps estimated from the simulated realizations reflect only the uncertainty verified by new survey sample data and do not reflect the uncertainty contained in the legacy soil map but unverified by sample data. In general, we conclude that Co-MCSS may provide a practical spatial statistical tool for revising categorical soil maps.

Finally, other related data, such as land cover/land use and discretized DEM-derived data (e.g., elevation), are often correlated with the spatial distributions of soil series and may also be incorporated as auxiliary information to improve the accuracy of soil mapping, especially when legacy soil maps are of low quality or unavailable and the survey sample data are very sparse. In this study, because we assumed that legacy soil maps were available and of high quality and only limited soil changes occurred, other auxiliary variables were not considered.

References

[1] P. Campling, A. Gobin, and J. Feyen, "Logistic modeling to spatially predict the probability of soil drainage classes," *Soil Science Society of America Journal*, vol. 66, no. 4, pp. 1390–1401, 2002.

[2] N. Bailey, T. Clements, J. T. Lee, and S. Thompson, "Modelling soil series data to facilitate targeted habitat restoration: a polytomous logistic regression approach," *Journal of Environmental Management*, vol. 67, no. 4, pp. 395–407, 2003.

[3] T. Hengl, N. Toomanian, H. I. Reuter, and M. J. Malakouti, "Methods to interpolate soil categorical variables from profile observations: lessons from Iran," *Geoderma*, vol. 140, no. 4, pp. 417–427, 2007.

[4] M. Debella-Gilo and B. Etzelmüller, "Spatial prediction of soil classes using digital terrain analysis and multinomial logistic regression modeling integrated in GIS: examples from Vestfold County, Norway," *Catena*, vol. 77, no. 1, pp. 8–18, 2009.

[5] C. J. Moran and E. N. Bui, "Spatial data mining for enhanced soil map modelling," *International Journal of Geographical Information Science*, vol. 16, no. 6, pp. 533–549, 2002.

[6] P. Scull, J. Franklin, and O. A. Chadwick, "The application of classification tree analysis to soil type prediction in a desert landscape," *Ecological Modelling*, vol. 181, no. 1, pp. 1–15, 2005.

[7] I. O. A. Odeh, A. B. McBratney, and D. J. Chittleborough, "Soil pattern recognition with fuzzy-c-means: application to classification and soil-landform interrelationships," *Soil Science Society of America Journal*, vol. 56, no. 2, pp. 505–516, 1992.

[8] P. A. Burrough, R. A. Macmillan, and W. Van Deursen, "Fuzzy classification methods for determining land suitability from soil profile observations and topography," *Journal of Soil Science*, vol. 43, no. 2, pp. 193–210, 1992.

[9] R. A. MacMillan, D. E. Moon, and R. A. Coupé, "Automated predictive ecological mapping in a Forest Region of B.C., Canada, 2001–2005," *Geoderma*, vol. 140, no. 4, pp. 353–373, 2007.

[10] M. F. P. Bierkens and P. A. Burrough, "The indicator approach to categorical soil data. II. Application to mapping and land use suitability analysis," *Journal of Soil Science*, vol. 44, no. 2, pp. 369–381, 1993.

[11] D. J. Brus, P. Bogaert, and G. B. M. Heuvelink, "Bayesian Maximum Entropy prediction of soil categories using a traditional soil map as soft information," *European Journal of Soil Science*, vol. 59, no. 2, pp. 166–177, 2008.

[12] C. Zhang and W. Li, "Regional-scale modelling of the spatial distribution of surface and subsurface textural classes in alluvial soils using Markov chain geostatistics," *Soil Use and Management*, vol. 24, no. 3, pp. 263–272, 2008.

[13] P. Goovaerts, "Stochastic simulation of categorical variables using a classification algorithm and simulated annealing," *Mathematical Geology*, vol. 28, no. 7, pp. 909–921, 1996.

[14] V. De Oliveira, "Bayesian prediction of clipped Gaussian random fields," *Computational Statistics and Data Analysis*, vol. 34, no. 3, pp. 299–314, 2000.

[15] J. Connolly, N. M. Holden, and S. M. Ward, "Mapping peatlands in Ireland using a rule-based methodology and digital data," *Soil Science Society of America Journal*, vol. 71, no. 2, pp. 492–499, 2007.

[16] S. J. Nield, J. L. Boettinger, and R. D. Ramsey, "Digitally mapping gypsic and natric soil areas using landsat ETM data," *Soil Science Society of America Journal*, vol. 71, no. 1, pp. 245–252, 2007.

[17] W. Li, "Markov chain random fields for estimation of categorical variables," *Mathematical Geology*, vol. 39, no. 3, pp. 321–335, 2007.

[18] W. Li, "Transiograms for characterizing spatial variability of soil classes," *Soil Science Society of America Journal*, vol. 71, no. 3, pp. 881–893, 2007.

[19] K. Abend, T. J. Harley, and L. N. Kanal, "Classification of binary random patterns," *IEEE Transactions on Information Theory*, vol. 11, no. 4, pp. 538–544, 1965.

[20] W. Li and C. Zhang, "A random-path markov chain algorithm for simulating categorical soil variables from random point samples," *Soil Science Society of America Journal*, vol. 71, no. 3, pp. 656–668, 2007.

[21] B. D. Hudson, "The soil survey as paradigm-based science," *Soil Science Society of America Journal*, vol. 56, no. 3, pp. 836–841, 1992.

[22] R. Kerry, P. Goovaerts, B. G. Rawlins, and B. P. Marchant, "Disaggregation of legacy soil data using area to point kriging for mapping soil organic carbon at the regional scale," *Geoderma*, vol. 170, pp. 347–358, 2012.

[23] A. J. Gray, J. W. Kay, and D. M. Titterington, "Empirical study of the simulation of various models used for images," *IEEE Transactions on Pattern Analysis and Machine Intelligence*, vol. 16, no. 5, pp. 507–513, 1994.

[24] W. Li and C. Zhang, "A single-chain-based multidimensional Markov chain model for subsurface characterization," *Environmental and Ecological Statistics*, vol. 15, no. 2, pp. 157–174, 2008.

[25] W. Li and C. Zhang, "Some further clarification on Markov chain random fields and transiograms," *International Journal of Geographical Information Science*, vol. 27, no. 3, pp. 423–430, 2013.

[26] J. Besag, "On the statistical analysis of dirty pictures (with discussions)," *Journal of the Royal Statistical Society B*, vol. 48, pp. 259–302, 1986.

[27] D. K. Pickard, "Unilateral Markov fields," *Advances in Applied Probability*, vol. 12, no. 3, pp. 655–671, 1980.

[28] J. Haslett, "Maximum likelihood discriminant analysis on the plane using a Markovian model of spatial context," *Pattern Recognition*, vol. 18, no. 3-4, pp. 287–296, 1985.

[29] A. Rosholm, *Statistical methods for segmentation and classification of images [Ph.D. thesis]*, Technical University of Denmark, Lyngby, Denmark, 1997.

[30] W. Li and C. Zhang, "A generalized Markov chain approach for conditional simulation of categorical variables from grid samples," *Transactions in GIS*, vol. 10, no. 4, pp. 651–669, 2006.

[31] W. Schwarzacher, "The use of Markov chains in the study of sedimentary cycles," *Journal of the International Association for Mathematical Geology*, vol. 1, no. 1, pp. 17–39, 1969.

[32] J. Lou, "Transition probability approach to statistical analysis of spatial qualitative variables in geology," in *Geologic Modeling and Mapping*, A. Forster and D. F. Merriam, Eds., pp. 281–299, Plenum Press, New York, NY, USA, 1996.

[33] G. E. Fogg, "Transition probability-based indicator geostatistics," *Mathematical Geology*, vol. 28, no. 4, pp. 453–477, 1996.

[34] S. F. Carle and G. E. Fogg, "Modeling spatial variability with one and multidimensional continuous-lag Markov chains," *Mathematical Geology*, vol. 29, no. 7, pp. 891–918, 1997.

[35] R. W. Ritzi, "Behavior of indicator variograms and transition probabilities in relation to the variance in lengths of hydrofacies," *Water Resources Research*, vol. 36, no. 11, pp. 3375–3381, 2000.

[36] W. Li and C. Zhang, "Linear interpolation and joint model fitting of experimental transiograms for Markov chain simulation of categorical spatial variables," *International Journal of Geographical Information Science*, vol. 24, no. 6, pp. 821–839, 2010.

[37] Soil Conservation Service, *Soil Survey-Iowa County, Wisconsin*, U.S. Government Printing Office, Washington, DC, USA, 1962.

[38] A. Soares, "Geostatistical estimation of multi-phase structures," *Mathematical Geology*, vol. 24, no. 2, pp. 149–160, 1992.

12

Simulation of Soil Temperature Dynamics with Models Using Different Concepts

Renáta Sándor and Nándor Fodor

Centre for Agricultural Research, Hungarian Academy of Sciences, 2462 Martonvásár, Hungary

Correspondence should be addressed to Nándor Fodor, fodornandor@rissac.hu

Academic Editors: A. Ferrante, R. M. Mian, T. Rennert, and J. Viers

This paper presents two soil temperature models with empirical and mechanistic concepts. At the test site (calcaric arenosol), meteorological parameters as well as soil moisture content and temperature at 5 different depths were measured in an experiment with 8 parcels realizing the combinations of the fertilized, nonfertilized, irrigated, nonirrigated treatments in two replicates. Leaf area dynamics was also monitored. Soil temperature was calculated with the original and a modified version of CERES as well as with the HYDRUS-1D model. The simulated soil temperature values were compared to the observed ones. The vegetation reduced both the average soil temperature and its diurnal amplitude; therefore, considering the leaf area dynamics is important in modeling. The models underestimated the actual soil temperature and overestimated the temperature oscillation within the winter period. All models failed to account for the insulation effect of snow cover. The modified CERES provided explicitly more accurate soil temperature values than the original one. Though HYDRUS-1D provided more accurate soil temperature estimations, its superiority to CERES is not unequivocal as it requires more detailed inputs.

1. Introduction

Soil temperature (T_{soil}) is one of the most important variables of the soil. It can significantly influence seed germination [1], plant growth [2], uptake of nutrients [3], soil respiration [4, 5], soil evaporation [6], and the intensity of physical [7], chemical [8, 9], and microbiological processes [10, 11] in the soil.

Solar radiation and air temperature are the main driving forces determining the soil temperature which is influenced by numerous other factors such as precipitation, soil texture, and moisture content as well as the type of surface cover (plant canopy, crop residue, snow, etc.) [12]. Yearly, monthly, or daily means of soil temperature measurements are frequently reported, but the variability of T_{soil} is similarly important [13]. In spite of this, at many meteorological stations only aboveground variables (e.g., air temperature) are observed, or the soil temperature sensors are installed at the station (close to the mast that supports other sensors and the data logger) and not at the plots of the experimental site which could make the measured data unrepresentative.

If soil temperature is not measured, several methods are available to calculate it using meteorological variables and other parameters. As the simple air-temperature-based methods (e.g., [14]) provided inadequate T_{soil} data, an improved formula was introduced that uses precipitation data [15] as well. There are three types of soil temperature models [16]: (1) empirical models that are based on statistical relationships between soil temperature at some depth and climatological and soil variables (e.g., [17]); (2) mechanistic models that focus on physical processes (radiative energy balance as well as sensible, latent, and ground-conductive heat fluxes) to predict the upper boundary temperature and estimate the temperature of deeper layers with Fourier's equation (e.g., [18]); (3) mixed empirical and mechanistic models that calculate the temperature of different soil layers based on physical principles of heat flow, but the boundary temperature at the soil surface must be provided empirically (e.g., [19]).

Since LAI (leaf area index) and soil water balance strongly influence the, soil temperature dynamics, soil temperature calculating methods function more precisely when those are integrated into hydrological [20, 21] or crop simulation models [22, 23]. The primary purpose of these models is to describe the processes of the very complex atmosphere-soil-plant system, including human activities, using mathematical tools and to simulate them with the help of computers.

The objectives of this paper are as follows: (1) presenting the effect of LAI on soil temperature; (2) comparison of an empirical and a mechanistic soil temperature model using measured data; (3) enhancing the performance of the empirical model.

2. Materials and Methods

Data of the agrometeorological station at Őrbottyán, Hungary were used in the study. The arenosol of the experiment site has the following characteristics: bulk density: 1.67 gcm^{-1}; organic matter content: 0.91%; CaCO$_3$ content: 5.1%; sand fraction: 86.3%; silt fraction: 8.3%; clay fraction: 5.4% [24]. Saturated hydraulic conductivity (Ks) and characteristic points of the soil water retention curve (SWRC) were measured with Guelph permeameter [25] and Eijkelkamp sand/kaolin box apparatus, respectively. pF-measurements were carried out with 100 cm^3 undisturbed samples taken in 5 replicates. The van Genuchten parameters [26] of the SWRC were determined with Soilarium [27]. The above parameters characterize the 0–20 cm layer of the soil. In the 20–60 cm layer, the parameter values are practically the same except for the organic matter content which gradually decreases to zero with depth.

Soil temperature sensors (thermistor type, ±0.5°C accuracy, 0.1°C resolution) were installed at 5 different depths (5, 10, 20, 40, and 60 cm) at the centre of each 10 × 15 m test plot of an experiment with 8 parcels realizing the combinations of the fertilized, nonfertilized, irrigated, and nonirrigated treatments in two replicates. Temperature data were recorded every 15 minutes. A meteorological station was installed next

to the experiment where precipitation, relative humidity, wind velocity and direction, global radiation, and air temperature were measured every 5 minutes in 2010 and 2011. In these two years, maize was grown at the site. LAI of every parcel in three-week intervals were determined by direct measurements. Three plants were cut out randomly at every observation time, and the area of the leaves was calculated with Montgomery's method [28].

Site-specific measured data were used as inputs for the CERES-Maize [29] crop simulation model as well as for the HYDRUS-1D [21] hydrological model. CERES is a daily-step deterministic model that simulates plant (assimilation, biomass accumulation, leaf area, dynamics and root growth) as well as soil (water, temperature and nutrient dynamics) processes using empirical equations. HYDRUS-1D is designed for simulating one-dimensional variably saturated water flow, heat movement, and the transport of solutes in the soil. It numerically solves the Richards equation for saturated-unsaturated water flow (including a sink term to account for water uptake by plant roots) and advection-dispersion type equations for heat and solute transport using Galerkin-type linear finite element schemes. Several studies proved the efficiency of both models [30, 31].

The soil temperature calculation module of CERES belongs to the empirical model group. When calculating the actual temperature (T_{soil}^i) at a given depth (x), this model takes into account that the upper soil layers absorb energy, and the heat needs time to reach the lower layers as in (1). The effect of the energy reaching the soil surface appears delayed and decreased in the lower soil layers. The extent of the delay and the decrease is a function of the actual average moisture content (Θ_{avg}^i) and the average bulk density of the topsoil (BD$_{avg}$). The model assumes a sinusoidal annual course of the soil surface temperature that is modified by an additive term of a five-day moving average of a factor described by (2) as follows:

$$T_{soil}^i(x) = T_{avg} + \overbrace{\left(\frac{T_{amp} \cdot \cos\left(0.0174 \cdot (i - I) + x \cdot f_1\left(\Theta_{avg}^i, BD_{avg}\right)\right)}{2} + DT^i\right)}^{Td} \cdot e^{x \cdot f_2(\Theta_{avg}^i, BD_{avg})}, \tag{1}$$

$$DT^i = \frac{\sum_{j=i-4}^{i}(1 - ALB) \cdot \left(T_{mean}^j + \left(T_{max}^j - T_{mean}^j\right) \cdot \sqrt{0.03 \cdot S_{rad}^j}\right)}{5} - T_{avg} - \frac{T_{amp} \cdot \cos(0.0174 \cdot (i - I))}{2}, \tag{2}$$

where i denotes the day of the year; I equals 200 on the northern hemisphere, while it is 20 on the southern hemisphere. ALB is the albedo of the surface, T_{avg} and T_{amp} denote the average temperature and the average temperature difference of the site. T_{mean}^i, T_{max}^i, and S_{rad}^i denote the daily mean and maximum temperature as well as the daily global radiation on the ith day of the year, respectively. The term Td

in (1) describes the delay of the effect of energy reaching the surface in deeper layers. The exponential term in (1) is more related to the heat capacity of the topsoil as it governs the decrease of the effect of the incoming energy at the surface in deeper layers.

The soil temperature calculation module of HYDRUS-1D belongs to the mechanistic model group. It numerically solves the convection-dispersion equation describing the one-dimensional heat transfer as follow: (3).

$$\frac{\partial C_s(\Theta)T}{\partial t} = \frac{\partial}{\partial x}\left(\lambda(\Theta)\frac{\partial T}{\partial x}\right) - C_w\frac{\partial qT}{\partial x} - C_w \cdot S \cdot T. \quad (3)$$

θ is the volumetric water content; λ denotes the apparent thermal conductivity of the soil. C_p and C_w are the volumetric heat capacities of the solid and the liquid phases, respectively. S is the sink term, and q is the Darcian fluid flux density. The apparent thermal conductivity can be expressed with (4) based on the work of de Marsily [32] as well as of Chung and Horton [33]:

$$\lambda(\Theta) = b_1 + b_2 \cdot \Theta + b_3 \cdot \Theta^{0.5} + \beta_t \cdot C_w|q|, \quad (4)$$

where β_t is the thermal dispersivity, while b_1, b_2, and b_3 are empirical parameters that can be estimated by using the sand, silt, and clay content of the soil.

Initial conditions of the water flow domain were measured with TDR (IMKO TRIME-FM3) tube access probe in 10 cm increments in three replicates. The initial soil temperature was set to uniform 10°C in the whole profile. HYDRUS-1D requires the setting of boundaries conditions for solving the flow equations. For water flow, atmospheric boundary conditions with surface runoff and free drainage were prescribed at the upper and lower boundary, respectively. For heat flow, the temperature values at both boundaries were provided in the model input file.

The parameters of CERES were calibrated by inverse modeling [34] so that the simulated LAI values would be in good agreement with the observed values (Figure 1). The obtained daily LAI values were used as inputs for HYDRUS-1D, as well. The measured and calculated soil temperature values were compared with simple graphical and statistical tools.

Originally, the user cannot alter the functions ((1) and (2)) of heat transport in CERES. Though it is an empirical model, it cannot be calibrated. In other words, it is postulated that it works for all soil types. A simple modification of one of its governing equations (1) is proposed to provide greater flexibility and the possibility of site-specific calibration for the following model: (5).

$$T^i_{soil}(x) = Td \cdot e^{c \cdot x \cdot f_2(\Theta^i_{avg}, BD_{avg})}. \quad (5)$$

By modifying the value of parameter c in (5), the amount of heat reaching the deeper soil layers could be adjusted. Though this parameter has no clear physical meaning, it most likely integrates the effect of soil organic matter, soil structure, and other implicit factors on soil-specific heat.

3. Results

Considerable differences in the leaf area indices were observed in different treatments of the experiment at the end of the canopy development in 2011 (Figure 2). Over 5°C, difference was observed in the daily maximum temperature at 5 cm depth, in the selected parcels. At 20 cm depth the observed difference was still explicitly greater (3.8°C) than

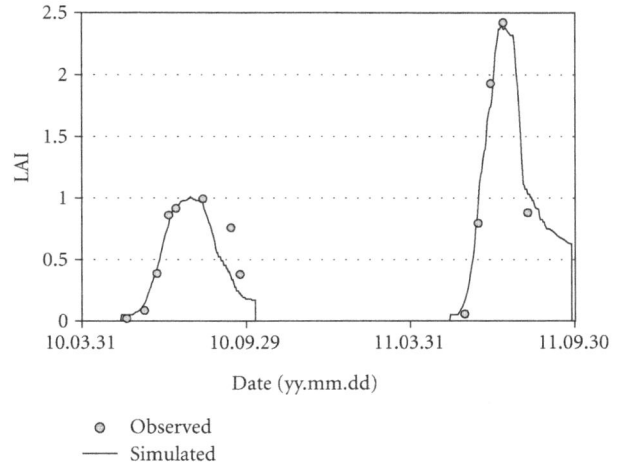

FIGURE 1: Observed and simulated leaf area index (LAI) values for the 1st parcel (fertilized, nonirrigated) of the experiment.

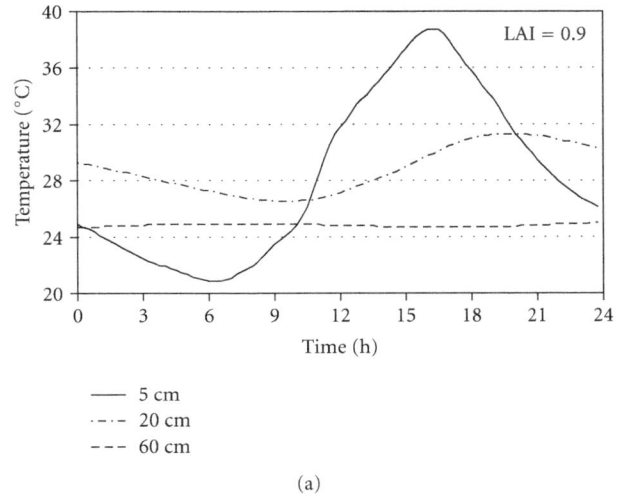

(a)

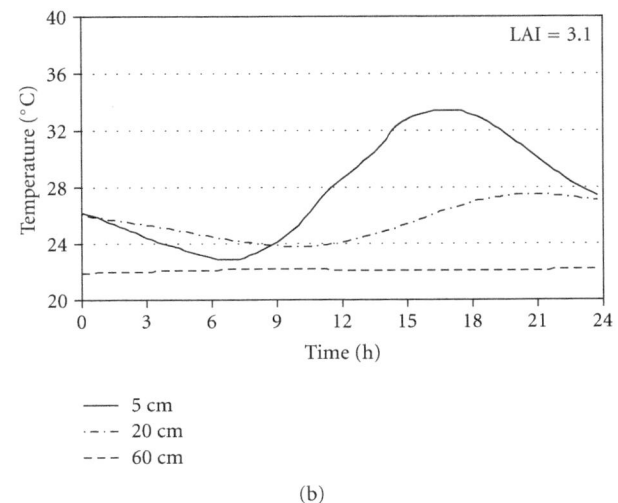

(b)

FIGURE 2: Effect of leaf are index (LAI) on the soil temperature dynamics at different depths in a non-fertilized (to the left) and in a fertilized parcel (to the right) on 14/07/2011 at Őrbottyán, Hungary.

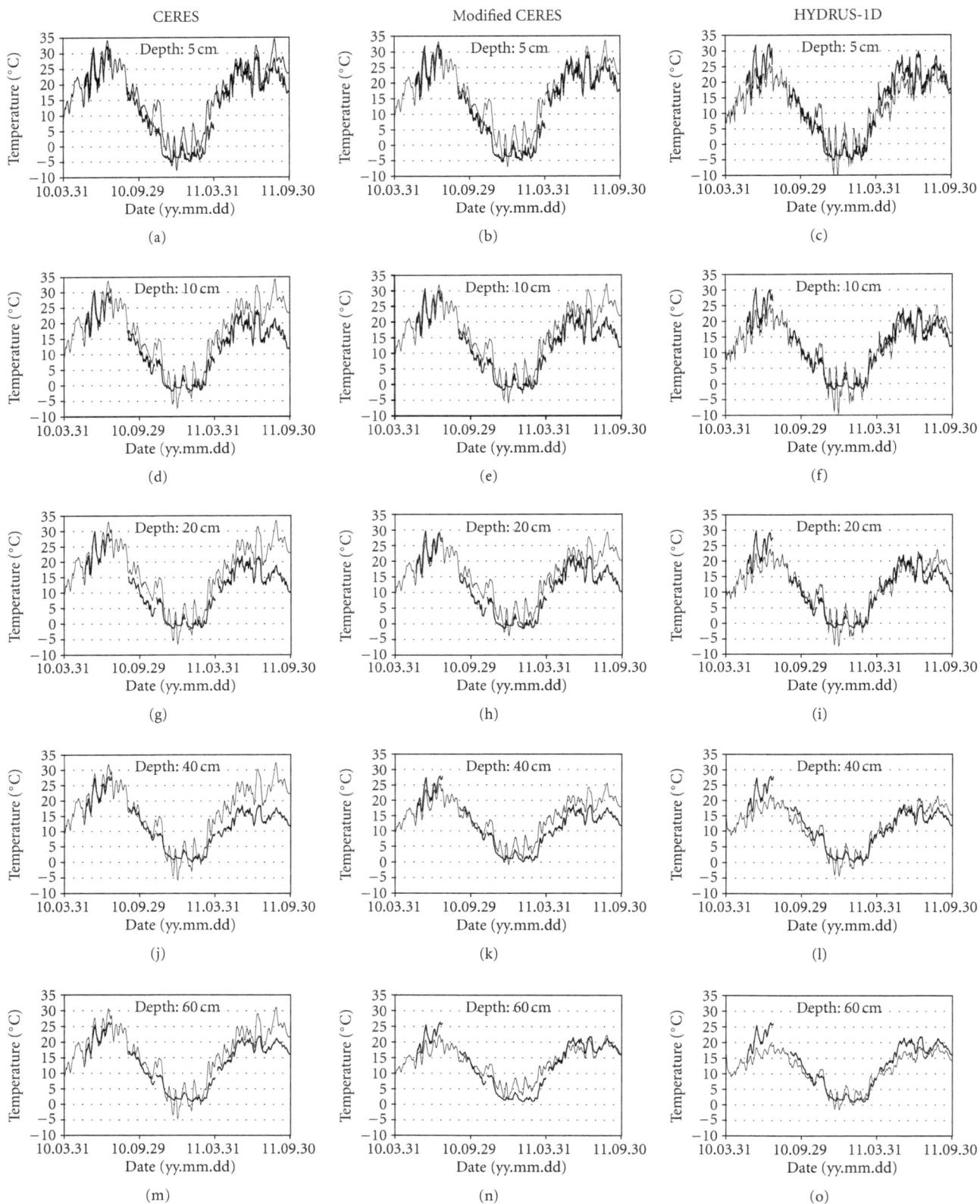

FIGURE 3: Series of measured and calculated soil temperature values at different depths, at Őrbottyán, Hungary. Thick lines—measured, thin lines—calculated.

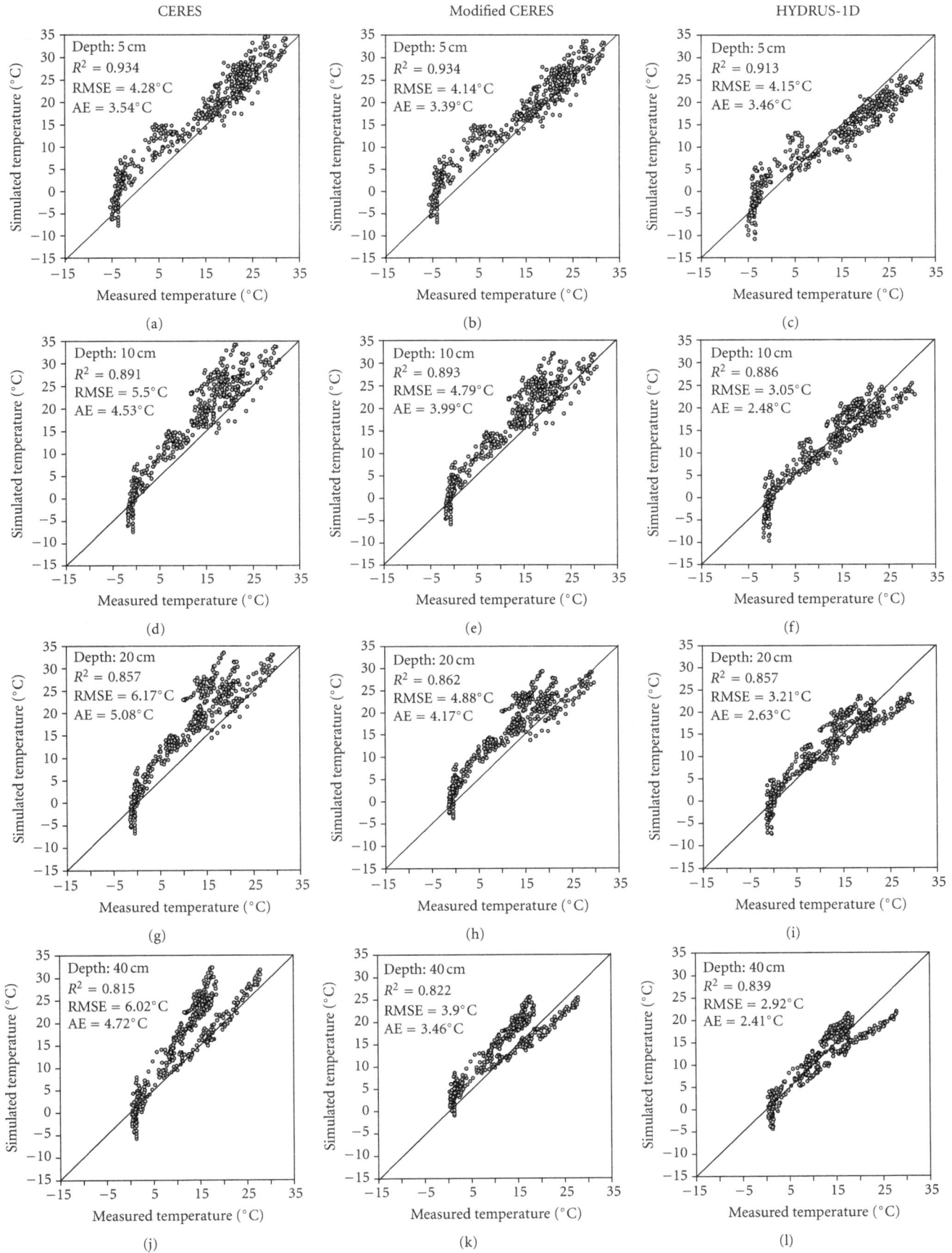

FIGURE 4: Continued.

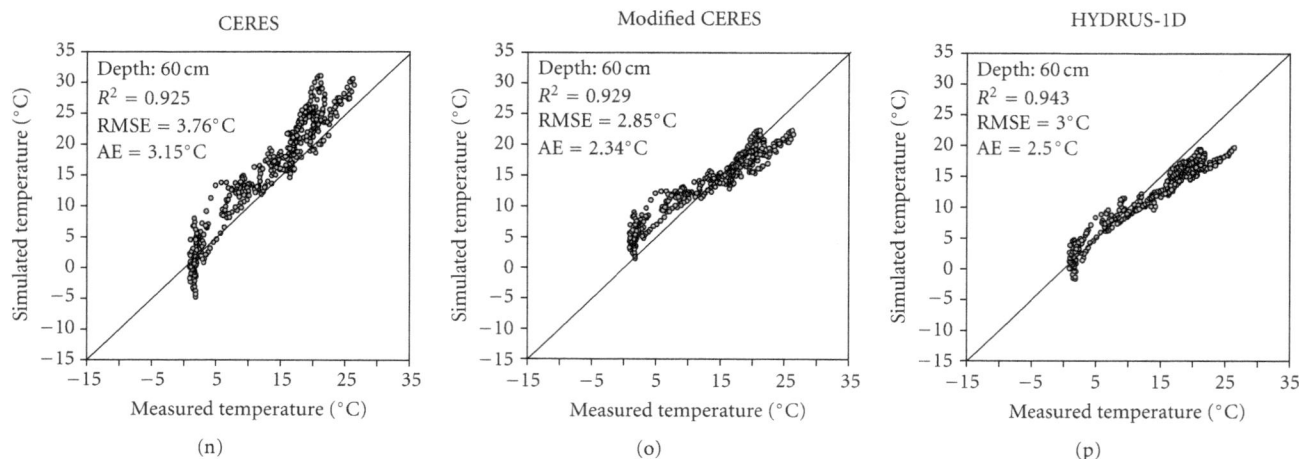

FIGURE 4: Comparison of the measured and calculated soil temperature values at different depths, at Őrbottyán, Hungary.

the measurement error. The peek temperature at this depth occurred 6 hours later (at sunset) than the peek of the air temperature.

The calibration of (5) resulted in $c = 4$ for the newly introduced parameter. When the series of measured and calculated temperature data were analyzed, it became obvious that the original CERES considerably underestimated the soil temperature of the deeper soil layers especially in the year 2011 (Figure 3). According to the error indicators (Figure 4), the modified CERES estimated better the T_{soil} than the original CERES especially in the deeper layers. The calculated values of HYDRUS-1D fit the best to the measured data (Figure 4). At the top and bottom layers, the modified CERES presented similar performance indicators than those of HYDRUS-1D.

HYDRUS-1D considerably underestimated the soil temperature of the upper layers in the frosty period when the average air temperature was $-7.1°C$ between 27/12/2010 and 04/01/2011, while the average observed and calculated T_{soil} was -3.4 and $-6.5°C$, respectively, at 5 cm depth.

During the winter period, all models overestimated the trends of temperature changes and resulted in more pronounced oscillations of soil temperature than that of the observed values.

4. Discussion

The comparison of the course of measured soil temperature at two parcels (different treatments of the experiment) on a summer day highlights the effect of canopy development status on soil temperature dynamics (Figure 2). The vegetation reduces both the average soil temperature and its diurnal amplitude; therefore, considering the LAI is important in modeling.

The measured soil temperature was not below $-5°C$ in the upper layers despite the fact that the average daily air temperature was permanently below $-7°C$ (some days the daily minimum was below $-15°C$) for longer periods in December 2010 and January 2011. During this period, the site was

covered with snow which reduces the effect of freezing since the depth of frost penetration is sensitive to the details of snow cover buildup [35]. As the models underestimated the actual soil temperature and overestimated the temperature oscillation within the winter period, it is obvious that all of them failed to account for the insulation effect of snow cover.

When the series of measured and calculated temperature data were analyzed, it became obvious that the original CERES considerably overestimated the summer soil temperature, while it underestimated T_{soil} during the winter. The minimum of the calculated soil temperature was below $-5°C$ at 60 cm depth, while the corresponding measured temperature was $+2°C$. The modified CERES let less heat to be transmitted to the deeper layers resulting in lower temperatures. The calculated average error of T_{soil} was reduced with almost 70% compared to the original CERES at 40 and 60 cm depths.

Though HYDRUS-1D gave the most accurate soil temperature estimations, it has to be noted that this model requires more detailed inputs than CERES does. For example, CERES calculates the leaf area development, while HYDRUS-1D requires LAI data to be provided as inputs. Furthermore, HYDRUS-1D requires some thermal properties of the soil (e.g., thermal conductivity) which was estimated from the available textural information in this study. This might explain the relative moderate performance of this model.

5. Conclusions

Two soil temperature models using different concepts were compared in this study. The simpler empirical model was enhanced by introducing an extra parameter in one of its governing equations. The experimental results clearly showed that crop cover significantly influences the soil temperature dynamics of the upper soil layers. Therefore, considering the LAI in model calculation is indispensable. The seasonal snow cover could significantly modify the freezing of soil as it builds up an isolating layer. The simulation of the effect of snow cover should be enhanced in the investigated models. The additional parameter proposed to modify the calculation of the CERES model provided greater flexibility and resulted in better performance, though the comparison was carried out only for a very sandy soil. Further studies should be conducted to investigate the capability of the modified CERES for simulating the heat transport of more structured soils with higher clay and organic matter contents. Though the more sophisticated HYDRUS-1D provided more accurate soil temperature estimations, its superiority to CERES is not unequivocal. The considerable input requirements of HYDRUS-1D may force the users to apply parameter estimation methods which most likely decrease the model accuracy.

Acknowledgment

The authors gratefully acknowledge the financial support of the Hungarian Scientific Research Foundation (K67672).

References

[1] G. Nabi and C. E. Mullins, "Soil temperature dependent growth of cotton seedlings before emergence," *Pedosphere*, vol. 18, no. 1, pp. 54–59, 2008.

[2] X. Liu and B. Huang, "Root physiological factors involved in cool-season grass response to high soil temperature," *Environmental and Experimental Botany*, vol. 53, no. 3, pp. 233–245, 2005.

[3] S. Dong, C. F. Scagel, L. Cheng, L. H. Fuchigami, and P. T. Rygiewicz, "Soil temperature and plant growth stage influence nitrogen uptake and amino acid concentration of apple during early spring growth," *Tree Physiology*, vol. 21, no. 8, pp. 541–547, 2001.

[4] R. D. Boone, K. J. Nadelhoffer, J. D. Canary, and J. P. Kaye, "Roots exert a strong influence on the temperature sensitivity of soil respiration," *Nature*, vol. 396, no. 6711, pp. 570–572, 1998.

[5] R. Inclán, D. De La Torre, M. Benito, and A. Rubio, "Soil CO_2 efflux in a mixed pine-oak forest in Valsaín (Central Spain)," *TheScientificWorldJournal*, vol. 7, no. 1, pp. 166–174, 2007.

[6] G. G. Katul and M. B. Parlange, "Estimation of bare soil evaporation using skin temperature measurements," *Journal of Hydrology*, vol. 132, no. 1–4, pp. 91–106, 1992.

[7] G. S. Rahi and R. D. Jensen, "Effect of temperature on soil-water diffusivity," *Geoderma*, vol. 14, no. 2, pp. 115–124, 1975.

[8] L. C. Paraíba, A. L. Cerdeira, E. F. Da Silva, J. S. Martins, and H. L. Da Costa Coutinho, "Evaluation of soil temperature effect on herbicide leaching potential into groundwater in the Brazilian Cerrado," *Chemosphere*, vol. 53, no. 9, pp. 1087–1095, 2003.

[9] X. Zhou, N. Persaud, D. P. Belesky, and R. B. Clark, "Significance of transients in soil temperature series," *Pedosphere*, vol. 17, no. 6, pp. 766–775, 2007.

[10] A. Olness, D. Lopez, D. Archer et al., "Factors affecting microbial formation of nitrate-nitrogen in soil and their effects on fertilizer nitrogen use efficiency," *TheScientificWorldJournal*, vol. 1, pp. 122–129, 2001.

[11] J. Pietikäinen, M. Pettersson, and E. Bååth, "Comparison of temperature effects on soil respiration and bacterial and fungal growth rates," *FEMS Microbiology Ecology*, vol. 52, no. 1, pp. 49–58, 2005.

[12] K. I. Paul, P. J. Polglase, P. J. Smethurst, A. M. O'Connell, C. J. Carlyle, and P. K. Khanna, "Soil temperature under forests: a simple model for predicting soil temperature under a range of forest types," *Agricultural and Forest Meteorology*, vol. 121, no. 3-4, pp. 167–182, 2004.

[13] B. Bond-Lamberty, C. Wang, and S. T. Gower, "Spatiotemporal measurement and modeling of stand-level boreal forest soil temperatures," *Agricultural and Forest Meteorology*, vol. 131, no. 1-2, pp. 27–40, 2005.

[14] V. R. Hasfurther and R. D. Burman, "Soil temperature modeling using air temperature as a driving mechanism," *Transactions of the American Society of Agricultural Engineers*, vol. 17, no. 1, pp. 78–81, 1974.

[15] D. Zheng, E. R. Hunt Jr., and S. W. Running, "A daily soil temperature model based on air temperature and precipitation for continental applications," *Climate Research*, vol. 2, no. 3, pp. 183–191, 1993.

[16] Y. Luo, R. S. Loomis, and T. C. Hsiao, "Simulation of soil temperature in crops," *Agricultural and Forest Meteorology*, vol. 61, no. 1-2, pp. 23–38, 1992.

[17] T. J. Toy, A. J. Kuhaida, and B. E. Munson, "The prediction of mean monthly soil temperature from mean monthly air temperature," *Soil Science*, vol. 126, pp. 96–104, 1978.

[18] C. H. M. van Bavel and D. I. Hillel, "Calculating potential and actual evaporation from a bare soil surface by simulation of concurrent flow of water and heat," *Agricultural Meteorology*, vol. 17, no. 6, pp. 453–476, 1976.

[19] P. J. Wierenga and C. T. De Wit, "Simulation of heat flow in soils," vol. 34, no. 6, pp. 845–848, 1970.

[20] H.-J. G. Diersch, *WASY Software FEFLOW 5.4 Finite Element Subsurface Flow and Transport Simulation System. Reference Manual*, WASY Institute for Water Resources Planning and Systems Research Ltd., Berlin, Germany, 2009.

[21] J. Šimůnek, M. T. Van Genuchten, and M. Šejna, "Development and applications of the HYDRUS and STANMOD software packages and related codes," *Vadose Zone Journal*, vol. 7, no. 2, pp. 587–600, 2008.

[22] N. Brisson, B. Mary, D. Ripoche et al., "STICS: a generic model for the simulation of crops and their water and nitrogen balances. I. Theory and parameterization applied to wheat and corn," *Agronomie*, vol. 18, no. 5-6, pp. 311–346, 1998.

[23] J. W. Jones, G. Hoogenboom, C. H. Porter et al., "The DSSAT cropping system model," *European Journal of Agronomy*, vol. 18, no. 3-4, pp. 235–265, 2003.

[24] N. Fodor, R. Sándor, T. Orfanus, L. Lichner, and K. Rajkai, "Evaluation method dependency of measured saturated hydraulic conductivity," *Geoderma*, vol. 165, pp. 60–68, 2011.

[25] W. D. Reynolds and D. E. Elrick, "In situ measurement of field-saturated hydraulic conductivity, sorptivity and the alpha-parameter using the Guelph permeameter," *Soil Science*, vol. 140, pp. 292–302, 1985.

[26] M. T. van Genuchten, "A closed form equation for predicting the hydraulic conductivity of unsaturated soils," *Soil Science Society of America Journal*, vol. 44, no. 5, pp. 892–898, 1980.

[27] N. Fodor and K. Rajkai, "Computer program (SOILarium 1.0) for estimating the physical and hydrophysical properties of soils from other soil characteristics," *Agrokémia és Talajtan*, vol. 60, pp. 327–340, 2011.

[28] E. G. Montgomery, "Correlation studies in corn. 24th Annual Report," *Agricultural Experiment Station of Nebraska*, pp. 108–159, 1911.

[29] J. T. Ritchie, U. Singh, D. C. Godwin, and W. T. Bowen, "Cereal growth, development and yield," in *Understanding Options for Agricultural Production*, G. Y. Tsuji, G. Hoogenboom, and

P. K. Thornton, Eds., pp. 79–98, Kluwer Academic Publishers, Dordrecht, The Netherlands, 1998.

[30] P. D. Jamieson, J. R. Porter, J. Goudriaan, J. T. Ritchie, H. Van Keulen, and W. Stol, "A comparison of the models AFR-CWHEAT2, CERES-Wheat, Sirius, SUCROS2 and SWHEAT with measurements from wheat grown under drought," *Field Crops Research*, vol. 55, no. 1-2, pp. 23–44, 1998.

[31] T. H. Skaggs, T. J. Trout, J. Šimůnek, and P. J. Shouse, "Comparison of HYDRUS-2D simulations of drip irrigation with experimental observations," *Journal of Irrigation and Drainage Engineering*, vol. 130, no. 4, pp. 304–310, 2004.

[32] G. de Marsily, *Quantitative Hydrogeology*, Academic Press, London, UK, 1986.

[33] S. O. Chung and R. Horton, "Soil heat and water flow with a partial surface mulch," *Water Resources Research*, vol. 23, no. 12, pp. 2175–2186, 1987.

[34] K. Soetaert and T. Petzoldt, "Inverse modelling, sensitivity and monte carlo analysis in R using package FME," *Journal of Statistical Software*, vol. 33, no. 3, pp. 1–28, 2010.

[35] L. E. Goodrich, "The influence of snow cover on the ground thermal regime.," *Canadian Geotechnical Journal*, vol. 19, no. 4, pp. 421–432, 1982.

13

The Effects of Excess Copper on Antioxidative Enzymes, Lipid Peroxidation, Proline, Chlorophyll, and Concentration of Mn, Fe, and Cu in *Astragalus neo-mobayenii*

P. Karimi,[1] R. A. Khavari-Nejad,[1,2] V. Niknam,[3] F. Ghahremaninejad,[1] and F. Najafi[1]

[1] *Faculty of Biological Sciences, Kharazmi University, Tehran 15719-14911, Iran*
[2] *Department of Biology, Faculty of Science, Islamic Azad University, Science and Research Branch, Tehran 14778-93855, Iran*
[3] *School of Biology and Center of Excellence in Phylogeny of Living Organisms, College of Science, University of Tehran, Tehran 14115-154, Iran*

Correspondence should be addressed to P. Karimi, parviz2125@gmail.com

Academic Editors: A. Bosabalidis, M. R. Cave, F. Darve, and H. Freitas

To probe the physiological and biochemical tolerance mechanisms in *Astragalus neo-mobayenii* Maassoumi, an endemic plant around the Cu-rich areas from the North West of Iran, the effect of different copper concentrations at toxic levels on this plant was investigated. Copper was applied in the form of copper sulfate ($CuSO_4 \cdot 5H_2O$) in four levels (0, 50, 100, and 150 μM). We observed no visible symptoms of Cu toxicity in this plant species. During the exposure of plants to excess copper, the antioxidant defense system helped the plant to protect itself from the damage. With increasing copper concentration, superoxide dismutase (SOD), peroxidase (POD), and catalase (CAT) activities increased in leaves and roots ($P < 0.001$) compared with that of the control group. The chlorophyll amount gradually declined with increasing Cu concentrations. However, reduction in the 50 μM level showed insignificant changes. Enhanced accumulation of proline content in the leaves was determined, as well as an increase of MDA content (oxidative damage biomarker) ($P < 0.001$). The results indicated that Cu contents in leaves and roots enhanced with increasing levels of Cu application. The Fe and Mn contents in both shoots and roots significantly decreased with increasing Cu concentration. Finally, the mechanisms of copper toxicity and copper tolerance in this plant were briefly discussed.

1. Introduction

Copper, an essential element for normal plant growth and metabolism [1, 2], plays a significant role in a number of physiological processes such as the photosynthetic and respiratory electron transport chains [3], nitrogen fixation, protein metabolism, antioxidant activity, cell wall metabolism, and hormone perception [2, 4, 5]. As a structural and catalytic component of proteins and enzymes, it is also well documented [6] and has been reported to be among the most toxic heavy metals [7]. However, when absorbed in excess quantities, Cu is highly toxic to plant growth potentially leading to physiological disorders that inhibit plant growth [8, 9]. It has been reported that excess Cu, at the cellular level, causes molecular damage to plants via the generation of reactive oxygen species (ROS) and free radicals [10]. Oxidative

stress by formation of ROS and oxidation of biomolecules such as lipids, proteins, nucleic acids, carbohydrates, and almost every other organic constituent of the living cell is an important aspect of Cu toxicity [11–13]. Plant cells can be protected from ROS by enzymatic defense mechanisms like superoxide dismutase (SOD), catalase (CAT), and peroxidase (POD) and nonenzymatic defense mechanisms like free amino acids especially proline, ascorbate, and glutathione and phenolic compounds [12, 14]. Free proline is known to accumulate under heavy metal exposure and considered to be involved in stress resistance [15]. In addition, Cu toxicity is related to disturbances in the uptake and transport of other mineral elements [16]. Less is known about the effects of Cu transport and uptake on Fe, Mn, Mg, and other mineral element assimilation. The induced deficiency of mineral content under excess copper from previous investigations is also available [16–20]. *Astragalus* with nearly 3000 species is generally considered the largest genus of vascular plants. Iran is one of the largest centers of diversity for the genus. It has nearly 750 species and an endemism rate of nearly 50% [21, 22]. It was determined how some physiological and biochemical parameters and Cu, Fe, and Mn concentration in roots and shoots were changed due to excess Cu in *Astragalus* plants grown in heavy metal soils constituting the flora of Iran Northwest.

2. Materials and Methods

2.1. Seeds Germination and Growth Conditions. *Astragalus* (*A. neo-mobayenii* Maassoumi) seeds were collected from Cu-rich areas (East Azerbaijan Province, Iran) and sterilized in 1% active sodium hypochlorite solution for 5 min, carefully washed by deionized water, and germinated on damp filter paper in the dark. Six-day seedlings were transferred to appropriate light conditions and supplied with 20%, 50%, and the whole Hoagland solution for 10 days. Seedlings were then cultivated in polyethylene pots containing perlite and vermiculite, and treatments were applied after three weeks. Seedlings were grown for 30 d in a growth chamber (greenhouse) at 65% constant relative humidity, 16/8 h day/night regime under $600\,\mu\,\text{mol}\,\text{m}^{-2}\,\text{s}^{-1}$ of light intensity, and day/night temperatures 25/20°C. Plants were supplied with the Hoagland nutrient solution (pH 6.2) which contained (macronutrients in mM) 1 KH_2PO_4, 2 $MgSO_4 \cdot 4H_2O$, 5 KNO_3, and 5 $Ca(NO_3)_2 \cdot 4H_2O$ and (micronutrients in μM) 9 $MnCl_2 \cdot 4H_2O$, 4.6 H_3BO_3, 0.8 $ZnSO_4 \cdot 7H_2O$, 0.3 $CuSO_4 \cdot 5H_2O$, and 0.1 $H_2MoO_4 \cdot H_2O$. Iron was supplied as Fe-EDTA (1.8 mM). Copper in four levels (0, 50, 100, and 150 μM) as $CuSO_4 \cdot 5H_2O$ was added to the nutrient solution. The experiment was conducted in four treatments with four replicates. 30 days after treatment, plants were harvested and used for physiological and biochemical analysis.

2.2. Photosynthetic Pigments' Analysis. Photosynthetic pigments (chlorophylls and carotenoids) were extracted by 80% acetone and centrifuged at 3000 g for 5 min [23]. Absorbance was determined in supernatant spectrophotometrically at 645 nm (Chl*b*), 663 nm (Chl*a*), and 470 nm (Car), and according to the Lichtenthaler and Wellburn formulae [24], pigment concentrations were calculated.

2.3. Enzyme Activity. The plant material (fresh weight) was homogenized on ice with 5 mL of 50 mmol sodium phosphate buffer (pH 7) including 0.5 mmol EDTA and 0.15 mol NaCl, in a mortar and pestle. The homogenate was centrifuged at 12000 g for 15 min at 4°C. The supernatant was used for enzyme assays. The activity of SOD was determined as described by Chen and Pan [25] in a 3 mL reaction mixture containing 50 mmol sodium phosphate buffer (pH 7), 10 mmol methionine, 1.17 mmol riboflavin, 56 mmol NBT, and 100 μL enzyme extract spectrophotometrically at 560 nm based on the photoreduction of nitroblue tetrazolium (NBT). The blue formazan produced by NBT photoreduction was measured by an increase in absorbance at 560 nm.

An SOD unit was defined as the amount of enzyme required to inhibit 50% of the NBT photoreduction.

The activity of CAT was determined as described by Havir and McHale [26] by a decrease in absorbance of the reaction mixture at 240 nm. The activity was assayed for 1 min in a reaction solution composed of 2.9 mL potassium phosphate buffer 50 mmol (2.85 mL, pH 7.0), H_2O_2 12.5 mmol (50 μL), and 100 μL of crude extract. The enzyme activity was calculated using the molar extinction coefficient of $36\,\text{M}^{-1}\,\text{cm}^{-1}$.

Peroxidase activity was determined based on an increase in absorbance at 470 nm as described by Sakharov and Ardilla [27]. The mixture composed of 2.8 mL guaiacol (3%), 100 μL H_2O_2 and 100 μL enzyme extract. A POD unit was defined as an increase in absorbance of 1.0 per min.

2.4. Determination of Lipid Peroxidation. Lipid peroxidation in roots was determined using thiobarbituric acid test by measurement of malondialdehyde level [28]. Roots were homogenized in 20% trichloroaceticacid (TCA) containing 0.5% thiobarbituric acid (TBA). The extracts were centrifuged at 10000 g for 15 min after incubation in 95°C water bath for 30 min and immediately ice bath. The amount of MDA-TBA complex was calculated by its specific absorbency at 532 nm in supernatant. Nonspecific absorbency at 600 nm was also subtracted [29]. The data was obtained as nm gr^{-1} FW using the extinction coefficient of $155\,\text{mM}^{-1}\,\text{cm}^{-1}$.

2.5. Proline Content. To estimate proline content of shoot, according to the Bates et al. [30] method, samples were homogenized in sulphosalicylic acid. The homogenate was filtered through Whatman's no. 1 filter paper. The filtrate was boiled for 1 hr after adding acetic acid and acid ninhydrin, and absorbance was taken at 520 nm wavelength.

2.6. Plant Sampling and Digestion. Roots and shoot were separated and washed by double-distilled water for at least four times. The samples were oven dried at 80°C for 48 hours and then milled by mixer. Homogenate powder was weighted (150 mg) and digested in 10 mL concentrated HNO_3 at 300°C heating plate. Cooled digests were diluted to 50 mL by double-distilled water and then filtrated by Whatman's no. 1 paper [31].

2.7. Metal Analysis. Metal contents of prepared samples were analyzed by ICP-OES spectroscopy (Varian VISTA-MPX) for manganese (Mn), copper (Cu), and (Fe). The metal concentrations were calculated as μg gr^{-1} DW.

2.8. Statistical Analysis. Statistical analysis were determined both based on one-way analysis of variance (ANOVA) and least significant difference (LSD) test with SPSS at significance levels of $P < 0.001$, $P < 0.01$, $P < 0.05$.

3. Results and Discussion

Total chlorophyll ($a + b$) content varied with Cu levels. With the increasing Cu concentration, the chlorophyll a and b

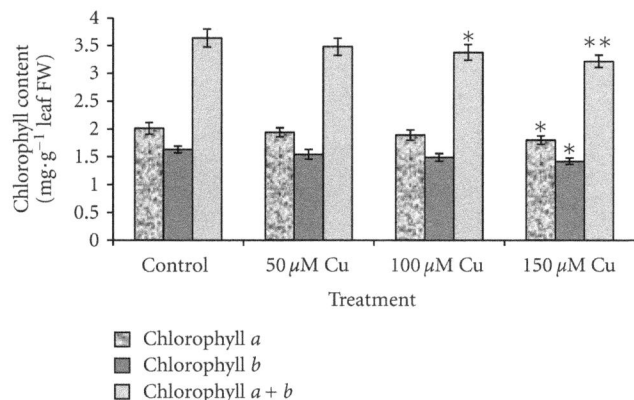

FIGURE 1: Chlorophyll *a* and chlorophyll *b* content and chlorophyll (*a* + *b*) in leaf tissues of *A. neo-mobayenii* grown in different concentrations of copper. Vertical bars represent standard error of the mean (n = 4). Asterisks indicate that the mean values are significantly different between treatments and control (*P < 0.05, **P < 0.01) according to LSD.

content decreased gradually. However, reduction in the 50 and 100 μM levels showed insignificant changes compared with that of the control group but showed significant change in 150 μM (P < 0.05) (Figure 1). Reduction of chlorophyll content in plants due to excess copper was also observed by Quzounidou [32]; Rama Devi and Prasad [33]; Monni et al. [34]; Xiong et al. [35]; Singh et al. [36]. It has been proposed that Cu at toxic concentration interferes with enzymes associated with chlorophyll biosynthesis and protein composition of photosynthetic membranes [37–39]. Also, possibility of Cu-induced Fe deficiency [16] and displacing Mg required for chlorophyll biosynthesis [40] have been proposed as a damage mechanism.

Figure 2 shows the changes of SOD activity in leaves and roots. No significant changes in SOD activity were observed in the leaves under 50 μM Cu concentration, while the activities showed significant increases (P < 0.001) under higher level of Cu concentration. Significant increases in root SOD activities under all treatments were observed (P < 0.05 or P < 0.001). As Cu concentration increased, the root CAT activity increased significantly (P < 0.01). The same result was observed in leaves as shown in Figure 3. Figure 4 shows increased POD activity in both leaves and roots concomitantly with increased Cu level. The increase in POD activity in both was significant (P < 0.001). The result of lipid peroxidation in root in the control and treatment groups is shown in Figure 5. MDA level in roots significantly increased with the increase of Cu concentration (P < 0.001). Our studied plant was endemic around the Cu-rich area and had adapted to contaminated soils by developing tolerance mechanisms to this metal stress. Many studies reported that internal protective responses to excess copper can vary among plant species and among different tissues [41]. It is well known that when copper is in excess, it catalyzes the formation of ROS and particularly, the highly toxic hydroxyl radicals from Haber-Weiss reaction [42], leading to an increase in MDA as biomarkers of oxidative damages. Hence,

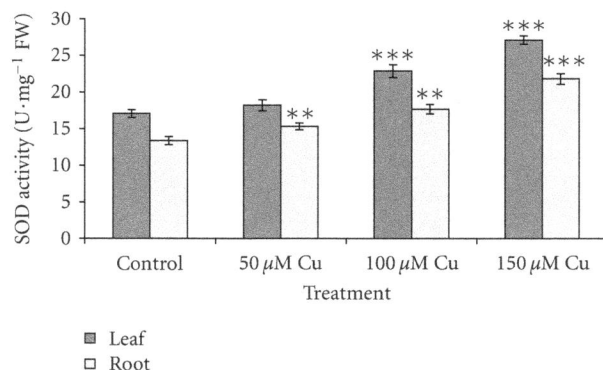

FIGURE 2: Effects of different concentrations of copper on super-oxide dismutase (SOD) activity in leaves and roots of *A. neo-mobayenii*. Vertical bars represent standard error of the mean (n = 4). Asterisks indicate that the mean values are significantly different between treatments and control (*P < 0.05, **P < 0.01, ***P < 0.001) according to LSD.

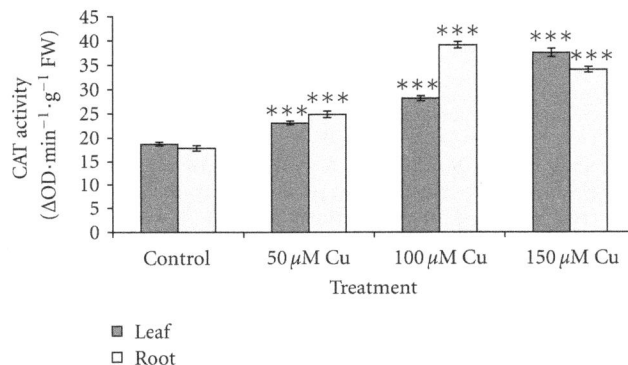

FIGURE 3: Effects of different concentrations of copper on catalase (CAT) activity in leaves and roots of *A. neo-mobayenii*. Vertical bars represent standard error of the mean (n = 4). Asterisks indicate that the mean values are significantly different between treatments and control (*P < 0.05, **P < 0.01, ***P < 0.001) according to LSD.

in response to the presence of excess Cu, plants increased the antioxidant responses due to increased generation of ROS. Accordingly, it was observed an excess Cu in plants inducing defense genes responsible for antioxidant enzymes, including SOD, POD, and CAT, which contribute to the removal of ROS [43–46]. SOD catalyzes the dismutation of superoxide into oxygen and hydrogen peroxide. The enhanced activity of catalase demonstrated that any hydrogen peroxide formed as a result of SOD activity was consumed by catalase and/or peroxidase. This indicated that these enzymes were known as a mediator of oxidative damage and might be sufficient to protect biomolecules of some parts of plants against ROS attack [13].

Figure 6 shows Cu-induced proline accumulation in shoots. The proline content increased substantially with increasing Cu concentrations (P < 0.001). This may be because synthesis of proline is considered to be one of the first metabolic responses to stress and acts osmoregulator,

TABLE 1: Effects of excess copper on Cu, Mn, and Mg contents of the shoots and roots of *A. neo-mobayenii.*

Cu (μM)	Shoot			Root		
	Cu (μg/g DW)	Mn (μg/g DW)	Fe (mg/g DW)	Cu (μg/g DW)	Mn (μg/g DW)	Fe (mg/g DW)
Control	12.32	42.9	147	8.14	24.67	93
50	23.69	42.72[NS]	146	14.73	26.12	85
100	31.92	39.14	119	18.12	26.89	67
150	44.58	36.12	106	24.29	25.09	53

Each value is the mean of the four replications.
All the values are significant at $P < 0.01$.
NS: nonsignificant.

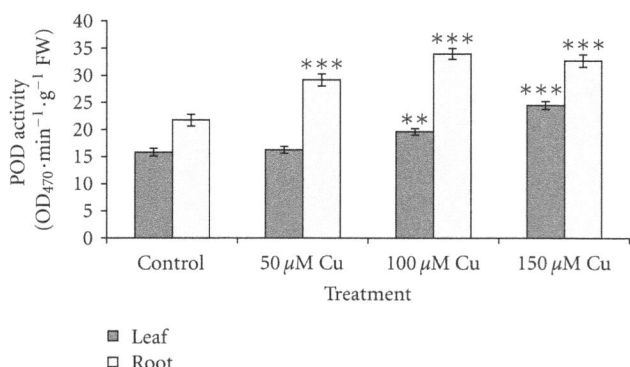

FIGURE 4: Effects of different concentrations of copper on peroxidase (POD) activity in leaves and roots of *A. neo-mobayenii.* Vertical bars represent standard error of the mean ($n = 4$). Asterisks indicate that the mean values are significantly different between treatments and control (*$P < 0.05$, **$P < 0.01$, ***$P < 0.001$) according to LSD.

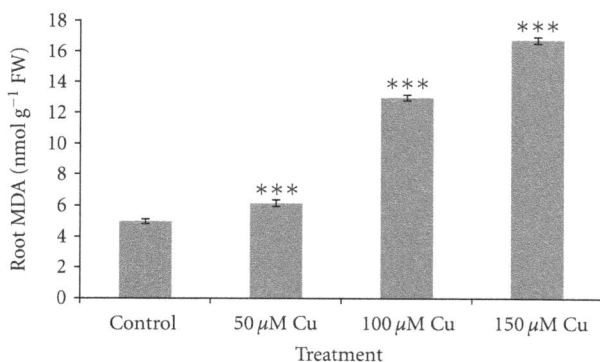

FIGURE 6: Proline contents in leaves of *A. neo-mobayenii* grown in different concentrations of copper. Vertical bars represent standard error of the mean ($n = 4$). Asterisks indicate that the mean values are significantly different between treatments and control (*$P < 0.05$, **$P < 0.01$, ***$P < 0.001$) according to LSD.

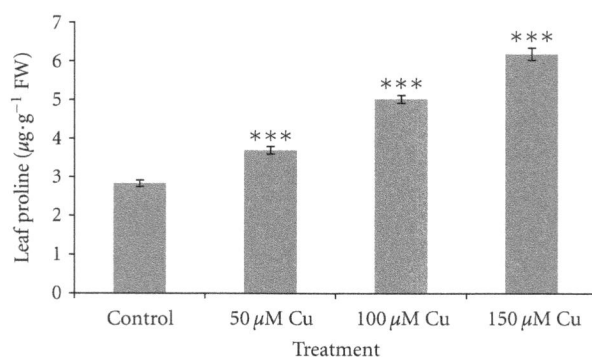

FIGURE 5: MDA levels in roots of *A. neo-mobayenii* grown in different concentrations of copper. Vertical bars represent standard error of the mean ($n = 4$). Asterisks indicate that the mean values are significantly different between treatments and control (*$P < 0.05$, **$P < 0.01$, ***$P < 0.001$) according to LSD.

stabilizer of protein synthesis, a metal chelator, and a hydroxyl radical scavenger [47–49].

The Cu content in shoots and roots increased significantly with an increase in the level of applied Cu. The accumulations in shoots were higher than that of roots in all treatments. Fe content in both shoots and roots reduced with increasing Cu concentration in the medium. However, a slight increase was observed in the lower level of applied Cu. The Mn content decreased insignificantly at higher levels of applied Cu. In roots increased levels of Mn were observed (Table 1). The results are in close conformity with the findings that an elevated copper application resulted in an increase in plant Cu content [8, 50–52]. In high concentration of copper application, the copper levels in leaves were above the threshold for copper toxicity [4]. On the other hand, normal growth of studied plants without any visible symptoms of Cu toxicity implied that this plant was tolerant to toxic levels of Cu. In addition, translocation of copper to the shoots was suggested as a strategy to explain the copper tolerance mechanism developed by plant in order to reduce copper stress. Thus according to the present study, this plant could be suitable for phytoextraction [53, 54].

Interference of Cu and Cd with the root uptake of mineral nutrients has been observed [55, 56]. Moreover, antagonistic effects of Cu and Fe have been suggested by many workers and often occur in plants grown under Cu toxicity [17, 57–59]. Also, competition of copper with Mn for transport sites in plasmalemma has been reported [60, 61]. In this study reduction of Mn with increasing levels of copper was observed. However, Mn contents in leaves did not drop below the critical deficiency range [4].

References

[1] R. K. Sharma and M. Agrawal, "Biological effects of heavy metals: an overview," *Journal of Environmental Biology*, vol. 26, no. 2, pp. 301–313, 2005.

[2] I. Yruela, "Copper in plants," *Brazilian Journal of Plant Physiologyogy*, vol. 17, no. 1, pp. 145–156, 2005.

[3] F. Van Assche and H. Clijters, "Effect of metals on enzyme activity in plants," *Plant Cell Environment*, vol. 13, pp. 195–206, 1990.

[4] H. Marschner, *Mineral Nutrition of Higher Plants*, Academic Press, London, UK, 1995.

[5] J. A. Raven, M. C. W. Evans, and R. E. Korb, "The role of trace metals in photosynthetic electron transport in O_2-evolving organisms," *Photosynthesis Research*, vol. 60, no. 2-3, pp. 111–149, 1999.

[6] M. Pilon, S. E. Abdel-Ghany, C. M. Cohu, K. A. Gogolin, and H. Ye, "Copper cofactor delivery in plant cells," *Current Opinion in Plant Biology*, vol. 9, no. 3, pp. 256–263, 2006.

[7] T. Li and Z. -T. Xiong, "A novel response of wild-type duckweed (Lemna paucicostata Hegelm.) to heavy metals," *Environmental Toxicology*, vol. 19, pp. 95–102, 2004.

[8] G. Ouzounidou, "Root growth and pigment composition in relationship to element uptake in Silene compacta plants treated with copper," *Journal of Plant Nutrition*, vol. 17, no. 6, pp. 933–943, 1994.

[9] V. Caspi, M. Droppa, G. Horváth, S. Malkin, J. B. Marder, and V. I. Raskin, "The effect of copper on chlorophyll organization during greening of barley leaves," *Photosynthesis Research*, vol. 62, no. 2-3, pp. 165–174, 1999.

[10] J. Liu, Z. Xiong, T. Li, and H. Huang, "Bioaccumulation and ecophysiological responses to copper stress in two populations of *Rumex dentatus* L. from Cu contaminated and non-contaminated sites," *Environmental and Experimental Botany*, vol. 52, no. 1, pp. 43–51, 2004.

[11] A. Murphy and L. Taiz, "Correlation between potassium efflux and copper sensitivity in 10 Arabidopsis ecotypes," *New Phytologist*, vol. 136, no. 2, pp. 211–222, 1997.

[12] O. Acar, I. Türkan, and F. Özdemir, "Superoxide dismutase and peroxidase activities in drought sensitive and resistant barley (*Hordeum vulgare* L.) varieties," *Acta Physiologiae Plantarum*, vol. 23, no. 3, pp. 351–356, 2001.

[13] R. Mittler, "Oxidative stress, antioxidants and stress tolerance," *Trends in Plant Science*, vol. 7, no. 9, pp. 405–410, 2002.

[14] S. Gao, R. Yan, M. Cao, W. Yang, S. Wang, and F. Chen, "Effects of copper on growth, antioxidant enzymes and phenylalanine ammonia-lyase activities in *Jatropha curcas* L. seedling," *Plant, Soil and Environment*, vol. 54, no. 3, pp. 117–122, 2008.

[15] F. B. Wu, F. Chen, K. Wei, and G. P. Zhang, "Effect of cadmium on free amino acid, glutathione and ascorbic acid concentrations in two barley genotypes (*Hordeum vulgare* L.) differing in cadmium tolerance," *Chemosphere*, vol. 57, no. 6, pp. 447–454, 2004.

[16] E. Pätsikkä, M. Kairavuo, F. Šeršen, E. M. Aro, and E. Tyystjärvi, "Excess copper predisposes photosystem II to photoinhibition in vivo by outcompeting iron and causing decrease in leaf chlorophyll," *Plant Physiologyogy*, vol. 129, no. 3, pp. 1359–1367, 2002.

[17] W. Schmidt, "Mechanisms and regulation of reduction-based iron uptake in plants," *New Phytologist*, vol. 141, no. 1, pp. 1–26, 1999.

[18] Y. Chen, J. Shi, G. Tian, S. Zheng, and Q. Lin, "Fe deficiency induces Cu uptake and accumulation in Commelina communis," *Plant Science*, vol. 166, no. 5, pp. 1371–1377, 2004.

[19] A. D. Rombolà, Y. Gogorcena, A. Larbi et al., "Iron deficiency-induced changes in carbon fixation and leaf elemental composition of sugar beet (Beta vulgaris) plants," *Plant and Soil*, vol. 271, no. 1-2, pp. 39–45, 2005.

[20] H. Lequeux, C. Hermans, S. Lutts, and N. Verbruggen, "Response to copper excess in *Arabidopsis thaliana*: impact on the root system architecture, hormone distribution, lignin accumulation and mineral profile," *Plant Physiologyogy and Biochemistry*, vol. 48, no. 8, pp. 673–682, 2010.

[21] J. M. Lock and K. Simpson, *Legumes of West Asia*, Kew, Royal Botanic Gardens, Canada, 1991.

[22] A. A. Maassoumi, *Astragalusin the Old World*, Tehran, Iran, 1998.

[23] D. Arnon, "Copper enzymes in isolated chloroplast: polyphenoloxidase in *Beta vulgaris*," *Plant Physiologyogy*, vol. 24, pp. 1–15, 1949.

[24] H. Lichtenthaler and A. Wellburn, "Determination of total carotenoids and chlorophyllaand *b* of leaf extracts in different solvents," *Biochemical Society Transactions*, vol. 603, pp. 591–592, 1983.

[25] C. N. Chen and S. M. Pan, "Assay of superoxide dismutase activity by combining electrophoresis and densitometry," *Botanical Bulletin of Academia Sinica*, vol. 37, no. 2, pp. 107–111, 1996.

[26] E. A. Havir and N. A. McHale, "Biochemical and developmental characterization of multiple forms of catalase in tobacco leaves," *Plant Physiology*, vol. 84, pp. 450–455, 1987.

[27] I. Y. Sakharov and G. B. Ardila, "Variations of peroxidase activity in cocoa beans during their ripening, fermentation and drying," *Food Chemistry*, vol. 65, no. 1, pp. 51–54, 1999.

[28] R. L. Heath and L. Packer, "Photoperoxidation in isolated chloroplasts—I. Kinetics and stoichiometry of fatty acid peroxidation," *Archives of Biochemistry and Biophysics*, vol. 125, no. 1, pp. 189–198, 1968.

[29] C. H. De Vos, R. Vooijs, H. Schat, and W. Ernst, "Cooper-induced damage to the permeability barrier in roots of *Silenecucubalus*," *Journal of Plant Physiologyogy*, vol. 135, pp. 165–169, 1989.

[30] L. S. Bates, R. P. Waldren, and I. D. Teare, "Rapid determination of free proline for water-stress studies," *Plant and Soil*, vol. 39, no. 1, pp. 205–207, 1973.

[31] N. Lavid, Z. Barkay, and E. Tel-Or, "Accumulation of heavy metals in epidermal glands of the waterlily (Nymphaeaceae)," *Planta*, vol. 212, no. 3, pp. 313–322, 2001.

[32] G. Ouzounidou, "The use of photoacoustic spectroscopy in assessing leaf photosynthesis under copper stress: correlation of energy storage to photosystem II fluorescence parameters and redox change of P700," *Plant Science*, vol. 113, no. 2, pp. 229–237, 1996.

[33] S. Rama Devi and M. N. V. Prasad, "Copper toxicity in Ceratophyllum demersum L. (Coontail), a floating macrophyte: response of antioxidant enzymes and antioxidants," *Plant Science*, vol. 138, no. 2, pp. 157–165, 1998.

[34] S. Monni, M. Salemaa, and N. Millar, "The tolerance of *Empetrum nigrum* to copper and nickel," *Environmental Pollution*, vol. 109, no. 2, pp. 221–229, 2000.

[35] Z. T. Xiong, C. Liu, and B. Geng, "Phytotoxic effects of copper on nitrogen metabolism and plant growth in Brassica pekinensis Rupr," *Ecotoxicology and Environmental Safety*, vol. 64, no. 3, pp. 273–280, 2006.

[36] D. Singh, K. Nath, and Y. K. Sharma, "Response of wheat seed germination and seedling growth under copper stress," *Journal of Environmental Biology*, vol. 28, no. 2, pp. 409–414, 2007.

[37] F. C. Lidon and F. S. Henriques, "Limiting step in photosynthesis of rice plants treated with varying copper levels," *Journal of Plant Physiologyogy*, vol. 138, pp. 115–118, 1991.

[38] W. Maksymiec, R. Russa, T. Urbanik-Sypniewska, and T. Baszynski, "Effect of excess Cu on the photosynthetic apparatus of runner bean leaves treated at two different growth stages," *Physiologia Plantarum*, vol. 91, no. 4, pp. 715–721, 1994.

[39] M. F. Quartacci, C. Pinzino, C. L. M. Sgherri, F. Dalla Vecchia, and F. Navari-Izzo, "Growth in excess copper induces changes in the lipid composition and fluidity of PSII-enriched membranes in wheat," *Physiologia Plantarum*, vol. 108, no. 1, pp. 87–93, 2000.

[40] H. Küpper, I. Šetlík, E. Šetliková, N. Ferimazova, M. Spiller, and F. C. Küpper, "Copper-induced inhibition of photosynthesis: limiting steps of in vivo copper chlorophyll formation in Scenedesmus quadricauda," *Functional Plant Biology*, vol. 30, no. 12, pp. 1187–1196, 2003.

[41] F. Passardi, C. Cosio, C. Penel, and C. Dunand, "Peroxidases have more functions than a Swiss army knife," *Plant Cell Reports*, vol. 24, no. 5, pp. 255–265, 2005.

[42] B. Halliwell and J. M. C. Gutteridge, "Oxygen toxicity, oxygen radicals, transition metals and disease," *Biochemical Journal*, vol. 219, no. 1, pp. 1–14, 1984.

[43] M. E. Alvarez and C. Lamb, "Oxidative burst mediated defense responses in plant disease resistance," in *Oxidative Stress and the Molecular Biology of Antioxidant Defenses*, J. G. Scandalios, Ed., pp. 815–839, Cold Spring Harbor Laboratory Press, New York, NY, USA, 1997.

[44] F. Navari-Izzo, M. F. Quartacci, C. Pinzino, F. DallaVecchia, and C. L. M. Sgherri, "Thylakoid-bound and stromal antioxidative enzymes in wheat treated with excess copper," *Physiologia Plantarum*, vol. 104, no. 4, pp. 630–638, 1998.

[45] M. Drazkiewicz, E. Skorzynska-Polit, and Z. Krupa, "Response of the ascorbate-glutathione cycle to excess copper in *Arabidopsis thaliana* (L.)," *Plant Science*, vol. 164, no. 2, pp. 195–202, 2003.

[46] H. Wang, X. Q. Shan, B. Wen, S. Zhang, and Z. J. Wang, "Responses of antioxidative enzymes to accumulation of copper in a copper hyperaccumulator of Commoelina communis," *Archives of Environmental Contamination and Toxicology*, vol. 47, no. 2, pp. 185–192, 2004.

[47] M. E. Farago and W. A. Mullen, "Plants which accumulate metals—part 4. A possible copper-proline complex from the roots of armeria maritima," *Inorganica Chimica Acta*, vol. 32, no. C, pp. L93–L94, 1979.

[48] S. Siripornadulsil, S. Traina, D. P. S. Verma, and R. T. Sayre, "Molecular mechanisms of proline-mediated tolerance to toxic heavy metals in transgenic microalgae," *Plant Cell*, vol. 14, no. 11, pp. 2837–2847, 2002.

[49] V. V. Kuznetsov and N. I. Shevyakova, "Stress responses of tobacco cells to high temperature and salinity. Proline accumulation and phosphorylation of polypeptides," *Physiologia Plantarum*, vol. 100, no. 2, pp. 320–326, 1997.

[50] V. Kumar, D. V. Yadav, and D. S. Yadav, "Effects of nitrogen sources and copper levels on yield, nitrogen and copper contents of wheat (Triticum aestivum L.)," *Plant and Soil*, vol. 126, no. 1, pp. 79–83, 1990.

[51] H. Panou-Filotheou, A. M. Bosabalidis, and S. Karataglis, "Effects of copper toxicity on leaves of oregano (*Origanum vulgare subsp. hirtum*)," *Annals of Botany*, vol. 88, no. 2, pp. 207–214, 2001.

[52] J. Cambrollé, E. Mateos-Naranjo, S. Redondo-Gómez, T. Luque, and M. E. Figueroa, "Growth, reproductive and photosynthetic responses to copper in the yellow-horned poppy, *Glaucium flavum Crantz*," *Environmental and Experimental Botany*, vol. 71, no. 1, pp. 57–64, 2011.

[53] D. C. McCain and J. L. Markley, "More manganese accumulates in maple sun leaves than in shade leaves," *Plant Physiologyogy*, vol. 90, pp. 1414–1421, 1989.

[54] J. Yoon, X. Cao, Q. Zhou, and L. Q. Ma, "Accumulation of Pb, Cu, and Zn in native plants growing on a contaminated Florida site," *Science of the Total Environment*, vol. 368, no. 2-3, pp. 456–464, 2006.

[55] D. T. Clarkson and U. Luttge, "Mineral nutrition: divalent cations, transport and compartmentation," *Progress in Botany*, vol. 51, pp. 93–100, 1989.

[56] R. B. Harrison, C. L. Henry, and D. Xue, "Magnesium deficiency in Douglas-fir and Grand fir growing on a sandy outwash soil amended with sewage sludge," *Water, Air, and Soil Pollution*, vol. 75, no. 1-2, pp. 37–50, 1994.

[57] C. D. Foy, R. L. Chaney, and M. C. White, "The physiology of metal toxicity in plants," *Annual Review of Plant Physiologyogy*, vol. 29, pp. 511–566, 1978.

[58] G. Ouzounidou, I. Ilias, H. Tranopoulou, and S. Karataglis, "Amelioration of copper toxicity by iron on spinach physiology," *Journal of Plant Nutrition*, vol. 21, no. 10, pp. 2089–2101, 1998.

[59] L. Lombardi and L. Sebastiani, "Copper toxicity in Prunus cerasifera: growth and antioxidant enzymes responses of in vitro grown plants," *Plant Science*, vol. 168, no. 3, pp. 797–802, 2005.

[60] B. F. Hulagur and R. T. Dangarwala, "Effect of zinc, copper and phosphorus fertilization on the uptake of iron, manganese and molybdenum by hybrid maize," *Madras Agricultural Journal*, vol. 69, pp. 11–16, 1982.

[61] F. C. Lidon and F. S. Henriques, "Copper toxicity in rice; a diagnostic criteria and its effect on Mn and Fe contents," *Soil Science*, vol. 154, no. 2, pp. 130–135, 1992.

14

Methane Production and Consumption in Loess Soil at Different Slope Position

Małgorzata Brzezińska, Magdalena Nosalewicz, Marek Pasztelan, and Teresa Włodarczyk

Institute of Agrophysics, Polish Academy of Sciences, Ulica Doświadczalna 4, 20290 Lublin, Poland

Correspondence should be addressed to Małgorzata Brzezińska, m.brzezinska@ipan.lublin.pl

Academic Editor: Hubert Hasenauer

Methane (CH_4) production and consumption and soil respiration in loess soils collected from summit (Top), back slope (Middle), and slope bottom (Bottom) positions were assessed in laboratory incubations. The CH_4 production potential was determined under conditions which can occur in the field (relatively short-term flooding periods with initially ambient O_2 concentrations), and the CH_4 oxidation potential was estimated in wet soils enriched with CH_4. None of the soils tested in this study emitted a significant amount of CH_4. In fact, the Middle and Bottom soils, especially at the depth of 20–40 cm, were a consistent sink of methane. Soils collected at different slope positions significantly differed in their methanogenic, methanotrophic, and respiration activities. In comparison with the Top position (as reference soil), methane production and both CO_2 production and O_2 consumption under flooding were significantly stimulated in the soil from the Middle slope position ($P < 0.001$), while they were reduced in the Bottom soil (not significantly, by 6 to 57%). All upper soils (0–20 cm) completely oxidized the added methane (5 kPa) during 9–11 days of incubation. Soils collected from the 20–40 cm at the Middle and Bottom slope positions, however, consumed significantly more CH_4 than the Top soil ($P < 0.001$).

1. Introduction

Methane (CH_4) is the most abundant hydrocarbon in the atmosphere, and it is an important greenhouse gas, which so far has contributed to an estimated 18–20% [1, 2] of postindustrial global warming. Methane has environmental impacts beyond those of a direct greenhouse gas, through atmospheric chemistry that enhances the abundance of tropospheric ozone (O_3) and decreases that of hydroxyl radicals (OH) and hence the atmospheric lifetime of many other pollutants [3]. The atmospheric CH_4 concentration has risen from the background level from 700 to 1782 ppb in 2006, and the growth rate in CH_4 concentration was changing considerably; the very large and interannual variations in CH_4 concentration remain unexplained and present an important challenge to the research community [4, 5]. Estimated surface CH_4 emissions reach 643 Tg year^{-1} [3]. Oxidation of atmospheric methane by well-drained soils accounts for about 10% [6] or 6% [4] of the global methane sink, that is about 30 Tg CH4 per year. Other CH_4 sinks are the

stratosphere (40 Tg year^{-1}) and tropospheric OH (445 Tg year^{-1}) [4].

Most methane on Earth is produced by Archaea through methanogenesis, the final step in fermentation of organic matter, which takes place in rice fields, the guts of animals, soils, wetlands, and landfills, as well as in freshwater and marine sediments. As a simple assumption, about 10–20% of reactive organic material buried in soils and sediments is converted to methane [1]. The potential impact of methane on future global warming and an important role of soils in sorption of this gas have led to many terrestrial studies of methods and techniques to quantify CH_4 flux at the soil-atmosphere interface [7]. Numerous experimental data on emission of greenhouse gases are used in modelling of the local and global gas emissions, while some models were developed to determine abatement strategies to meet restrictions on emission and/or deposition levels at the least cost [8].

Soil saturation with water has dramatic consequences for gas diffusion processes in soil (as gases diffuse 10,000

TABLE 1: Basic characteristics of loess soils at three slope positions.

Slope position	Soil depth (cm)	C_{org} (%)	C_{inorg} (%)	pH (H_2O)	Sand	Silt (%)	Clay
Top	0–20	0.97	0.690	8.11	33.6	62.1	4.3
	0–40	0.57	0.002	8.14	16. 2	77.4	6.3
Middle	0–20	2.14	0.006	7.60	27.1	68.7	4.2
	20–40	1.42	0.275	7.91	27.5	68.3	4.2
Bottom	0–20	0.92	0.657	7.88	30.9	64.9	4.2
	20–40	0.39	0.393	7.56	34.1	61.6	4.3

faster in air than in water). Consequently, one of the main effects of flooding is a lower pool of available O_2 [9, 10] and a several-fold change in the activity of the oxidoreductases—intracellular enzymes involved in the oxidative metabolism of soil microorganisms [11]. Conventional knowledge states that water-saturated systems like wetlands (swamps, marshes) and paddy soils (rice fields) are net contributors of CH_4 to the atmosphere, whereas upland soils (with the exception of landfills) are generally sinks for CH_4 [12]. However, significant methane emission from field soils may also occur after normal precipitation if the soils remain saturated for a long enough period, since water occupation of soil voids may cause oxygen deficiency and development of reducing conditions. Even in unsaturated conditions, there may be anaerobic microsites capable of evolving methane. Little is known, however, about methane emission when usually well-drained soils become flooded for a short period [13]. In fact, soils can act as a source and a sink for CH_4, depending on their air-water conditions [7, 14].

Soil properties are a product of soil-forming factors including landscape variability, agroecosystem management, and climatic factors. Numerous studies were performed to measure the effect of landscape position and land management on physical, chemical and biological soil properties [15–21]. Soils developed from loess are fertile and show high erodibility [22]. Soil erosion results in heavy differentiation of a soil cover with natural pedons being reduced or overbuilt. Both eroded and colluvial soils differ from uneroded soils not only in morphological features but also in particle-size and pore distributions, organic matter content and plant nutrients, water retention, and bulk density [23]. Loess soils are among the most susceptible to the drop in redox potential under anaerobic conditions, which is followed by a rapid reduction of the oxidized inorganic soil components [24]. In consequence, periodical soil hypoxia changes soil respiration, which plays a fundamental role in the metabolism of the soil biota and promotes development of methanogenic microorganisms.

The objective of this study was to compare the CH_4 production and CH_4 consumption in slightly eroded loess soils taken at the summit, back slope, and bottom of a hill. The experiment was performed in laboratory under controlled temperature and air-water conditions. Initially, ambient O_2 concentrations were present in both flooded and wet soil incubations. Our intention was to determine the soil potential for methane production under conditions which occur in field (relatively short-term flooding periods for methanogenic activity) and for methane oxidation (soil enriched with CH_4).

2. Materials and Methods

2.1. Site and Soil Description. A loessial agricultural basin of the Ciemięga River (near Lublin, south-east part of Poland) is a region of the water erosion risk, including sediment transport and nutrient runoff, and is under intensive agricultural use [24, 25]. Soil samples were collected near Baszki village from two depths (0–20 cm and 20–40 cm) and three slope positions: at the summit (Top), back slope (Middle), and slope bottom (Bottom).

The slope is about 15 m high and 60 m long and is covered by natural grass vegetation; it is at the distance of about 150 m from the river. The annual precipitation in this region is 570 mm, and the average annual temperature is +7.5°C [25]. The basic characteristic of the tested brown loess soil (Eutric Cambisol) is shown in Table 1.

2.2. Incubation Experiment. For methanogenic activity measurements, 20 g portions of air-dry soils were placed into 60 cm^3 glass vessels and flooded with 15 cm^3 of distilled water. All the vessels were tightly closed with rubber stoppers and aluminium caps, and the flooded soils were incubated at 25°C for 28 days.

For methanotrophic activity measurements, 10 g portions of air-dried soils were placed into 60 cm^3 glass vessels and 5 cm^3 of distilled water was added. All the vessels were tightly closed with rubber stoppers and aluminium caps, and wet soils were enriched with 5% (v/v) CH_4 (5 kPa). The soil samples were incubated at 20°C for 21 days.

Initially, ambient O_2 concentrations were present in both incubations (20.5% v/v). Our intention was to determine the potential of soils for methane production under field conditions, with relatively short-term flooding periods, and for methane oxidation after soil enrichment with CH_4.

2.3. Methods. The concentrations of gases in the headspace were measured with gas chromatographs Shimadzu GC-14B and GC-14A (Japan) equipped with a flame ionization detector (FID) and a thermal conductivity detector (TCD), respectively. Methane was detected by the FID detector at 150°C. The gas components were separated on a column packed with a Porapak Q maintained at 80°C, and the temperature of the injector was 150°C. Carbon dioxide and O_2 were detected by TCD with the use of two 2 m columns

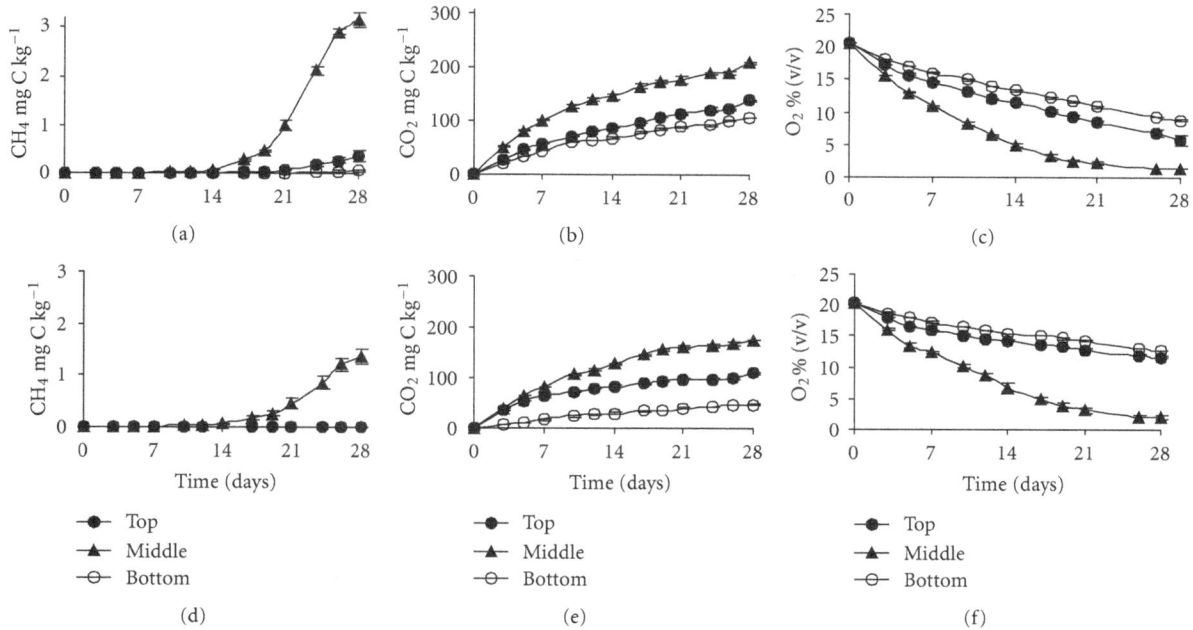

FIGURE 1: Changes of CH_4, CO_2 and O_2 over time in loess soils collected from three slope positions and incubated under flooding (methanogenic potential). Top: summit, Middle: back slope, Bottom: bottom of the slope. Upper graphs (a–c) upper soil depth of 0–20 cm; lower graphs (d–f) lower soil depth of 20–40 cm. (a) and (d) cumulative CH_4 production; (b) and (e) cumulative CO_2 production; (c) and (f) changes in O_2 in the headspace. Points represent triplicate-means with standard error.

(3.2 diameter), one packed with Porapak Q (for CO_2) and the other packed with Molecular Sieve 5A (for O_2) with He as a carrier gas flowing at a rate of 40 cm³ min⁻¹. The temperatures of the column and detector were 40°C and 60°C, respectively. The detector responses were calibrated using certified gas standards (Air Products) containing 20.9% O_2 in N_2 and 4% CH_4 or 1% CO_2 in He [26, 27].

Soil properties were determined by standard methods. The organic (C_{org}) and inorganic (C_{inorg}) carbon was determined using a TOC-VCPH analyzer (Shimadzu, Japan). The particle size distribution was measured with the laser diffraction method [28]. All measurements were done in triplicate, and the results were expressed on an oven-dry weight basis (105°C, 24 h).

2.4. Calculations. The concentrations of gases were corrected for solubility in water by using published values of the Bunsen absorption coefficient [29]. The rates of methane production and consumption were calculated by linear regression of increases and decreases, respectively, in CH_4 concentrations against incubation time, using at least three consecutive measurements with a regression coefficient (r^2) of >0.9, and expressed in mg CH_4-C per kg of oven-dry soil per day [30, 31]. The total cumulative CH_4 production and CH_4 consumption were determined in each sample by the difference in the headspace CH_4 concentration at the beginning and end of the assay period [32]. The rates of CH_4 production and consumption and the total amount of CH_4 produced and consumed were used as a measure of the methanogenic and methanotrophic activity (potential), respectively. Similarly, the total amounts of CO_2 produced

and O_2 consumed were calculated to describe the respiration activity (and expressed as mg CO_2-C kg⁻¹ and O_2% (v/v), resp.). Final amounts of CH_4, CO_2, and O_2 were assessed by Student's "t" test to determine the significance of the differences in gas production or consumption between soils. Correlations between total gases produced or consumed over time, and organic carbon in soils collected from different slope positions were tested with regression analysis.

3. Results

3.1. Methanogenic Activity of Soils from Different Slope Positions. Position of soil in the slope strongly affected the capacity of CH_4 production. Methane was produced in flooded soils after a 17-day lag (Figure 1). The highest methanogenic activity was observed in the Middle soil. During 28-day incubation, the upper 0–20 cm soil evolved 3.17 mg CH_4-C kg⁻¹ soil at a rate of 0.304 mg CH_4-C kg⁻¹ d⁻¹ (Figure 1(a), Table 2). Soil sampled at the Top position produced only 0.359 mg CH_4-C kg⁻¹, while the Bottom soil evolved even less than 0.06 mg CH_4-C kg⁻¹.

Deeper soil layers (20–40 cm) showed significantly lower methanogenic activity ($P < 0.001$) with the highest production in the soil from the Middle position 1.35 mg CH_4-C kg⁻¹ and much lower in the other soils: less than 0.002 mg CH_4-C kg⁻¹ (Figure 1(d)).

The tested soils showed relatively large differences in their respiration under flooding. Both CO_2 evolution and O_2 consumption were more intensive in the upper rather than deeper soil layers (Figures 1(b) and 1(c)). The Middle soil produced 207.6 mg CO_2-C kg⁻¹, and consumed 19.14%

TABLE 2: CH_4 production, CO_2 evolution and O_2 uptake of soils collected from three slope positions, and incubated for 28 days under flooding (average values $\pm$ standard error, $n = 3$).

Slope position	Soil depth (cm)	CH₄ production		CO₂ evolution (mg C kg⁻¹)	O₂ uptake % (v/v)
		Total (mg C kg⁻¹)	Rate (mg C kg⁻¹ d⁻¹)		
Top	0–20	0.3595 ± 0.118	0.0316	138.1 ± 2.56	14.61 ± 0.85
	20–40	0.0017 ± 0.001	0.0001	110.2 ± 2.19	8.73 ± 0.04
Middle	0–20	3.1679*** ± 0.140	0.3042	207.6*** ± 2.34	19.14** ± 0.10
	20–40	1.3538*** ± 0.129	0.1162	174.6*** ± 0.48	18.49*** ± 0.29
Bottom	0–20	0.0584 ns ± 0.011	0.0065	106.7** ± 0.13	11.65** ± 0.12
	20–40	0.0018 ns ± 0.001	0.0003	47.8*** ± 0.59	7.71 ns ± 0.13

*, **, ***, different from the Top position (reference soil) at $P < 0.05$, $P < 0.01$ and $P < 0.001$, respectively, according to Student's t-test; ns—not significant difference.

(v/v). At the beginning of CH_4 evolution in this soil after 17 days of incubation, there was only 3.21% (v/v) O_2 left in the headspace. At the end of incubation, O_2 was hardly depleted (1.36% v/v in the headspace). The Top and Bottom soils consumed 14.6 and 11.6% (v/v) O_2, respectively, which yielded the final O_2 concentration in the headspace of 5.88 and 8.85% (v/v), respectively.

The subsurface-flooded soils showed some lower respiration (Figures 1(e) and 1(f)). An exception was the Middle soil, which produced as much as 174.6 mg CO_2-C kg⁻¹, while it consumed 18.5% (v/v) O_2 (2% v/v O_2 left in the headspace). The other soils followed the tendency observed in CH_4 production; thus they evolved less CO_2 and consumed less O_2 (<111 mg CO_2-C kg⁻¹ and <8.8% (v/v), resp.). In most cases, the CO_2 produced and O_2 consumed in the course of methanogenic incubation in the Middle and Bottom soils significantly differed from those in Top soil (Table 2).

3.2. Methanotrophic Activity of Soils from Different Slope Positions. The capacity of CH_4 consumption in the upper soils was slightly modified by the slope position (Figure 2(a)). All the soils collected at the 0–20 cm depth consumed completely the added methane (100%) after a 3-4 day lag. The highest methanotrophic activity was observed in the Middle soil, which rapidly utilized the whole gas between 4 and 9 day, at a rate of -20.66 mg CH_4-C kg⁻¹ d⁻¹ (Table 3). CH_4 consumption in the Top and Bottom samples lasted some longer, until the 11th incubation day (16.08 and 17.58 mg CH_4-C kg⁻¹ d⁻¹, resp.).

Deeper soil layers showed greater variation in CH_4 oxidation capacity (Figure 2(d)). The highest activity was shown, again, by the soil collected from Middle position, which completely utilized 121.3 mg CH_4-C kg⁻¹ between the 3rd and 9th incubation days (at a rate of -12.26 mg CH_4-C kg⁻¹ d⁻¹). The Bottom soil consumed 116.49 mg CH_4-C kg⁻¹ (92% of initial) at a rate of -7.95 mg CH_4-C kg⁻¹ d⁻¹, whereas the Top soil oxidized only 14% of added CH_4, at a low rate of -0.799 mg CH_4-C kg⁻¹ d⁻¹ ($P < 0.001$) (Table 3).

In general, CO_2 production and O_2 consumption followed the tendencies in CH_4 uptake (Figure 2). The Middle soil of both depths showed an apparent increase in CO_2, and a decrease in O_2 in the headspace after 4 days of incubation, when CH_4 oxidation started. Total CO_2 evolution

of 198.9 mg C kg⁻¹ was, however, similar to that measured in the Top upper soil, which consumed significantly less CH_4 (Table 3). The amount of CO_2 evolved by the Middle upper soil was significantly higher than that produced in the Bottom soil (151.9 mg C kg⁻¹, $P < 0.001$). Among the soils collected from deeper layers, the Middle soil produced more CO_2 and consumed more O_2 than the other soils, with accumulation of 224.1 mg CO_2-C kg⁻¹ and utilization of 13.6% (v/v) O_2, $P < 0.001$ (Figure 2).

3.3. Relations between Measured Soil Properties. The amount of methane produced in flooded soils showed a close relationship with the amount of organic C modified by the soil position in the slope (Figure 3(a)).

Similar significant relations were observed for CO_2 produced and O_2 consumed during incubation of flooded and wet soils (i.e., in the course of methane production and oxidation, resp.) versus C_{org} (Figures 3(a) and 3(b)). Such correlations for methane oxidation were not shown.

4. Discussion

Position in the landscape affects the accumulation and redistribution of water, nutrients, sediments, and organic matter. Soils on ridges and upper slopes will tend to loose soil and organic matter that will tend to accumulate on lower slopes and in depressions. Generally, soils in lower-slope positions will tend to have a wetter moisture regime for a longer period [33], while soil O_2 concentrations may decrease significantly from ridges to valleys [34]. Methane emission from low-slope positions may be observed already one or three days after summer rainfall, depending on the intensity of precipitation [13]. It has been assumed that, in well-aerated soils, CH_4 production in anaerobic microsites could be an important source of methane for methane oxidizing bacteria [35]. Little is known, however, about methane emission when usually well-drained soils become flooded for a short period [13]. The characteristics of CH_4 oxidizing and producing communities and the factors which affect these characteristics as well as CH_4 transport determine the magnitude of the surface CH_4 flux to the atmosphere [36]. Our studies with loess soil collected from different slope positions and incubated under laboratory conditions showed that the slope position significantly affected the soil

TABLE 3: CH$_4$ consumption, CO$_2$ evolution, and O$_2$ uptake in soils collected from three slope positions and incubated with 5 kPa methane for 21 days (average values $\pm$ standard error, $n = 3$).

Slope position	Soil depth (cm)	CH$_4$ consumption			CO$_2$ evolution (mg C kg^{-1})	O$_2$ uptake % (v/v)
		Total (mg C kg^{-1})	% of initial CH$_4$	Rate (mg C kg^{-1} d^{-1})		
Top	0–20	130.84 ± 2.31	100	−16.08	186.7 ± 11.4	11.14 ± 0.53
	20–40	17.80 ± 2.62	14	−0.799	83.9 ± 2.86	2.55 ± 0.11
Middle	0–20	121.36* ± 0.01	100	−20.66	198.9 ns ± 3.57	11.77 ns ± 0.34
	20–40	131.71* ± 0.80	100	−12.26	224.1*** ± 5.26	13.63*** ± 0.32
Bottom	0–20	130.96 ns ± 43.7	100	−17.58	151.9 ns ± 5.34	9.44 ns ± 0.38
	20–40	116.49*** ± 12.2	92	−7.954	104.1** ± 3.14	6.99*** ± 0.50

*, **, ***, different from the Top (reference) soil at $P < 0.05$, $P < 0.01$, and $P < 0.001$, respectively, according to Student's t-test; ns—not significant difference.

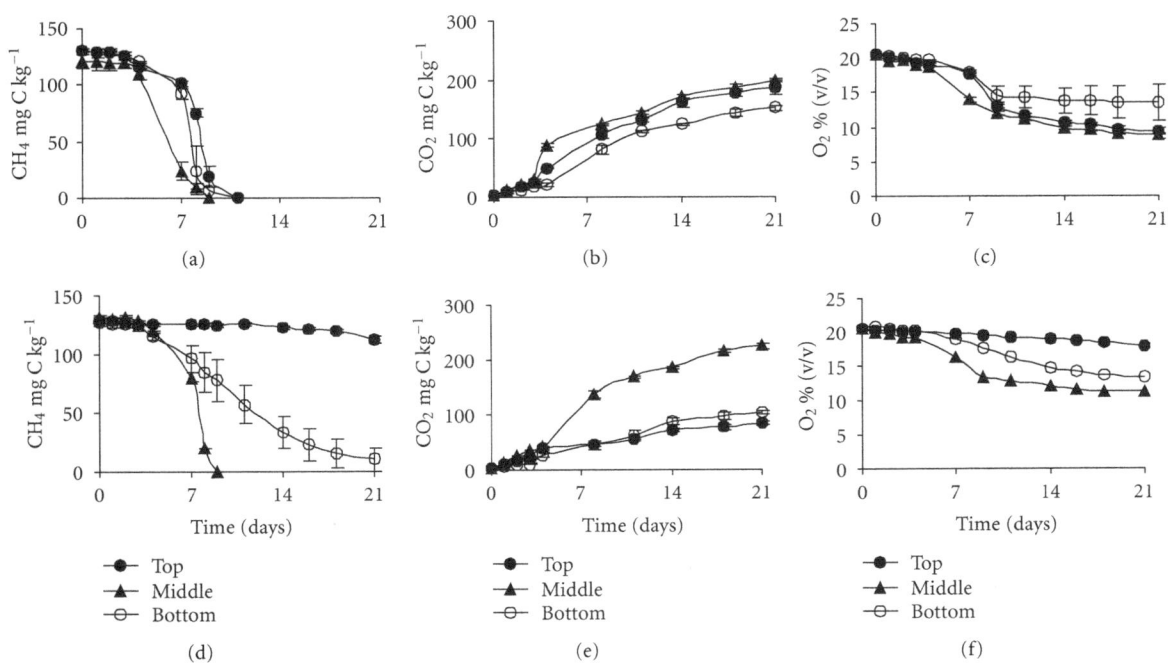

FIGURE 2: Changes of CH$_4$, CO$_2$, and O$_2$ over time in loess soils collected from three slope positions and incubated with added methane, 5 kPa (methanotrophic potential). Top: summit, Middle: back slope, Bottom: bottom of the slope. Upper graphs (a–c) upper soil depth of 0–20 cm; lower graphs (d–f) lower soil depth of 20–40 cm. (a) and (d) cumulative CH$_4$ production; (b) and (e) cumulative CO$_2$ production; (c) and (f)—changes in O$_2$ in the headspace. Points represent triplicate-means with standard error.

potential for CH$_4$ production and oxidation and modified soil respiration involved in CH$_4$ transformations.

We compared soils from the Middle and Bottom positions with the soils at the summit (Top) position, which may be regarded as a reference soil. The methane production potential of the soil from the Middle position significantly increased (9-fold in the upper soil layer and even more in the deeper layer $P < 0.001$), while at the Bottom position CH$_4$ production did not changed significantly (Table 2). In turn, the methane oxidation potential was unchanged in the 0–20 cm layer (as all upper soils depleted all added CH$_4$) but strongly increased in the 20–40 cm soil layers and form the level of 14% in the Top soil, to 92–100% in the other soils ($P < 0.05$ or $P < 0.001$).

In the course of methanogenesis, soil respiration underwent modification similar to that observed for the methanogenic potential. In the Middle slope position, CO$_2$ production and O$_2$ uptake were significantly stimulated as compared with the Top soil (by 50% and 30%, resp., $P < 0.001$ and $P < 0.01$). Soil collected at the Bottom position showed lower respiration than the Top soil (both CO$_2$ and O$_2$ less by about 20%, $P < 0.01$). In the deeper soil layer, the changes were generally more pronounced.

Changes in soil respiration in the course of methanotrophy were apparently dependent on soil depth. In comparison with the Top site, the upper Middle and Bottom soils were not changed, as all soils consumed comparable amounts of methane. However, soils sampled from a depth of 20–40 cm

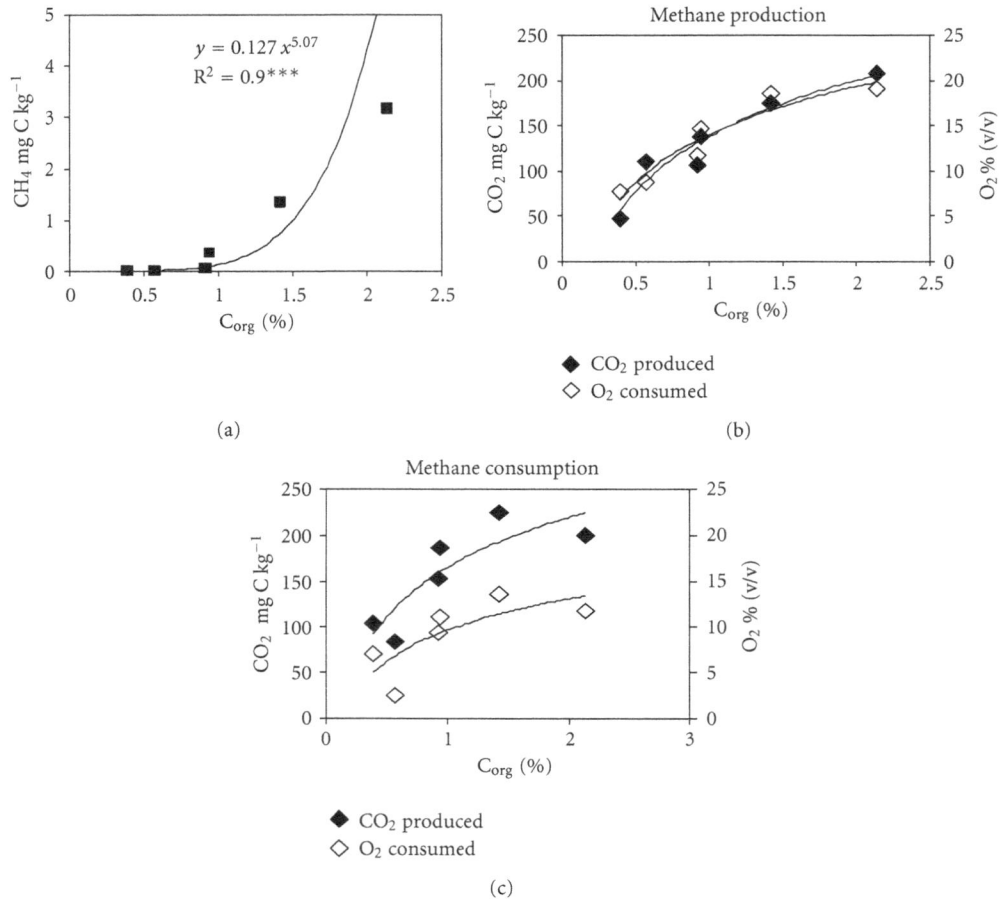

(a)

(b)

(c)

FIGURE 3: Relationships between total gases produced or consumed over time and organic carbon (C_{org}) in soils collected from different slope positions. (a) CH_4 produced in flooded soils versus C_{org}; (b) CO_2 produced and O_2 consumed in flooded soils versus C_{org} ($y = 7.60 \cdot Ln(x) + 14.1$, $R^2 = 0.92***$) and ($y = 88.4 \cdot Ln(x) + 138.7$, $R^2 = 0.93***$); respectively, (c) CO_2 produced and O_2 consumed in wet soils enriched with CH_4 soils versus C_{org} ($y = 78.1 \cdot Ln(x) + 165.2$, $R^2 = 0.74***$) and ($y = 4.38 \cdot Ln(x) + 4.60$, $R^2 = 0.92*$), respectively. Points present mean values. *** and *, $P < 0.001$ and $P < 0.05$, respectively.

respired at a significantly higher rate than the Top soil (up to 5 times) ($P < 0.001$).

In the experiments of [21], water-stable aggregates were significantly different among landscape positions and decreased from lower > middle > summit landscape position. However, in some soils, the landscape effect was insignificant, for example, for enzyme activities or emission of CH_4 and CO_2 [19, 21]. Fang et al. [17] observed that neither potential net N mineralization nor nitrification was differentiated by the slope position, nor was accumulative emissions of N_2O or CO_2 from incubated soils in laboratory; in contrast, the ability to oxide CH_4 appeared to decrease from the bottom to the top.

It is well known that methane fluxes are strongly regulated by the presence or absence of methanotrophs (CH_4 oxidizers), which are generally found in the upper (0–20 cm) soil [37]. On the other hand, methanogens (CH_4 producers) use labile carbon compounds that were produced in the root zone and are less abundant with increasing distance from the soil surface, and rates of potential CH_4

production decline with depth below the aerobic zone [38]. However, deeper soil also contributes to CH_4 emission. In our experiment, both upper (0–20 cm) and lower (20–40 cm) soil layers of the Middle position evolved CH_4. The process started relatively fast, after 17 days of flooded incubation at 25°C, and CH_4 reached 3.16 mg CH_4-C kg^{-1} (0.35% v/v of CH_4 in the headspace) over 28-day incubation. Similarly, Mayer and Conrad [39] observed a rapid increase in CH_4 production within 25 days of flooding an upland agricultural soil and a forest soil. It is possible that, in our experiment, CH_4 emissions did not occur in the other soils since O_2, which is the most thermodynamically favourable electron acceptor [29, 37], was still in the headspace. Methanogenesis is evidently inhibited by O_2 this is apparent from field studies that show no overlap in the depth distributions of O_2 penetration in soils or sediments and net CH_4 production [38]. However, the lack of CH_4 production in the presence of O_2 in situ may be due to a combination of factors, of which O_2 toxicity is just one. For example, methanogens are more sensitive to desiccation than O_2 exposure in a paddy soil [40]. In our experiment, methane formation in the Middle soil was preceded by relatively rapid O_2 consumption, which created soil hypoxia and allowed methanogens to develop. However, CH_4 was also detected in the upper Top soil on day 21 (0.008% v/v in the headspace), although there was 8.4% (v/v) O_2 in the headspace.

Probably, the mechanisms of the slope position effect on methanogenic and methanotrophic activities are complex. Some decline in soil pH was observed for the Middle and Bottom soils as compared with the Top soil (from pH 8.1 to 7.6–7.9, Table 1). However, most methanogenic communities seem to be dominated by neutrophilic species [38]; in a survey of 68 methanogenic species conducted by Garcia et al. [41], most species grew best in a pH range from 6 to 8. Evidently, in our experiment, the soil position changed the sand, silt, and clay content at the depth of 20–40 cm (Table 1). The clay fraction content decreased from 6.3% in the Top soil to 4.2 and 4.3% in the Middle and Bottom soils, respectively. Similarly, the silt content decreased from 77.4% to 68.7 and 61.6%, respectively. A change in the sand content (an increase from 16.3% in the Top position to 27.5 and 34.5% in the Middle and Bottom positions, resp.) probably stimulated CH_4 oxidation due to better draining and easier gas diffusion in soil containing more sand, because O_2 is necessary for monooxygenase enzyme which catalyzes methane oxidation [42]. Coarse-textured soils have been documented as supporting CH_4 oxidation by enhancing gas diffusion (CH_4 and O_2) into the soil [43].

Paluszek and Żembrowski [44] present their findings from a long-term study designed to explore effect of accelerated erosion on soil properties in a loessial landscape. They observed that slight, moderate, and severe erosion has an adverse effect on soil physical properties. The clay content and bulk density in Ap horizons of eroded soils are on the increase whereas the content of organic matter, content of water-stable aggregates, field water capacity, and retention of water useful for plants decrease. In the consequence, soil porosity, air capacity, and air permeability deteriorate. By contrast, in very severely eroded soils whose Ap horizons developed from carbonate loess, pore-size distribution, field water capacity and retention of water useful for plants are favourable and comparable to those in noneroded soils [44].

High organic C content in the Middle soil, as compared with the Top soil (Table 1), stimulated methane production under soil hypoxia. Probably, small differences in C_{org} between Top and Bottom soils may explain insignificant differences in methanogenic activity. Nevertheless, high correlation coefficients obtained for relationships between C_{org} and produced CH_4, evolved CO_2, and consumed O_2 confirm their universal character (Figure 3). However, better explanation of the changes observed in our experiment needs more information on the properties of tested soils at different slope positions.

Acknowledgment

The paper was partly financed by the State Committee for Scientific Research, Poland (Grants no. N N310 043838). The authors greatly appreciate Professor R. Dębicki (Maria Curie-Skłodowska University in Lublin, Poland) who inspired us to perform these studies.

References

[1] K. Knittel and A. Boetius, "Anaerobic oxidation of methane: progress with an unknown process," *Annual Review of Microbiology*, vol. 63, pp. 311–334, 2009.

[2] Q. Zhuang, J. M. Melack, S. Zimov, K. M. Walter, C. L. Butenhoff, and M. Aslam K Khalil, "Global methane emissions from wetlands, rice paddies, and lakes," *Eos*, vol. 90, no. 5, pp. 37–38, 2009.

[3] M. J. Prather and J. Hsu, "Coupling of nitrous oxide and methane by global atmospheric chemistry," *Science*, vol. 330, no. 6006, pp. 952–954, 2010.

[4] D. Fowler, K. Pilegaard, M. A. Sutton et al., "Atmospheric composition change: ecosystems-atmosphere interactions," *Atmospheric Environment*, vol. 43, no. 33, pp. 5193–5267, 2009.

[5] Intergovernmental Panel on Climate Change, "Climate change 2007: the scientific basis," Contribution of Working Group 1 to the Fourth Assessment Report of the Intergovernmental Panel on Climate Change, Cambridge University Press, Cambridge, UK, 2007.

[6] E. Topp and E. Pattey, "Soils as sources and sinks for atmospheric methane," *Canadian Journal of Soil Science*, vol. 77, no. 2, pp. 167–178, 1997.

[7] A. S. K. Chan and T. B. Parkin, "Effect of land use on methane flux from soil," *Journal of Environmental Quality*, vol. 30, no. 3, pp. 786–797, 2001.

[8] C. Brink, E. van Ierland, L. Hordijk, and C. Kroeze, "Cost-effective emission abatement in europe considering interrelations in agriculture," *TheScientificWorldJournal*, vol. 1, pp. 814–821, 2001.

[9] W. Stępniewski, "Aeration of soils and plants," in *Encyclopedia of Agrophysics*, J. Gliński, J. Horabik, and J. Lipiec, Eds., pp. 8–12, Springer Science+Business Media B.V., Dordrecht, The Netherlands, 1st edition, 2011.

[10] W. Stępniewski, Z. Stępniewska, R. P. Bennicelli, and J. Gliński, *Oxygenology in Outline*, Institute of Agrophysics of the Polish Academy of Sciences, Lublin, Poland, 2005.

[11] M. Brzezińska, Z. Stępniewska, and W. Stępniewski, "Soil oxygen status and dehydrogenase activity," *Soil Biology and Biochemistry*, vol. 30, no. 13, pp. 1783–1790, 1998.

[12] A. S. K. Chan and T. B. Parkin, "Methane oxidation and production activity in soils from natural and agricultural ecosystems," *Journal of Environmental Quality*, vol. 30, no. 6, pp. 1896–1903, 2001.

[13] F. L. Wang and J. R. Bettany, "Methane emission from Canadian prairie and forest soils under short term flooding conditions," *Nutrient Cycling in Agroecosystems*, vol. 49, no. 1–3, pp. 197–202, 1997.

[14] M. Brzezińska, "Impact of treated wastewater on biological activity and accompanying processes in organic soils (field and model experiments)," *Acta Agrophysica*, vol. 131, pp. 7–164, 2006 (Polish).

[15] S. Abrishamkesh, M. Gorji, and H. Asadi, "Long-term effects of land use on soil aggregate stability," *International Agrophysics*, vol. 25, no. 2, pp. 103–108, 2011.

[16] A. Akbarzadeh, R. T. Mehrjardi, H. G. Refahi, H. Rouhipour, and M. Gorji, "Using soil binders to control runoff and soil loss in steep slopes under simulated rainfall," *International Agrophysics*, vol. 23, no. 2, pp. 99–109, 2009.

[17] Y. Fang, P. Gundersen, W. Zhang et al., "Soil-atmosphere exchange of N2O, CO2 and CH 4 along a slope of an evergreen broad-leaved forest in southern China," *Plant and Soil*, vol. 319, no. 1-2, pp. 37–48, 2009.

[18] A. Ferrero and J. Lipiec, "Determining the effect of trampling on soils in hillslope-woodlands," *International Agrophysics*, vol. 14, no. 1, pp. 9–16, 2000.

[19] C. N. Gacengo, C. W. Wood, J. N. Shaw, R. L. Raper, and K. S. Balkcom, "Agroecosystem management effects on greenhouse gas emissions across a coastal plain catena," *Soil Science*, vol. 174, no. 4, pp. 229–237, 2009.

[20] M. T. Rashid and R. P. Voroney, "Field-scale application of oily food waste and nitrogen fertilizer requirements of corn at different landscape positions," *Journal of Environmental Quality*, vol. 34, no. 3, pp. 963–969, 2005.

[21] R. P. Udawatta, R. J. Kremer, B. W. Adamson, and S. H. Anderson, "Variations in soil aggregate stability and enzyme activities in a temperate agroforestry practice," *Applied Soil Ecology*, vol. 39, no. 2, pp. 153–160, 2008.

[22] J. Rejman, "Effect of water and tillage erosion on transformation of soils and loess slopes," *Acta Agrophysica*, vol. 136, pp. 5–91, 2006 (Polish).

[23] R. Turski, J. Paluszek, and A. Slowinska-Jurkiewicz, "The effect of erosion on the spatial differentiation of the physical properties of Orthic Luvisols," *International Agrophysics*, vol. 6, no. 3-4, pp. 123–136, 1992.

[24] P. Gliński and Z. Stępniewska, "Changes of redox properties in slightly eroded loess soil," *Acta Agrophysica*, vol. 4, no. 3, pp. 669–686, 2004.

[25] P. Gliński and R. Dębicki, "Degradation of loessial soils in the Ciemięga river basin," *Acta Agrophysica*, vol. 23, pp. 39–46, 1999.

[26] M. Brzezińska, P. Rafalski, T. Włodarczyk, P. Szarlip, and K. Brzeziński, "How much oxygen is needed for acetylene to be consumed in soil?" *Journal of Soils and Sediments*, vol. 11, no. 7, pp. 1142–1154, 2011.

[27] T. Włodarczyk, W. Stępniewski, and M. Brzezińska, "Dehydrogenase activity, redox potential, and emissions of carbon dioxide and nitrous oxide from Cambisols under flooding conditions," *Biology and Fertility of Soils*, vol. 36, no. 3, pp. 200–206, 2002.

[28] M. Ryzak and A. Bieganowski, "Determination of particle size distrubution of soil using laser diffraction—comparison with areometric method," *International Agrophysics*, vol. 24, no. 2, pp. 177–181, 2010.

[29] J. Gliński and W. Stępniewski, *Soil Aeration and Its Role for Plants*, CRC Press, Boca Raton, Fla, USA, 1985.

[30] T. E. Freitag and J. I. Prosser, "Correlation of methane production and functional gene transcriptional activity in a peat soil," *Applied and Environmental Microbiology*, vol. 75, no. 21, pp. 6679–6687, 2009.

[31] X. Xu and K. Inubushi, "Temperature effects on ethylene and methane production from temperate forest soils," *Chinese Science Bulletin*, vol. 54, no. 8, pp. 1426–1433, 2009.

[32] J. Gulledge and J. P. Schimel, "Moisture control over atmospheric CH4 consumption and CO2 production in diverse Alaskan soils," *Soil Biology and Biochemistry*, vol. 30, no. 8-9, pp. 1127–1132, 1998.

[33] B. Murphy, B. Wilson, and A. Rawson, "Development of a soil carbon benchmark matrix for central west NSW," in *Proceedings of the 19th World Congress of Soil Science. Soil Solutions for a Changing World*, Brisbane, Australia, August 2010.

[34] W. L. Silver, A. E. Lugo, and M. Keller, "Soil oxygen availability and biogeochemistry along rainfall and topographic gradients in upland wet tropical forest soils," *Biogeochemistry*, vol. 44, no. 3, pp. 301–328, 1999.

[35] R. Conrad, "Soil microbial processes and the cycling of atmospheric trace gases," *Advances in Microbial Ecology*, vol. 14, pp. 207–250, 1995.

[36] M. A. Bradford, P. Ineson, P. A. Wookey, and H. M. Lappin-Scott, "Role of CH4 oxidation, production and transport in forest soil CH4 flux," *Soil Biology and Biochemistry*, vol. 33, no. 12-13, pp. 1625–1631, 2001.

[37] A. Matson, D. Pennock, and A. Bedard-Haughn, "Methane and nitrous oxide emissions from mature forest stands in the boreal forest, Saskatchewan, Canada," *Forest Ecology and Management*, vol. 258, no. 7, pp. 1073–1083, 2009.

[38] J. P. Megonigal, M. E. Hines, and P. T. Visscher, "Anaerobic metabolism: linkages to trace gases and aerobic processes," in *Biogeochemistry*, W. H. Schlesinger, Ed., pp. 317–424, Elsevier-Pergamon, Oxford, UK, 2004.

[39] H. P. Mayer and R. Conrad, "Factors influencing the population of methanogenic bacteria and the initiation of methane production upon flooding of paddy soil," *FEMS Microbiology Ecology*, vol. 73, no. 2, pp. 103–111, 1990.

[40] S. Fetzer, F. Bak, and R. Conrad, "Sensitivity of methanogenic bacteria from paddy soil to oxygen and desiccation," *FEMS Microbiology Ecology*, vol. 12, no. 2, pp. 107–115, 1993.

[41] J.-L. Garcia, B. K. C. Patel, and B. Ollivier, "Taxonomic, phylogenetic, and ecological diversity of methanogenic Archaea," *Anaerobe*, vol. 6, no. 4, pp. 205–226, 2000.

[42] X. K. Xu and K. Inubushi, "Ethylene oxidation, atmospheric methane consumption, and ammonium oxidation in temperate volcanic forest soils," *Biology and Fertility of Soils*, vol. 45, no. 3, pp. 265–271, 2009.

[43] A. Saari, P. J. Martikainen, A. Ferm et al., "Methane oxidation in soil profiles of Dutch and Finnish coniferous forests with different soil texture and atmospheric nitrogen deposition," *Soil Biology and Biochemistry*, vol. 29, no. 11-12, pp. 1625–1632, 1997.

[44] J. Paluszek and W. Żembrowski, "Improvement of the soils exposed to erosion in a loessial landscape," *Acta Agrophysica*, vol. 164, pp. 5–159, 2008 (Polish).

15

Impact of Long-Term Forest Enrichment Planting on the Biological Status of Soil in a Deforested Dipterocarp Forest in Perak, Malaysia

D. S. Karam,[1] A. Arifin,[1, 2] O. Radziah,[3, 4] J. Shamshuddin,[3] N. M. Majid,[1] A. H. Hazandy,[1, 2] I. Zahari,[5] A. H. Nor Halizah,[5] and T. X. Rui[1]

[1] Department of Forest Production, Faculty of Forestry, Universiti Putra Malaysia, 43400 Serdang, Selangor, Malaysia
[2] Laboratory of Sustainable Bioresource Management, Institute of Tropical Forestry and Forest Products, Universiti Putra Malaysia, 43400 Serdang, Selangor, Malaysia
[3] Department of Land Management, Faculty of Agriculture, Universiti Putra Malaysia, 43400 Serdang, Selangor, Malaysia
[4] Laboratory of Food Crops and Floriculture, Institute of Tropical Agriculture, Universiti Putra Malaysia, 43400 Serdang, Selangor, Malaysia
[5] Forestry Department Peninsular Malaysia, Jalan Sultan Salahuddin, 50660 Kuala Lumpur, Malaysia

Correspondence should be addressed to A. Arifin, arifin_soil@yahoo.com

Academic Editor: Tom Jensen

Deforestation leads to the deterioration of soil fertility which occurs rapidly under tropical climates. Forest rehabilitation is one of the approaches to restore soil fertility and increase the productivity of degraded areas. The objective of this study was to evaluate and compare soil biological properties under enrichment planting and secondary forests at Tapah Hill Forest Reserve, Perak after 42 years of planting. Both areas were excessively logged in the 1950s and left idle without any appropriate forest management until 1968 when rehabilitation program was initiated. Six subplots (20 m × 20 m) were established within each enrichment planting (F1) and secondary forest (F2) plots, after which soil was sampled at depths of 0–15 cm (topsoil) and 15–30 cm (subsoil). Results showed that total mean microbial enzymatic activity, as well as biomass C and N content, was significantly higher in F1 compared to F2. The results, despite sample variability, suggest that the rehabilitation program improves the soil biological activities where high rate of soil organic matter, organic C, N, suitable soil acidity range, and abundance of forest litter is believed to be the predisposing factor promoting higher population of microbial in F1 as compared to F2. In conclusion total microbial enzymatic activity, biomass C and biomass N evaluation were higher in enrichment planting plot compared to secondary forest. After 42 years of planting, rehabilitation or enrichment planting helps to restore the productivity of planted forest in terms of biological parameters.

1. Introduction

Malaysia is a country rich in biodiversity of which natural forest is a home for thousands of flora and fauna [1]. However, the need for development and urbanization catalysed by the pressure of rising human population has made vast area of natural forests cleared up to cultivate new area for housing and wood productions. Liebig et al. [2] stated that the fertility of soil proportionally change with time catalyzed by natural phenomena and human activities. Hence, deforestation of natural forest leads to soil degradation, which proceeds rapidly under tropical climatic conditions [3, 4]. Forest rehabilitation is believed to be one of the best ways to overcome and lower down the demand for woody and nonwoody products from natural forest. Besides that, forest plantation also supports the shortage of wood supply, while sustaining world ecosystem [3]. In addition, forest plantation is also known as an alternative way to restore degraded sites to its original condition and sustains its soil fertility [5, 6]. Insam [7] found that soil fertility and its management are the most crucial part to evaluate a particular site of soil ecological area which gives a preview of the site's environmental management and the extent of success for a particular forest rehabilitation program which can only be identified through its soil fertility evaluation.

Enrichment planting is one of important technique used in forest rehabilitation [8, 9]. Montagnini et al. [10] defined enrichment planting as the introduction of valuable species to degraded forests without the elimination of valuable individual which already existed at that particular site. Adjers et al. [11] summarized that there are total of 25857 ha of forest plantation had been planted through enrichment planting technique in Peninsular Malaysia. *Shorea acuminata, S. leprosula, Dryobalanops aromatica,* and *D. oblongifolia* are among the favorite species planted in Peninsular Malaysia [12]. While for secondary forest, it is a forest area which has regrown trees after major disruption and disturbance such as fire and deforestation. Normally, the regeneration of plants species in secondary forests are done naturally by itself without any forest treatment given for a period of long time till the effect of disturbance is no longer noticed.

It is undeniable that soil microorganism is the major agents in promoting nutrient cycling including carbon (C), nitrogen (N), phosphorus (P), and sulphur (S). Furthermore, Gaspar et al. [13] concluded that soil microbial biomass comprises 1–4% and 2–6% of total organic C and N in soil, respectively. Rapid turnover of microbial activities in soil is dependent on the changes occurring in the surrounding environment such as climate change, disturbance, and pollutant toxicity [14, 15] which made microbial activity a good sensitive indicator [16] for soil fertility evaluation. Islam and Weil [17] also stressed the importance of including microbial biomass evaluation to describe the status of fertility and quality of soil at a particular study site.

Enzymatic activities are also one of the important evaluation aspects for determining soil fertility. They play a vital role in the organic residues degradation, humic substance synthesis, pollutant degradation, and nutrient cycles in soil [18]. Fluorescein diacetate (FDA) hydrolysis assay provides a reliable estimation of overall microbial activity in soil [19] and is widely used to analyse bacterial and fungal enzymatic activities [20, 21]. In addition, FDA analysis is considered as nonspecific because it is hydrolysed by various types of enzymes which include protease, esterase, and lipase [13, 21]. Heal and Maclean [22] found that approximately 90% of the energy transfer cycle in the soil was via microbial decomposer, and total microbial activity illustrates a general measurement of the organic matter turnover. Behera and Sahani [5] stated the importance of including biological studies, such as the evaluation of microbial biomass in land evaluations, because they provide a better indication of changes or degradation in forest soils than carbon and nitrogen analyses. Vásquez-Murrieta et al. [23] also stated that the key factors regulating and maintaining continuous supplies of nutrients in the soil for plant uptake are circulated by soil microbes. Soil fertility evaluation primarily focuses on the physicochemical properties in order to describe the growth performance of particular tree species at the plantation without taking into account the importance of soil biological properties as sensitive indicator to the changes occurring in the soil [24]. Hence, the objective of this study was to provide information and compare soil biological properties under enrichment planting and secondary forests after 42 years (as for 2010) of planting at Bukit Tapah Forest Reserve, Perak, Malaysia.

2. Materials and Methods

2.1. Description of the Study Site. The study was carried out in enrichment planting (N 04.179394° E 101.31998°) and secondary forest (N 04.17336° E 101.31974°) at Bukit Tapah Forest Reserves, Perak (Figure 1) on 21st until 23rd July 2011. The mean annual rainfall and temperature are 2,417 mm and 24.5°C, respectively. The soils in this study area are classified as Ultisols, which are considered as highly weathered due to large amount of low-activity clays associated with high Al saturation [3]. All of the tree species of *Shorea leprosula, S. bracteolata,* and *S. macroptera* planted were done on 2nd February 1968, and the age of the trees was 42 years old in 2010, while adjacent secondary forest was left idle to undergo natural regeneration without any reforestation activity. Compartment 13 of Bukit Tapah is one of the 10 compartments that was gazetted for enrichment planting at Perak South District, Malaysia. About 1,185 hectares out of 64,984 hectares of Bukit Tapah Forest Reserve were converted to enrichment planting program of which compartment 13 covers 87.2 hectares of the forest reserves. The purpose of enrichment planting done at this area is to replace and curtail this particular area which had undergone excessive logging before 1968.

The size of the poly bags used to plant the seedlings was 10 cm × 15 cm × 23 cm. Twenty-six thousand five hundred and forty-four saplings were planted with 304 saplings per hectare, and the rates of survival recorded in 1970 found that only 9,158 trees managed to grow well and survive with resulting in 105 saplings per hectare, respectively. *Shorea leprosula, S. parvifolia, S. bracteolata,* and *S. macroptera* were the main species of Dipterocarpaceae planted in compartment 13 enrichment planting plot. The trees were planted on a 10 m × 3 m grid.

2.2. Experimental Design and Soil Sampling. This study used a completely randomized design. Enrichment planting and secondary forest plots were designated as F1 and F2, respectively. Six subplots were demarcated in each plot in order to serve as replicates. Six soil samples were randomly collected at depths of 0–15 cm and 15–30 cm in each subplot. The samples were then mixed together to form a composite sample for each soil depth range. Hence, 12 composite samples (six from soil depth 0–15 cm and six from soil depth 15–30 cm) were collected from each plot for the analysis. The composite samples were kept in UV-sterilized polyethylene bags at 0°–4°C.

2.3. Total Microbial Population. Spread-plate technique or direct count of colony forming unit was used to evaluate the estimation of microbial population [25, 26]. Nutrient agar was used for bacterial culture. Dilution factor of 10^{-2}, 10^{-3},

Impact of Long-Term Forest Enrichment Planting on the Biological Status of Soil in a Deforested Dipterocarp
Forest in Perak, Malaysia

129

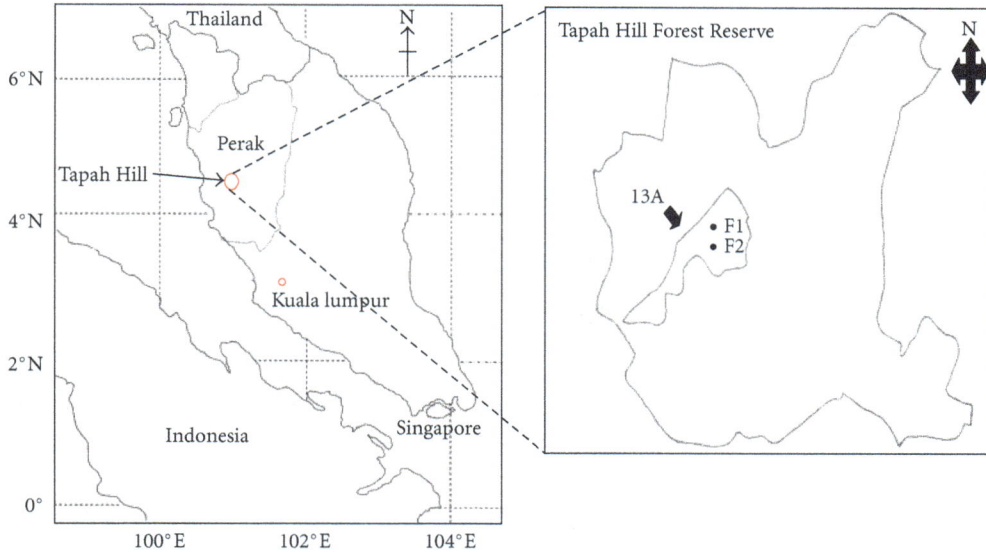

FIGURE 1: Enrichment planting (F1) and secondary forest (F2) plots at Tapah Hill Forest Reserve, Perak, Malaysia (Scale 1 : 20 000).

and 10^{-4} was found to be suitable for colony calculation after few pilot test carried out to standardize the dilution factor for every population counts. The number of colony forming units per gram soil was calculated using the following equation:

number of colony forming units/g of dry weight soil

$$= \frac{[(\text{mean plate count})(\text{dilution factor})]}{(\text{dry weight soil, initial dilution})}, \tag{1}$$

where dry weight soil = (Weight of moist soil, initial dilution blank) × [(1 − % moisture soil sample)/100]. The results were expressed in $\log_{10} g^{-1}$ soil.

2.4. Microbial Enzymatic Activity.
Fluorescein diacetate (FDA) hydrolysis assay illustrated by Sánchez-Monedero et al. [18] and Gagnon et al. [27] was used to evaluate microbial enzymatic activity.

2.5. Microbial Biomass Analysis.
Soil microbial biomass C (MBC) and N (MBN) were extracted using rapid chloroform fumigation extraction described by Witt et al. [28]. Soil MBC analysed by wet dichromate oxidation [23] and calculation for biomass C is as below:

$$MBC = \frac{\left(C_{fumigated} - C_{control}\right)}{kEC}. \tag{2}$$

The chloroform-labile C pool was calculated as the difference between samples of un-fumigated and fumigated C which is proportional to MBC, where kEC is soil specifically estimated as 0.38 [29].

Soil MBN was determined using Kjeldahl digestion and distillation technique [30, 31]. The calculation for biomass N is

$$MBN = \frac{\left(N_{fumigated} - N_{control}\right)}{kEN}. \tag{3}$$

The chloroform-labile N pool was calculated as the difference between samples of un-fumigated and fumigated N which is proportional to MBN, where kEN is soil specifically estimated as 0.54 [32].

2.6. Measurement of Soil Organic Matter, Organic C, Total N, Soil Acidity, Bulk Density, and Moisture Content.
Soil organic matter and organic C were determined using loss on ignition method [33] total N via Kjeldahl digestion [31], and soil acidity was elucidated in a 1 : 2.5 of soil : distilled water suspension using a glass electrode [34, 35]. Bulk density was determined using the disturbed soil technique, and the gravimetric method was used to measure soil moisture content.

3. Statistical Analysis

Student's t-test was used to compare the differences between the mean values for microbial population, enzymatic activity, biomass C, biomass N, and selected physicochemical properties for samples collected at the same depths in the adjacent plots. Pearson correlation analysis was used to detect the correlation between microbial biomass C with organic matter and microbial biomass N with total N. SPSS version 16.0 was used for the statistical analysis.

4. Results

There were no significant differences ($P \leq 0.05$) between F1 ($2.96 \pm 0.04 \log_{10} g^{-1}$ soil) and F2 ($2.87 \pm 0.06 \log_{10} g^{-1}$ soil) for microbial population count (Figure 2). The total mean of microbial population count for 15–30 cm depth for both plots was too low and was excluded from the final results to avoid bias.

Microbial enzymatic activities were significantly different ($P \leq 0.05$) for both F1 and F2 at each depth (Figure 3).

Total mean of microbial enzymatic activity rate was 24.45 $\pm$ 0.65 $\mu g\,g^{-1}$ soil 0.5 h^{-1} and 22.91 $\pm$ 0.53 $\mu g\,g^{-1}$ soil 0.5 h^{-1} under F1 and F2 at 0–15 cm depths while, for 15–30 cm depth, F1 and F2 enzymatic activity rate each 22.25 $\pm$ 0.49 $\mu g\,g^{-1}$ soil 0.5 h^{-1} and 17.91 $\pm$ 1.73 $\mu g\,g^{-1}$ soil 0.5 h^{-1}, respectively.

MBC rate was significantly higher ($P \leq 0.05$) in F1 compared to F2 at the same soil depths (Figure 4). Total mean of MBC rate for each F1 and F2 at 0–15 cm and 15–30 cm depths was 465 $\pm$ 105 $\mu g\,g^{-1}$ soil, 325 $\pm$ 58 $\mu g\,g^{-1}$ soil, 158 $\pm$ 66 $\mu g\,g^{-1}$ soil, and 124 $\pm$ 35 $\mu g\,g^{-1}$ soil, respectively.

F1 and F2 were significantly different ($P \leq 0.05$) where F1 contained higher rate of MBN compared to F2 at 0–15 cm and 15–30 cm soil depths (Figure 5). Total mean of MBN rate for each F1 and F2 plots was 239 $\pm$ 8 $\mu g\,g^{-1}$ soil and 162 $\pm$ 18 $\mu g\,g^{-1}$ soil at 0–15 cm depth and 134 $\pm$ 12 $\mu g\,g^{-1}$ soil and 78 $\pm$ 11 $\mu g\,g^{-1}$ soil at 15–30 cm depth, respectively. There were no significant differences of ratio of MBC/MBN between F1 and F2 plots ($P \leq 0.05$) (Figure 6). MBC/MBN ratio for F1 at 0–15 cm and 15–30 cm depths was 1.91 $\pm$ 0.41 and 2.48 $\pm$ 0.48. In contrast, F2 exhibits a lower MBC/MBN ratio of 1.03 $\pm$ 0.45 at 0–15 cm depth and 1.84 $\pm$ 0.49 at 15–30 cm depth.

Soil organic matter and organic C were significantly different ($P \leq 0.05$) for both F1 and F2 at 0–15 cm and 15–30 cm depths (Table 1). Soil acidity does not show any significant difference for both plots at the same soil depths. At 0–15 cm, there were significant differences in bulk density and moisture content compared to F2. However, there were no significant differences detected between F1 and F2 at 15–30 cm.

Table 2 shows the results of the Pearson correlation for selected chemical and biological properties in both plots. There were no linear relationship detected between microbial biomass C and organic matter for both plots at the same soil depths. Besides that, microbial biomass N and total N also do not show any linear relationship between the same soil depths. Correlation analysis of organic matter content and MBC/MBN ratios showed no strong relationship.

5. Discussion

Microbial population count between enrichment planting and secondary forest show a proportional in microbial growth, and this situation could be catalysed by the abundance of forest litter available on the forest floor which promotes microbial decomposing activity to take place and increase soil fertility [5, 36].

Microbial enzymatic activity was found to be higher in 0–15 cm depth compared to the lower depth, and also greater in enrichment planting compared to secondary forest. This activity is probably facilitated by the thicker and greater abundance of forest litter available, which enhances microbial decomposing processes. Higher content of organic matter in enrichment planting as compared with secondary forest contributes to the higher enzymatic activity. Smith and Paul [16] justified that microbial activity has been proven to be a "sensitive indicator" to illustrate changes in soil organic matter. The higher microbial activity of the

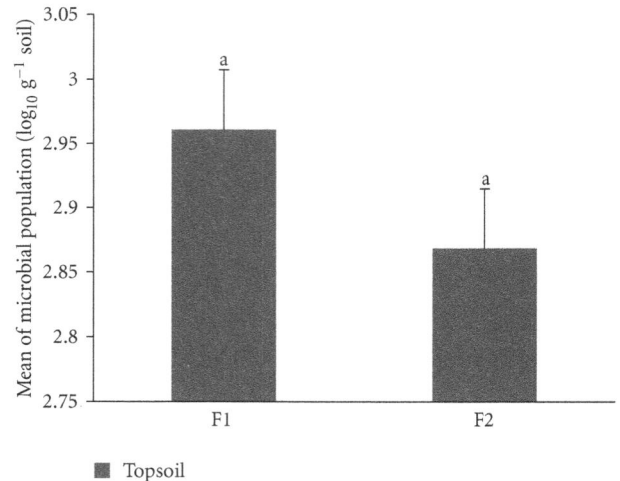

FIGURE 2: Total mean microbial population at F1 and F2 plots. Different letters indicate significant difference between means of the same soil depths at enrichment planting (F1) compared to secondary forest (F2) plots, using the Student's t-test ($P \leq 0.05$) (bars are means, whiskers indicate standard error).

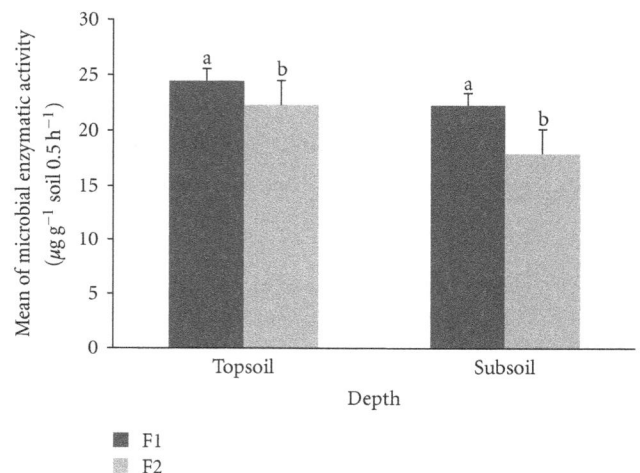

FIGURE 3: Total mean microbial enzymatic activity at F1 and F2 plots. Different letters indicate significant difference between means of the same soil depths at enrichment planting (F1) compared to secondary forest (F2) plots, using the Student's t-test ($P \leq 0.05$) (bars are means, whiskers indicate standard error).

enrichment plot at 0–15 cm may also be due to the high moisture content, which, along with surrounding humidity, enhances the microorganism cycles in the soil. Moreover, low soil compaction in the enrichment plot would also provide better air and water penetration in the soil to allow macro- and microorganisms to thrive and undergo necessary daily biochemical processes.

Greater amount of organic matter in enrichment planting is a valuable indication of greater amount of MBC. Islam and Weil [17] suggested that abundance and thickness of the layer of litter on the forest floor promotes high decomposing processes by soil microorganism. In addition,

Impact of Long-Term Forest Enrichment Planting on the Biological Status of Soil in a Deforested Dipterocarp Forest in Perak, Malaysia

131

TABLE 1: Selected soil physicochemical properties of enrichment planting (F1) and secondary forest (F2) plots.

Parameters	F1	F2	P value
	0–15 cm depth		
Organic matter (%)	16.99 ± 0.84^a	12.12 ± 0.35^b	0.001947
Organic carbon (%)	9.86 ± 0.49^a	7.03 ± 0.47^b	0.001947
Total nitrogen (%)	1.55 ± 0.09^a	1.11 ± 0.09^b	0.006318
pH-H_2O	4.36 ± 0.11^a	4.19 ± 0.05^a	0.348473
Bulk density (g cm^{-3})	1.16 ± 0.01^a	1.24 ± 0.02^b	0.007088
Moisture content (%)	26.33 ± 0.61^a	20.50 ± 1.91^b	0.015656
	15–30 cm depth		
Organic matter (%)	14.29 ± 0.35^a	11.27 ± 0.78^b	0.005467
Organic carbon (%)	8.29 ± 0.20^a	6.54 ± 0.45^b	0.005466
Total nitrogen (%)	0.81 ± 0.05^a	0.77 ± 0.10^a	0.713792
pH-H_2O	4.42 ± 0.10^a	4.23 ± 0.08^b	0.059146
Bulk density (g cm^{-3})	1.22 ± 0.01^a	1.26 ± 0.02^a	0.153677
Moisture content (%)	23.33 ± 0.49^a	19.17 ± 2.60^a	0.146512

Note: different letters each row indicate significant differences between the means of soil properties at both depths at enrichment planting (F1) or secondary forest (F2) plots using the Student's t-test ($P < 0.05$).

TABLE 2: Pearson correlation analysis results comparing microbial biomass C (MBC) with organic matter (OM), microbial biomass N (MBN) with total N (TN), and OM with MBC/MBN ratio for both plots at the same soil depths.

Soil depth (cm)	MBC versus OM		MBN versus TN		OM versus MBC/MBN ratio	
	P value	r^2	P value	r^2	P value	r^2
F1 (0–15)	0.197	0.708	0.087	0.749	0.830	0.113
F1 (15–30)	0.091	0.864	0.603	−0.271	0.667	−0.226
F2 (0–15)	0.947	−0.036	0.120	0.702	0.202	−0.606
F2 (15–30)	0.215	−0.593	0.939	0.040	0.146	−0.670

Note: F1: enrichment planting; F2: secondary forest.

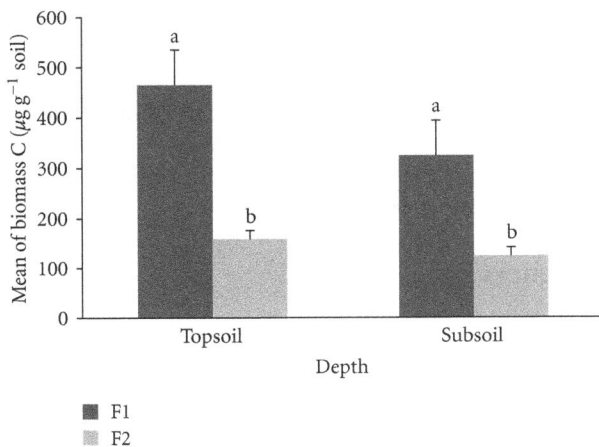

FIGURE 4: Total mean microbial biomass C at F1 and F2 plots. Different letters indicate significant difference between means of the same soil depths at enrichment planting (F1) compared to secondary forest (F2) plots, using the Student's t-test ($P \leq 0.05$) (bars are means, whiskers indicate standard error).

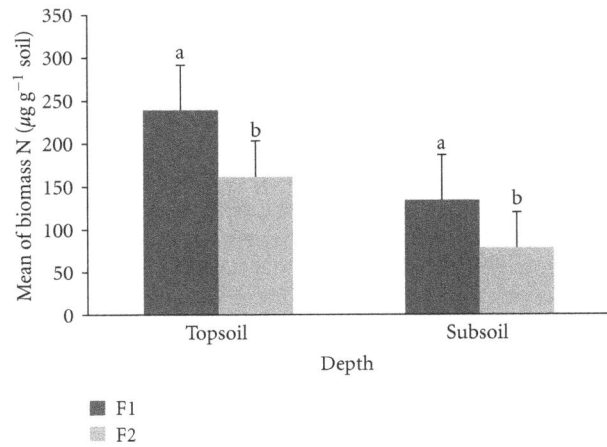

FIGURE 5: Total mean microbial biomass N at F1 and F2 plots. Different letters indicate significant difference between means of the same soil depths at enrichment planting (F1) compared to secondary forest (F2) plots, using the Student's t-test ($P \leq 0.05$) (bars are means, whiskers indicate standard error).

Powlson et al. [37] claimed the sensitivity posed by labile C is proportional to the limitation of soil microbial biomass, and this affects organic C aggradation.

MBN in enrichment planting is greater compared to secondary forest for both soil depths. Higher MBN could be due to the higher total N availability possessed by enrichment

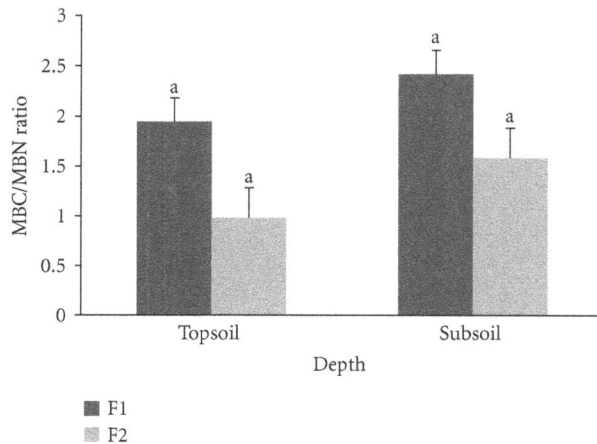

FIGURE 6: Ratio of microbial biomass C to microbial biomass N (MBC/MBN) between the same soil depths at enrichment planting (F1) and secondary forest (F2) plots. Different letters indicate significant difference between means of the same soil depths comparing enrichment planting (F1) to secondary forest (F2) plots, using the Student's t-test ($P \leq 0.05$) (bars are means, whiskers indicate standard error).

planting compared to secondary forest. Kandeler et al. [38] observed that increase in microbial N might be reflected by the competition between microorganism and plants in limited N ecosystem condition. Hence, these results proved that changes in N whether it increases or decreases will catalyze the level of MBN as what we can observe at enrichment planting and secondary forest, respectively.

Variation of MBC/MBN ratio between enrichment planting and secondary forest shows the qualitative changes occurring in the soil biological composition [5]. The ratio of MBC and MBN was found to be proportional due to the same gradient level for both enrichment planting and secondary forest and [39] explained that reasonably high soil organic substrate and low total N compared to organic C at both sites are believed to be similar. Arifin et al. [3] and Carter [40] evaluated that the restoration of soil organic substrate in soil also depends on the carrying capacity, solum type, climate, and land usage management of soil. Likewise, the vast diversity of the organic substrate production in enrichment planting which promotes and sustains the food chain in soil contributes to sustaining an ideal amount of microbial biomass per unit soil [39].

The high acidity at both plots could be due to the formation of decomposition byproducts such as humic and fulvic acids [5], which decrease soil pH. At both forests, abundance of forest litters provides suitable medium for soil macro- and microdecomposer to break down forest litter constituent to release macro- and micronutrients to the soil to increase the soil fertility. However, soil microorganisms in tropical dry environment are found to be able to withstand high acidic condition in the soil as long as the pH does not decrease to the point where H^+ ions begin to form precipitation products [41].

6. Conclusion

Total microbial enzymatic activity, biomass C, and N were found to be higher in enrichment planting plots compared to secondary forest. The abundance of organic substrate and increased soil acidity play important roles in the biological properties at both sites. The soil biological properties in enrichment planting were found to be improved compared to secondary forest after 42 years of planting. It is recommended that further research be done to determine the most sensitive microorganisms that caused the changes in the soil. Biological components of soils help in increasing the fertility of the soils by enhancing the retention capacity of nutrients for plant uptake and, thus, promoting the soil fertility and productive capability especially in the tropical environment condition. Further research must be conducted to identify the microorganisms that are most influential in soil changes. Biological properties of soil help increase fertility by enhancing the retention of nutrients available for plant uptake, thus, promoting soil fertility and productivity, especially in tropical environments. In conclusion, 42 years of forest enrichment planting using indigenous dipterocarp species led to recovery or restoration of soil biological properties to levels higher than observed in secondary forest. Therefore, forest enrichment planting by the Forestry Department Peninsular Malaysia effectively increased the productivity and fertility of soil in previously degraded forestland.

Acknowledgments

The authors wish to thank Perak South District, Department of Forestry, Perak, Malaysia that allowed us to carry out the research project. This study was financially supported by the Fundamental Research Grant Scheme (FRGS) and Research University Grant Scheme (RUGS) from the Ministry of Higher Education of Malaysia (MOHE) through the Universiti Putra Malaysia (UPM), Malaysia. They also like to express their gratitude to all Department of Forestry, Perak staff who helped us with the fieldwork.

References

[1] D. K. Singh, F. Abood, H. Abdul-Hamid, Z. Ashaari, and A. Abdu, "Boric acid toxicity trials on the wood borer *Heterobostrychus aequalis* waterhouse (coleoptera: Bostrychidae)," *American Journal of Agricultural and Biological Science*, vol. 6, no. 1, pp. 84–91, 2011.

[2] M. A. Liebig, G. E. Varvel, J. W. Doran, and B. J. Wienhold, "Crop sequence and nitrogen fertilization effects on soil properties in the Western Corn Belt," *Soil Science Society of America Journal*, vol. 66, no. 2, pp. 596–601, 2002.

[3] A. Arifin, S. Tanaka, S. Jusop et al., "Assessment on soil fertility status and growth performance of planted dipterocarp species in Perak, Peninsular Malaysia," *Journal of Applied Sciences*, vol. 8, no. 21, pp. 3795–3805, 2008.

[4] M. K. Jarecki and R. Lal, "Crop management for soil carbon sequestration," *Critical Reviews in Plant Sciences*, vol. 22, no. 6, pp. 471–502, 2003.

[5] N. Behera and U. Sahani, "Soil microbial biomass and activity in response to Eucalyptus plantation and natural regeneration on tropical soil," *Forest Ecology and Management*, vol. 174, no. 1–3, pp. 1–11, 2003.

[6] J. C. Dagar, A. D. Mongia, and N. T. Singh, "Degradation of tropical rain forest soils upon replacement with plantations and arable crops in Andaman and Nicobar islands in India," *Tropical Ecology*, vol. 36, no. 1, pp. 89–101, 1995.

[7] H. Insam, "Developments in soil microbiology since the mid 1960s," *Geoderma*, vol. 100, no. 3-4, pp. 389–402, 2001.

[8] M. Schulze, "Technical and financial analysis of enrichment planting in logging gaps as a potential component of forest management in the eastern Amazon," *Forest Ecology and Management*, vol. 255, no. 3-4, pp. 866–879, 2008.

[9] J. L. Doucet, Y. L. Kouadio, D. Monticelli, and P. Lejeune, "Enrichment of logging gaps with moabi (*Baillonella toxisperma* Pierre) in a Central African rain forest," *Forest Ecology and Management*, vol. 258, no. 11, pp. 2407–2415, 2009.

[10] F. Montagnini, B. Eibl, L. Grance, D. Maiocco, and D. Nozzi, "Enrichment planting in overexploited subtropical forests of the Paranaense region of Misiones, Argentina," *Forest Ecology and Management*, vol. 99, no. 1-2, pp. 237–246, 1997.

[11] G. Adjers, S. Hadengganan, J. Kuusipalo, K. Nuryanto, and L. Vesa, "Enrichment planting of dipterocarps in logged-over secondary forests: effect of width, direction and maintenance method of planting line on selected Shorea species," *Forest Ecology and Management*, vol. 73, no. 1–3, pp. 259–270, 1995.

[12] M. S. Safa, Z. Ibrahim, and A. Abdu, "Potentialities of new line planting technique of enrichment planting in Peninsular Malaysia: a review of resources sustainability and economic feasibility," *Munich Personal RePec Archive*, Paper no. 10889, 2004.

[13] M. L. Gaspar, M. N. Cabello, R. Pollero, and M. A. Aon, "Fluorescein diacetate hydrolysis as a measure of fungal biomass in soil," *Current Microbiology*, vol. 42, no. 5, pp. 339–344, 2001.

[14] H. A. Ajwa, C. J. Dell, and C. W. Rice, "Changes in enzyme activities and microbial biomass of tallgrass prairie soil as related to burning and nitrogen fertilization," *Soil Biology and Biochemistry*, vol. 31, no. 5, pp. 769–777, 1999.

[15] C. W. Rice, T. B. Moorman, and M. Beare, "Role of microbial biomass carbon and nitrogen in soil quality," in *Methods for Assessing Soil Quality*, J. W. Doran and A. J. Jones, Eds., vol. 46, pp. 203–215, SSSA Special Public, Madison, Wis, USA, 1996.

[16] J. L. Smith and E. A. Paul, "The significance of soil microbial estimations," in *Soil Biochemistry*, G. Stotzky and J. M. Bollard, Eds., vol. 6, pp. 357–396, Marcell Dekker, New York, NY, USA, 1990.

[17] K. R. Islam and R. R. Weil, "Land use effects on soil quality in a tropical forest ecosystem of Bangladesh," *Agriculture, Ecosystems and Environment*, vol. 79, no. 1, pp. 9–16, 2000.

[18] M. A. Sánchez-Monedero, C. Mondini, M. L. Cayuela, A. Roig, M. Contin, and M. De Nobili, "Fluorescein diacetate hydrolysis, respiration and microbial biomass in freshly amended soils," *Biology and Fertility of Soils*, vol. 44, no. 6, pp. 885–890, 2008.

[19] M. González-Pérez, L. Martin-Neto, L. A. Colnago et al., "Characterization of humic acids extracted from sewage sludge-amended oxisols by electron paramagnetic resonance," *Soil and Tillage Research*, vol. 91, no. 1-2, pp. 95–100, 2006.

[20] E. R. Ingham and D. A. Klein, "Relationship between fluorescein diacetate-stained hyphae and oxygen utilization, glucose utilization, and biomass of submerged fungal batch cultures," *Applied and Environmental Microbiology*, vol. 44, no. 2, pp. 363–370, 1982.

[21] J. Schnurer and T. Rosswall, "Fluorescein diacetate hydrolysis as a measure of total microbial activity in soil and litter," *Applied and Environmental Microbiology*, vol. 43, no. 6, pp. 1256–1261, 1982.

[22] O. W. Heal and S. F. Maclean Jr., "Comparative productivity in ecosystem-secondary productivity," in *Unifying Concepts in Ecology*, W. H. van Dobben and R. H. Lowe-McConell, Eds., pp. 89–108, Dr. W. Junk B.V. Publishers, The Hague, The Netherlands, 1975.

[23] M. S. Vásquez-Murrieta, B. Govaerts, and L. Dendooven, "Microbial biomass C measurements in soil of the central highlands of Mexico," *Applied Soil Ecology*, vol. 35, no. 2, pp. 432–440, 2007.

[24] A. Pérez-Piqueres, V. Edel-Hermann, C. Alabouvette, and C. Steinberg, "Response of soil microbial communities to compost amendments," *Soil Biology and Biochemistry*, vol. 38, no. 3, pp. 460–470, 2006.

[25] M. Cycoń and Z. Piotrowska-Seget, "Changes in bacterial diversity and community structure following pesticides addition to soil estimated by cultivation technique," *Ecotoxicology*, vol. 18, no. 5, pp. 632–642, 2009.

[26] K. Sleytr, A. Tietz, G. Langergraber, and R. Haberl, "Investigation of bacterial removal during the filtration process in constructed wetlands," *Science of the Total Environment*, vol. 380, no. 1–3, pp. 173–180, 2007.

[27] V. Gagnon, F. Chazarenc, Y. Comeau, and J. Brisson, "Influence of macrophyte species on microbial density and activity in constructed wetlands," *Water Science and Technology*, vol. 56, no. 3, pp. 249–254, 2007.

[28] C. Witt, J. L. Gaunt, C. C. Galicia, J. C. G. Ottow, and H. U. Neue, "A rapid chloroform-fumigation extraction method for measuring soil microbial biomass carbon and nitrogen in flooded rice soils," *Biology and Fertility of Soils*, vol. 30, no. 5-6, pp. 510–519, 2000.

[29] E. D. Vance, P. C. Brookes, and D. S. Jenkinson, "An extraction method for measuring soil microbial biomass C," *Soil Biology and Biochemistry*, vol. 19, no. 6, pp. 703–707, 1987.

[30] C. B. Craft, E. D. Seneca, and S. W. Broome, "Loss on ignition and kjeldahl digestion for estimating organic carbon and total nitrogen in estuarine marsh soils: calibration with dry combustion," *Estuaries and Coasts*, vol. 14, no. 2, pp. 175–179, 1991.

[31] A. H. Simonne, E. H. Simonne, R. R. Eitenmiller, H. A. Mills, and C. P. Cresman, "Could the dumas method replace the Kjeldahl digestion for nitrogen and crude protein determinations in foods?" *Journal of the Science of Food and Agriculture*, vol. 73, no. 1, pp. 39–45, 1997.

[32] P. C. Brookes, A. Landman, G. Pruden, and D. S. Jenkinson, "Chloroform fumigation and the release of soil nitrogen: a rapid direct extraction method to measure microbial biomass nitrogen in soil," *Soil Biology and Biochemistry*, vol. 17, no. 6, pp. 837–842, 1985.

[33] P. Ahmadpour, A. M. Nawi, A. Abdu et al., "Uptake of heavy metals by *Jatropha curcas* L. Planted in soils containing Sewage sludge," *American Journal of Applied Sciences*, vol. 7, no. 10, pp. 1291–1299, 2010.

[34] M. H. Akbar, O. H. Ahmed, A. S. Jamaluddin et al., "Differences in soil physical and chemical properties of rehabilitated and secondary forests," *American Journal of Applied Sciences*, vol. 7, no. 9, pp. 1200–1209, 2010.

[35] B. T. Saga, O. H. Ahmed, A. S. Jamaluddin et al., "Selected soil morphological, mineralogical and sesquioxide properties

of rehabilitated and secondary forests," *American Journal of Environmental Sciences*, vol. 6, no. 4, pp. 389–394, 2010.

[36] K. Chander, S. Goyal, and K. K. Kapoor, "Microbial biomass dynamics during the decomposition of leaf litter of poplar and eucalyptus in a sandy loam," *Biology and Fertility of Soils*, vol. 19, no. 4, pp. 357–362, 1995.

[37] D. S. Powlson, P. C. Prookes, and B. T. Christensen, "Measurement of soil microbial biomass provides an early indication of changes in total soil organic matter due to straw incorporation," *Soil Biology and Biochemistry*, vol. 19, no. 2, pp. 159–164, 1987.

[38] E. Kandeler, A. R. Mosier, J. A. Morgan et al., "Response of soil microbial biomass and enzyme activities to the transient elevation of carbon dioxide in a semi-arid grassland," *Soil Biology and Biochemistry*, vol. 38, no. 8, pp. 2448–2460, 2006.

[39] A. R. Barbhuiya, A. Arunachalam, H. N. Pandey, K. Arunachalam, M. L. Khan, and P. C. Nath, "Dynamics of soil microbial biomass C, N and P in disturbed and undisturbed stands of a tropical wet-evergreen forest," *European Journal of Soil Biology*, vol. 40, no. 3-4, pp. 113–121, 2005.

[40] M. R. Carter, "Soil quality for sustainable land management: organic matter and aggregation interactions that maintain soil functions," *Agronomy Journal*, vol. 94, no. 1, pp. 38–47, 2002.

[41] J. L. Faulwetter, V. Gagnon, C. Sundberg et al., "Microbial processes influencing performance of treatment wetlands: a review," *Ecological Engineering*, vol. 35, no. 6, pp. 987–1004, 2009.

16

Ectomycorrhizal Influence on Particle Size, Surface Structure, Mineral Crystallinity, Functional Groups, and Elemental Composition of Soil Colloids from Different Soil Origins

Yanhong Li, Huimei Wang, Wenjie Wang, Lei Yang, and Yuangang Zu

The Key Laboratory of Forest Plant Ecology Ministry of Education, Harbin, Heilongjiang 150040, China

Correspondence should be addressed to Huimei Wang; whm0709@163.com and Wenjie Wang; wjwang225@hotmail.com

Academic Editors: A. Roldán Garrigós and J. Viers

Limited data are available on the ectomycorrhizae-induced changes in surface structure and composition of soil colloids, the most active portion in soil matrix, although such data may benefit the understanding of mycorrhizal-aided soil improvements. By using ectomycorrhizae (*Gomphidius viscidus*) and soil colloids from dark brown forest soil (a good loam) and saline-alkali soil (heavily degraded soil), we tried to approach the changes here. For the good loam either from the surface or deep soils, the fungus treatment induced physical absorption of covering materials on colloid surface with nonsignificant increases in soil particle size ($P > 0.05$). These increased the amount of variable functional groups (O–H stretching and bending, C–H stretching, C=O stretching, etc.) by 3–26% and the crystallinity of variable soil minerals (kaolinite, hydromica, and quartz) by 40–300%. However, the fungus treatment of saline-alkali soil obviously differed from the dark brown forest soil. There were 12–35% decreases in most functional groups, 15–55% decreases in crystallinity of most soil minerals but general increases in their grain size, and significant increases in soil particle size ($P < 0.05$). These different responses sharply decreased element ratios (C : O, C : N, and C : Si) in soil colloids from saline-alkali soil, moving them close to those of the good loam of dark brown forest soil.

1. Introduction

At the global scale, soil degradation, including soil erosion, is a potential threat to food security, and phytorehabilitation measures for controlling soil degradation are a popular and urgent topic of research [1, 2]. Ectomycorrhizal (ECM) fungi and associated symbionts can promote the growth of plants and increase their tolerance to unfavorable soil conditions such as nutrient deficiency or heavy metal pollution [3–5]. More than 5000 fungi can form ECM symbionts with over 2000 woody plants [6], showing the importance of ectomycorrhizae in plant-soil interactions. Commercially available mycorrhizal inocula, which consist of a single fungus species, are currently used for afforestation and grassland recovery [7]. ECM fungi enhance the growth and fitness of plants [8–11] by providing them with mineral and organic nutrients from the soil matrix and by protecting the carbohydrates

and organic compounds that are stored in the roots from pathogenic organisms [12]. Improvements in root length, soil P utilization efficiency, and disease and stress resistance, as well as enhanced soil nutrient availability, have been reported [13, 14]. Recent studies have been performed on the identification of ECM fungi [15–18], interactions among various fungi and their effects on soil pollution rehabilitation [17, 19], and the underlying genetic basis for ECM functions [20–22]. These studies have provided a sound basis for understanding the mechanisms of the interaction between ECM fungi and various plants. Mycorrhizal fungi can also directly stabilize soil both through their hyphal network and through the secretion of glue-like chemicals [7]. However, the rarity of studies on the interaction between soil particles and ECM fungi hinders a full understanding of the function of fungi in soil health maintenance and soil physical texture formation.

Owing to the profound heterogeneity among soil samples from field sampling campaigns, it is difficult to visualize the interaction between ECM fungi and soil aggregates at the microscale of millimeters or nanometers. Soil colloids are generally considered to be particles with effective diameters of around 10 nm to 10 μm, with the smallest colloids just larger than dissolved macromolecules and the largest colloids being those that resist settling once suspended in soil pore water [23–25]. Soil colloids are the most active portion of the soil and largely determine the physical and chemical properties of the soil. Organic colloids are more reactive chemically and generally have a greater influence on soil properties per unit weight than inorganic colloids. Inorganic colloids of clay minerals are usually crystalline (although some are amorphous) and usually have a characteristic chemical and physical configuration. These features of soil colloids made it more suitable to study the interaction between ECM fungi and soil particles in indoor laboratory tests compared to bulk soils with heterogeneous composition [25]. In this paper, one aim was to find conformational changes in fungal extracts induced by the interaction with soil colloidal particles.

Previous papers have described good methods for identifying the surface structure and composition of soil colloids. To examine surface changes and microstructure changes in soil colloids after the addition of soil conditioner, Wang et al. used atomic force microscopy (AFM) to characterize 3D structural changes and used scanning electron microscopy (SEM) to find 2D surficial changes [26]. Laser particle-size analyzers have been used in a variety of studies with soil samples [27–30]. X-ray powder diffraction (XRD) is a nondestructive and rapid analytical technique primarily used for phase identification of soil minerals that can provide information regarding grain size and relative crystallinity, for example, interactions between clay minerals and organic matter in relation to carbon sequestration [31, 32]. Infrared spectroscopy is a well-established technique for the identification of chemical compounds and specific functional groups in compounds and, thus, is a useful tool for soil applications [33–35]. X-ray photoelectron spectroscopy (XPS) has the advantage of being able to detect all elements in the soil (except for H and He) and provides much valuable information on the composition of, and bonding state of elements in, surface and near-surface layers of many minerals [36–38]. The above methods may be useful for clarifying the influence of ECM fungus extracts on the surface structure and composition of soil colloids.

To reveal the underlying mechanism of the impact of ECM fungi on soil particles, soil colloids were extracted from 3 types of soils, including dark brown forest soil (deep and surface layers) and saline-alkaline soil in a grassland (surface layer). After fungal extract treatment, the laser particle-size analyzer, AFM, SEM, XRD, IR, and XPS were used to characterize the structural and surface changes. We hypothesized that ECM fungi would have different effects on soil colloids from different origins, and this could contribute to the improvement of degraded soil and the formation of healthy soil structure. The main aim of this study was to investigate the relationship between ECM fungi and soil and reveal the mechanism underlying soil improvement.

2. Material and Methods

2.1. Preparation of Soil Colloids. Soil samples were collected from the top soil (0–20 cm) in a typical saline-alkali region of the Songnen Plain ($45°59'55''$N, $124°29'48''$E). As a nonsaline loam control, we used dark brown soil from the surface layer (0–20 cm) and a deeper layer (60–80 cm) from the Experimental Forest Farm of Northeast Forestry University ($45°43'6''$N, $126°37'54''$E).

The soil colloids were separated according to [25, 26] as follows. One gram of air-dried soil was fully dispersed in 100 mL of ultrapure water in a 250 mL beaker. The suspension was allowed to stand undisturbed for more than 24 h. Sands and silts in the soil sample were gradually deposited at the bottom of the beaker, whereas the soil colloids were left in the suspension indicated as a turbid solution. The upper suspension was carefully decanted to a centrifuge tube and then centrifuged at 12000 rpm for 10 min. The precipitates were dissolved with 25 mL of ultrapure water and regarded as the soil colloidal solution.

2.2. Preparation of Fungus Extract. *Gomphidius viscidus* was sampled from Inner Mongolia, China, for laboratory testing. It is an important mycorrhizal fungus in coniferous forests and belongs to the Gomphidiaceae family of the Agaricales [40]. It was grown in modified Melin-Norkrans (MMN) medium: $CaCl_2$, 0.05 g; NaCl, 0.025 g; KH_2PO_4, 0.5 g; $(NH_4)_2HPO_4$, 0.25 g; $MgSO_4 \cdot 7H_2O$, 0.15 g; $FeCl_3$ (1%), 1.2 mL; thiamine HCl, 0.2 mL; malt extract, 3 g; glucose, 10 g; stock solution of micronutrients (contents per liter: H_3BO_3, 2.86 g; $MnCl_2$, 1.81 g; $ZnSO_4$, 0.22 g; $CuSO_4$, 0.08 g; and $NaMoO_4$, 0.02 g), 1 mL; ultrapure water, 1000 mL; and 15 g of agar in the case of agar media. The pH of the media was adjusted to 5.45–5.55 before autoclaving (121°C, 0.1 Mpa, and 20 min). The medium was dispersed aseptically in a 10 cm culture dish and stored in a 4°C refrigerator. The strain was inoculated on the MMN solid culture medium, training repeatedly until no other bacteria were produced [41].

Next, liquid medium was used, and 200 mL aliquots of the medium were dispensed into 250 mL beakers and autoclaved at 117°C for 20 min. Each Erlenmeyer flask was inoculated with fungal colonies of *G. viscidus* and then cultured on a shaking table. After 14 days, the fungal hyphae were used to extract fungal solution. First, culture media were filtered using a 4-layer gauze. After being cleaned repeatedly with ultrapure water, the mycelium was ground in a mortar and centrifuged (12000 rpm) at 4°C for 10 min, and the supernatant was collected as the fungus extract solution.

2.3. Particle Size and Surface Structure Observation. After the addition of fungus extract solution to soil colloids from different origins, the particle size and surface structure were determined with a laser particle-size analyzer (ZetaPALS, USA), an atomic force microscope (AZ, USA), and a scanning electron microscope (Quanta 200, FEI, USA). Methods were revised from [26, 30], and detail was as follows.

For the laser size analysis, 10 μL of fungus extract was added to 4 mL of soil colloid solution (about 1 mg soil colloids), and the fungus extract was replaced by 10 μL of ultrapure water in the control group. Twelve hours after mixing, the mixture was analyzed using the laser particle-size analyzer. For AFM observation, 10 μL of fungus extract was added to 4 mL of soil colloid solution, and the fungus extract was replaced by 10 μL of ultrapure water in the control group. One hour after mixing, centrifugation (12000 rpm) at 25°C for 10 min was used to precipitate the soil colloids. This colloid precipitate was then dissolved with 4 mL of ultrapure water, dropped on the mica surface, and air-dried before AFM imaging. The AFM images of the soil colloids were obtained in tapping mode by using the PicoPlus II AFM system from Molecular Imaging (MI) Corporation (AZ, USA). For SEM observation, the same procedures as those for AFM observation were used, and the soil colloidal solution was dropped on the sample stage and then air-dried before SEM imaging (Quanta 200, FEI, USA). The samples were sputter-coated with a thin layer of gold-palladium (5–10 nm, 25 mA, and 3 min) at room temperature by using a sputter coater before the examination.

2.4. Crystal Structure Diffraction of Soil Minerals. The preparation of treatments and controls was the same as that for AFM, and SEM and X-ray powder diffraction (XPD) observation were carried out according to [32]. After being air-dried, X-ray powder diffraction patterns were collected in transmission by using an X-ray powder diffractometer (D/Max 2200, Rigaku, Japan) with a rotating anode (Philips) and Cu Kα_1 radiation generated at 30 mA and 40 kV. The range of 2θ diffraction angles examined was 10–40° with steps of 0.02° and a measuring time of 0.3 s per step.

2.5. Observations of Functional Group Change. The preparation of treatments and controls was the same as that for AFM, and SEM and infrared spectrum (IR) measurements were carried out according to [33–35]. The samples were diluted with 1% KBr mixing powder and separately pressed to obtain self-supporting disks. Tablets for IR measurements were prepared by pressing the powder mixture at a load of 8 tons for 8 min. The IR spectrum was obtained by a compact Fourier transform infrared spectrophotometer (IR Affinity-1, SHIMADZU, Japan) and recorded across a wave number range of 4000–500 cm^{-1} at a resolution of 4 cm^{-1}.

2.6. Analysis of the Atomic Concentration of Elements. A K-Alpha spectrometer equipped with a concentric hemispherical analyzer in the standard configuration (Thermo Scientific, USA) was used in this analysis with the method revised from [38]. After the preparation of control and treatment samples (the same as that for SEM and AFM), the soil colloidal solution was dropped on a cleaned HOPG surface and then air-dried before examination. The vacuum system consisted of a turbomolecular pump and a titanium sublimation pump. The residual pressure before the analysis was lower than 10^{-7} Pa. The X-ray source was AlKα, and it was run at 30 mA and 80 kV. The incident angle was 49.1°, and the emission angle was 0° with respect to the sample's surface normal. All the spectra were obtained in digital mode. The wide-scan spectra were acquired from 1000 eV to 0 eV. Sample charging was corrected by comparing all binding energies to the adventitious carbon at 285.0 eV. Detailed spectra were processed using CasaXPS software (V2.3.12, Casa Software Ltd., UK). An iterated Shirley-Sherwood background subtraction was applied before peak fitting using a nonlinear least-squares algorithm. The atomic concentration of elements was calculated using the software.

2.7. Data Analysis. In the analysis of XRD data, the original data were rectified using the Jade program to eliminate Kα and then obtain the XRD pattern for a sample. The upper area (a_c), which was separated with the smooth curve connecting each point of minimum intensity, corresponded to the crystalline portion, and the lower area was the background containing the amorphous portion (a_b). The Jade program was used to calculate grain size and relative crystallinity (relative crystallinity = $a_c/(a_c + a_b)$) [42] and check the effect of the addition of fungus extracts. The major components of the soil were obtained in conjunction with the database and according to the books written by Xie [43] and Marc and Jacques [44].

Six functional groups [45] were selected for IR analysis, as shown in Figure 5: 3750–3300 cm^{-1} is O–H stretching of structural OH; 1200–970 cm^{-1} is Si–O–Si stretching; 950–820 cm^{-1} is O–H bending of structural OH; 2970–2820 cm^{-1} is aliphatic C–H stretching; 1750–1630 cm^{-1} is C=O stretching of carboxylic acids, amides, and ketones; and 1650–1360 cm^{-1} is carbonates.

3. Results

3.1. Results Obtained Using the Laser Particle-Size Analyzer. The particle size and distribution of soil colloids from dark brown forest soil (surface and deep layers) and saline-alkali soil with and without the addition of fungus extracts are shown in Figure 1. After the addition of the fungus extract, the particle size of soil colloids from dark brown forest soil in the surface layer was larger than that in the control, and similar results were found for deep soils. However, the differences were not significant between the treatment and the control ($P > 0.05$). In contrast, the addition of the fungus extract increased the particle size of colloids from saline-alkaline soil from 472 ± 11.3 nm to 502 ± 4.0 nm, and this 6.29% increase was statistically significant ($P < 0.05$) (Figure 1).

3.2. AFM Results. As shown in Figure 2, soil colloids from the surface and deep layers of dark brown forest soil showed similar tendencies. More soil colloids aggregated together and became larger than those in the control. Moreover, more covering materials were found on the surface of these colloids than those in the control samples. Contrary to the soil colloids from the dark brown forest soil, soil colloids from the saline-alkali soil were dispersed after the addition of fungus extracts, and the interaction between the different particles was not as dense as that in the control (Figure 2).

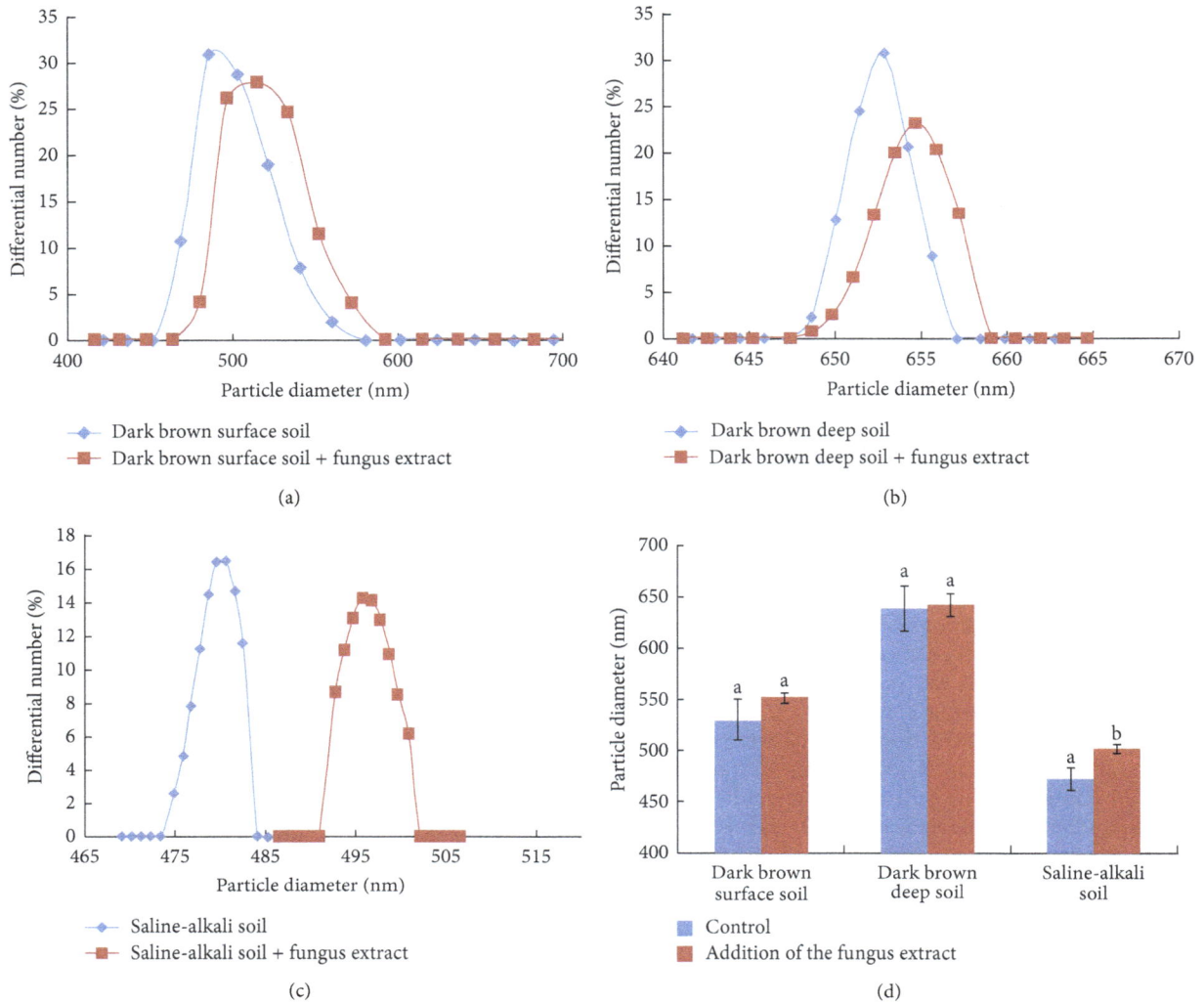

FIGURE 1: Particle diameter changes in soil colloids with and without fungus extract addition.

3.3. SEM Results. The results of SEM were similar to those of AFM (Figure 3). After the addition of the fungus extract, the edges of colloid particles from the dark brown forest soil became much smoother, the small gap between soil colloid particles became invisible, and some covering materials seemed to fill these gaps and made the edges less distinct. This tendency was more evident in surface soils than in deep soils (Figures 3(a)–3(d)).

However, results from the saline-alkali soil were different from those from the dark brown forest soil; relatively larger particles with more distinct and acute edges were observed after the addition of the fungus extract (Figures 3(e) and 3(f)).

3.4. XRD Results. Both the surface and deep layers of dark brown forest soil had 4 obvious diffraction peaks located at 12.3°, 17.8°, 25.0°, and 26.7°. These peaks are indicative of 3 kinds of minerals: kaolinite (7.15 Å, 3.56 Å), hydromica (4.98 Å), and quartz (3.34 Å) (Figures 4(a) and 4(b)). Soil colloids from saline-alkali soil had 6 diffraction peaks, located

at 12.3°, 17.8°, 18.6°, 25.0°, 26.7°, and 29.3° indicating that the main soil mineral composition was kaolinite (7.15 Å, 3.56 Å), hydromica (4.98 Å), vermiculite (4.74 Å), quartz (3.34 Å), and calcite (3.03 Å) [44] (Figure 4(c)).

The addition of the fungus extract increased the relative crystallinity of soil colloids from both the surface and deep layers of dark brown forest soil (Figures 4(a) and 4(b)). Kaolinite increased by 300%, hydromica by 40–70%, and quartz by 83–157%. The grain size of hydromica and quartz both increased by 3–25%, but the kaolinite was reduced by 50–77%.

Unlike with dark brown forest soil, the addition of fungus extracts decreased the relative crystallinity of soil colloids from saline-alkali soil, except for quartz (a 17% increase) (Figure 4(c)). The largest reduction was 59% in kaolinite, while the decreases for hydromica, vermiculite, and calcite ranged from 15% to 25%. In most cases, the grain size of these soil minerals increased after the addition of the fungus extract, except for quartz (reduced 22%) (Figure 4(c)). Grain size increased by about 7% in vermiculite and calcite, while

Ectomycorrhizal Influence on Particle Size, Surface Structure, Mineral Crystallinity, Functional Groups, and Elemental
Composition of Soil Colloids from Different Soil Origins

139

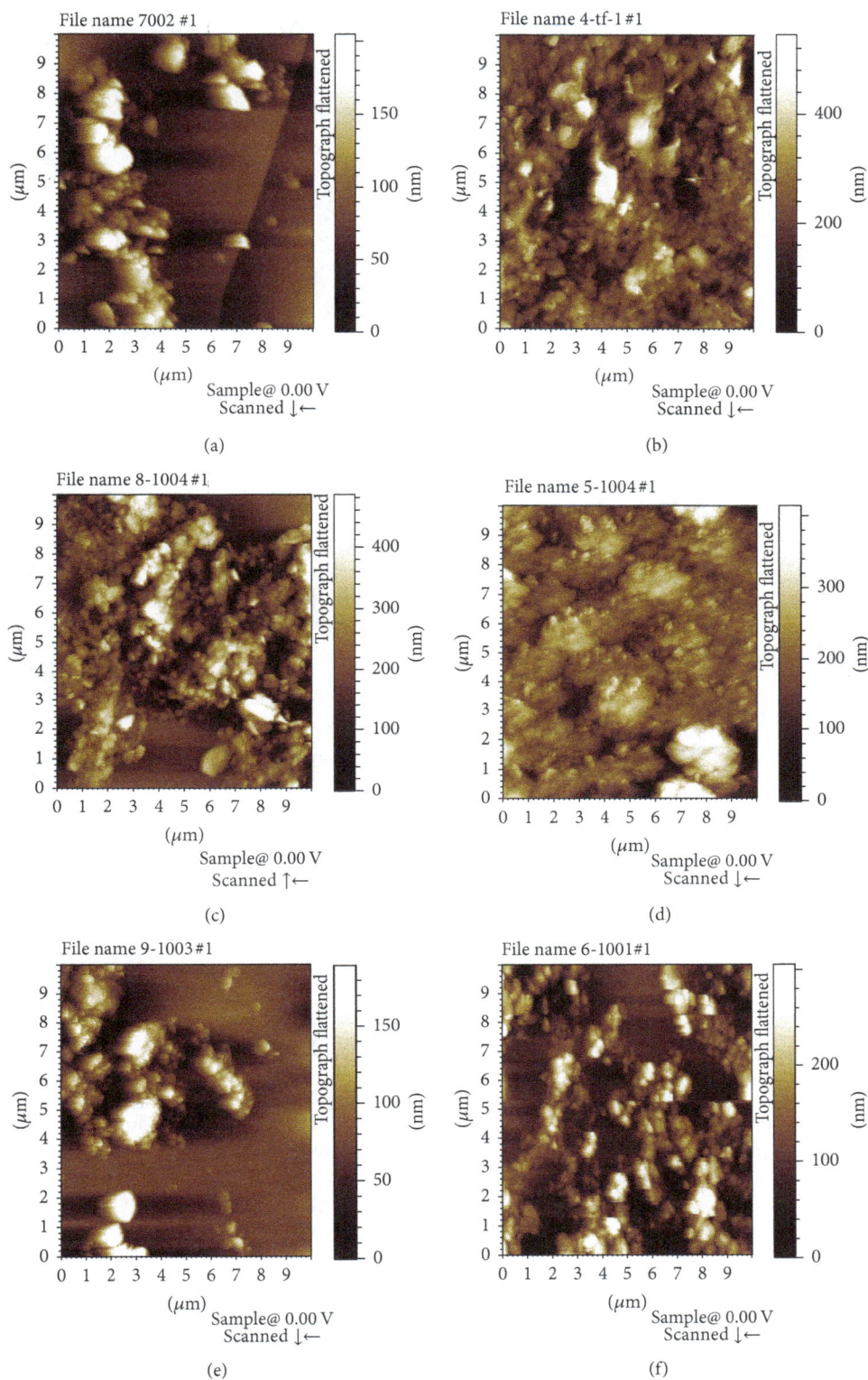

FIGURE 2: Atomic force microscopy images of soil colloids with (b, d, and f) and without (a, c, and e) fungus extract addition. (a) Colloids
from the surface layer of dark brown forest soil; (b) colloids from the surface layer of dark brown forest soil + fungus extract; (c) colloids
from the deep layer of dark brown forest soil; (d) colloids from the deep layer of dark brown forest soil + fungus extract; (e) colloids from
saline-alkali soil; (f) colloids from saline-alkali soil + fungus extract.

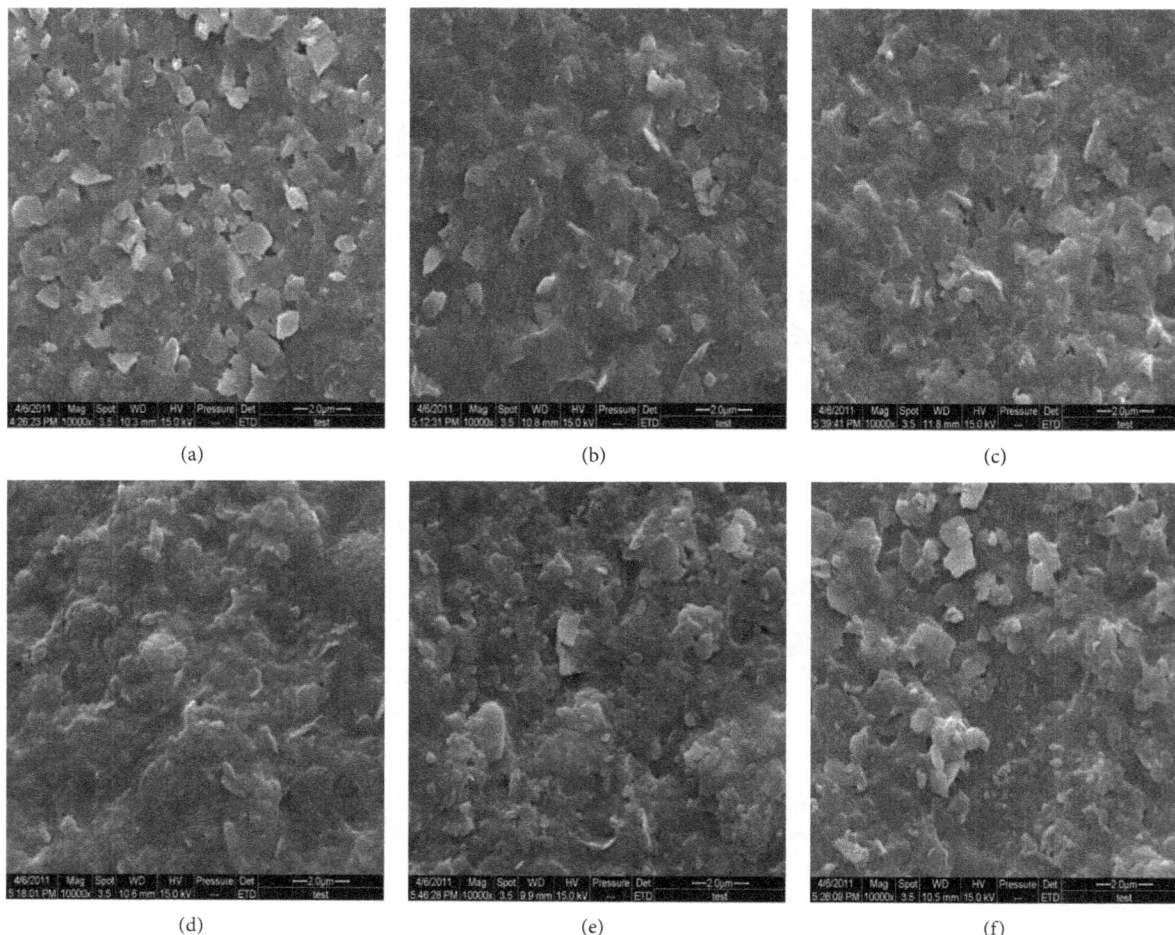

FIGURE 3: Scanning electron microscopy images of soil colloids with (b, d, and f) and without (a, c, and e) fungus extract addition. The labels are the same as those for Figure 2.

much larger increases (over 57%) were found in hydromica and kaolinite (Figure 4(c)).

3.5. IR Results. The addition of the fungus extract to soil colloids from the dark brown forest soil slightly reduced the amount of stretching of the COO⁻ and carbonate functional groups (about 10%) but increased O–H bending, increased stretching in most of the studied functional groups, including C–H, Si–O–Si, and O–H (all about 10–20%), and slightly increased (5%) C=O stretching (Figure 5(a)). Similar tendencies but slight differences in the size of changes were found in deep soil than in surface soil (Figure 5(b)).

Most functional group traits in the surface and deep layers of dark brown forest soil increased. However, a completely different pattern was found in saline-alkali soil (Figure 5(c)). In contrast with dark brown forest soil, the addition of fungus extracts to soil colloids from saline-alkali soil reduced the traits of most functional groups from 10% to 35% (Figure 5(c)). Functional group traits that decreased included O–H bending, C=O stretching, Si–O–Si stretching, O–H stretching, COO⁻ stretching, and carbonate stretching, with the exception of C–H stretching (a 56% increase) (Figure 5(c)).

3.6. XPS Results. Semiquantitative analysis of variable elements with and without the addition of the fungus extract was performed using XPS (Figure 6). In the case of soil colloids from the surface layer of dark brown forest soil, the addition of the fungus extract induced <5% changes in all elements, for example, a 5% increase in C1s and <5% decreases for all O1s, Si2p, N1s, and Ca2p (Figure 6(a)). Changes in variable elements in the deep soil due to the addition of fungus extracts were more evident than those in the surface layers (Figures 6(a) and 6(b)). The changes in C1s, O1s, and Si2p were less than 5%, while 6–9% decreases in N1s and Ca2p were observed (Figure 6(b)).

Compared to the dark brown forest soil, addition of the fungus extract to the saline-alkali soil caused large reductions in variable elements (Figure 6(c)). C1s decreased by 21%, Ca2p by 10%, and O1s, Si2p, and N1s by 5%.

Stoichiometric changes induced by fungus extract addition were also found in the ratios among different elements (Table 1). In the case of the surface layer of dark brown forest soil, the ratios of C : N, Si : Ca, and C : Ca increased by 7–14%. In the case of the deep layer of dark brown forest soil, changes were also mainly found in C : N (6.5%), Si : Ca (20.5%), and C : Ca (17.1%). Stoichiometric changes were much more

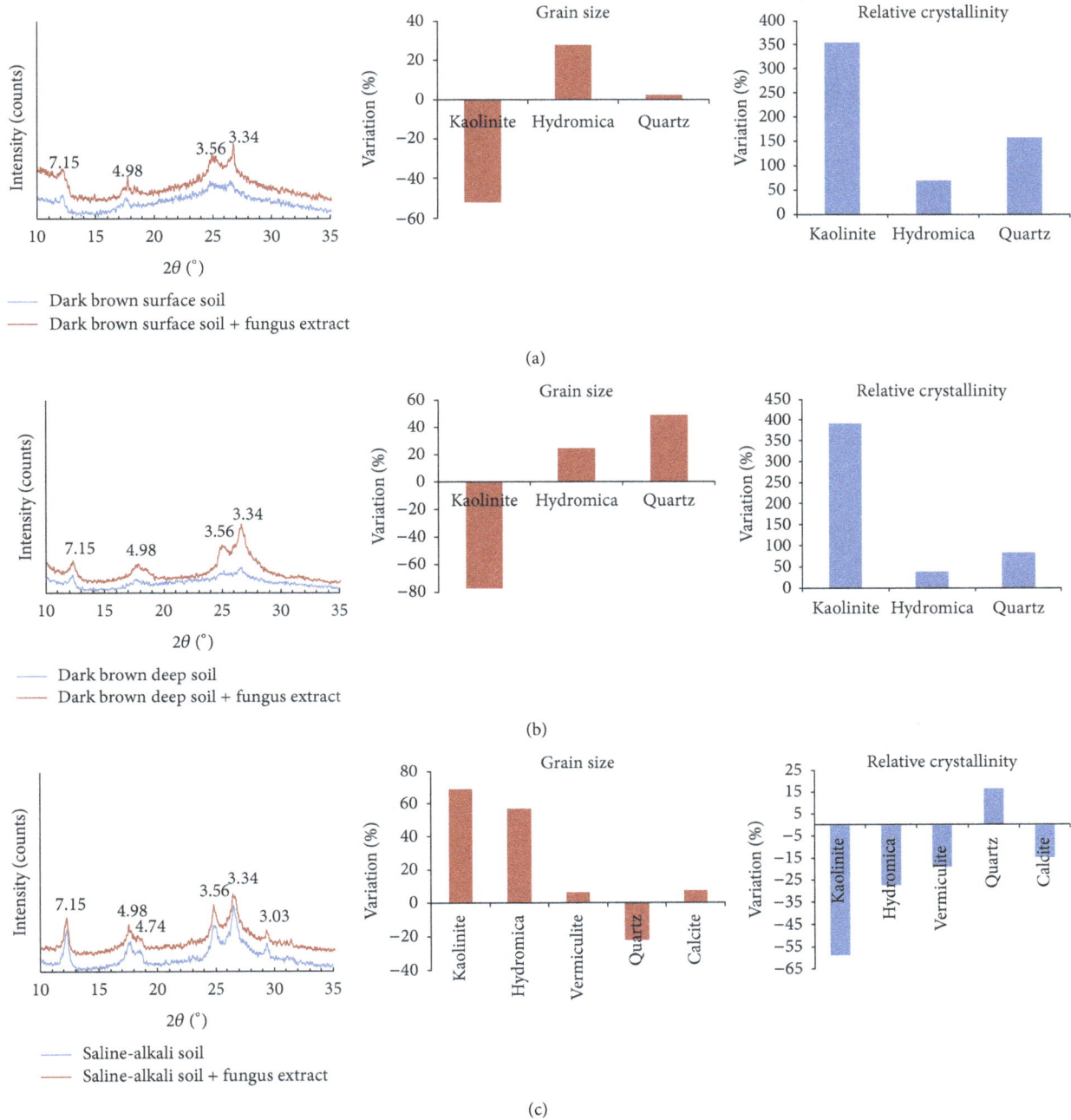

FIGURE 4: X-ray powder diffraction results with and without fungus extract addition. (a) Colloids from the surface layer of dark brown forest soil; (b) colloids from the deep layer of dark brown forest soil; (c) colloids from saline-alkali soil.

evident in saline-alkali soil than in dark brown forest soil. Over 25% decreases were found in $C:O$, $C:N$, and $C:Si$, and a 12.7% decrease was found in $C:Ca$. The $Si:Ca$ ratio increased by 16.39% (Table 1).

4. Discussion

Heavy soil degradation is common in China, and rehabilitation via vegetation recovery is mainly conducted in degraded regions, such as the saline-alkali soil region in the Songnen Plain [26]. Symbiotic associations between tree roots and ECM fungi play important roles in promoting the growth of these plants, and this promotion can be affected by variation in the strains of ECM fungi [46] and their interactions between biotic and abiotic factors [47]. ECM can aid the recovery of degraded soil both through the direct absorption of variable pollutants [17, 19] and indirect protection from rehabilitated vegetation [41]. In a previous study of ECM fungi, the main focus was on the relationship

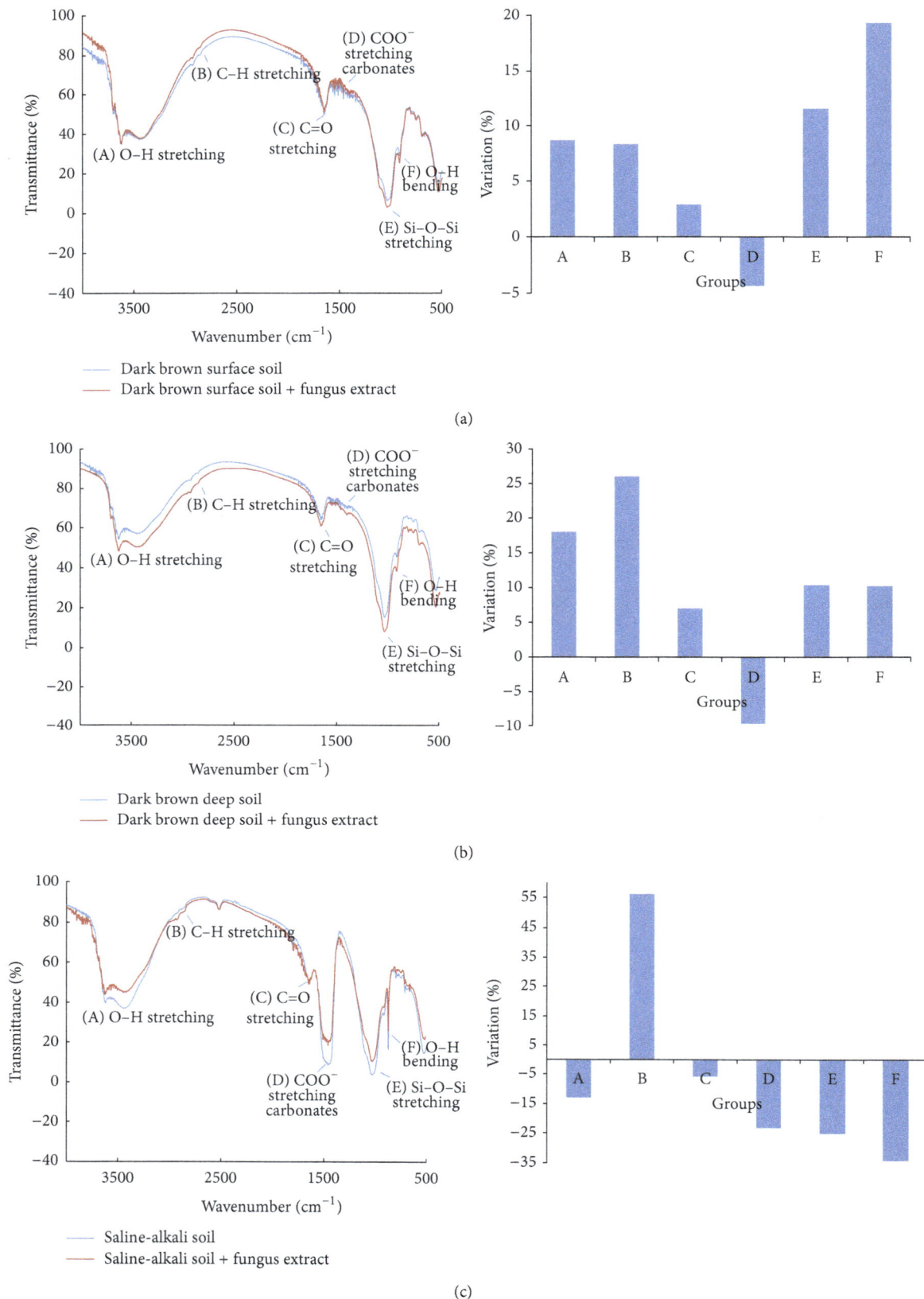

FIGURE 5: Infrared spectrum results after fungus extract addition. The labels are the same as those for Figure 4.

Ectomycorrhizal Influence on Particle Size, Surface Structure, Mineral Crystallinity, Functional Groups, and Elemental Composition of Soil Colloids from Different Soil Origins

143

(a)

(b)

(c)

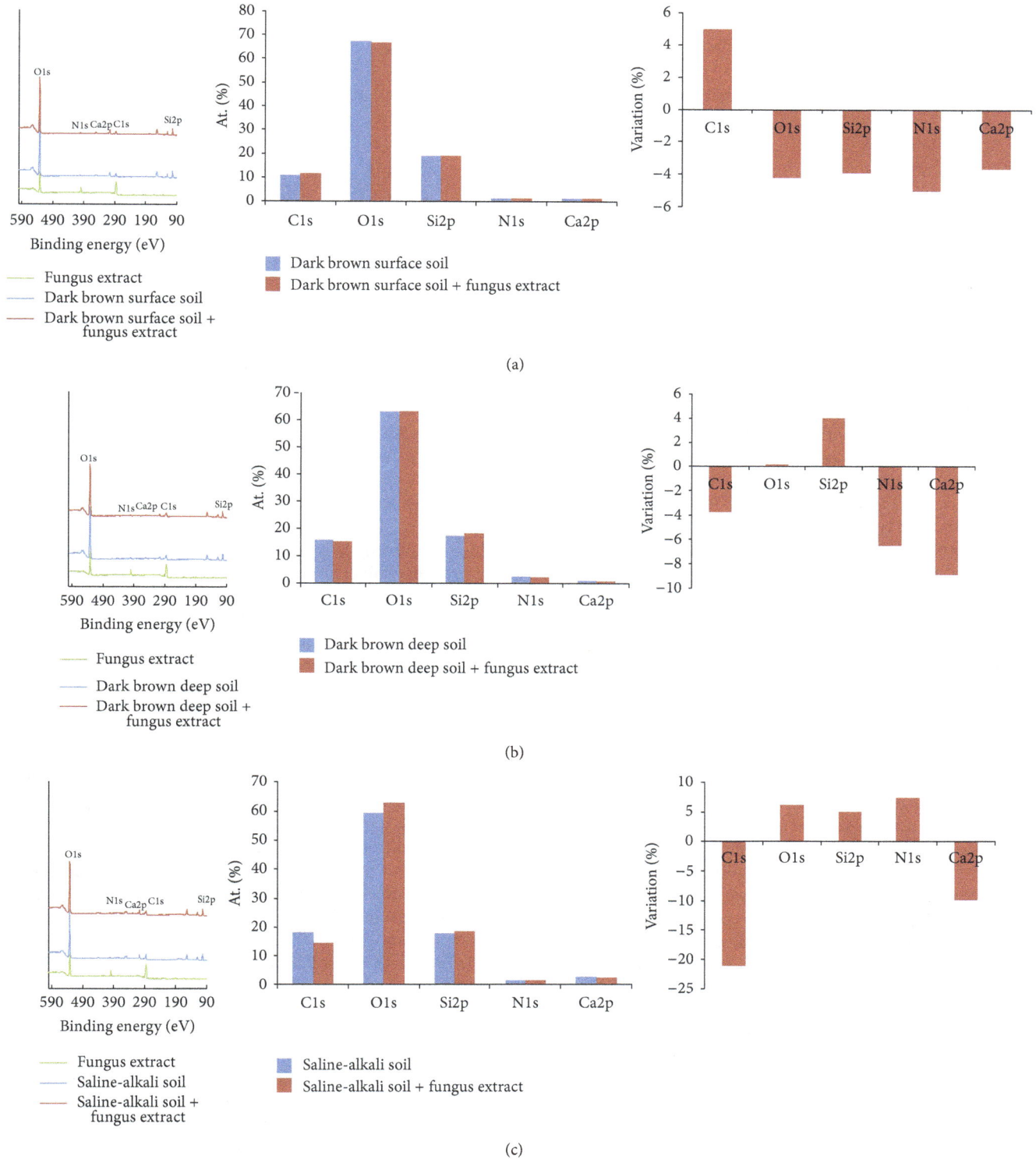

FIGURE 6: X-ray photoelectron spectroscopy results with and without fungus extract addition. The labels are the same as those for Figure 4.

between ECM and plants. In the case of the ECM studied in this paper (*Gomphidius viscidus*), reviewed data have found significant increase in biomass of variable trees together with more soil nutrient absorption (Figure 7): that is, growth of biomass, ground diameter, lateral root and height of larch, pine, and oak increased from 20% to 40%, and much higher increase (50%–60%) in N, P absorption was found. This result indicates that soil nutrient absorption increase should be a basis for the biomass increase after ECM infection. However, few studies paid attention to the underlying mechanism of the

TABLE 1: Results from X-ray photoelectron spectroscopy of the variation in element ratios with and without the addition of the fungus extract.

	Dark brown forest soil—surface			Dark brown forest soil—deep			Saline-alkali soil		
	Control	Addition of the fungus extract	Variation%	Control	Addition of the fungus extract	Variation%	Control	Addition of the fungus extract	Variation%
C : O	0.16	0.18	9.58	0.25	0.24	−3.95	0.31	0.23	−25.68
C : N	7.30	8.10	10.96	6.30	6.50	3.17	12.00	8.80	−26.67
C : Si	0.57	0.62	8.77	0.90	0.84	−6.67	1.03	0.77	−25.24
Si : Ca	13.60	13.60	0.00	17.90	20.50	14.53	6.10	7.10	16.39
C : Ca	7.70	8.40	9.09	16.20	17.10	5.56	6.30	5.50	−12.70

interaction between ECM and the soil matrix. The findings in this paper showed that the ECM influences on soil colloids should be important aspect of degraded soil improvement.

Dark brown forest soil and saline-alkali soil are two types of soil that are widespread in northeastern China. The first is an example of loam with good physicochemical properties [48], while the latter is notorious for poor physicochemical properties because of long-term degradation from human disturbance [26]. In the case of dark brown forest soil from surface and deep layers, both AFM and SEM images revealed that viscous materials wrapped around soil particles, filled some gaps among particles, and induced smoother surfaces with unclear edges (Figures 2(a)–2(d) and Figures 3(a)–3(d)). High absorption capacity is the basis for adhesive material absorption on soil colloids. This absorption has been reported previously [23]. Yan et al. reported that the maximum adsorption capacity (q_0) for fine soil colloids ranged from 169.6 to 203.7 $\mu g\,mg^{-1}$ [49], which was higher than that for coarse soil colloids (81.0–94.6 $\mu g\,mg^{-1}$). Thus, physical absorption instead of chemical reactions possibly occurred in dark brown forest soil.

However, saline-alkali soil was different from dark brown forest soil. There were no clear adhesive layers on soil colloids, and smoother edges were found after the addition of fungus extracts (Figures 3(e) and 3(f)). Significantly larger particle sizes (Figure 1) and relatively loose interactions between different soil particles were observed in saline-alkali soil colloids (Figures 2(e) and 2(f)). Saline-alkali soil has a clay texture with tiny soil particles, high pH, and rich carbonates (especially calcium carbonate) [26]. This soil has poor soil physical structure with limited soil aggregates, and soil colloid surfaces are overloaded by Na^+ [48]. Wang et al. (2011) reported that the addition of the soil conditioner HPMA induced flocculation of soil colloids with looser structure and larger aggregates [26]. The findings on AFM, SEM images, and soil particle size indicate that the function of the fungus extract is like that of the soil conditioner (e.g., HPMA [26]), and a chemical reaction instead of physical adhesion may occur between the fungus extract and soil colloids. The mechanism of saline-alkali soil improvement by HPMA conditioner is to activate inorganic calcium in the soil via a chemical reaction between calcium carbonate and HPMA, as well as the exchange of Ca^{2+} and excessive Na^+ on the surface of soil colloids [26]. A similar chemical reaction is possibly activated after the addition of the fungus extract

to soil colloids from saline-alkali soils, owing to the fact that organic acids were present in the fungus extract solution.

It is worth mentioning that carbonate functional groups appear obviously reduced by 25%, perhaps because saline-alkali soil is basically an alkaline soil with a pH of 9-10 and includes more carbonate ($CaCO_3$) in calcite, while the fungus extract is generally acidic; therefore, combining them may cause some chemical reaction and lead to calcium carbonate being dispersed. Dissolution of calcium carbonate could lead to more Ca^{2+} being adsorbed on the surface of soil colloids, which would tend to increase soil colloid dispersion and form a more loose soil colloid apparent structure [26], as shown in Figure 2.

Measuring differences in mineral composition on the surface of soil colloids and in mineral crystallinity may help explain the changes in particle diameter and surface structure as shown in Figures 1–3. The techniques of XRD, IR, and XPS are commonly used for mineral crystallinity observation [42, 43], functional group identification in soils [44], and elemental composition changes [50]. These methods could be used to examine mineral crystallinity changes, variable functional group changes, and the elemental composition of soil colloids, together with surface and particle size changes after the addition of the fungus extract.

XRD data proved that fungus extract treatment could change the relative crystallinity and grain size of variable soil minerals and showed differences between dark brown forest soil and saline-alkali soil (Figure 4). In the case of dark brown forest soil, addition of fungus extracts 30% ~ 400% increased crystallinity of kaolinite, hydromica, and quartz, while 17% ~58% decreases were found in saline-alkali soil (except for a 15% increase for quartz) (Figure 4). These data were well-matched with the observations from SEM and AFM; that is, soil colloid particle edges became blurred with covering materials in dark brown forest soil, while soil colloids from saline-alkali soil tended to be dispersed with relatively loose interactions (Figures 2–3). By using a similar XRD method, Zheng and Zhao (2011) found that soil clay minerals (illite, kaolinite, and montmorillonite) are very sensitive to disturbance, and their changes could affect soil fertility [51]. Our results showed that ECM fungi in soil may favor this process via their effect on the relative crystallinity and grain size of soil minerals.

IR can provide detailed information about variable functional groups [44], and IR data from this study showed

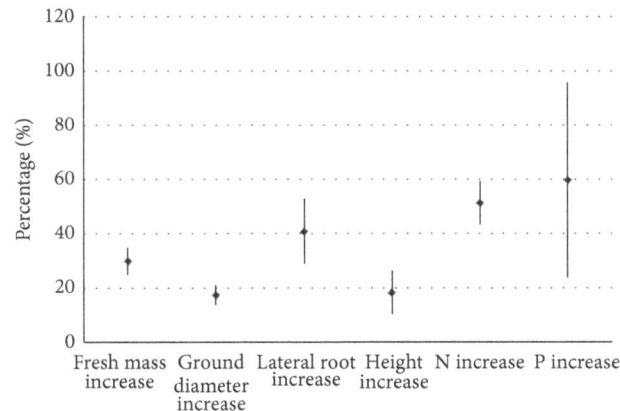

FIGURE 7: ECM infection influences on plant growth and biomass N, P absorption [39]. The ECM is the same species, *Gomphidius viscidus*, used in this study, and plant species are *Larix kaempferi*, *Pinus tabulaeformis*, and *Quercus liaotungensis*. The percentage in y-axis is the improvement of each parameter relative to the untreated control (without ECM infection).

ECM fungus could induce evidently different responses between saline-alkali soil and dark brown forest soil. Traits of functional groups such as O–H stretching and bending, C–H stretching, and Si–O–Si stretching in soil colloids of dark brown forest soil were increased by 8–27% after the addition of the fungus extract (Figures 5(a) and 5(b)). However, all the functional group traits in saline-alkali soil were reduced by about 10–35%, except for C–H stretching. The changes in functional group composition (Figure 4) and mineral crystallization (Figure 5) in soil colloids were accompanied by relative changes in element composition (Figure 6 and Table 1), too. Compared with very slight changes in dark brown forest soil, addition of the fungus extract to degraded soil (saline-alkali soil) majorly reduced C1s (21%) and made the C : O, C : N, C : Si, and C : Ca element ratios more similar to those of high-quality dark brown forest soil (Table 1). ECM inoculum has been commonly used during forestry plantings and native prairie restorations to enhance tree and plant growth [11].

The present study clarifies the mycorrhizal-soil interaction. Field studies indicate that mycorrhizal inocula benefit native plant production and establishment in severely degraded areas [11, 52], and commercial mycorrhizal-soil conditioners are now available for vegetation recovery [7]. The interactions between ECM fungus extracts and soil colloids observed in this study indicate that there are relatively fast physical and chemical reactions between fungi and tiny soil particles and that the clarification of this process may promote the understanding of the function of ECM in soil nutrient exploitation and soil health maintenance.

5. Conclusions

The addition of fungus extract to soil colloids could significantly affect the surface structure, mineral crystallization, functional groups, and elemental composition of colloids, although the effects were different for dark brown forest soil (a good loam) and saline-alkali soil (a heavily degraded soil). For dark brown forest soil, physical absorption at

the surface of soil colloids together with increases in functional group traits such as O–H stretching and bending, C–H stretching, C=O stretching, and Si–O–Si stretching and the relative crystallinity of kaolinite, hydromica, and quartz were observed after the addition of fungus extracts. In the case of degraded saline-alkali soil, the particle diameter and grain size of variable soil minerals were sharply increased, and remarkable reductions were found in the relative crystallization of variable minerals and most functional groups (5 out of 6), showing that chemical reactions instead of physical absorption possibly occurred after the addition of the fungus extract. As good loams and degraded soils are typical in northeastern China, the findings in this paper will improve the understanding of the mechanisms for ECM-aided soil improvements, especially for highly degraded soil.

Conflict of Interests

No conflict of interests is needed to be declared for all authors.

Acknowledgments

This study was supported financially by China's Ministry of Science and Technology (2011CB403205), China's National Foundation of Natural Sciences (31170575), China's Postdoctoral Foundation (201003406 and 20080430126), and Basic Research Fund for National Universities from Ministry of Education of China (DL12DA03).

References

[1] R. Lal, "Soil degradation by erosion," *Land Degradation & Development*, vol. 12, no. 6, pp. 519–539, 2001.

[2] S. Kapur and E. Akca, "Global assessment of land degradation," in *Encyclopedia of Soil Science*, R. Lal, Ed., pp. 296–306, Marcel Dekker, New York, NY, USA, 2002.

[3] C. Baum, K. Hrynkiewicz, P. Leinweber, and R. Meißner, "Heavy-metal mobilization and uptake by mycorrhizal and

nonmyrrhizal willows (*Salix × dasyclados*)," *Journal of Plant Nutrition and Soil Science*, vol. 169, no. 4, pp. 516–522, 2006.

[4] K. Hrynkiewicz, C. Baum, and P. Leinweber, "Mycorrhizal community structure, microbial biomass P and phosphatase activities under *Salix polaris* as influenced by nutrient availability," *European Journal of Soil Biology*, vol. 45, no. 2, pp. 168–175, 2009.

[5] A. G. Khan, C. Kuek, T. M. Chaudhry, C. S. Khoo, and W. J. Hayes, "Role of plants, mycorrhizae and phytochelators in heavy metal contaminated land remediation," *Chemosphere*, vol. 41, no. 1-2, pp. 197–207, 2000.

[6] X. M. Hua, *Trees Mycorrhizal Research*, Chinese Science and Technology Press, Beijing, China, 1995.

[7] K. M. Vogelsang, J. D. Bever, M. Griswold, and P. A. Schultz, "The use of mycorrhizal fungi in erosion control applications," Final Report for Caltrans 65A0070, California Department of Transportation, Sacramento, Calif, USA, 2004.

[8] Y. Li, S. Q. Wang, and W. P. Ju, "Effects of mycorrhizal fungi on pine seedling root disease prevention and cure," *Journal of Liaoning Forestry Science and Technology*, vol. 4, pp. 21–22, 2000.

[9] X. F. Yan and Q. Wang, "Effects of ectomycorrhizal inoculation on the seeding growth of QuercusLiaotungensis," *Acta Phytoecologica Sinica*, vol. 26, no. 6, pp. 701–707, 2002.

[10] Q. Ma, J. G. Huang, and J. B. Jiang, "Effect of inoculating with the ectotrophic mycorrhizal epiphyte on the *Pinus massoniana* seedling growth," *Journal of Fujian Forestry Science and Technology*, vol. 32, no. 2, pp. 85–88, 2005.

[11] B. M. Ohsowski, J. N. Klironomos, K. E. Dunfield, and M. Harta Miranda, "The potential of soil amendments for restoring severely disturbed grasslands," *Applied Soil Ecology*, vol. 60, pp. 77–83, 2012.

[12] D. J. Read, "Mycorrhizas in ecosystems," *Experientia*, vol. 47, no. 4, pp. 376–391, 1991.

[13] N. L. Carly, E. D. Laura, and H. Jonathan, "Mycorrhizae and soil phosphorus affect growth of *Celastrus orbiculatus*," *Biological Invasions*, vol. 13, no. 10, pp. 2339–2350, 2011.

[14] K. Bojarczuk and B. Kieliszewska-Rokicka, "Effect of ectomycorrhiza on Cu and Pb accumulation in leaves and roots of silver birch (*Betula pendula* Roth.) seedlings grown in metal-contaminated soil," *Water, Air and Soil Pollution*, vol. 207, no. 1-4, pp. 227–240, 2010.

[15] C. Voiblet, S. Duplessis, N. Encelot, and F. Martin, "Identification of symbiosis-regulated genes in *Eucalyptus globulus*—*Pisolithus tinctorius* ectomycorrhiza by differential hybridization of arrayed cDNAs," *The Plant Journal*, vol. 25, no. 2, pp. 181–191, 2001.

[16] A. Aučina, M. Rudawska, T. Leski, D. Ryliškis, M. Pietras, and E. Riepšas, "Ectomycorrhizal fungal communities on seedlings and conspecific trees of *Pinus mugo* grown on the coastal dunes of the Curonian Spit in Lithuania," *Mycorrhiza*, vol. 21, no. 3, pp. 237–245, 2011.

[17] K. Hrynkiewicz, G. Dabrowska, C. Baum, K. Niedojadlo, and P. Leinweber, "Interactive and single effects of ectomycorrhiza formation and *Bacillus cereus* on metallothionein *MT1* expression and phytoextraction of Cd and Zn by willows," *Water, Air & Soil Pollution*, vol. 223, no. 3, pp. 957–968, 2012.

[18] T. Leski and M. Rudawska, "Ectomycorrhizal fungal community of naturally regenerated European larch (*Larix decidua*) seedlings," *Symbiosis*, vol. 56, no. 2, pp. 45–53, 2012.

[19] D. Zimmer, C. Baum, P. Leinweber, K. Hrynkiewicz, and R. Meissner, "Associated bacteria increase the phytoextraction of cadmium and zinc from a metal-contaminated soil by mycorrhizal willows," *International Journal of Phytoremediation*, vol. 11, no. 22, pp. 200–213, 2009.

[20] G. K. Podila, J. Zheng, S. Balasubramanian et al., "Fungal gene expression in early symbiotic interactions between *Laccaria bicolor* and red pine," *Plant and Soil*, vol. 244, no. 1-2, pp. 117–128, 2002.

[21] A. L. Quéré, D. P. Wright, B. Söderström, A. Tunlid, and T. Johansson, "Global patterns of gene regulation associated with the development of ectomycorrhiza between birch (*Betula pendula* Roth.) and *Paxilus involutus* (Batsch) fr," *Molecular Plant-Microbe Interacionst*, vol. 18, no. 7, pp. 659–673, 2005.

[22] G. Heller, A. Adomas, G. Li et al., "Transcriptional analysis of *Pinus sylvestris* roots challenged with the ectomycorrhizal fungus *Laccaria bicolor*," *BMC Plant Biology*, vol. 8, article 19, 2008.

[23] D. M. Zhou, D. J. Wang, L. Cang, X. Hao, and L. Chu, "Transport and re-entrainment of soil colloids in saturated packed column: effects of pH and ionic strength," *Soils Sediments*, vol. 11, no. 3, pp. 491–503, 2011.

[24] N. M. DeNovio, J. E. Saiers, and J. N. Ryan, "Colloid movement in unsaturated porous media: recent advances and future directions," *Vadose Zone Journal*, vol. 3, no. 2, pp. 338–351, 2004.

[25] C. D. Barton and A. D. Karathanasis, "Influence of soil colloids on the migration of atrazine and zinc through large soil monoliths," *Water, Air and Soil Pollution*, vol. 143, no. 1–4, pp. 3–21, 2003.

[26] W. J. Wang, H. S. He, Y. G. Zu et al., "Addition of HPMA affects seed germination, plant growth and properties of heavy saline-alkali soil in Northeastern China: comparison with other agents and determination of the mechanism," *Plant and Soil*, vol. 339, no. 1, pp. 177–191, 2011.

[27] C. C. Muggler, T. H. Pape, and P. Buurman, "Laser grain-size determination in soil genetic studies. 2. Clay content, clay formation, and aggregation in some Brazilian oxisols," *Soil Science*, vol. 162, no. 3, pp. 219–228, 1997.

[28] L. Beuselinck, G. Govers, and J. Poesen, "Assessment of micro-aggregation using laser diffractometry," *Earth Surface Processes and Landforms*, vol. 24, no. 1, pp. 41–49, 1999.

[29] T. M. Zobeck, "Rapid soil particle size analyses using laser diffraction," *Applied Engineering in Agriculture*, vol. 20, no. 5, pp. 633–639, 2004.

[30] J. R. Campbell, "Limitations in the laser particle sizing of soils," in *Advances in Regolith*, pp. 38–42, 2003.

[31] S. Fontaine, S. Barot, P. Barré, N. Bdioui, B. Mary, and C. Rumpel, "Stability of organic carbon in deep soil layers controlled by fresh carbon supply," *Nature*, vol. 450, no. 8, pp. 277–280, 2007.

[32] B. D. Lee, S. K. Sears, R. C. Graham, C. Amrhein, and H. Vali, "Secondary mineral genesis from chlorite and serpentine in an ultramafic soil toposeqnence," *Soil Science Society of America Journal*, vol. 67, no. 4, pp. 1309–1317, 2003.

[33] G. Haberhauer and M. H. Gerzabek, "FTIR-spectroscopy of soils-characterisation of soil dynamic processes," *Trends in Applied Spectroscopy*, vol. 3, pp. 103–109, 2001.

[34] C. T. Johnston and Y. O. Aochi, "Fourier transform infrared and raman spectroscopy," in *Methods of Soil Analysis. Part 3*, J. M. Bartels and J. M. Bigham, Eds., pp. 269–321, Soil Science Society of America, American Society of Agronomy, Madison, Wis, USA, 1996.

Ectomycorrhizal Influence on Particle Size, Surface Structure, Mineral Crystallinity, Functional Groups, and Elemental
Composition of Soil Colloids from Different Soil Origins

147

[35] R. J. Cox, H. L. Peterson, J. Young, C. Cusik, and E. O. Espinoza, "The forensic analysis of soil organic by FTIR," *Forensic Science International*, vol. 108, no. 2, pp. 107–116, 2000.

[36] M. Soma, H. Seyama, N. Yoshinaga, B. K. G. Theng, and C. W. Childs, "Bonding state of silicon in natural ferrihydrites by X-ray photoelectron spectroscopy," *Clay Science Society of Japan*, vol. 9, no. 6, pp. 385–391, 1996.

[37] C. W. Childs, K. Inoue, H. Seyama, M. Soma, B. K. G. Theng, and G. Yuan, "X-ray photoelectron spectroscopic characterization of silica springs allophane," *Clay Minerals*, vol. 32, no. 4, pp. 565–572, 1997.

[38] G. Yuan, M. Soma, H. Seyama, B. K. G. Theng, L. M. Lavkulich, and T. Takamatsu, "Assessing the surface composition of soil particles from some Podzolic soils by X-ray photoelectron spectroscopy," *Geoderma*, vol. 86, no. 3-4, pp. 169–181, 1998.

[39] Y. H. Li, B. Y. Zhang, Y. Jiang, and W. J. Wang, "The effect of organic and inorganic carbon on extracellular enzyme activity of acid phosphatase and proteases in three kinds of fungal hyphae," *Bulletin of Botanical Research*, vol. 33, no. 4, pp. 404–409, 2013.

[40] Y. H. Dong, C. Hao, Y. R. Niu, and Q. J. Chen, "The ecological elementary research of *chroogomphisrutillus*," *Chinese Agricultural Science Bulletin*, vol. 26, no. 7, pp. 191–194, 2010.

[41] F. Q. Yu and P. G. Liu, "Reviews and prospects of the ectomycorrhizal research and application," *Acta Ecologica Sinica*, vol. 22, no. 22, pp. 2217–2226, 2002.

[42] B. Xu, J. M. Man, and C. X. Wei, "Methods for determining relative crystallinity of plant starch X-ray powder diffraction spectra," *Chinese Bulletin of Botany*, vol. 47, no. 3, pp. 278–285, 2012.

[43] P. R. Xie, *Soil Chemistry Mineralogy Properties in Northeast of China*, Science Press, Beijing, China, 2010.

[44] P. Marc and G. E. Jacques, *Handbook of Soil Analysis*, Springer, Berlin, Germany, 2006.

[45] D. L. Spark, *Methods of Soil Analysis Part 3: Chemical Methods*, Soil Science Society of America, 1996.

[46] S. E. Smith and D. J. Read, *Mycorrhizal Symbiosis*, Academic Press, London, UK, 1997.

[47] C. Baum, Y. K. Toljander, K. U. Eckhardt, and M. Weih, "The significance of host-fungus combinations in ectomycorrhizal symbioses for the chemical quality of willow foliage," *Plant and Soil*, vol. 323, no. 1-2, pp. 213–224, 2009.

[48] HLJTR. Soil management bureau and soil census office of Heilongjiang Province, China. *Soil of Heilongjiang Province*. China Agriculture, Beijing, China, 1993.

[49] J. L. Yan, G. X. Pan, L. Q. Li, G. Quan, C. Ding, and A. Luo, "Adsorption, immobilization, and activity of β-glucosidase on different soil colloids," *Journal of Colloid and Interface Science*, vol. 348, no. 2, pp. 565–570, 2010.

[50] B. Stenberg, R. A. ViscarraRossel, A. M. Mouazen, and J. Wetterlind, *Visible and Near Infrared Spectroscopy in Soil Science*, Academic Press, Burlington, Mass, USA, 2010.

[51] Q. F. Zheng and L. P. Zhao, "Effect of agricultural use on clay minerals and nutrient of black soil in Northeast of China," *Journal of Jilin Agricultural Sciences*, vol. 36, no. 5, pp. 29–32, 2011.

[52] S. R. Matias, M. C. Pagano, F. C. Muzzi et al., "Effect of rhizobia, mycorrhizal fungi and phosphate-solubilizing microorganisms in the rhizosphere of native plants used to recover an iron ore area in Brazil," *European Journal of Soil Biology*, vol. 45, no. 3, pp. 259–266, 2009.

17

Jatropha curcas L. Root Structure and Growth in Diverse Soils

Ofelia Andrea Valdés-Rodríguez,[1] **Odilón Sánchez-Sánchez,**[2] **Arturo Pérez-Vázquez,**[1] **Joshua S. Caplan,**[3] **and Frédéric Danjon**[4,5]

[1] *Colegio de Postgraduados, Campus Veracruz 421, 91690 Veracruz, VER, Mexico*
[2] *Centro de Investigaciones Tropicales, UV 91110 Xalapa, VER, Mexico*
[3] *Department of Ecology, Evolution and Natural Resources, Rutgers, The State University of New Jersey, New Brunswick, NJ 08901, USA*
[4] *INRA, UMR1202 BIOGECO, 33610 Cestas, France*
[5] *Université de Bordeaux, UMR1202 BIOGECO, 33610 Cestas, France*

Correspondence should be addressed to Arturo Pérez-Vázquez; parturo@colpos.mx

Academic Editors: J. Aherne, F. Bastida, and K. Wang

Unlike most biofuel species, *Jatropha curcas* has promise for use in marginal lands, but it may serve an additional role by stabilizing soils. We evaluated the growth and structural responsiveness of young *J. curcas* plants to diverse soil conditions. Soils included a sand, a sandy-loam, and a clay-loam from eastern Mexico. Growth and structural parameters were analyzed for shoots and roots, although the focus was the plasticity of the primary root system architecture (the taproot and four lateral roots). The sandy soil reduced the growth of both shoot and root systems significantly more than sandy-loam or clay-loam soils; there was particularly high plasticity in root and shoot thickness, as well as shoot length. However, the architecture of the primary root system did not vary with soil type; the departure of the primary root system from an index of perfect symmetry was 14 ± 5% (mean ± standard deviation). Although *J. curcas* developed more extensively in the sandy-loam and clay-loam soils than in sandy soil, it maintained a consistent root to shoot ratio and root system architecture across all types of soil. This strong genetic determination would make the species useful for soil stabilization purposes, even while being cultivated primarily for seed oil.

1. Introduction

Jatropha curcas L. has received a great deal of attention for its potential as a biofuel crop due to the high oil content of its seeds and because it can grow in soils with low nutrient content or water availability and on thin or steeply sloping soils [1, 2]. *J. curcas* seedlings are known to have consistent root system architecture, with a prominent vertical taproot and four lateral roots branching at equal angles (90°). The structural characteristics of *J. curcas* roots may therefore provide soil resistance to water and wind erosion in some sites, while simultaneously providing seeds for biofuel production [3].

One problem in considering *J. curcas* for projects in degraded soils is that its response to varying soil conditions has not been quantitatively evaluated. There are indications that *J. curcas* may alter its growth patterns in response to suboptimal conditions. For example, it is capable of shedding its leaves during prolonged dry periods [4, 5]. However, Heller [6], who made qualitative observations of the species in the African continent, reported that *J. curcas* grows well even on gravelly, sandy, and saline soils. Although not based on quantitative data, his observations are still referenced frequently in efforts to promote *J. curcas* as a biofuel crop [1, 5]. In Mexico and Central America, where *J. curcas* is native, reports also state that it is normally found in marginal soils of low nutrient content [7, 8]. There are suggestions that the plant grows better in sandy and loamy (i.e., aerated) soils than in clayey soils [9, 10]. Clay soils are reportedly less suitable because they limit root system development, especially when they are saturated [10, 11]. However, Valdes et al. [12] found that *J. curcas* could be more productive in sandy-loam and clay-loam soils than in sandy soils.

While the basic patterns described in the literature on *J. curcas* may be accurate, the response of root structure to different soil conditions has never been evaluated directly, and aboveground responses are mainly based on observational studies. Knowledge of how *J. curcas* root system architecture varies across a range of soil types will facilitate an evaluation of its suitability for revegetation in soil conservation efforts, will be relevant for biofuel purposes, and may also help determine if both aims can be achieved simultaneously. The objective of this study was to quantitatively describe the shoot and root structural variation of *J. curcas* seedlings in three different soils that are characteristic of the Mexican tropics.

2. Materials and Methods

2.1. Biological Material. Native Mexican seeds of *J. curcas* were collected in Papantla, in southeastern Mexico (20.2558° N, 97.2600° W, 77 masl) during August 2010. Seeds were selected from the middle of their weight distribution for sowing; average ± standard deviation (SD) measures were mass: 758 ± 97 mg, length: 8.4 ± 1.0 mm, width: 10.4 ± 0.50 mm, and thickness: 9.0 ± 0.5 mm.

2.2. Soil Selection. Soils were selected based in their textural characteristics and because they represented prominent soils of the eastern Mexican tropics. The sandy soil was an arenosol, the sandy-loam was a regosol, while the clay-loam was a phaeozem; typologies were based on previous research performed in the region [13]. Sandy-loam and clay-loam soils were obtained from the premises of the Colegio de Postgraduados in Veracruz (19.1954° N, 96.3389° W), while sandy soil was obtained from a dune near the city of Veracruz (19.2093° N, 96.2597° W). The upper 50 cm of soil was collected and homogenized; one subsample (500 g) was taken from each soil type for physical and chemical analyses. Textural characterization was performed following Bouyoucos [14] and classified according to NRCS [15]; bulk density was estimated by the gravimetric method. Analysis of pH was conducted using an electronic potentiometer in a 1 : 1 slurry, organic matter content was determined by the Walkley-Black method, extractable phosphorus was determined following Olsen and Sommers [16], and exchangeable calcium and magnesium concentrations were determined using methods based on Diehl et al. [17], all adapted for Mexican soils [18].

2.3. Experimental Conditions. The experiment was conducted outdoors in Veracruz, Mexico (19.1988° N, 96.1522° W, 2 masl) and was carried out using a completely randomized design, with 15 replicates per soil type (clay-loam, sandy-loam, and sand; n = 45 plants). Seeds were sown in early September 2010 and were uprooted three months after germination (when they were in the juvenile life stage). The maximum, minimum, and average temperatures recorded at a local meteorological station (Skywatch Geos no. 11) during the period were 29.2, 19.4, and 23.7°C, respectively. The average relative humidity was 75.3%.

One seed was sown per pot, which consisted of a black polyethylene bag (40 cm diameter × 50 cm length) filled with the assigned soil. The soil in each bag was watered to field capacity daily to maintain near-constant moisture levels in all containers. Average irrigation provided per pot was approximately 310 mm (sand), 666 mm (sandy-loam), or 597 mm (clay-loam) in total through the experimental period. Pots with sandy soil received less water because of the lower water requirements of these plants.

2.4. Aboveground Measurements. At the conclusion of the experiment (three months after germination), we measured shoot length, leaf number, and diameter at the root collar. We also calculated the area of the largest leaf on each plant based on the model obtained by Liv et al. [19] for *J. curcas* (Figure 1(a)):

$$\text{Leaf Area} = 0.84 * (t * l)^{0.99}, \tag{1}$$

where t = leaf cross-sectional length and l = leaf longitudinal length.

Stem volume (V) was calculated assuming that the stem was composed of two conical frustums: one extending from the root crown to the widest point on the stem and the other extending from the widest point to the attachment point of the most basal leaf (Figure 1(b)):

$$V = \frac{\pi}{3} * L_1 * \left[\left(\frac{R_c}{2} \right)^2 + \left(\frac{d_{\max}}{2} \right)^2 + (R_c * d_{\max}) \right] + \frac{\pi}{3}$$
$$* L_2 * \left[\left(\frac{d_{\min}}{2} \right)^2 + \left(\frac{d_{\max}}{2} \right)^2 + (d_{\min} * d_{\max}) \right], \tag{2}$$

where R_c is root collar diameter; $d_{\max}$ is stem diameter at its widest point; $d_{\min}$ is stem diameter at the attachment point of the most basal leaf; L_1 is length from R_c to $d_{\max}$; L_2 is length from $d_{\max}$ to $d_{\min}$.

2.5. Uprooting. Plants were uprooted using methods that previous experience showed to be optimal for the various soil textures. Plants in sandy and sandy-loam soils were uprooted while the root zone was sprayed with water at low pressure. Plants in clay-loam soils were watered to 50% of the soil's saturation level and uprooted without the use of sprayed water.

2.6. Identification and Digitization of the Root Structure. The primary coarse root structure of *J. curcas* includes the taproot and four main lateral roots; these are all present within 24 hours of germination (Figure 2). The architecture of this set of five roots was encoded in 3D using methods adapted from Reubens et al. [3]. The taproot and the four primary lateral roots were encoded in terms of length (measuring tape, 1.0 mm precision), diameter (at bases and tips with a caliper, 0.01 mm precision), and orientation in the X, Y, and Z planes (at the bases and at 20 cm from their bases with a protractor, 1° precision). Secondary roots that emerged from any of the five

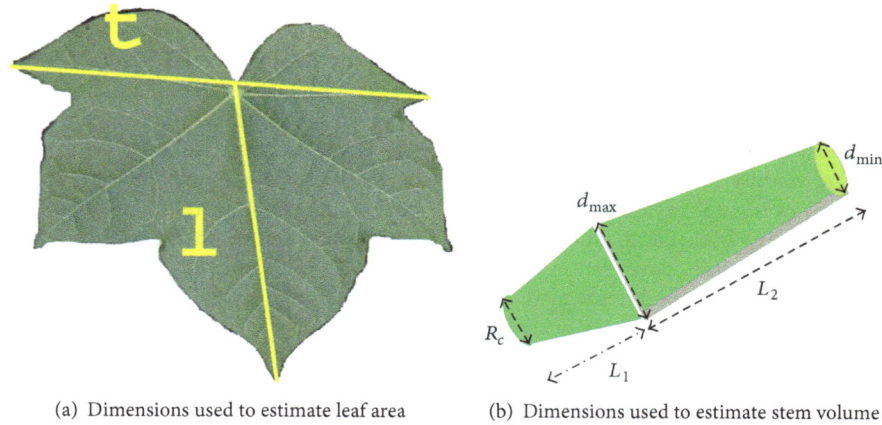

(a) Dimensions used to estimate leaf area

(b) Dimensions used to estimate stem volume

FIGURE 1: Leaf area and stem volume calculations.

FIGURE 2: *Jatropha curcas* seedling 24 hr after germination. Note the radicle and four lateral roots which comprise its fundamental root structure.

primary roots and had a diameter thicker than 2.0 mm were also recorded. The soil level at the center of the stem base was considered the initial reference $(0, 0, 0)$, while one of the four primary lateral roots was selected to define zero azimuth (Figure 3). Root segments ended at a branching point or where there was an abrupt change of growth direction. The above data were organized as Multi-scale Tree Graphs (MTGs), which are specialized databases for three-dimensional plant structure [20]. AMAPmod software version 2.2.30 [21] was used to derive architectural characteristics from the MTGs. Leaf, stem, and root dry masses were measured (analytical balance, 0.001 g precision) after oven drying at 70°C for 72 hr.

2.7. Modeling the Root Structure. In the idealized case, the four primary lateral roots of *J. curcas* would originate at the same vertical position along the stem, be symmetrically distributed in the horizontal plane, have the same diameters, and have the same inclinations. The consistency with which plants conformed to this idealized root structure was evaluated using a model that considers five estimators or indexes that range from zero to one, where zero is the perfect conformation to the idealized model and one represents

maximal deviation from the model (modified from Reubens et al. [3], Figure 4).

With respect to the horizontal plane, we considered the symmetry in the angular distribution of the four primary lateral roots (β_{symm}):

$$\beta_{symm}$$
$$= \frac{[\text{abs}(b_{1\text{-}2}-90°)+\text{abs}(b_{2\text{-}3}-180°)+\text{abs}(b_{3\text{-}4}-270°)]}{540°},$$

(3)

where b_{ij} is the horizontal angle between two neighboring primary lateral roots i and j (Figure 4(a)). $\beta_{symm} = 0$ if all the roots are distributed at 90° intervals and 1 if all the roots extend from a single point.

We also evaluated consistency in the basal diameter of the four primary lateral roots (D_{symm}):

$$D_{symm} = \frac{(\sum(d_{max}-d_i)/\sum d_{1\text{-}4})}{3},$$

(4)

where d_i is the basal diameter of the ith primary lateral root (Figure 4(b)); $\sum d_{1\text{-}4}$ is the sum of the four primary lateral root diameters; and d_{max} is the maximum diameter of the four primary laterals. $D_{symm} = 0$ if all roots have the same diameter and 1 if there is only one lateral root.

In this study, instead of considering oblique roots, as in Reubens et al. [3], we considered the consistency in the angle of the four primary lateral roots below the horizontal surface (their inclinations, θ_i), for the root within the ZRT (Zone of Rapid Taper, as defined by Danjon et al. [22]).

With respect to the vertical plane, we considered the symmetry in the angular deviation from the horizontal (θ_{symm}):

$$\theta_{symm} = \frac{(\sum \text{abs}(\theta_{max}-\theta_i)/\sum \theta_{1\text{-}4})}{3},$$

(5)

where θ_i is the angle of the ith primary lateral root below horizontal surface within the ZRT (Figure 4(c)); θ_{max} is the

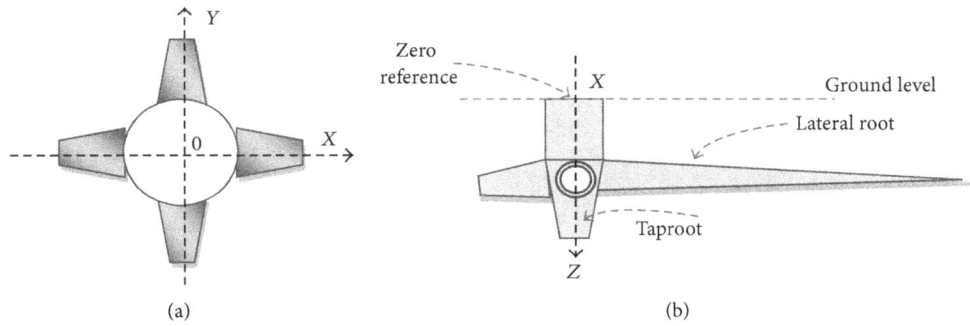

FIGURE 3: Coarse root structure. (a) Top view, (b) Lateral view.

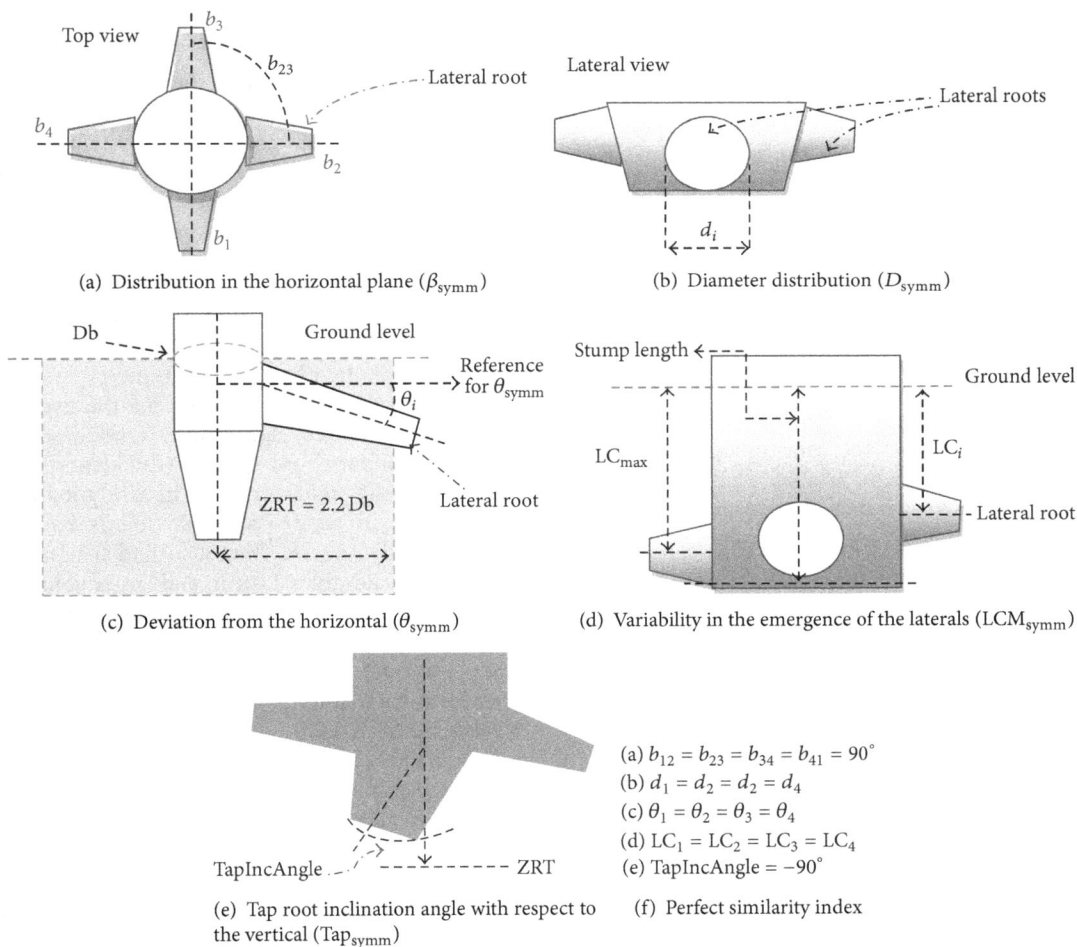

(a) Distribution in the horizontal plane (β_{symm})

(b) Diameter distribution (D_{symm})

(c) Deviation from the horizontal (θ_{symm})

(d) Variability in the emergence of the laterals (LCM_{symm})

(e) Tap root inclination angle with respect to the vertical (Tap_{symm})

(a) $b_{12} = b_{23} = b_{34} = b_{41} = 90°$
(b) $d_1 = d_2 = d_2 = d_4$
(c) $\theta_1 = \theta_2 = \theta_3 = \theta_4$
(d) $LC_1 = LC_2 = LC_3 = LC_4$
(e) $TapIncAngle = -90°$

(f) Perfect similarity index

FIGURE 4: Parameters considered for calculating the similarity index and its component measurements.

maximum inclination of the four primary roots. $\theta_{symm} = 0$ if all roots have the same inclination angle; 1 represents the highest difference between angles.

We also calculated the variability in the position of emergence of the four primary laterals from the taproot (LCM_{symm}). To do this we considered the length from the root collar (at the level of the soil surface) to the branching point of each of the four primary lateral roots (LC_i, Figure 4(d)):

$$LCM_{symm} = \left[\frac{\sum (LC_{max} - LC_i)}{3} \right] * (\text{Stump Length}), \quad (6)$$

where LC_{max} is the longest LC_i, and Stump is the portion of the taproot from which the four main lateral roots branch [22] (Figure 4(d)). LCM_{symm} = 0 if all main laterals originate at the same point and 1 if roots originate from opposite extremes of the stump. Note that this definition of LCM_{symm} differs from that of Reubens et al. [3], insofar as they considered a departure from the fixed value of 2.5 cm for all seedlings in their study. As this value depends on the soil type and how deeply the seed was sown, we only evaluated the similarity of the length to root base collar (LC) from each lateral root.

The final similarity measurement we considered was the angle from which the taproot deviated from a vertical line (Tap_{symm}):

$$Tap_{symm} = \frac{[(-90°) - TapIncAngle]}{90°}, \tag{7}$$

where TapIncAngle is the inclination angle between the taproot and the vertical at the level of the root stump (Figure 4(e)). Tap_{symm} = 0 if the taproot is vertically oriented and 1 if the taproot is horizontally oriented.

We computed a composite metric for the degree to which *J. curcas* plants adhered to the idealized model plant (SI):

$$SI = \frac{\left(\beta_{symm} + D_{symm} + \theta_{symm} + LCM_{symm} + Tap_{symm}\right)}{5}. \tag{8}$$

SI = 0 for root systems perfectly matching the model and 1 for complete lack of adherence.

We used an index of phenotypic plasticity (PI) [23] to quantify the magnitude of the morphological response to varying soil types. For each variable, PI uses the mean response for individuals grown in each treatment to evaluate the greatest change displayed by the species among treatments:

$$PI = \frac{(Maximum\ value - Minimum\ value)}{Maximum\ value}. \tag{9}$$

PI ranges from 0 to 1, with 1 representing the greatest possible plasticity.

2.8. Statistical Analysis. Differences in parameter means among soil types were statistically compared using one-way analysis of variance (ANOVA) in SigmaPlot 10.0. Tests of residual normality and equal variance were conducted. *Post hoc* comparisons were made for normally distributed parameters with a Tukey test, while non-normally distributed parameters were analyzed with Dunn's Method, all with a 95% confidence level.

3. Results

3.1. Soil Analysis. All substrates were found to be slightly alkaline. However, the sandy soil had very low organic matter content, being 2–4% of the amount in the other soils (Table 1). The sand also had 10–26% of the P, 23–44% of the Ca, and 29–53% of the Mg found in other soils (Table 1).

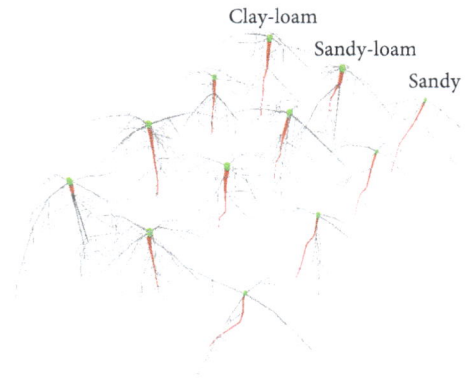

FIGURE 5: Digitized *Jatropha curcas* root systems grown in three different soils.

3.2. Above- and Belowground Response to Soil Types. Plants grown in the sandy-loam and clay-loam soils had, on average, approximately twice the height and three times wider collar diameter than plants grown in sandy soil. Stem volumes, numbers of leaves, and leaf areas were more than five times greater for plants grown in these soils than for plants grown in sandy soil (Table 2). All plants grown in sandy soil survived, but 62% were completely defoliated by the conclusion of the experiment; none of those grown in sandy-loam or clay-loam soils lost all leaves. Stem slenderness ratio (height over root collar diameter) did not differ among soils, indicating that the seedlings were not plastic in this trait.

Root diameters and volumes for the five primary roots in *J. curcas* differed significantly among soil types ($P < 0.001$). Secondary root growth (thickening) was lower for plants in sandy soil than those in sandy-loam or clay-loam soils. Roots in sandy- and clay-loams had similar basal and apical diameters. They also had a greater number of branches thicker than 2.0 mm and larger volumes than those in sandy soil. All taproots in sandy-loam and clay-loam soils developed secondary roots thicker than 2.0 mm, whereas only 13% of the taproots in sandy soil developed such roots. However, root lengths did not differ significantly among treatments ($P > 0.05$).

Stem mass, leaf mass, and root system mass were lower at the conclusion of the three-month growing period in sandy soil than in clay-loam or sandy-loam soils (Table 2, Figure 5). Despite these differences, allocation of biomass was greater to stems than to roots in all three soil types (Table 3). Within root systems, the greatest proportion of biomass and volume was allocated to taproots (Table 4). The uppermost 10 cm of soil contained the majority of root volume (Figure 6).

3.3. Root Structure, Similarity Indices, and Plasticity to Soil. Root system symmetry index scores were typically 0.146 ± 0.05 (mean ± SD); mean SI did not statistically differ among soil types (Table 4). Of the main parameters defining *J. curcas* root structure, taproot inclination, and primary lateral root distribution (β) had the lowest plasticity index scores (0.05 and 0.06, respectively). Biomass allocation to roots and the inclination angles (θ), length, and apical diameters of the five

TABLE 1: Soil characteristics for the three soil types in which *J. curcas* seedlings were grown for three months.

Soil type	Texture (%) Sand	Silt	Clay	pH	Bulk density $(g\,cm^{-3})$	Organic matter $(g\,kg^{-1})$	P $(g\,kg^{-1})$	Ca $(mmol\,kg^{-1})$	Mg $(mmol\,kg^{-1})$
Sand	96.0	2.5	1.5	7.81	1.56	1.68	0.01	77.17	154.35
Sandy-loam	66.0	21.0	13.0	7.26	1.47	39.00	0.05	175.40	294.66
Clay-loam	30.0	35.0	35.0	7.43	1.26	72.62	0.12	329.74	519.17

TABLE 2: Aboveground parameters in *J. curcas* seedlings grown in three different soils.

Soil type	Stem length (mm)	Root collar diameter (mm)	Stem slenderness $(cm\,cm^{1})$	Stem volume (cm^3)	Number of leaves	Leaf area (cm^2)
Sand	209.4 ± 26.6^b	12.1 ± 1.6^b	17.5 ± 2.7^a	21.71 ± 8.9^b	0.5 ± 0.8^b	29.5 ± 0.2^b
Sandy-loam	380.2 ± 88.7^a	23.4 ± 3.3^a	15.9 ± 2.9^a	118.78 ± 45.7^a	7.0 ± 2.6^a	223.0 ± 5.4^a
Clay-loam	361.6 ± 72.8^a	23.1 ± 3.0^a	15.6 ± 1.7^a	114.11 ± 46.8^a	6.8 ± 2.8^a	178.0 ± 54.9^a

[a,b] Means within a column which do not share the same letter are significantly different ($P < 0.05$).

TABLE 3: Average $\pm$ SD dry matter allocation in *J. curcas curcas* grown in three different soil types.

Soil type	Total biomass (g)	Stem, total^{-1}	Leaves, total^{-1}	Root, total^{-1}
Sand	3.17 ± 1.24^b	0.77 ± 0.10^a	0.03 ± 0.04^b	0.20 ± 0.07^a
Sandy-loam	29.59 ± 9.81^a	0.63 ± 0.06^a	0.19 ± 0.06^a	0.18 ± 0.03^a
Clay-loam	30.01 ± 11.01^a	0.66 ± 0.10^a	0.17 ± 0.04^{ab}	0.17 ± 0.04^a

[a,b] Means within a column which do not share the same letter are significantly different ($P < 0.05$).

TABLE 4: Average $\pm$ SD below-ground parameters in *J. curcas* seedlings grown in three different soils.

Parameter	Units	Sand	Sandy-loam	Clay-loam	PI
Root length	cm				
Total		115.9 ± 17.0^b	132.0 ± 18.2^{ab}	144.6 ± 38.6^a	0.20
Taproot		36.1 ± 85.1^a	39.7 ± 95.9^a	41.1 ± 72.1^a	0.12
Four main laterals		27.7 ± 8.5^a	32.3 ± 10.1^a	37.5 ± 14.6^a	0.26
Basal diameter	mm				
Taproot		8.2 ± 1.3^b	20.1 ± 4.4^a	18.4 ± 3.2^a	0.59
Four main laterals		2.9 ± 0.4^b	4.8 ± 1.9^a	5.2 ± 1.3^a	0.44
Apex diameter	mm				
Taproot		0.70 ± 0.25^a	0.63 ± 0.17^a	0.70 ± 0.15^a	0.10
Four main laterals		0.41 ± 0.11^b	0.53 ± 0.09^{ab}	0.59 ± 0.19^a	0.31
Number of roots > 2.0 mm thick		5.13 ± 0.35^b	12.09 ± 5.85^a	13.89 ± 5.84^a	0.63
Root mass					
Total	g	0.61 ± 0.20^b	5.33 ± 1.74^a	5.34 ± 3.02^a	0.89
Taproot	%	74.40 ± 9.85^b	85.48 ± 6.97^a	75.78 ± 7.05^b	
Coarse root structure					
TapIncAng	deg	-89.36 ± 4.48^a	-85.21 ± 3.72^a	-88.67 ± 6.12^a	0.05
θ	deg	-20.71 ± 4.79^a	-18.59 ± 3.05^a	-17.94 ± 3.31^a	0.13
LCM	cm	1.07 ± 0.41^a	1.36 ± 0.50^a	1.0 ± 0.38^a	0.26
β	deg	89.3 ± 4.40^a	94.7 ± 3.7^a	91.3 ± 6.1^a	0.06
Similarity indexes					
β_{symm}		0.02 ± 0.02^a	0.04 ± 0.03^a	0.03 ± 0.03^a	
D_{symm}		0.24 ± 0.10^a	0.29 ± 0.09^a	0.25 ± 0.14^a	
θ_{symm}		0.27 ± 0.10^a	0.34 ± 0.11^a	0.35 ± 0.12^a	
LCM_{symm}		0.12 ± 0.06^a	0.16 ± 0.13^a	0.11 ± 0.07^a	
Tap_{Symm}		0.04 ± 0.04^a	0.04 ± 0.03^a	0.04 ± 0.03^a	
SI		0.13 ± 0.05^a	0.15 ± 0.07^a	0.16 ± 0.03^a	

[a,b] Means within a column which do not share the same letter are significantly different ($P < 0.05$).

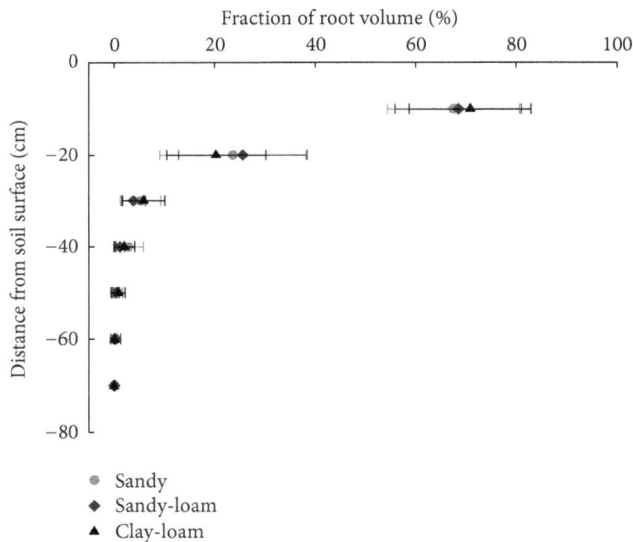

FIGURE 6: Root volume distribution by depth in sandy, sandy-loam and clay-loam soils.

primary coarse roots also showed low plasticity (PI < 0.31). However, the number of secondary lateral roots thicker than 2 mm, as well as root mass, was highly plastic ($P > 0.63$) (Table 4).

4. Discussion

4.1. Growth and Mass Distribution. *J. curcas* had a significant growth response to soil conditions. In sandy soil, it displayed characteristics typical of plants grown in arid conditions, including reduction of leaf area and defoliation. These responses reduce transpirational surface area and are common adaptations of species with photosynthetic stems, such as *J. curcas* [24]. Additionally, the low nutrient availability of the sandy soil strongly reduced stem and leaf growth [25, 26]. Higher biomass allocation to stems over leaves and roots, regardless of soil type, indicates that this ratio is strongly genetically determined. This lack of plasticity may be adaptive because stem tissue is used for water storage by seedlings of *J. curcas*, allowing them to survive during dry periods [5]. Another pattern that remained consistent across soil types was that the largest fraction of the mass was allocated to roots in the uppermost 10 cm of the surface (Figure 4) and to the taproot (73% in sand, 87% in sandy-loam, and 76% in clay-loam) (Table 3). The root architecture of one-month-old seedlings grown in sandy soil was previously described by Reubens et al. [3]; they had 50% of their root volume allocated to the taproot. Taken together, these results suggest that there is an increase in the mass and volume of taproot as compared to lateral roots over time. Enlarged taproots, in combination with consistently shallow lateral roots, indicate that seedlings search simultaneously for resources in deep and shallow soil.

Although the clay-loam soil used in this experiment had the highest nutrient content, there were no differences in growth parameters or biomass compared to the sandy-loam, which had a lower nutrient content (Tables 2 and 3). This result is contrary to that of Patolia et al. [27], who reported greater biomass production under elevated nutriment conditions. In this study, it is likely that plant roots were more easily able to obtain nutrients from the sandy-loam than from the clay-loam because of the high soil aeration requirements of *J. curcas*. It is also possible that nutrient levels in our sandy-loam soil were near the ideal levels to which this species is adapted at this stage of growth [28].

Stem growth rates of 4.1 mm day^{-1} recorded in sandy-loam and clay-loam soils were similar to growth rates measured by Jimu et al. [29] in clay soils and under similar temperatures. However, the low aboveground development in sandy soil found in this study (2.3 mm day^{-1}) contrasts with claims that *J. curcas* can grow well under semiarid conditions and sandy soils [6, 9]. Higher stem and leaf growth rates in sandy soil have been reported before [30] but under periodic water irrigation with amendments of N, P, K, Ca, and Mg. Low levels of N and P in our sandy soil probably contributed to leaf loss and slow stem growth rates [31]. Maintenance of root length in sandy soil, despite extreme reductions in root mass, indicates that plants retain a capacity for soil exploration under limiting nutrient conditions. Similar patterns were reported by Achten et al. [30] under extreme drought stress. This strategy may serve to improve foraging outcomes for soil resources. Each of the five primary roots in *J. curcas* took on a strongly herring-bone branching structure. This structure is highly efficient in soil exploration [32] and is indicative of *J. curcas*' adaptation to well-drained and nutrient-poor soils [33, 34].

4.2. Root Structure and Plasticity to Soil Type. As observed previously by Reubens et al. [3], we found similar symmetry index values among soils. Primary lateral root and taproot inclination angles were also similar among the three soils, suggesting that the arrangement and structure of *J. curcas* root systems are strongly determined by genetics and only weakly affected by environmental conditions, such as the soil textures used in this experiment. Having prominent lateral roots with a symmetrical radial distribution and consistent diameters provides balanced anchorage to *J. curcas* plants; this root structure can tolerate forces originating from varying directions and maintain stability. Low plasticity in stem allocation, root allocation, and root structure (Tables 3 and 4) indicates that these characteristics are also strongly determined by genetics and are minimally influenced by soil conditions. Maintenance of higher mass in stems than in roots, independent of the soil condition, may also indicate that *J. curcas* is a species that evolved to store resources in the stem and thereby avoid physiological stress in extreme environmental conditions [23]. Positioning lateral roots near the soil surface is a characteristic of plants adapted to arid climates [24]. Therefore, this species could be established in sites with limited nutrient and water resources, although growth rates and seed production under these circumstances could be extremely low.

The fact that the primary root system structure of *J. curcas* (a long, thick taproot with four, nearly perpendicular lateral roots) was not plastic in response to soil type indicates that its large lateral roots are able to stabilize superficial soils, while its large taproot can provide reinforcement across planes of weakness, for example, along the flanks of potential slope failures [22, 35]. Therefore, this plant will reliably reinforce soils in which it is planted by increasing the shear and tensile strength of the rooting zone [36]. Additionally, *J. curcas* has been shown to raise the macroaggregate stability and organic matter content of the soils in which it grows [37], ensuring that precipitation infiltrates rather than runs off and that a minimal amount of soil erodes.

5. Conclusions

J. curcas seedlings developed well in both sandy-loam and clay-loam soils. In sandy soil, its growth was reduced significantly, though plants were still able to survive and maintain a favorable root-shoot relationship. These characteristics would allow the plant to survive under a wide variety of soil conditions, making it well suited for preventing soil erosion. Although its growth, seed production, and performance for erosion control could be lower in poor soils, *J. curcas* cultivation programs could not only serve as a source of income generation, but could also improve the quality of soils in the long run.

References

[1] K. Li, W. Y. Yang, L. Li, C. H. Zhang, Y. Z. Cui, and Y. Y. Sun, "Distribution and development strategy for *Jatropha curcas* L. in Yunnan Province, Southwest China," *Forestry Studies in China*, vol. 9, no. 2, pp. 120–126, 2007.

[2] S. K. Behera, P. Srivastava, R. Tripathi, J. P. Singh, and N. Singh, "Evaluation of plant performance of *Jatropha curcas* L. under different agro-practices for optimizing biomass—a case study," *Biomass and Bioenergy*, vol. 34, no. 1, pp. 30–41, 2010.

[3] B. Reubens, W. M. J. Achten, W. H. Maes et al., "More than biofuel? *Jatropha curcas* root system symmetry and potential for soil erosion control," *Journal of Arid Environments*, vol. 75, no. 2, pp. 201–205, 2011.

[4] J. C. Dagar, O. S. Tomar, Y. Kumar, H. Bhagwan, R. K. Yadav, and N. K. Tyagi, "Performance of some under-explored crops under saline irrigation in a semiarid climate in northwest India," *Land Degradation and Development*, vol. 17, no. 3, pp. 285–299, 2006.

[5] R. Brittaine and L. Lutaladio, *Jatropha: A Smallholder Bioenergy Crop: The Potential for Pro-Poor Development*, Integrated Crop Management, Food and Agriculture Organization of the United Nations, Rome, Italy, 2010.

[6] J. Heller, *Physic Nut. Jatropha Curcas L.*, Promoting the Conservation and Use of Underutilized and Neglected Crops, Institute of Plant Genetics and Crop Plant Research, Gatersleben/ International Plant Genetic Resources Institute, Rome, Italy, 1996.

[7] J. Martínez-Herrera, P. Siddhuraju, G. Francis, G. Dávila-Ortíz, and K. Becker, "Chemical composition, toxic/antimetabolic constituents, and effects of different treatments on their levels, in four provenances of *Jatropha curcas* L. from Mexico," *Food Chemistry*, vol. 96, no. 1, pp. 80–89, 2006.

[8] Semarnat-INE-UNAM-CIECO, "Análisis integrado de las tecnologías, el ciclo de vida y la sustentabilidad de las opciones y escenarios para el aprovechamiento de la bioenergía en México," Final Report, Instituto Nacional de Ecología, México, Mexico, 2008.

[9] V. K. Gour, "Production practices including post harvest management of *Jatropha curcas*," in *Proceedings of the Biodiesel Conference Towards Energy Independence—Focus on Jatropha*, B. Singh, R. Swaminathan, and V. Ponraj, Eds., pp. 223–251, Hyderabad, India, June 2006.

[10] W. M. J. Achten, L. Verchot, Y. J. Franken et al., "Jatropha biodiesel production and use," *Biomass and Bioenergy*, vol. 32, no. 12, pp. 1063–1084, 2008.

[11] K. D. Ouwens, G. Francis, Y. J. Franken et al., *Position Paper on Jatropha curcas. State of the Art, Small and Large Scale Project Development*, FACT Foundation, Wageningen, The Netherlands, 2007.

[12] R. O. A. Valdes, S. O. Sanchez, V. A. Perez, and B. R. Ruiz, "Soil texture effects on the development of *Jatropha curcas* seedlings—Mexican variety 'Piñon manso'," *Biomass Bioenery*, vol. 35, pp. 3529–3536, 2011.

[13] S. C. A. Ortiz and C. C. J. Lopez, *Los Suelos del Campus Veracruz*, Colegio de Postgraduados, Campus Veracruz, Veracruz, Mexico, 2000.

[14] G. J. Bouyoucos, "Directions for making mechanical analysis of soil by hydrometer method," *Soil Science*, vol. 42, no. 3, pp. 225–230, 1936.

[15] National Resources Conservation Services (NRCS), "Soil Texture Calculator," 2010, http://soils.usda.gov/technical/aids/investigations/texture/.

[16] S. R. Olsen and L. E. Sommers, "Phosphorus," in *Methods of Soil Analysis*, A. L. Page, Ed., pp. 420–422, American Society of Agronomy, Madison, Wis, USA, 1982.

[17] H. Diehl, C. A. Goetz, and C. C. Hach, "The Versenate titration for total hardness," *American Water Works Association Journal*, vol. 42, pp. 40–48, 1950.

[18] Secretaría de Medio Ambiente y Recursos Naturales (Semarnat), "Norma Oficial Mexicana NOM-021-SEMARNAT-2000 que establece las especificaciones de fertilidad, salinidad y clasificación de suelos, estudio, muestreo y análisis," *Diario Oficial (Segunda Sección)*, 2002 (Spanish), México, Mexico, http://www.profepa.gob.mx/innovaportal/file/3335/1/nom-021-semarnat-2000.pdf.

[19] S. S. Liv, L. S. do Vale, and B. N. E. de Macedo, "A simple method for measurement of *Jatropha curcas* leaf area," *Revista Brasileira de Oleaginosas e Fibrosas*, vol. 11, no. 1, pp. 9–14, 2007.

[20] C. Godin, Y. Guedon, and E. Costes, "Exploration of plant architecture databases with the AMAPmod software illustrated on an apple-tree hybrid family," *Agronomie*, vol. 19, no. 3-4, pp. 163–184, 1999.

[21] CIRAD/INRA-UMR, "Exploring and Modeling Plant Architecture. Software for Windows v 2.2.30," La Recherche Agronomique pour le Développement/Inventeurs du Monde Numérique, Montpellier, France, 2006.

[22] F. Danjon, T. Fourcaud, and D. Bert, "Root architecture and wind-firmness of mature Pinus pinaster," *New Phytologist*, vol. 168, no. 2, pp. 387–400, 2005.

[23] F. Valladares, S. J. Wright, E. Lasso, K. Kitajima, and R. W. Pearcy, "Plastic phenotypic response to light of 16 congeneric shrubs from a panamanian rainforest," *Ecology*, vol. 81, no. 7, pp. 1925–1936, 2000.

[24] L. Ci and X. Yang, "Biological and technical approaches to control windy desertification," in *Desertification and Its Control*, pp. 35–426, Higher Education Press, Beijing, China, 2010.

[25] D. Sánchez and J. Aguirreolea, "Transporte del agua y balance hídrico en la planta," in *Fundamentos de Fisiología Vegetal*, J. Azcón-Bieto and J. Talón, Eds., pp. 45–63, McGraw-Hill, Barcelona, Spain, 2000.

[26] S. Tracy, C. R. Black, J. A. Roberts et al., "Quantifying the impact of soil compaction on root system architecture in tomato (Solanumlycopersicum) by X-ray micro-computed tomography," *Ann Bot*, vol. 110, no. 2, pp. 511–519, 2012.

[27] J. S. Patolia, A. Ghosh, J. Chikara, D. R. Chaudharry, D. R. Parmar, and H. M. Bhuva, "Response of *Jatropha curcas* grown on wasted land to P and N," in *Proceedings of the Expert Seminar on Jatropha curcas L. Agronomy and Genetics*, FACT Foundation, Wageningen, The Netherlands, March 2007.

[28] J. P. Grime, "Estrategias primarias en la fase establecida," in *Estrategias de adaptación de las plantas y procesos que controlan la vegetación*, pp. 19–75, Limusa, Mexico City, Mexico, 1989.

[29] L. Jimu, I. W. Nyakudya, and C. A. T. Katsvanga, "Establishment and early field performance of *Jatropha curcas* L. at Bindura University farm, Zimbabwe," *Journal of Sustainable Development in Africa*, vol. 10, no. 4, pp. 445–469, 2009.

[30] W. M. J. Achten, W. H. Maes, B. Reubens et al., "Biomass production and allocation in *Jatropha curcas* L. seedlings under different levels of drought stress," *Biomass and Bioenergy*, vol. 34, no. 5, pp. 667–676, 2010.

[31] I. Bonilla, "Introducción a la nutrición mineral de las plantas. Los elementos minerales," in *Fundamentos de Fisiología Vegetal*, J. Azcon-Bieto and J. Talón, Eds., pp. 83–97, McGraw-Hill, Mexico City, Mexico, 2000.

[32] F. Danjon and B. Reubens, "Assessing and analyzing 3D architecture of woody root systems, a review of methods and applications in tree and soil stability, resource acquisition and allocation," *Plant and Soil*, vol. 303, no. 1-2, pp. 1–34, 2008.

[33] T. Koike, M. Kitao, A. M. Quoreshi, and Y. Matsuura, "Growth characteristics of root-shoot relations of three birch seedlings raised under different water regimes," *Plant and Soil*, vol. 255, no. 1, pp. 303–310, 2003.

[34] L. Qu, A. M. Quoreshi, and T. Koike, "Root growth characteristics, biomass and nutrient dynamics of seedlings of two larch species raised under different fertilization regimes," *Plant and Soil*, vol. 255, no. 1, pp. 293–302, 2003.

[35] A. Stokes, C. Atger, A. G. Bengough, T. Fourcaud, and R. C. Sidle, "Desirable Plant root traits for protecting natural and engineered slopes against landslides," *Plant and Soil*, vol. 324, no. 1, pp. 1–30, 2009.

[36] Y. Zhou, D. Watts, Y. Li, and X. Cheng, "A case study of effect of lateral roots of Pinus yunnanensis on shallow soil reinforcement," *Forest Ecology and Management*, vol. 103, no. 2-3, pp. 107–120, 1998.

[37] J. O. Ogunwole, D. R. Chaudhary, A. Ghosh, C. K. Daudu, J. Chikara, and J. S. Patolia, "Contribution of *Jatropha curcas* to soil quality improvement in a degraded Indian entisol," *Acta Agriculturae Scandinavica B*, vol. 58, no. 3, pp. 245–251, 2008.

18

Influence of Long-Term Thinning on the Biomass Carbon and Soil Respiration in a Larch (*Larix gmelinii*) Forest in Northeastern China

Huimei Wang,[1] Wei Liu,[1,2] Wenjie Wang,[1] and Yuangang Zu[1]

[1] Key Laboratory of Forest Plant Ecology, Ministry of Education, Northeast Forestry University, Harbin 150040, China
[2] College of Art and Landscape, Jiangxi Agricultural University, Nanchang 330045, China

Correspondence should be addressed to Wenjie Wang; wjwang225@hotmail.com

Academic Editors: F. Darve and A. Roldán Garrigós

Thinning management is used to improve timber production, but only a few data are available on how it influences ecosystem C sink capacity. This study aims to clarify the effects of thinning on C sinks of larch plantations, the most widespread forests in Northeastern China. Both C influx from biomass production and C efflux from each soil respiration component and its temperature sensitivity were determined for scaling-up ecosystem C sink estimation: microbial composition is measured for clarifying mechanism for respiratory changes from thinning treatment. Thinning management induced $6.23 \, \text{mol C m}^{-2} \, \text{yr}^{-1}$ increase in biomass C, while the decrease in heterotrophic respiration (R_h) at the thinned sites ($0.9 \, \text{mol C m}^{-2} \, \text{yr}^{-1}$) has enhanced 14% of this biomass C increase. This decrease in R_h was a sum of the 42% decrease ($4.1 \, \text{mol C m}^{-2} \, \text{yr}^{-1}$) in litter respiration and $3.2 \, \text{mol C m}^{-2} \, \text{yr}^{-1}$ more CO_2 efflux from mineral soil in thinned sites compared with unthinned control. Increases in temperature, temperature sensitivity, alteration of litters, and microbial composition may be responsible for the contrary changes in R_h from mineral soil and litter respiration, respectively. These findings manifested that thinning management of larch plantations could enhance biomass accumulation and decrease respiratory efflux from soil, which resulted in the effectiveness improvement in sequestrating C in forest ecosystems.

1. Introduction

Carbon dynamics in forest plantations have become a hot topic for forest ecological researches due to their important role in global climate change [1, 2]. In recent Kyoto Protocol negotiations, it was agreed that carbon sequestration in intensively managed plantation forests could be used to offset industrial carbon efflux [2, 3], and this was highlighted in the Marrakesh Accords. To determine the optimal management techniques for these forests, the effects of commonly used management practices (e.g., selective thinning) on ecosystem carbon sinks need to be assessed. Thinning was originally performed to obtain larger diameter and higher quality timber, and more recently, an increasing number of studies have investigated the biomass productivity. The effects of thinning on biomass carbon accumulation have varied between studies [4–7], due to differences in thinning intensity and the length of time after thinning practice was carried out

[7, 8]. Estimation of the effect of thinning on biomass carbon accumulation should be surveyed and analyzed.

China is home to the world's largest plantation forests (over 62 Mha), and selective thinning is one of the main management practices used in this region. Larch forests are widely distributed around the world at a latitude of 60 degrees north, and about 4.5 Mha of larch forests is distributed in Northeastern China [9, 10]. Comparison in boreal and temperate region indicates that larch forest in Northeastern China is more productive than that of Siberian forests, boreal evergreen, and south Boreas sites and similar to that of Europe Russia forest, boreal deciduous, and temperate evergreen forests [11]. Therefore, quantification of the influence of long-term thinning on the ecosystem carbon budget in these forests is of both scientific and economic significances [3, 5, 6].

Owing to the large storage, a small change in soil carbon from soil heterotrophic respiration changes can reverse the direction of forest ecosystem balance [12]. Instead of biomass carbon alone, net ecosystem productivity (the difference between annual biomass carbon increase (NPP) and soil heterotrophic respiration (R_h)) is often used to assess ecosystem carbon sink capacity as well as in international carbon trading [3, 13]. To calculate net ecosystem productivity, R_h needs to be distinguished from root autotrophic respiration, and few partitioned data is available for thinning management of forests. This makes it difficult to evaluate the effects of thinning on the ecosystem carbon sink. Given that a decrease of R_h after thinning practices, enhancement of forest carbon sink could be observed via decreasing CO_2 efflux from soil. Or else, the biomass carbon gain from NPP may be offset by the increase in R_h. A long-term respiration separating R_h (litter decomposition and microbial respiration in mineral soil) and autotrophic respiration from roots may facilitate the quantification of R_h changes owing to thinning management [12, 14, 15].

Forest management has been shown to have profound but inconsistent influences on soil respiration [16–18]. This inconsistent effect may be related to the composition of the microbial community [19, 20], thermal condition alteration [21], temperature sensitivity changes [14, 15], and as litter amount and composition. Removal trees and shrubs in thinning management may possibly alter soil respiration and its contribution to different components, and as the decomposer, the changes in soil microbes are important [21]. A full check on soil microbial composition will help the understanding of underlying mechanism of thinning effects on soil carbon processes [14, 21].

The aims of this study were (1) to quantify the effects of long-term thinning on ecosystem carbon sink capacity via survey of biomass carbon and soil respiratory efflux and (2) to check the microbial and thermal changes for soil respiratory alteration from thinning treatment.

2. Materials and Methods

2.1. Study Site and Experimental Design. The study was conducted at Laoshan Experimental Station ($127°34'41''$ E, $45°20'45''$ N). The larch plantation (*L. gmelinii*) being studied was afforested in 1969 at an initial planting density of 3300 plants·ha^{-1}. Thinning was performed on 3 occasions following afforestation: at 11 years (1980), 20 years (1989), and 25 years (1994). The removal of the first and third thinnings was approximately 200 tree·ha^{-1}, while the second thinning was approximately 300 tree·ha^{-1}. During thinning, weak larches and other competing shrubs and saplings under the canopy were removed. The unthinned site is a long-term permanent plot located in the same larch plantation. This site was established in 1978 and has no artificial thinning except dead tree movement for timber utilization. Three replicating pairs of thinned and unthinned treatments (ca. 20 m * 20 m) were selected in this paper. For each replicating pair, thinned plot and unthinned plot are neighbor for securing the data reliability. Area of the thinned site (2.5 ha) and unthinned site (0.5 ha) is approximately 3 ha.

2.2. Soil Respiration Partitioning and Environmental Parameters. Owing to its simple and cost effective and realistic respiration partitioning [22, 23], the trenched box method was used to partition respiration by autotrophic roots from respiration by heterotrophic soil microbes and litter decomposition. Each trenched box was 50 cm × 50 cm × 50 cm deep and settled permanently during the measurements. Soil respiration was measured using a Li-6400 system (LI-COR Inc., USA), and measurements were taken at least 12 hours after PVC collars (inner diameter = 10.2 cm, height = 5 cm) insertion to avoid soil disturbance from affecting the result. At each unthinned and thinned site, 4 trenched boxes were settled to make respiration measurements. Within each trenched box, microbial respiration in the mineral soil excluding recognizable litters and roots (R_m) was measured, while outside the trenched box, soil respiration excluding recognizable litters ($R_{-litter}$) and total soil respiration (R_t) were measured. Root respiration can be calculated as the difference between $R_{-litter}$ and R_m, while respiration from litter decomposition is the difference between R_t and $R_{-litter}$. Duration of the measurement was in the growing season (end of April to early Oct) from April 2005 to September 2007. The measurements were carried out one time per month.

At both the thinned and unthinned sites, a thermometer probe (Li-6400 system) was used to measure soil temperature at a depth of 5 cm at the same time as soil respiration measurements were taken. Continuous soil temperature data were also recorded at 30 min intervals using a thermo Recorder mini Rt-21s (Espec, Japan) from 2005 to 2007. These data were used to compare thermal conditions at the thinned and unthinned sites and were scaled-up to estimate whole ecosystem respiration.

2.3. Tree Growth and Biomass-Related Parameters. An inventory of the arbor layer species larch (*L. gmelinii*), birch (*Betula platyphylla*), and ash (*Fraxinus mandshurica*) was carried out at both the thinned and unthinned sites. Tree height (*H*) and diameter at breast height (DBH) were recorded, and tree density was calculated for each species. Similarly, measurements of the height, basal diameter, and tree density of 7 understory shrubs and saplings (*Syringa amurensis*, *F. mandshurica*, *Ulmus propinqua*, *Tilia amurensis*, *Corylus heterophylla*, *Pinus koraiensis*, and *Acer mono*) were also carried out using 4 quadrats (5 m × 5 m) at both the thinned and unthinned sites. Following the arbor layer censuses, 10 to 18 different-sized individuals of each species were harvested to determine the biomass of each of plant parts (leaves, stems, branches, and roots); no such distinction between different organs was made for the understory species. Oven-dried biomass (108°C) was used to determine the allometric relationship between biomass and DBH$^2 H$ for each species (see Table s1 in Supplementary Materials available online at http://dx.doi.org/10.1155/2013/865645). These relationships were then used to determine the difference in total biomass between the thinned and unthinned sites.

2.4. Soil Microbial Carbon and Microbial Composition. Soil microbial carbon was measured using the chloroform

TABLE 1: Long-term thinning effects on the growth and biomass C accumulation of species found in larch plantations. Values in parentheses are standard deviations.

Parameters	Treatment	Arbor layer				Shrubs and saplings	Total
		Larix gmelinii	Betula platyphylla	Fraxinus mandshurica	Arbor total		
Diameter (cm)	Thinned	21.7 (4.7)b	19.2 (2.4)a	18.1 (9.2)a	—	1.34 (1.57)a	—
	Unthinned	18.0 (5.3)a	21.8 (3.8)a	14.2 (6.7)a	—	2.12 (2.06)a	—
Density (No·hm^{-2})*	Thinned	950 (45)a	38 (15)a	63 (16)a	1050 (40)a	10500 (550)a	—
	Unthinned	1063 (67)b	100 (18)b	100 (24)b	1263 (60)b	7700 (450)b	—
Mean height (m)	Thinned	18.5 (1.9)a	17.5 (3.4)a	16.6 (6.5)a	—	1.68 (1.22)a	—
	Unthinned	17.8 (2.0)a	18.1 (4.3)a	13.8 (6.8)a	—	2.08 (1.46)a	—
Biomass (Mg·ha^{-1})	Thinned	216.0 (12.1)a	7.1 (3.1)b	5.0 (1.3)b	231.1 (13.2)a	1.52 (0.73)a	232.6
	Unthinned	156.4 (13.1)b	15.1 (4.2)a	12.9 (3.2)a	191.2 (15.1)b	2.51 (0.63)a	193.7
Biomass C (g C m^{-2})	Thinned	10800 (605)a	355 (155)a	250 (38)b	11555 (660)a	76 (36)a	11630
	Unthinned	7820 (655)b	756 (210)b	646 (160)a	9560 (755)b	126 (31)a	9686
Difference (g C m^{-2})	Thinned-Unthinned	2980	−400	−396	1994	−50	1944
Annual C change (g C m^{-2} yr^{-1})#		114.6	−15.4	−15.2	76.7	−1.9	74.8

*Tree density for arbor layer species was calculated for trees of DBH > 10 cm, while for shrubs and saplings it was calculated for plants of basal diameter > 0.4 cm. #The annual C change is the difference between thinned and unthinned divided by the total thinning duration (26 years).

(CHCL$_3$) fumigation method, which was first proposed by Jenkinson and Powlson [24] and subsequently revised by Lin et al. [25]. Six soil samples were collected from a depth of 0–10 cm from both the thinned and unthinned sites in summer (August). Enumeration of soil bacteria, fungi, and actinomycetes was carried out using a plate counting method [26]. The number of colony-forming units (CFU) was then counted, and a CFU number per unit of fresh soil was calculated.

2.5. Data Analysis. The relationship between soil respiration and soil temperature was expressed by the exponential relationship: $R_s = R_0 e^{bT}$, where R_s is the measured soil respiration rate (R_m, $R_{-litter}$, and R_t), T is the measured soil temperature, and R_0 and b are the best-fitting coefficients. R_0 is theoretical soil respiration rate at $0°C$. Q_{10} was then calculated using the expression $\exp(10 * b)$, to assess temperature sensitivity [15]. The best-fitted equation from thinned sites and unthinned sites was used to scale-up respiration of each component (mineral soils, litters and roots) from the continuous measurement of soil temperatures.

The differences of diameter, tree height, biomass of trees between different treatments, respiration between different treatments as well as among measured seasons were statistical analyzed by SPSS 17.0.

3. Results

3.1. Impact of Long-Term Thinning on Forest Biomass Carbon. Long-term thinning had a significant effect on tree size and density and most parameters for larch were statistically significant ($P < 0.05$) (Table 1). DBH and H of larch trees were, on average, 3.7 cm and 0.7 m larger, respectively, at the thinned site than at the unthinned site. Similarly, DBH and

H of ash were 3.9 cm and 2.8 m larger, respectively, at the thinned site than at the unthinned site. In contrast, DBH and H of birch were 2.6 cm and 0.6 m smaller at the thinned site than at the unthinned site. Three arbor species were generally present at a higher density at the unthinned site than at the thinned site, and overall, total density was 20% higher at the unthinned site (1263 individuals·ha^{-1}) than at the thinned site (1050 individuals·ha^{-1}). The basal diameter and height of shrubs and saplings were 0.78 cm and 0.4 m smaller, respectively, at the thinned site than at the unthinned site (Table 1).

The allometric relationships (Table s1) and inventory data were used to calculate differences in biomass accumulation between the thinned and unthinned sites (Table 1). The biomass of the arbor layer at the unthinned site (191.2 Mg·ha^{-1}) was lower than that at the thinned site (231.1 Mg·ha^{-1}) ($P < 0.05$), while the biomass of the shrub layer at the unthinned site (2.51 Mg·ha^{-1}) was approximately 1.0 Mg·ha^{-1} higher than that at the thinned site ($P > 0.05$). By summing the biomass of the 2 layers, total biomass at the unthinned site was 38.9 Mg·ha^{-1} (approximately 20%) lower than that at the thinned site. In total, arbor biomass carbon at the thinned site was 1994 g C m^{-2} higher than that at the unthinned site ($P < 0.05$), while understory carbon at the thinned site was 50 g C m^{-2} lower than that at the unthinned site ($P > 0.05$) (Table 1). By dividing the differences between the thinned site and the unthinned site by 26 years (duration of long-term thinning), we can determine the annual changes owing to the thinning treatment, about 74.8 g C m^{-2} yr^{-1} increase in total biomass carbon (Table 1).

3.2. Influence of Long-Term Thinning on Microbial Carbon and Composition. Microbial carbon at the thinned site was 8% lower than that at the unthinned site (Figure 1). Similarly,

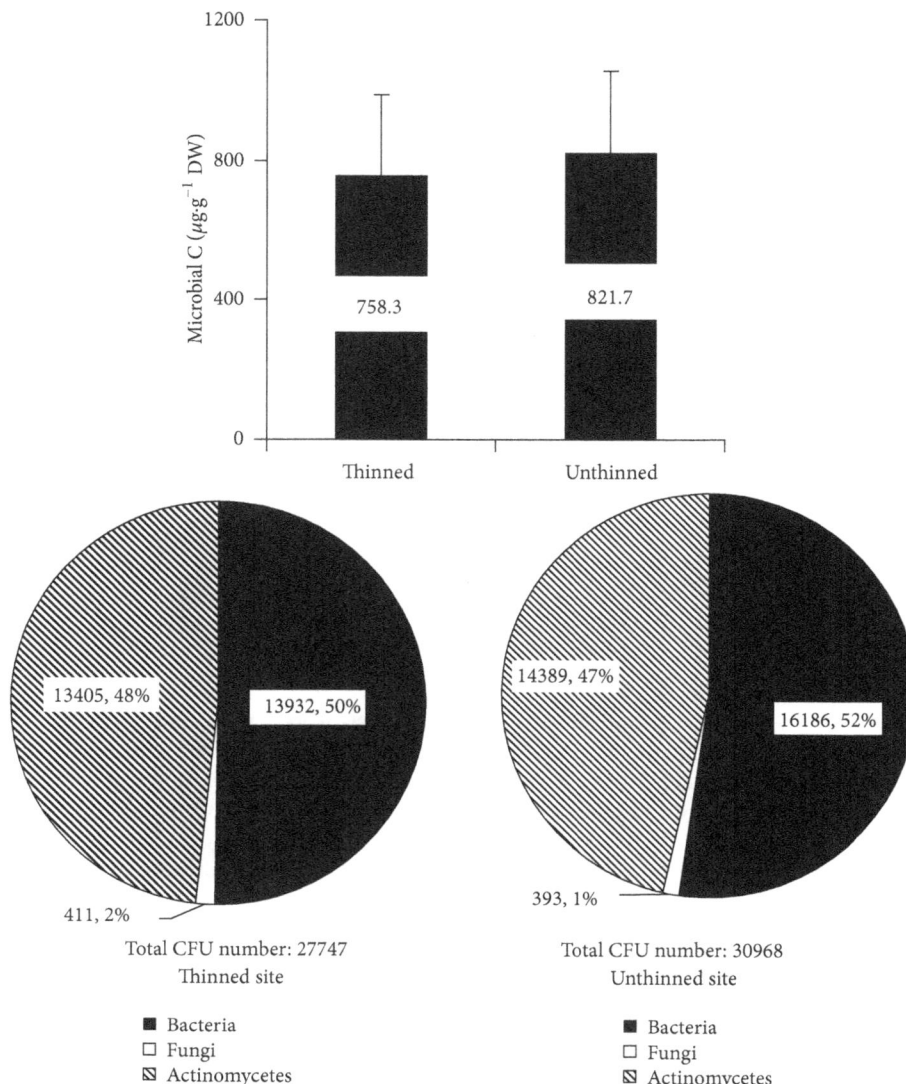

FIGURE 1: The effect of long-term thinning on microbial carbon and the microbial composition of bacteria, fungi, and actinomycetes.

total CFU number at the thinned site was 90% of that at the unthinned site (Figure 1). The main differences observed in the soil microbe composition at the thinned site were 1% increase in the percentage of actinomycetes and fungi but a 2% decrease in the percentage of bacteria (Figure 1).

3.3. Influence of Long-Term Thinning on Soil Temperature and Respiration Rates. The effect of long-term thinning on soil temperature can be observed by comparing the continuous measurements and discrete measurements taken at the thinned and unthinned sites (Figure 2). On average across the 3-year period, the discrete soil temperature measurements were 0.38°C higher at the thinned site (13.14°C) than at the unthinned site (12.76°C). In the growing season (May to Oct) of 2005 and 2007, the continuous measurement data also showed 0.25°C higher soil temperature at the thinned site, while in the nongrowing season, contrary tendency was

observed (−0.42°C at the unthinned site while −0.71°C at the thinned site) (Figure 2).

Thinning had a varied effect on the different soil respiration components, with no general pattern being observed (Figure 3). Nine out of the 18 measurements showed that R_m was higher at the thinned site than at the unthinned site, while the other half showed the opposite pattern (Figure 3(a)). For both $R_{-litter}$ and R_t, most of the measurements (11 out of 18 for each) showed that the unthinned site had higher respiration rates than the thinned site, while the remainder indicated that the unthinned site had lower or similar respiration rates compared to the thinned site (Figures 3(b) and 3(c)).

By classifying the measurements by season, it was found that the respiration of soil microbes peaked in summer. In spring and autumn, soil microbes contributed more to the total soil respiration at the thinned site (73.6% and 69.2%, resp.) than at the unthinned site (55.9% and 65.2%), while levels were similar during the summer (55.2% and

Influence of Long-Term Thinning on the Biomass Carbon and Soil Respiration in a Larch (Larix gmelinii) Forest in Northeastern China

161

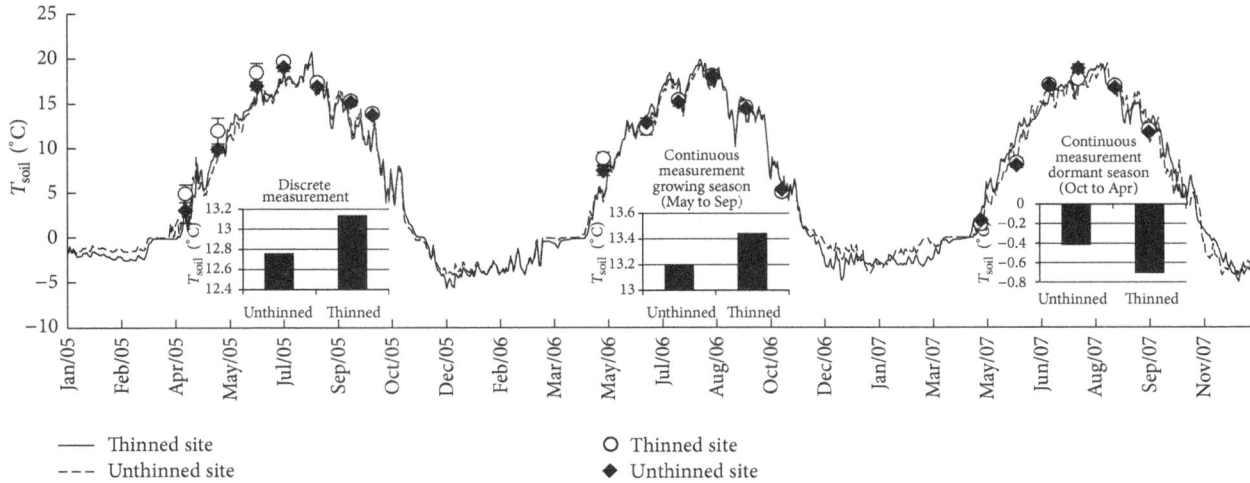

FIGURE 2: Differences in soil temperature (daily mean value) between thinned and unthinned sites. Line data were the daily mean values of continuously recorded temperatures measured using an RT21s thermometer, while the scattered data were discrete measurements that were taken at the same time as soil respiration was measured.

54.5%) (Table s2). This seasonal difference resulted in the R_m contribution being 7% higher at the thinned site (65.8%) than at the unthinned site (58.8%), on average. In contrast, root respiration was generally higher at the unthinned site than that at the thinned site throughout the year, contributing by 15.2% of total respiration compared with 9.1% at the thinned site. The 3-year average litter respiration was the same at the unthinned site and the thinned site (0.69 μmol m^{-2} s^{-1}). However, its contribution to total respiration was slightly higher at the unthinned site in spring and autumn but lower in summer. These resulted in a slightly higher overall average contribution at the unthinned site (26.2%) than at the thinned site (25.2%) (Table s2).

3.4. Influence of Long-Term Thinning on Temperature Respiration Relationships and Q_{10} Values.

There were significant exponential relationships between soil temperature and R_m, $R_{-litter}$, and R_t (Figure 4). There was a steady increase in b values for R_m, $R_{-litter}$, and R_t at the thinned site compared with the unthinned site (Figure 4). Q_{10} value for R_m at the unthinned site (2.15) is 10% lower than that at the thinned site (2.37). Higher Q_{10} values for $R_{-litter}$ and R_t at the thinned site were also observed; however, the percentage was less than 5% (Figure 4). R_0 value for R_m was almost the same between thinned and unthinned sites. A 7% higher R_0 for $R_{-litter}$ but a 6% lower R_0 for R_t at the thinned site were observed (Figure 4).

3.5. Influence of Long-Term Thinning on Annual CO_2 Efflux from Soil Microbes, Roots, and Litter.

Based on the continuous soil temperature data (Figure 2) and the exponential relationships (Figure 4), the respiration from soil microbes, roots, and litter was scaled-up (Figure s1). Generally, microbial respiration was higher at the thinned site, while root respiration was similar between the thinned and unthinned sites. However, respiration from litter decomposition was

generally higher at the unthinned site than that at the thinned site (Figure s1).

The annual total for each component of soil respiration was calculated from Figure s1 (Table 2). Although almost no change in total respiration was found; however its distribution in different components was altered (Table 2). Heterotrophic respiration from mineral soils at the thinned site was 3.0 to 3.4 mol m^{-2} yr^{-1} higher than that at unthinned site, while respiration from litter decomposition at thinned site was 4.0 to 4.2 mol m^{-2} yr^{-1} lower than that at unthinned site. Summing of these two components of heterotrophic respiration, thinned treatment decreased 0.7–1.2 mol m^{-2} yr^{-1} (averaged at 0.9 mol m^{-2} yr^{-1}) in total heterotrophic respiration. Autotrophic respiration from roots in thinned site was 0.7–0.9 mol m^{-2} yr^{-1} higher than that at the unthinned site (Table 2).

4. Discussion

Forest management practices such as tending and thinning can dramatically affect stand biomass and volume of harvested timber [4, 8, 27]. In this paper, average analysis showed that biomass carbon increased by 6.23 mol C m^{-2} yr^{-1} as the result of the long-term thinning (Table 1). Larch plantations are widespread in China and other northern hemisphere countries, and thinning is a common tending practice. At 2012, over 4.5 million hectares of land in Northeastern China was covered by larch forests (2 Mha in Heilongjiang Province, Liaoning Province, and Jilin Province [10] and 2.5 Mha in Daxinganling district of the Inner Mongolia autonomous region [9]). Based on our findings, approximately 12.1 Tg CO_2 (total area 4.5 Mha × annual C sink increase, 6.23 mol m^{-2} yr^{-1} as a result of the thinning practice) could be captured annually by these managed forests compared with unthinned forests. During the same period (2005–2007), total industrial CO_2 efflux was 670 Tg CO_2 on

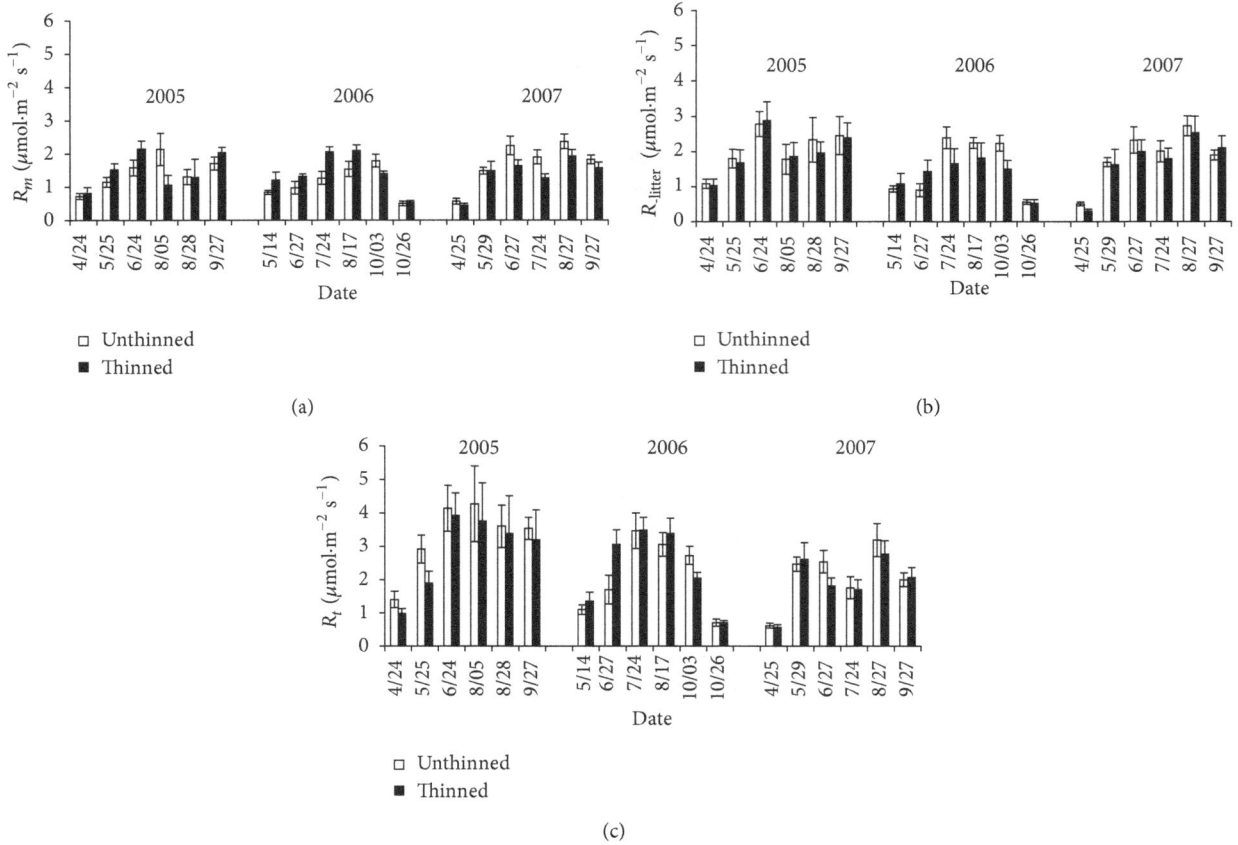

FIGURE 3: Comparison of R_m, $R_{-litter}$, and R_t between the unthinned and thinned sites from 2005 to 2007. (a) R_m, (b) $R_{-litter}$, and (c) R_t.

TABLE 2: Annual effluxes from different components of soil respiration at the thinned and unthinned sites.

Year	Items	Heterotrophic R						Autotrophic R from roots		Total	
		From mineral soil		From litters		Subtotal					
		Unthinned	Thinned	Unthinned	Thinned	Unthinned	Thinned	Unthinned	Thinned	Unthinned	Thinned
2005	Amount (mol m^{-2} yr^{-1})	28.7	32.1	10.1	5.9	38.8	38.0	8.0	8.8	46.8	46.8
	Difference	3.4		−4.2		−0.8		0.8		0.0	
2006	Amount (mol m^{-2} yr^{-1})	27.7	30.7	9.7	5.5	37.4	36.2	7.6	8.3	44.9	44.6
	Differences	3.0		−4.2		−1.2		0.7		−0.3	
2007	Amount (mol m^{-2} yr^{-1})	27.7	30.9	9.6	5.6	37.3	36.5	7.5	8.4	44.8	44.8
	Differences	3.2		−4.0		−0.7		0.9		0.0	
Mean	Amount (mol m^{-2} yr^{-1})	28.0	31.2	9.8	5.7	37.8	36.9	7.7	8.5	45.5	45.4
	Differences	3.2		−4.1		−0.9		0.8		−0.1	

average, and the annual increase was as high as 60 Tg CO_2 in Northeastern China [28]. Thus, the management of larch forests in this region alone could offset 2% of this industrial CO_2 efflux and 20% of its annual increase of this industrial efflux.

However, feasibility of this carbon increase used in trade-off industrial emission depends on the ecosystem carbon sink, instead of biomass carbon alone [3, 13]. Thus, the influences on soil heterotrophic respiration should be numerated, owing to that ecosystem carbon sink equals the differences

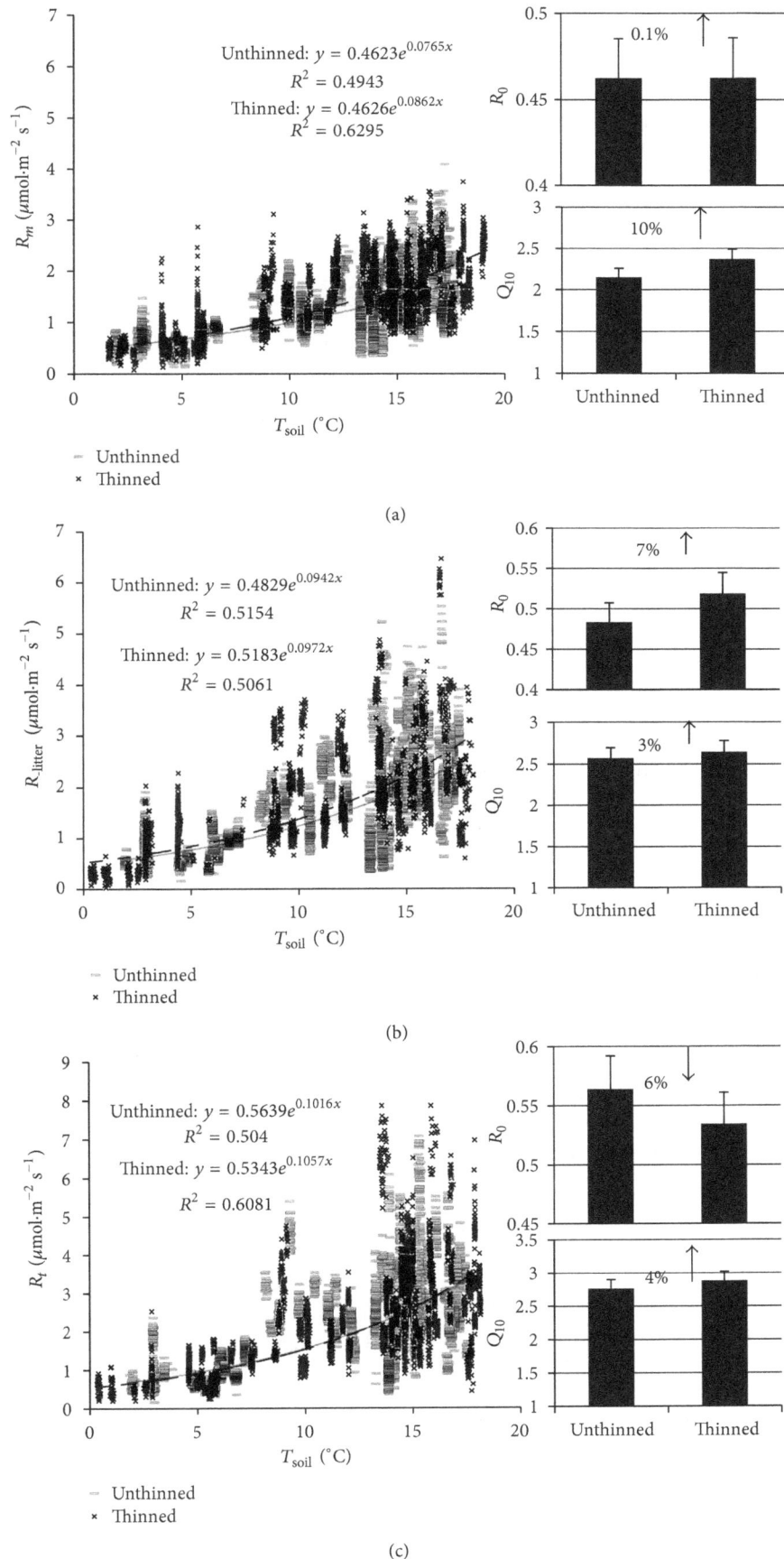

(a)

(b)

(c)

FIGURE 4: Comparison of the temporal response of R_m, $R_{-litter}$, and R_t between the unthinned and thinned sites from 2005 to 2007.

between biomass productivity and soil heterotrophic respiration. Although total CO_2 efflux from soil was similar in the thinned site ($45.4 \, \text{mol m}^{-2} \, \text{yr}^{-1}$) and the unthinned site ($45.5 \, \text{mol m}^{-2} \, \text{yr}^{-1}$), the percentage of heterotrophic respiration and autotrophic respiration were altered (Table 2). On average, heterotrophic respiration of the thinning practice has decreased by $0.9 \, \text{mol m}^{-2} \, \text{yr}^{-1}$. This has resulted in a 14% increase in total ecosystem carbon sink at the thinned site. Therefore, calculation of ecosystem carbon sink proved that thinned treatment could enhance carbon sink size via increase in biomass and decrease in respiratory efflux. Thinning practices is a common method for managing forest [4, 27]. Its effects on biomass and soil respiration were studied previously, although conclusions are variable [2, 8, 16]. In our study, we proved that light thinning (tree density decreased from 1263 trees·ha^{-1} to 1050 trees·ha^{-1} following 3 selective thinned treatments) could be a practical silvicultural strategy to offset industrial CO_2 emission fulfilling the requirements of the Kyoto Protocol [3, 13].

In the past, the effect of thinning on soil respiration has been found to vary. For example, investigations in a young *Pinus ponderosa* plantation demonstrated a 13% reduction in soil respiration in the first year after thinning [16]. Similar reductions have been found in other forests [1, 29]. In contrast, some other studies have found increased soil respiration in thinned stands [16, 30]. In this study, we found no overall differences in R_m, $R_{-litter}$, and R_t in any season using the instant measurements from 2005 to 2007 (Figures 3 and 4). Similarly, there were no differences in R_t between the unthinned and thinned sites when the annual data were used (Figure s1, Table 2). However, obvious alternation in its contrition in variable components was observed; that is, microbial respiration in the mineral soil at the thinned site increased, while litter respiration decreased (Table 2). The sum of these 2 components finally resulted in a decrease in total heterotrophic carbon efflux (Table 2).

These differences in respiration between the long-term thinned and unthinned sites should be related to variations in thermal condition, temperature sensitivity of respiration (Q_{10} values), and the soil microbial composition [12]. At the thinned site, the increased soil temperature (Figure 2) and increased temperature sensitivity of R_m (Figure 4) are responsible for the higher level of heterotrophic respiration from mineral soil (Table 2). On the other hand, the decrease in litter respiration at thinned site (Table 2) should be in accordance with the less return from litter both in total amount and in diversity, as shown in Table 1. Organic matter decomposition depends on specific soil microbial communities [19]. The difference in total microbial biomass decrease (biomass C and CFU data) as well as composition alteration (Figure 1) resulted in changes in the level of litter decomposition, too.

Forest thinning is one basic silviculture method to improve timber quality and productivity; this study gave a case study for the influences of thinning on ecosystem carbon sink. Just as mentioned by many previous studies, biomass C changes during forest managements are the most important item that needs to be fully considered, and our study also confirmed this. However, underground soil C dynamics should be considered in C sink evaluation, particularly in region (like northeastern China) with highly intensive agricultural exploitation [31]. In short-term instant measurement, soil respiration and its heterotrophic components showed diversified seasonal changes (Figure 3); thus instant measurement is difficult to identify the long-term effect on soil C efflux. In this study, annual sum of respiration and its components can be used to find the difference between thinning and control treatment, and data in replicating years showed similar pattern.

5. Conclusion

Our findings support the suggestion that long-term thinning of larch plantation in Northeastern China can improve forest carbon sink through both increase in biomass carbon and decrease in soil heterotrophic CO_2 efflux, although the short-term instant measurement may show diversified result. Scaled-up data manifested that biomass increase could trade off 21% of the annual increase of total emission of local industry, and inclusion of soil respiration can give another 14% increase in the sink size. These findings support the use of thinning practices in larch plantation management in Northeastern China for improving ecosystem carbon sink capacity.

Authors' Contribution

Huimei Wang contributed to data analysis, laboratory assay, and paper preparation; Wei Liu contributed to field data measurement; Wenjie Wang contributed to experiment design and paper preparation and revision; Yuangang Zu provided long-term data for thinning history and experiment design.

Acknowledgments

This study was supported financially by China's and the Ministry of Science and Technology (2011CB403205), the National Natural Science Foundation of China (31170575), and the basic research fund for national universities from Ministry of Education of China (DL12DA03).

References

[1] M. F. Laporte, L. C. Duchesne, and I. K. Morrison, "Effect of clearcutting, selection cutting, shelterwood cutting and microsites on soil surface CO_2 efflux in a tolerant hardwood ecosystem of northern Ontario," *Forest Ecology and Management*, vol. 174, no. 1–3, pp. 565–575, 2003.

[2] R. Jandl, M. Lindner, L. Vesterdal et al., "How strongly can forest management influence soil carbon sequestration?" *Geoderma*, vol. 137, no. 3-4, pp. 253–268, 2007.

[3] J. H. Borden, G. Marland, B. Schlamadinger et al., "'Kyoto forests#8217; and a broader perspective on management," *Science*, vol. 290, no. 5498, pp. 1895–1896, 2000.

[4] X. B. Dong, "The impacts of cutting intensity on the growth of larch forest," *Journal of Northeast Forestry University*, vol. 29, pp. 44–47, 2001.

Influence of Long-Term Thinning on the Biomass Carbon and Soil Respiration in a Larch (Larix gmelinii)
Forest in Northeastern China

165

[5] C. Kim, Y. Son, W. K. Lee, J. Jeong, and N. J. Noh, "Influences of forest tending works on carbon distribution and cycling in a *Pinus densiflora* S. et Z. stand in Korea," *Forest Ecology and Management*, vol. 257, no. 5, pp. 1420–1426, 2009.

[6] J. Novak, M. Slodicak, and D. Dusek, "Thinning effects on forest productivity and site characteristics in stands of *Pinus sylvestris* in the Czech Republic," *Forest Systems*, vol. 20, no. 3, pp. 464–474, 2011.

[7] S. Y. Tian, "Modeling water, carbon, and nitrogen dynamics for two drained pine plantations under intensive management practices," *Forest Ecology and Management*, vol. 264, pp. 20–36, 2012.

[8] P. Nilsen and L. T. Strand, "Thinning intensity effects on carbon and nitrogen stores and fluxes in a Norway spruce (*Picea abies* (L.) Karst.) stand after 33 years," *Forest Ecology and Management*, vol. 256, no. 3, pp. 201–208, 2008.

[9] Y. J. Sun, J. Zhang, A. H. Han, X. J. Wang, and X. J. Wang, "Biomass and carbon pool of *Larix gmelini* young and middle age forest in Xing'an Mountains Inner Mongolia," *Acta Ecologica Sinica*, vol. 27, no. 5, pp. 1756–1762, 2007.

[10] Z. H. Sun, G. Z. Jin, and C. C. Mu, *Study on the keeping long-term productivity of* Larix olgensis *plantation*, Science Press, Beijing, China, 2009.

[11] W. J. Wang, W. J. ZU, H. M. Wang, Y. Matsuura, K. Sasa, and T. Koike, "Plant biomass and productivity of *Larix gmelinii* forest ecosystems in Northeast China: intra- and inter-species comparison," *Eurasian Journal of Forest Research*, vol. 8, no. 1, pp. 21–41, 2005.

[12] E. A. Davidson and I. A. Janssens, "Temperature sensitivity of soil carbon decomposition and feedbacks to climate change," *Nature*, vol. 440, no. 7081, pp. 165–173, 2006.

[13] E. D. Schulze, C. Wirth, and M. Heimann, "Managing forests after Kyoto," *Science*, vol. 289, no. 5487, pp. 2058–2059, 2000.

[14] P. J. Hanson, N. T. Edwards, C. T. Garten, and J. A. Andrews, "Separating root and soil microbial contributions to soil respiration: a review of methods and observations," *Biogeochemistry*, vol. 48, no. 1, pp. 115–146, 2000.

[15] R. D. Boone, K. J. Nadelhoffer, J. D. Canary, and J. P. Kaye, "Roots exert a strong influence on the temperature sensitivity of soil respiration," *Nature*, vol. 396, no. 6711, pp. 570–572, 1998.

[16] J. W. Tang, Y. Qi, M. Xu, L. Misson, and A. H. Goldstein, "Forest thinning and soil respiration in a ponderosa pine plantation in the Sierra Nevada," *Tree Physiology*, vol. 25, no. 1, pp. 57–66, 2005.

[17] M. F. Selig, J. R. Seiler, and M. C. Tyree, "Soil carbon and CO_2 efflux as influenced by the thinning of loblolly pine (*Pinus taeda* L.) plantations on the Piedmont of Virginia," *Forest Science*, vol. 54, no. 1, pp. 58–66, 2008.

[18] B. W. Sullivan, T. E. Kolb, S. C. Hart, J. P. Kaye, S. Dore, and M. Montes-Helu, "Thinning reduces soil carbon dioxide but not methane flux from southwestern USA ponderosa pine forests," *Forest Ecology and Management*, vol. 255, no. 12, pp. 4047–4055, 2008.

[19] A. P. C. Houston, S. Visser, and R. A. Lautenschlager, "Microbial processes and fungal community structure in soils from clear-cut and unharvested areas of two mixedwood forests," *Canadian Journal of Botany*, vol. 76, no. 4, pp. 630–640, 1998.

[20] S. Maassen, H. Fritze, and S. Wirth, "Response of soil microbial biomass, activities, and community structure at a pine stand in northeastern Germany 5 years after thinning," *Canadian Journal of Forest Research*, vol. 36, no. 6, pp. 1427–1434, 2006.

[21] Y. G. Zu, W. J. Wang, H. M. Wang, W. Liu, S. Cui, and T. Koike, "Soil CO_2 efflux, carbon dynamics, and change in thermal conditions from contrasting clear-cut sites during natural restoration and uncut larch forests in northeastern China," *Climatic Change*, vol. 96, no. 1, pp. 137–159, 2009.

[22] P. Rochette, L. B. Flanagan, and E. G. Gregorich, "Separating soil respiration into plant and soil components using analyses of the natural abundance of carbon-13," *Soil Science Society of America Journal*, vol. 63, no. 5, pp. 1207–1213, 1999.

[23] M. S. Lee, K. Nakane, T. Nakatsubo, and H. Koizumi, "Seasonal changes in the contribution of root respiration to total soil respiration in a cool-temperate deciduous forest," *Plant and Soil*, vol. 255, no. 1, pp. 311–318, 2003.

[24] D. S. Jenkinson and D. S. Powlson, "The effects of biocidal treatments on metabolism in soil-V. A method for measuring soil biomass," *Soil Biology and Biochemistry*, vol. 8, no. 3, pp. 209–213, 1976.

[25] Q. M. Lin, Y. G. Wu, and H. L. Liu, "Modification of fumigation extraction method for measuring soil microbial biomass carbon," *Chinese Journal of Applied Ecology*, vol. 18, pp. 63–66, 1999.

[26] A. Pandey and L. M. S. Palni, "The rhizosphere effect of tea on soil microbes in a Himalayan monsoonal location," *Biology and Fertility of Soils*, vol. 21, no. 3, pp. 131–137, 1996.

[27] A. Juodvalkis, L. Kairiukstis, and R. Vasiliauskas, "Effects of thinning on growth of six tree species in north-temperate forests of Lithuania," *European Journal of Forest Research*, vol. 124, no. 3, pp. 187–192, 2005.

[28] H. Li, "Evolution and decomposition analysis of industrial CO_2 emissions in Northeast China during the period 1995–2009," *Resources Science*, vol. 34, pp. 309–315, 2012.

[29] D. E. Toland and D. R. Zak, "Seasonal patterns of soil respiration in intact and clear-cut northern hardwood forests," *Canadian Journal of Forest Research*, vol. 24, no. 8, pp. 1711–1716, 1994.

[30] Y. Peng and S. C. Thomas, "Soil CO_2 efflux in uneven-aged managed forests: temporal patterns following harvest and effects of edaphic heterogeneity," *Plant and Soil*, vol. 289, no. 1-2, pp. 253–264, 2006.

[31] W. Wen-Jie, Q. Ling, Z. Yuan-Gang et al., "Changes in soil organic carbon, nitrogen, pH and bulk density with the development of larch (*Larix gmelinii*) plantations in China," *Global Change Biology*, vol. 17, no. 8, pp. 2657–2676, 2011.

19

Current Status of Trace Metal Pollution in Soils Affected by Industrial Activities

Ehsanul Kabir,[1,2] Sharmila Ray,[1] Ki-Hyun Kim,[1] Hye-On Yoon,[3] Eui-Chan Jeon,[1] Yoon Shin Kim,[4] Yong-Sung Cho,[4] Seong-Taek Yun,[5] and Richard J. C. Brown[6]

[1] Department of Environment and Energy, Sejong University, Seoul 143-747, Republic of Korea
[2] Department of Farm Power and Machinery, Bangladesh Agricultural University, Mymensingh, Bangladesh
[3] Korea Basic Science Institute, Seoul Center, Anamdong, Seoul 136-713, Republic of Korea
[4] Institute of Environmental and Industrial Medicine, Hanyang University, Seoul 133-791, Republic of Korea
[5] Department of Earth and Environmental Sciences, Korea University, Seoul 136-701, Republic of Korea
[6] Analytical Science Division, National Physical Laboratory, Hampton Road, Teddington, TW11 0LW, UK

Correspondence should be addressed to Ki-Hyun Kim, khkim@sejong.ac.kr

Academic Editors: M. B. Amran, I. Ciucanu, and S. O. Fakayode

There is a growing public concern over the potential accumulation of heavy metals in soil, owing to rapid industrial development. In an effort to describe the status of the pollutions of soil by industrial activities, relevant data sets reported by many studies were surveyed and reviewed. The results of our analysis indicate that soils were polluted most significantly by metals such as lead, zinc, copper, and cadmium. If the dominant species are evaluated by the highest mean concentration observed for different industry types, the results were grouped into Pb, Zn, Ni, Cu, Fe, and As in smelting and metal production industries, Mn and Cd in the textile industry, and Cr in the leather industry. In most cases, metal levels in the studied areas were found to exceed the common regulation guideline levels enforced by many countries. The geoaccumulation index (I_{geo}), calculated to estimate the enrichment of metal concentrations in soil, showed that the level of metal pollution in most surveyed areas is significant, especially for Pb and Cd. It is thus important to keep systematic and continuous monitoring of heavy metals and their derivatives to manage and suppress such pollution.

1. Introduction

Industrial pollution has been and continues to be a major cause of environmental degradation. Numerous studies have already demonstrated that areas in close proximity to industrial activities are marked by noticeable contamination of air, soil, and water [1–3]. Hence, such activities can affect the air we breathe, the water we use, and the soil we stand on and can ultimately lead to illness and/or harm to the residents in the affected area.

Among various toxic substances released by industrial activities, heavy metals have been seen as a key marker because they may be analysed effectively and consistently in most environmental matrices. Unlike organic pollutants which may degrade to less harmful components as a result of biological or chemical processes, metals are not degradable by natural processes especially when elemental metallic content is considered [4]. The effects of metal pollution on local environments and organisms may therefore be substantial and long lasting in spite of extensive remediation efforts [4]. In fact, lead, cadmium, copper, manganese, and so forth have been commonly chosen as representative metals for which their concentrations in the environment may be used as reliable indices of environmental pollution [5].

In most parts of the world, large quantities of trace metals are directly discharged to nearby land and into surface waters. This activity adversely affects the quality of air, soil, and ground water, such that it becomes a subject of serious concern worldwide [6–8]. In recent years, many governments and policy makers have continued to strive for a more comprehensive understanding of environmental health hazards due to intensive industrial activities in order to inform future policy and abatement legislation [9]. In this paper, we intend to provide the results of a review of a survey of environmental pollution caused by industrial activities. Through an in-depth analysis of basic methodologies and relevant databases, we provide some insights into the fundamental aspects of metal pollution associated with industrial activities with a major emphasis on the soil matrix.

2. Status of Data Availability of Metal Pollution between Different Criteria: Matrix Types (Air, Soil, and Water) and Industry Types

During the past few decades, industrial activities have increased greatly around the world with rapid economic growth. This has been accompanied by severe environmental pollution. Many studies have been carried out to evaluate the status of industrial pollution and its environmental impacts in a broad and aggregative manner. There are, however, very few studies that describe the impact of such pollution with respect to the spread of key pollutants across various environmental media. In order to build a database to assess the basic features of heavy metal pollution due to industrial activities, we conducted a literature survey of major articles dealing with this topic that have been published since 1996 (a total of 61 references). If these metal pollution data are sorted by matrix type, 42 of the articles dealt with the soil phase. The remaining ones dealt with the air (11) and water (8) phases. As the classification of the data sets surveyed is important, we used criteria provided by the International Standard Industrial Classification (ISIC) division of the United Nations (Table 1) [10]. The literature reviewed was chosen so as to provide a representative sample across industries, geographic area, and measured concentrations. It covered a range of industries including smelters, mining and metal (ME + MI), chemical and petrochemical (CE + PE), tannery (LE), ceramic and cement (NM), textile (TA), and industrial complexes (containing multiple generic process) (IC) of the studies in each category numbering 17, 7, 4, 3, 2, and 8, respectively. In case of air ($n = 11$), four case studies were mainly conducted near chemical and petrochemical (CE + PE) types, while all others concerned various industrial complexes ($n = 7$). Metal pollution in the water phase was investigated mainly in the brewery (BE = 4), tannery (LE = 2), and textile industries (TE = 2 cases). Figure 1 shows the frequency of data availability for this survey as a function of industry type and between the different media.

3. Comparison of Experimental Approaches for Data Acquisition

3.1. Sampling. In this study, considering the availability of data, our analysis concentrated on the soil matrix. In this respect, we analyzed the basic methodological approaches employed for data acquisition in all the selected references. Based on this analysis, we evaluated the fundamental features of metal pollution in soil layers resulting from industrial activities. The basic information concerning the methodologies for sampling, sample treatment, and analysis for metal

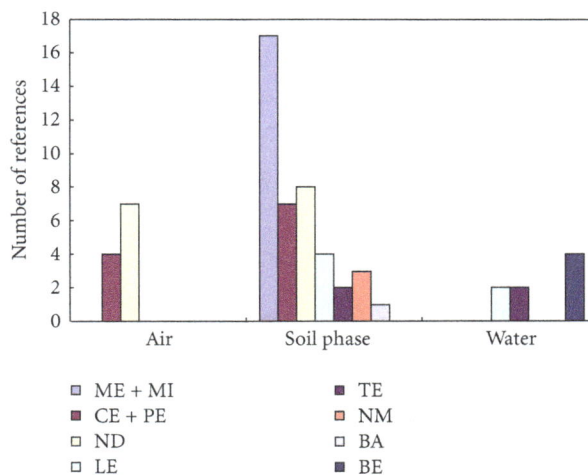

FIGURE 1: Source frequency of soil metal pollution data across different environmental matrices and with respect to different industry types ($n = 61$ references for comparison: refer to Table 1 for acronyms).

content is summarized in Table 2. In order to investigate the metal load onto soil, the distance between the sampling sites and the emission sources is a critical parameter. In general, there is a progressive decrease of metal concentrations with increasing distance from the source. Most of the studies reported soil sampling within a radius of 50 m to 2 km from industrial sources [15, 23, 30]. The collection of soil samples has commonly been made by random sampling [11] or grid sampling [14]. As seen in Table 2, grid sampling based on 1 km × 1 km squares has been adopted most commonly in many of these studies [12, 16, 22]. There are, however, some exceptions like the use of 20 m × 20 m grids (e.g. [28]). Most of the studies considered employing vertical sampling in the range of 0–5 cm [11] to 0–20 cm [33]. Soil samples were typically collected with a stainless steel spatula or auger and kept in PVC packages until analysis.

The metal data in soils derived from random and grid sampling are often used to represent an entire area. In this respect, grid sampling can be particularly useful when prior knowledge of the likely spatial variability is limited. This technique also avoids any sampling bias that could result from the collection of an unrepresentative average sample as a result high portion of subsamples from the same region. Two subtypes of grid sampling such as gridcell and gridpoint have commonly been employed (e.g., [34]).

3.2. Sample Preparation. Soil samples were commonly dried at room temperature and sorted via sieving (e.g., using a 2 mm sieve). They were then mineralized in a single acid like HNO_3 within Teflon bombs in a microwave digester [12]. In many studies, however, authors preferred to use mixtures of various acids like (1) HCl, HNO_3, and H_2O_2 ([11], (2) HCl, HNO_3, and HF [14], (3) HCl, HNO_3, HF, and $HClO_4$ [16], (4) HCl, HF, and $HClO_4$ [22], and (5) aqua regia [33]). Reagent blanks and standard reference soil samples were also analyzed to reduce experimental biases and properly validate

TABLE 1: Classification of industry types for the analysis of soil metal pollution: the International Standard Industrial Classification (ISIC) codes.

Order	Section	Division	Short mane	Groups	Industry Code[a]
1		Beverages	BE	Beverages	110
2		Textiles	TE	Spinning, weaving, and finishing of textiles	131
3				Other textiles	139
4		Leather and related products	LE	Tanning and dressing of leather, luggage, handbags, saddlery and harness, dressing, and dyeing	151
5		Paper and paper products	PA	Paper and paper products	170
6	Manufacturing	Coke and refined petroleum products		Refined petroleum products	192
7		Chemicals and chemical products	CE	Chemicals, fertilizers and nitrogen compounds, plastics, and synthetic rubber in primary forms	201
8		Pharmaceutical products and pharmaceutical preparations	PH	Pharmaceuticals, medicinal chemical, and botanical products	210
9		Nonmetallic mineral products	NM	Glass and glass products	231
10				Nonmetallic mineral products	239
11		Metals	ME	Iron and steel industry	241
12				Precious and other nonferrous metals industry	242
13		Machinery and equipment	MA	General-purpose machinery	281
14		Electrical equipment	BA	Batteries and accumulators	272
15		Transport equipment	TR	Building of ships and boats	301
16		Furniture	FU	Furniture	310
17	Mining and quarrying	Extraction of crude petroleum and natural gas	PE	Extraction of crude petroleum	061
18		Mining of metal ores	MI	Mining of iron ores	071
19				Mining of non-ferrous metal ores	072

[a] Source: ISIC Rev. 4 structure (2008, United Nations Statistics Division) [10].

each extraction method [33]. The digestion procedure used will depend on the species in the soil requiring digestion and the final analytical step. For instance, chromates may require more extreme digestion conditions, such as HF, to properly dissolve the chromium present. However, analysis techniques such as ICP-MS prefer final solutions with relatively low ionic contents and so HF (subsequently neutralized with HBO_3) may cause instrumental drift during analysis [13].

3.3. Instrumental Detection. Flame/atomic absorption spectrometer (FAAS/AAS) is the dominant technique employed for metal analysis in soil [19, 22, 23, 27, 32]. In FAAS, either an air/acetylene or a nitrous oxide/acetylene flame is used to evaporate the solvent and dissociate the sample into its component atoms. The atoms of interest absorb light from a hollow cathode lamp (selected for the target element) as it passes through the cloud of atoms produced by the atomization process. The amount of absorbed light is measured and used to calculate the concentration of each metal of

interest. Compounds of the alkali and transition metals can all be atomized with good efficiency yielding typical FAAS detection limits in the sub-ppm range.

AAS is also used in combination with a graphite furnace (GF) mode—known as GF-AAS [12, 25]. It is essentially the same as flame AA, except the flame is replaced by a small, electrically heated graphite tube, or cuvette, which is heated up to 3000°C to generate the cloud of atomized sample. The higher atom density and longer residence time in the tube yield improved detection limits (DLs) for GF-AAS in the sub-ppb range, which is by 3 orders of magnitude superior to flame AAS. However, because of the temperature limitation and the use of graphite cuvettes, the analytical performance for refractory elements is still somewhat limited. The techniques also exhibit a lower throughput than many of the more recent mass spectrometry (MS) techniques; this is because it is only able to determine one element at a time, unlike MS methods which can determine a range of elements at once.

TABLE 2: Sampling and analytical procedures of metals in soil matrix reported between different studies.

| Order | Study area | No. of samples | Sample collection method | | | Method Instrument | Reference materials used | Reference |
			Period	Soil depth (cm)	Grid			
1	Albania	21	Jul '95	0–15	—	ICP-AES	CRMs-BCR 142R	[11]
2	Algeria	119	Jan-Feb '06	—	1 × 1 km	FAAS[a], GFS[b]	CRM 141 R, SRM 2709 SJS	[12]
3	Australia	25	Dec '00–Feb '01	0–5 and 5–20	—	ICP-MS	CRMs-AGAL 10, AGAL 11	[13]
4	Belgium	27	'93–'98	0–20	10 m interval	ICP-MS, ICP-AES	CRMs-SRM2710, GBW07411, GBW07311 & SRM1633	[14]
5	Bulgaria	14	Jun '04	—	—	XRF	—	[15]
6	Kosovo	82	Jun–Nov '02	—	1×1 km	XRF, ICP-MS	CRMs-BR, DR-N, GH, DTS-1, SDO-1, AGV-2, NIST 2709, 2710	[16]
7	Greece	30	'93–'94	—	—	INAA[c] & AAS	IAEA SOIL-7, Pepperbush SRM	[17]
8	Italy	280	—	0–5	—	ICP-MS	—	[18]
9	Jordan	3	Sept '03	0–20 and 20–40	—	AAS	—	[19]
10	Peru	6	—	0–10	—	ICP-ES	—	[20]
11	Slovenia	40	—	0–25	—	FAAS	WEPAL 2004.3.3	[21]
12	UK	70	—	0–15	1 km interval	FAAS	—	[22]
13	Bangladesh	53	Nov '95	0–5 and 5–15	—	AAS	—	[23]
14	India	30	'02–'04	—	—	XRF	SO-1, SO-2, SO-3, SO-4, NGRI-D, NGRI-U	[24]
15	Spain	24	Winter '07	0–5	—	ICP-MS & AAS-GF	CRM 052	[25]
16	Spain	27	Winter '05	0–3	—	ICP-MS	Lobster hepatopancreas	[26]
17	Turkey	29	Sep '03	—	—	FAAS	SRM - BCR-701	[27]
18	China	50	—	0–10	20 m × 20 m	XRF	GSS, GRS, GSD, SO-1, NIST-2709, NIST-2710, NIST-2711	[28]
20	Jordan	31	—	0–10 and 10–20	—	AAS	NBS SRMs	[29]
21	Pakistan	38	Oct '03–Dec '03	1–5	—	AAS	SRM-SR-96	[30]
22	Serbia	59	Jun '99–Mar '00	0–5	—	FAAS	—	[31]
23	Damascus	51	Summer '98	0–20	—	FAAS	—	[32]

[a]FAAS—Flame atomic absorption spectrometer.
[b]GFS—Graphite furnace spectrometry.
[c]Instrumental Neutron Activation Analysis with gamma-ray spectroscopy.

Many authors have also used inductively coupled plasma with mass spectrometry (ICP-MS) or atomic emission spectrometry (ICP-AES) for the simultaneous analysis of multiple metals [14, 16, 20, 33]. For ICP/ICP-AES analysis, the system uses temperatures as high as 10,000°C to atomize even the most refractory elements with high efficiency. As a result, DLs for these systems can be orders of magnitude lower (typically at the 1–10 parts-per-billion level) than FAAS techniques. The ICP method can simultaneously screen for up to 60 elements in a single sample run of less

than one minute, without any degradation of precision or detection limits. If run in a "sequential mode" ICPs can provide analytical results for about five elements per minute [35].

ICP-MS is a multielement technique that also uses an ICP plasma source to dissociate the sample into its constituent atoms or ions [35]. However, in this case, detection focuses on the ions themselves rather than the light that they emit. The ions are extracted from the plasma and passed into the mass spectrometer, where they are separated based on their atomic mass-to-charge ratio by a quadrupole or magnetic sector analyzer. In terms of DLs, ICP-MS can produce the best results (typically 1–10 ppt), followed by GF AAS (usually in the sub-ppb range), ICP-AES (of the order of 1–10 ppb), and FAAS (in the sub-ppm range). Being a mass spectrometric technique, ICP-MS also enables quantification by isotope dilution strategies for poly-isotopic elements, which can produce highly accurate results. However, mass spectrometric techniques may also suffer from isobaric mono- and polyatomic interferences which unless properly corrected can bias results [36].

Nondestructive methods for analysis such as X-ray fluorescence (XRF) and instrumental neutron activation analysis (INAA) have also been used commonly in many recent studies [16, 17, 24, 28]. In INAA, the sample is bombarded with neutrons, causing the elements to form radioactive isotopes. As the radioactive emissions and radioactive decay paths for each element are well known, one can determine the concentrations of the elements based on the information of spectral emission. This type of application is highly advantageous in that it does not destroy the sample and is generally matrixindependent, although often very difficult to calibrate accurately.

3.4. Quality Assurance (QA). QA procedures are made to ensure that the approach is properly validated, under control at all stages, and employed appropriately. Validation of analytical methodologies via the measurement of matrix-certified reference materials is good examples of a necessary QA activity. Another important performance criteria in QA/QC is the method detection limit (MDL). The MDL is usually defined as the lowest quantity or concentration of a component that can be reliably detected for a given analytical method [55].

Standard reference materials (SRMs—generally NIST in the USA) or certified reference materials (CRMs—the term used by most other National Measurement Institutes (NMIs)) are materials used to check the quality and traceability of results and can be divided into two categories: calibrants and matrix RMs (as mentioned above). The former is mono- or multielemental standard solutions used for calibrating instruments and ensuring traceability in measurement results, while the latter is a material of a similar matrix to the sample being analyzed which has been certified for homogeneity and its content of relevant analytes (e.g., lead in dust). Matrix RMs are used for method validation, rather than for calibration.

Most of the reference materials used in soils have been formulated by several NMIs like National Institute of Standards and Technology (NIST), USA, Institute for Reference Materials and Measurements (IRMM, Belgium), and so forth. In the industrial soils of Baoji city, China, soil standards NIST-2709, 2710, and 2711 were used along with Canadian certified reference materials such as SO-1, 2, 3, and 4 [28]. In the industrial regions of Kayseri, Turkey, BCR-701 was reported as a method validation tool [21]. Two certified reference materials, CRM 141 R and SRM 2709 SJS, were also used to develop method for the urban soils of Algeria [12] while in the industrial soils of Kosovo, CRMs in rock

forms (BR, DR-N, and GH) produced by Centre National de la Recherche Scientifique, CRNS (Notre Dame des Pauvres, France), and DTS-1, SDO-1, and AGV-2 produced by the United States Geological Survey, USGS (Denver, Colorado, USA) have been employed to develop suitable analytical methods [16].

4. In Depth Analysis of Soil Metal Pollution by Industry Types

As one of the major indices of environmental pollution, trace element concentration data in the soil phase affected by industrial activities complied in this study is listed in Table 3. The comparable data sets for the air and water, although much more limited relative to soil matrix, are also provided in Table 4. In light of differences in the relative abundance of metal pollution data between these different matrices, we conducted a detailed analysis of the impact of industrial activities on environmental metal pollution by focusing mainly on soil media.

Trace metals occur naturally in soils as a result of diverse geological processes such as chemical reaction and erosion of underground geological materials [56]. Beside these natural sources, industrial activities can supply a considerable quantity of metals to soil [57]. A large number of industrial activities produce wastes and contaminants that reach the soil through direct disposal, spills, leaks, atmospheric deposition from air, and other pathways [25]. Hence, enhanced metal levels (e.g., Cu, Zn, Pb, Co, Ni, Cd, As, and others) in soil media have been reported from in and around several industrial sites. As one of the dominant transportation routes of heavy elements, atmospheric emissions have commonly been designated as the main route of metallic accumulation in surface soils via their subsequent deposition, along with other transport routes like waste water discharge [58].

4.1. Mining, Smelter, and Metallurgical Industries. Severe metal pollution has been reported from areas surrounding mines and smelters in many countries [33, 59–63]. High levels of metals discharged from mine wastes may cause adverse environmental effects, because they can be dispersed into nearby agricultural soils and stream systems and taken up by food crops [64]. Among the 17 references dealing with metal pollution in mining and smelter sites, 4 elements (Cu, Cd, Pb, and Zn) stood out as the most common choice of target analyte. However, the data for other metals (e.g., Mn, Ni, Cr, Fe, and As), although not as common as the above 4 elements, are also available in many studies (Table 3). For this industry type, noticeably high concentrations of Pb, Zn, Mn, and Cr are frequently detected in the soil samples analysed. The relative ordering of the metal concentrations in these soils, if compared in terms of their median values, can be arranged in the following descending order: Mn > Zn > Pb > Cr > Ni > Cu > As > Fe > Cd (Table 5).

Fairly high concentrations of Mn, Zn, and Cr (e.g., 443, 68.3, and 98.7 mg kg^{-1}, resp.) were found near metal industries in Thessaloniki [17]. In general, all elements determined in their study were comparable to the levels normally determined in clay soils worldwide [17]. Enhanced

TABLE 3: Summary of trace element concentrations in soils affected by diverse industrial activities.

Order	Industry type	Industry group short name[a]	ISIC industry code[a]	Location City	Country	Trace element concentration (mg kg^{-1})									Reference
						Pb	Zn	Mn	Ni	Cn	Cr	Fe	Cd	As	
1	Iron smelter		241	Kremikovtzi	Bulgaria	—	—	—	111	61.8	—	—	2.10	—	[15]
2	Pb and Zn smelter		242	Zhuzhou	China	472	907	—	—	77.0	—	—	11.9	38.0	[28]
3	Zn Smelter		242	Celje	Slovenia	865	230	—	49.0	708	—	—	289	37.0	[21]
4	Smelter		241, 242	Wales	UK	136	183	—	—	92.0	—	—	4.9	—	[22]
5	Smelter		241, 242	Reppel-Bocholt	Belgium	899	137	246	94.0	801	—	—	12.0	—	[14]
6	Smelter		241, 242	Kosovska Mitrovica	Kosovo	973	560	—	214	56.2	418	—	7.2	87.1	[16]
7	Metal industry		241, 242	Potenza	Italy	87.7	62.8	—	—	26.1	—	—	1.31	18.3	[18]
8	Metal industry		241, 242	Annaba	Algeria	53.1	67.5	355	—	39.0	30.9	—	0.44	—	[12]
9	Metal industry	ME + MI	241, 242	Karak ind. Estate (S)	Jordan	15.7	4.9	—	3.70	1.60	—	23.4	—	—	[29]
10	Copper smelter, steel industry		241, 242	Port Kembla	Australia	20	42.0	—	—	49.0	12.0	—	—	3.20	[33]
11	Metal recycling, steel production		241, 242	Murcia	Spain	18.3	21.8	229	17.7	7.40	19.4	—	0.20	—	[37]
12	Ferrous and non-ferrous metal smelters, iron and steel manufacturing, metal recovery, manganese ore treatment, scrap, metal incineration		241, 242	Thessaloniki	Greece	24.2	68.3	443	—	26.8	98.7	37.1	0.22	36.1	[17]
13	Copper mines		072	Canchaque	Peru	156	235	714	801	—	—	—	183	279	[20]
14	Smelters, mining		071, 072, 241, 242	Elbasan	Albania	80	61.0	—	447	14.0	491	—	3.00	—	[11]
15	Smelters, mining		071, 072, 241, 242	Rubik	Albania	135	250	—	66.0	111	256	—	9.00	—	[11]
16	Smelters, mining		071, 072, 241, 242	Munelle	Albania	103	111	—	54.0	73.0	91.0	—	2.00	—	[11]
17	Smelters, mining		071, 072, 241, 242	Tharsis	Spain	85.0	92.0	654	—	47.0	—	—	—	37.0	[38]
18	Petrochemical		061	Tarragona	Spain	29.5	—	297	—	—	20.4	—	0.19	4.71	[25]
19	Petrochemical		061	Almowra	Spain	—	—	—	27.6	—	—	—	2.13	—	[39]
20	Petrochemical		061	Tarragona	Spain	36.3	—	—	—	—	13.8	—	0.21	5.50	[26]
21	Petrochemical	CE + PE	061	Catalonia	Spain	37.8	—	—	—	—	16.5	—	0.21	6.51	[40]
22	Chemical		201	Tarragona	Spain	24.5	—	248	—	—	14.8	—	0.17	6.87	[25]
23	Chemical		201	Valencia	Spain	22.2	—	—	—	—	17.5	—	0.16	6.24	[40]
24	Sulphuric acid-producing industry		201	Dhaka	Bangladesh	1.03	126	277	88.2	63.5	—	—	0.53	—	[23]
31	Textile industries	TE	131,139	Dhaka	Bangladesh	56.4	207	382	51.1	164	—	—	0.48	—	[23]
32	Textile industries		131,39	Haridwar	India	191	—	668	—	109	568	308	83.6	—	[41]
27	Tannery industries		151	Dhaka	Bangladesh	68.1	290	39.7	5.20	17.6	—	—	1.26	—	[23]
28	Tannery industries	LE	151	Haridwar	India	—	—	097	—	0.04	744	37.7	0.04	—	[41]
29	Tannery industries		151	Peshawar	Pakistan	4.66	2.38	5.29	—	—	29.9	18.0	0.60	—	[30]
30	Tannery industries		151	Damascus	Syria	17.0	103	—	39.0	34.0	57.0	—	—	—	[32]

TABLE 3: Continued.

Order	Industry type	Industry group short name[a]	ISIC industry code[a]	Location City	Country	Trace element concentration (mg kg^{-1}) Pb	Zn	Mn	Ni	Cn	Cr	Fe	Cd	As	Reference
33	Cement factory		239	Qadissiya	Jordan	55.0	45.0	204	39.0	2.90	22.2	—	5.00	—	[29]
25	Ceramic industries	NM	239	Dhaka	Bangladesh	28.6	287	217	50.1	38.4	—	—	0.33	—	[23]
26	Ceramic industries		239	Castellón	Spain	229	—	345	36.5	66.0	29.7	—	72.0	—	[42]
34	Battery manufacturing	BA	272	Baoji	China	268	169	—	39.3	62.3	100	—	—	—	[28]
35	Textile, plastic, furniture, industries		131, 201, 310	Thane-Belapur	India	—	191	—	184	105	521	—	—	—	[43]
36	Chemicals, dyes, textile, paint industries		131, 201, 202	Rajasthan	India	293	—	—	136	298	240	—	—	—	[43]
37	Paint, battery, textile, milling and chemical industries		131, 201, 272	Lagos	Nigeria	—	—	—	24.7	106	—	—	13.4	—	[44]
38	Ceramic, paper products, petrochemical industries		061, 170, 239	Belgrade	Serbia	37.7	103	—	66.6	40.0	—	—	—	—	[31]
39	Chemicals, pharmaceutical, batteries	ND	201, 210, 272	Hyderabad	India	65.0	313	—	45.0	193	433	—	0.70	33.0	[24]
40	Chemicals, furniture, printing, leather, textile Brewery, steel/metal works, paints, pharmaceuticals,		131, 151, 201, 310	Kayseri	Turkey	74.8	112	—	44.9	36.9	29.0	—	2.53	—	[27]
41	packaging, food processing, textiles and plastic products industries		110, 131, 139, 201, 241	Lagos	Nigeria	143	247	283	17.0	25.6	26.6	—	2.90	—	[45]
42	Tannery, textile, fertilizer, rerolling and casting, chemicals paints plastics		131, 139, 151, 201, 241	Jajmau, Unnao	India	38.3	159.9	—	—	42.9	265	—	—	—	[46]

[a] Refer to Table 2.
[b]ND—not defined due to the complexity of industry types.

TABLE 4: Trace element concentration in (a) air and (b) water phases affected by industrial activities.

(a) Air phase

Order	Industry type	Industry group Short name[a]	ISIC industry code[a]	Location City	Location Country	Trace element concentrations (ng m^{-3}) Pb	Zn	Mn	Ni	Cu	Cr	Fe	Cd	As	Reference
1	Petrochemical		061	Tarragona	Spain	1.22	—	4.70	—	—	2.19	—	0.05	0.12	[25]
2	Petrochemical	CE + PE	061	Rio de Janeiro	Brazil	8.90	25.1	1.50	—	12.1	—	15.9	0.30	—	[47]
3	Chemical and petrochemical		061, 201	Rio de Janeiro	Brazil	15.9	2124	16.0	2.10	22.0	2.40	775	0.40	—	[48]
4	Chemical		201	Tarragona	Spain	3.20	—	8.50	—	—	2.19	—	0.23	0.69	[25]
5	Machinery, metallurgical, petrochemistry, pulp and paper, textile		131, 170, 241, 242, 281,	Sihwa	Rrpublic of Korea	321	657	161	—	—	69.6	1203	7.10	7.70	[49]
6	Textile, metallurgical, paper		131, 241, 242, 281,	Banwol	Rrpublic of Korea	346	1272	86.3	—	—	21.0	1218	6.30	6.30	[49]
7	Metal recycling, steel production		241, 242	Pohang	Rrpublic of Korea	93.8	389	245	—	—	23.9	5423	1.95	—	[50]
8	Ceramic, paper products, petrochemical	ND[b]	061, 170, 239	Onda	Spain	164	203	7.00	—	—	3.00	432	—	—	[40]
9	Oil refineries, cement, shipyards, steel		192, 239, 241, 301	Elefsina	Greece	71.1	—	21.1	—	—	—	—	14.5	3.70	[51]
10	Oil refineries, steel mill		192, 241	Capana	Argentina	70.6	36.9	21.2	—	—	2.83	552	0.27	2.76	[44]
11	Ferrous and non-ferrous metal manufacturing, fertilizer production, manganese ore treatment, scrap, metal incineration, oil refining		192, 201, 241, 242	Thessaloniki	Greece	140	830	84.0	—	59.0	7.50	—	2.70	3.60	[17]

[a]Refer to Table 2.
[b]ND—not defined due to the complexity of industry types.

(b) Water phase

Phase	Industry type	Industry group short name[a]	ISIC industry code[a]	Location City	Location Country	Trace element concentration (mg L^{-1}) Pb	Zn	Mn	Ni	Cu	Cr	Fe	Cd	As	Reference
1	Brewery		110	Ibadan	Nigeria	—	0.33	0.04	—	0.60	0.01	0.18	—	—	[52]
2	Brewery	BE	110	Benin	Nigeria	—	0.17	0.03	—	0.46	0.01	0.38	—	—	[52]
3	Brewery		110	Lagos	Nigeria	—	0.22	—	—	0.54	0.02	0.15	—	—	[52]
4	Brewery		110	Ibadan	Nigeria	0.19	2.40	—	—	0.36	0.21	—	0.11	—	[53]
5	Tannery	LE	151	Peshawar	Pakistan	0.08	0.13	0.087	—	—	0.09	0.14	0.014	—	[30]
6	Tannery		151	Haridwar	India	—	—	0.02	—	—	0.93	0.84	0.007	—	[41]
7	Textile	TE	131,139	Kaduna	Nigeria	—	0.31	1.18	—	1.16	—	2.14	—	—	[54]
8	Textile		131,139	Haridwar	India	—	—	0.7	—	0.05	—	2.95	—	—	[41]

[a]Refer to Table 1.

concentrations (mg kg^{-1}) of Mn (652), Pb (85), Zn (92), and Cu (47) were also found in soils surrounding the mining and smelting areas in Tharsis, Spain [38]. An investigation covering eight smelters and mining sites in Albania reported exceedingly high concentrations of one or more metals [11]. These authors reported the maximum concentrations (mg kg^{-1}) of metals in soil dry matter (DM): Cd (14), Cr (3,865), Cu (1,107), Ni (3,579), Pb (172), and Zn (2,495). If one refers to the report of Kabata-Pendias and Pendias [65], the measurements of Shallari et al. [11] appear to be toxic with the observed levels harmful for plant growth.

Borgna et al. [16] measured 12 trace elements (As, Cd, Co, Cr, Cu, Ni, Pb, Sb, Th, Tl, U, and Zn) in topsoils from the smelter site in the K. Mitrovica area, Kosovo. They reported considerably elevated median values (mg kg^{-1}) for Pb, Zn, and Cu of 294, 196, and 37.7, respectively. Borgna et al. [16] also noticed that metal levels caused by mining activities decreased significantly with soil depth. Such pollution activities, therefore, basically affected the upper soil layer between 0 and 50 cm [66]. The analysis of vertical soil profiles generally showed an accumulation of trace metals towards the surface soil due to the outputs of mining, smelter, and metallurgical industries [61, 67, 68]. However, there is also contrasting evidence of vertical metal distribution patterns. For instance, near an iron smelter in Bulgaria, Schulin et al. [15] found that the differences in metal levels between topsoil and subsoil samples were generally small. These authors concluded that the observed vertical profiles of soil metal levels in the study area should primarily be related to geogenic processes. Martley et al. [33] was also unable to find any differences in metal concentrations at soil layers between 0–5 and 5–20 cm, except for Pb and Zn. Similar partitioning results for trace elements were also seen from soils around the Harjavalta smelter, Finland [69], and in Tharsis, Spain [38].

Considering horizontal distributions, metal concentrations in soil usually decrease with increasing distance from the mining or smelter site, generally following an exponential or negative-power decay function. In most cases, metal concentrations in topsoil layers significantly exceeded those of background levels (background levels being considered as those many kilometers from the smelter facilities). For instance, contamination was found to be detectable up to 33 km for a copper smelter or up to 217 km for a zinc smelter in Canada [70]. It was found that As was generally transported over long distances relative to other elements (e.g., Pb, Zn, etc.) [68, 71]. The reason for such enhanced distribution of As may come as a result of its higher volatility and extended atmospheric residence time. Hence, As is less likely to deposit very near the point source from which it was emitted [72]. However, smelter emissions generally depend on a variety of factors including the mass of emitted contaminants, their particle size distributions, stack height, meteorological conditions such as wind speed and direction, and topography [67]. Garmash [73] found that concentrations of zinc, lead, and cadmium were an order of magnitude lower in soil at iron smelter than from a nearby the zinc smelter as a function of the different emissions profile from each plant.

4.2. Chemical and Petrochemical Industries.
Chemical and petrochemical industries have been identified as large emitters of not only metals but also a wide variety of pollutants (e.g., volatile organic compounds (VOCs), polycyclic aromatic hydrocarbons (PAHs), polychlorinated biphenyls (PCBs), etc.) [25]. As these pollutants are also commonly recognized as markers of environmental pollution, they have also been identified as the cause of some adverse health effects in workers and people living nearby [25]. In Table 3, it is shown that Cd, Pb, Cr, and As are the most abundant trace metals found in soil samples around chemical and petrochemical sites. The release of metals such as Pb, As, or Cr may occur in refining operations and from the burning of residual oils [74]. In contrast, Mn, Ni, Zn, and Cu were determined only at 3, 2, 1, and 1 sites, respectively, from the 7 references examined in this study (Table 3). As shown in Table 5, the mean concentrations (mg kg^{-1}) of Pb, Cr, As, and Cd resulting from this industry type were 25.2, 16.6, 5.97, and 0.51, respectively.

In a study conducted near a sulphuric acid plant in Bangladesh, very high concentrations (mg kg^{-1}) of Zn (126), Mn (277), Ni (88.2) and Cu (63.5) were detected in soil samples [23]. The status of metal pollution in soil near the sulphuric acid production facilities should directly depend on the quantity of waste material discharged onto the land. Furthermore, many trace metals are likely to be deposited by localized acid rain, once emitted to air from the industry in question [23]. There is also a strong possibility of soil acidification as a result of SO$_2$ emissions from the acid-producing industries [75]. Low pH values from the sulphuric acid production facility are also likely to help increase solubility and mobility of metals in the soil.

4.3. Textile Industries.
Textile industries can act as one of the major sources of metal pollution in the environment [41]. There is evidence that significant amounts of trace metals have been released into the surrounding soil from textile industries (Table 3). In one of the previous studies conducted in Bangladesh, mean soil concentrations of Pb, Zn, Mn, Ni, and Cd in the vicinity of textile industries were found to be 56.4, 207, 382, 51.1, 164 mg kg^{-1}, respectively [23]. All metals except Ni were detected in 18 soil samples collected near textile industrial facilities with their mean values being (mg kg^{-1}) of 191 (Pb), 668 (Mn), 109 (Cu), 586 (Cr), 380 (Fe), and 83.6 (Cd) [41]. In addition, chromium and cadmium in the soils contaminated with textile effluent in Tamil Nadu, India were reported to be in the range of 55.4–180 and 0.2–5.8 mg kg^{-1}, respectively [76]. Although relatively high levels of lead are generally seen in soil samples contaminated by textile industries, conversely lead concentrations in textile effluents were below the detection limit. The presence of lead in soil was thus ascribed to its airborne transport and subsequent deposition from automobiles and other industries in the area.

4.4. Leather Industries.
Solid and liquid wastes emanating from the tanning industry are known to contain various toxic trace metals [68]. In most developing countries, tannery effluents are directly discharged to nearby land where they

TABLE 5: Statistical summary of trace element concentration (in mg kg^{-1}) in soils affected by industrial activities.

Industry group short name	Statistical parameter	Trace elements								
		Pb	Zn	Mn	Ni	Cu	Cr	Fe	Cd	As
ME + MI	Mean	258	190	440	186	137	177	30.3	37.6	67.0
	Median	95	102	399	80.0	52.6	94.9	30.3	3.95	37.0
	SD	343	234	205	252	244	189	9.69	86.7	89.0
	Min	15.7	4.90	229	3.70	1.60	12.0	23.4	0.20	3.20
	Max	973	907	714	801	801	491	37.1	289	279
	N	16	16	6	10	16	8	2	14	8
CE + PE	Mean	25.2	126	274	57.9	63.5	16.6	—	0.51	5.97
	Median	25.2	—	274	57.9	—	16.6	—	0.51	5.97
	SD	13.4	—	24.6	42.9	—	2.57	—	0.72	0.86
	Min	1.03	—	248	27.6	—	13.8	—	0.16	4.71
	Max	37.8	—	297	88.2	—	20.4	—	2.13	6.87
	N	6	1	3	2	1	5	0	7	5
TE	Mean	124	207	525	51.1	136	568	308	42.0	—
	Median	124	—	525	—	136	—	—	42.0	—
	SD	95.2	—	202	—	38.7	—	—	58.8	—
	Min	56.4	—	382	—	109	—	—	0.48	—
	Max	191	—	668	—	164	—	—	83.6	—
	N	2	1	2	1	2	1	1	2	0
LE	Mean	29.9	132	15.3	22.1	17.2	277	27.9	0.63	—
	Median	17.0	103	5.29	22.1	17.6	57.0	27.9	0.60	—
	SD	33.6	146	21.2	23.9	17.0	405	13.9	0.61	—
	Min	4.66	2.38	0.97	5.20	0.04	29.9	18.0	0.04	—
	Max	68.1	290	39.7	39.0	34.0	744	37.7	1.26	—
	N	3	3	3	2	3	3	2	3	0
NM	Mean	104	166	255	41.9	35.8	26.0	—	25.8	—
	Median	55.0	166	217	39.0	38.4	26.0	—	5.00	—
	SD	109	171	77.9	7.24	31.6	5.30	—	40.1	—
	Min	28.6	45.0	204	36.5	2.90	22.2	—	0.33	—
	Max	229	287	345	50.1	66.0	29.7	—	72.0	—
	N	3	2	3	3	3	2	0	3	0
BA	Value	268	169	—	39.3	62.3	100	—	—	—
ND	Mean	109	188	283	74.0	106	252	—	4.89	—
	Median	69.9	176	—	45.0	74	253	—	2.72	—
	SD	98.2	81.2	—	62.3	95.6	203	—	5.76	—
	Min	37.7	103	—	17.0	25.6	26.6	—	0.70	—
	Max	293	313	—	184	298	521	—	13.4	—
	N	6	6	1	7	8	6	0	4	0
All	Mean	158	180	312	106	106	176	84.8	21.6	42.8
	Median	65.0	131	280	49.6	52.6	44.0	37.1	2.00	25.7
	SD	247	180	208	168	176	217	125	59.8	71.8
	Min	1.03	2.38	0.97	3.70	0.04	12.0	18.0	0.04	3.20
	Max	973	907	714	801	801	744	308	289	279
	N	37	30	18	26	34	26	5	33	14

adversely affect the quality of both soil and ground water [23]. According to our evaluation, Cr showed the highest values followed by Zn, amongst the trace metals reported in this category (Table 5). This finding can be ascribed to the fact that chromium salts are the most widely used tanning substances [32, 77]. Only a fraction of the chromium salts are actually consumed during leather processing, and most of the salt is discharged as liquid effluent [30]. The mean concentrations (mg kg^{-1}) of trace metals were determined to be Cr (744), Zn (0.97), Cu (0.04), Fe (37.7), and As (0.04) in soil samples in the vicinity of leather industries in India [41]. On the other hand, higher iron and copper levels were seen (4837–6311 and 7.20–20.5 mg kg^{-1}, resp.) in soils affected by tannery effluent in Mexico [78]. Cr was also reported to range from 155 to 568 mg kg^{-1} in tannery waste contaminated soil in the Vellore district of Tamil Nadu, India [5].

4.5. Non-Metallic Mineral Industries. Nonmetallic mineral industries such as cement, ceramic, and battery manufacturing facilities can act as a major source of trace metal pollution in soil. In a study made in Jordan, it was suggested that cement factory emissions might represent the most important pollution source in the area investigated [29]. High Pb, Zn, Mn, and Ni concentrations (e.g., 55, 45, 204, and 39 mg kg^{-1}, resp.) were recorded in the soil samples close to the cement plant [29]. This may reflect the fact that the process and production of cement industry require a substantial amount of energy (supplied by burning fossil fuel) and traffic activity in and around the plant [79]. Kashem and Singh [23] found Ni and Zn in excess of tolerable levels, set as 50 mg kg^{-1} and 290 mg kg^{-1}, respectively, in the soil samples of ceramic industry sites in Bangladesh. High levels (mg kg^{-1}) of Pb (268) and Zn (169) were also found near battery manufacturing facilities, which are suspected to pollute the soil in the industrial area of Baoji city, China [28]. It was noted that Pb is the main primary raw material used for battery production, while the geochemical behavior of Pb, Zn, and Mn is known to be very similar in most natural processes [80].

5. Evaluation of Metal Pollution between Soils Affected by Different Industry Types

To learn more about the relative dominance of a given metal between different industry types, we evaluated our soil metal data by various classification criteria (Table 5). Considering the metal concentration levels determined for different industry types, the overall mean values for different metals can be arranged in the following descending manner: Mn > Zn > Cr > Pb > Ni > Cu > Fe > As > Cd (Table 5). Across the different industry types, the highest average metal concentrations (6 out of 9 studies) were observed from smelter and metal industries. However, the highest mean Mn and Cd concentrations were found from the textile industry, while the highest Cr from leather industry studies (Figure 2). In a study comparing textile and leather industries, Deepali and Gangwar [41] found chromium content to be 23.6% lower in soil contaminated by textile effluent than that contaminated by tannery effluent. Textile industry effluent

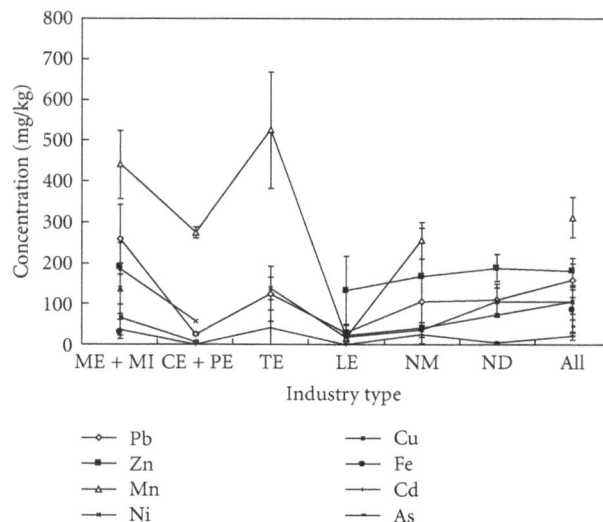

FIGURE 2: Comparison of soil metal levels affected by different industry types.

showed an excess concentration when compared to that from the tannery industry of 87.8% (Fe), 99.9% (Mn), and 99.9% (Cd), while soils affected by tannery industry were characterized by excess Cr. In another study comparing textile and tannery industries in Bangladesh, higher concentrations of Mn, Ni, and Cu were found in soil samples due to textile activity [23]. On the other hand, Pb and Zn showed the opposite trend, that is, higher values from tannery than textile industry samples.

The rapid growth in industrial activities has increased the pressure on environmental sustainability. For a healthy and balanced environment, several regulations have been set out by different governments recognizing the need for improved environmental management. In case of soil, the regulations are set on total metals in the soil, and as such it is important to consider bioavailability of the metals in relation to soil physiochemical properties (such as organic matter content, clay content, cation exchange capacity, etc.). The regulations for the maximum allowable soil metal levels established in different countries are presented in Table 6. However, there are large variations in regulatory metal limits in different countries. In this discussion, we will consider mainly regulatory values which are common to a large number of countries. If our database of soil pollution (Table 3) is examined in this respect, Pb is found to exceed the maximum allowable level (e.g., 100 mg kg^{-1} in Table 6) on many occasions as a result of the metal and mining industries (8 out of 17 studies) and on a single occasion from textile, ceramic, and battery manufacturing industries. Much more emphasis has been placed on lead contamination in soils in recent years, as it exerts very toxic effects on humans and animals. Lead enters human or animal metabolism either via the food chain or by intake of soil dust [93]. In addition to the sources mentioned above, battery production and scrap battery recovery facilities, thermal power plants, and iron-steel industries are commonly found to be major industrial sources

TABLE 6: Regulatory levels for soil metals established between different countries.

Order	Country	Concentration (mg kg^{-1})							Reference
		Pb	Zn	Ni	Cu	Cr	Cd	As	
1	Australia	—	200	60	60	50	—	20	[81]
2	Japan	—	—	—	125	—	1	15	[82]
3	Taiwan	100	300	—	150	200	4	20	[83]
4	Turkey	150	500	—	100	—	5	—	[84]
5	EU	300	300	75	140	150	3	—	[85]
6	Netherlands	150	500	100	100	250	5	—	[86]
7	Spain	300	450	112	210	150	3	—	[87]
8	Germany	100	300	50	100	100	3	20	[88]
9	France	100	300	50	100	150	2	20	[89]
10	UK	550	280	35	140	600	4	10	[90]
11	USA	150	300	31	45	212	2	5.6	[91]
12	Canada	500	500	100	100	250	5	—	[92]

of Pb [94]. The concentration of Zn measured from a (Pb and Zn) smelter in China, a metal industry facility in Kosovo, and an industrial complex in India averaged as 907, 560, and 313 mg kg^{-1}, respectively (Table 3). Hence, these observed Zn levels also exceeded commonly allowable concentration levels (e.g., 300 mg kg^{-1} in Table 6). The mean concentration of Ni was much higher than its allowable limit (e.g., 50 mg kg^{-1}) close to a number of industries such as mining and metal, chemical, textile, ceramic, and industrial complexes. In the case of Cu, concentrations above the regulatory value (e.g., 100 mg kg^{-1}) were mainly seen from textile and metal industries. In textile processing, a number of heavy metals (especially Cu, Cr, Zn, etc.) are used in dying and printing processes [95]. As such, the mean concentrations of Cr and Cd in metal, textile, leather, and industrial complexes exceeded the guideline values (i.e., 150 and 3 mg kg^{-1}, resp.). The contents of Cr and Cd in topsoil were reported to increase due to the release of various industrial wastes such as tannery wastes, electroplating sludges, leather manufacturing wastes, and so forth [94]. In contrast, exceedance of the regulatory guideline level for As (i.e., 20 mg kg^{-1}) was found only in the case of mining and metal industries.

6. Assessment of Geoaccumulation (I_{geo}) Index

In order to learn more about the level of metal pollution in soil and around industrialized areas, we employed a common approach to estimate the enrichment of metal concentrations in soil relative to background concentration, by computing the geoaccumulation index (I_{geo}) [96, 97]. This method allows assessment of the extent of metal pollution into seven classes based on the increasing numerical value of the index as follows: (1) $I_{geo} \leq 0$: practically uncontaminated, (2) $0 < I_{geo} < 1$: uncontaminated to moderately contaminated, (3) $1 < I_{geo} < 2$: moderately contaminated, (4) $2 < I_{geo} < 3$: moderately to heavily contaminated (5) $3 < I_{geo} < 4$: heavily contaminated (6) $4 < I_{geo} < 5$: heavily to extremely

contaminated, and (7) $5 \leq I_{geo}$: extremely contaminated [96]. This index can be derived by the following equation:

$$I_{geo} = \log_2 \left(\frac{Cn}{1.5Bn} \right), \quad (1)$$

where Cn is the measured concentration of the trace element in the soil samples and Bn is the geochemical background value in the earth's crust. The constant 1.5 allows us to account for the natural fluctuations in the content of a given substance in the environment and the possible influence of nonlocalized anthropogenic sources. Results for calculation of the geo-accumulation index using our survey soil data are summarized in Table 7. If we calculate the average I_{geo} values of the trace metals in our soil database, they can be arranged in ascending order: Zn (0.81), Ni (1.26), Cu (1.28), As (1.49), Cr (1.60), Pb (2.06), and Cd (3.60). According to this geo-accumulation index calculation, it can be seen that the extent of soil pollution is the least significant for such metals as Mn and Fe, despite the influence of industrial activities. This indicates that these two metals are commonly derived from natural (geogenic) processes. It is also found that Zn, Ni, and Cu remain below class 3 level, while the highest I_{geo} were recorded for Cr, As, Pb, and Cd such as 3.57, 3.84, 5.02, and 9.33, respectively (Table 7). Among the metals, higher I_{geo} values were found for Pb and Cd that are mainly in association with mining and metal industries. The relative ordering of I_{geo} values within our study is comparable to those reported in a number of previous studies. In one of the previous studies in the Gebze industrial area, Turkey, the highest I_{geo} values for metals were found as 10.2 (Cd), 8.38 (Pb), 6.64 (Zn), 4.77 (As), 3.63 (Cu), 3.52 (Mn), and 3.49 (Cr) [94]. Around a cement factory in Ghana, the I_{geo} of Ca, Cu, Mn were found 1.21, 1.36, and 2.96, respectively [98].

7. Conclusions

To learn more about the effect of industrial activities on environmental pollution, we performed a comprehensive

TABLE 7: Comparison of the geoaccumulation index calculated for the soil bound metals affected by industrial activities.

Order	Industry type	Industry group short name[a]	Country	Index of geoaccumulation (I_{geo})								
				Pb	Zn	Mn	Ni	Cu	Cr	Fe	Cd	As
1	Iron smelter		Bulgaria	—[a]	—	—	0.12	—	—	—	2.22	—
2	Pb and Zn smelter		China	3.98	2.67	—	—	0.19	—	—	4.72	0.96
3	Zn Smelter		Slovenia	4.85	0.69	—	—	3.39	—	—	9.33	0.92
4	Smelter		UK	2.18	0.36	—	—	0.45	—	—	3.44	—
5	Smelter		Belgium	4.91	—	—	—	3.57	—	—	4.74	—
6	Smelter		Kosovo	5.02	—	—	1.07	—	1.63	—	4.00	2.16
7	Metal industry		Italy	1.55	—	—	—	—	—	—	1.54	—
8	Metal industry		Algeria	0.82	—	—	—	—	—	—	—	—
9	Metal industry	ME + MI	Jordan	—	—	—	—	—	—	—	—	—
10	Copper smelter, steel industry		Australia	—	—	—	—	—	—	—	—	—
11	Metal recycling, steel production		Spain	—	—	—	—	—	—	—	—	—
12	Ferrous and non-ferrous metal smelters, iron and steel manufacturing, metal recovery, manganese ore treatment, scrap, metal incineration		Greece	—	—	—	—	—	—	—	—	0.89
13	Copper mines		Peru	2.38	0.72	—	2.97	—	—	—	8.67	3.84
14	Smelters, mining		Albania	1.42	—	—	2.13	—	1.86	—	2.74	—
15	Smelters, mining		Albania	2.17	0.81	—	—	0.71	0.92	—	4.32	—
16	Smelters, mining		Albania	1.78	—	—	—	0.11	—	—	2.15	—
17	Smelters, mining		Spain	1.50	—	—	—	—	—	—	—	0.92
18	Petrochemical		Spain	—	—	—	—	—	—	—	—	—
19	Petrochemical		Spain	—	—	—	—	—	—	—	—	—
20	Petrochemical		Spain	0.28	—	—	—	—	—	—	—	—
21	Petrochemical	CE + PE	Spain	0.33	—	—	—	—	—	—	—	—
22	Chemical		Spain	—	—	—	—	—	—	—	—	—
23	Chemical		Spain	—	—	—	—	—	—	—	—	—
24	Sulphuric acid producing industry		Bangladesh	—	—	—	—	—	—	—	0.24	—
31	Textile industries	TE	Bangladesh	0.91	0.54	—	—	1.28	—	—	0.09	—
32	Textile industries		India	2.67	—	—	—	0.69	2.07	—	7.54	—
27	Tannery industries		Bangladesh	1.18	1.03	—	—	—	—	—	1.49	—
28	Tannery industries	LE	India	—	—	—	—	—	2.46	—	—	—
29	Tannery industries		Pakistan	—	—	—	—	—	—	—	0.42	—

TABLE 7: Continued.

Order	Industry type	Industry group short name[a]	Country	Index of geoaccumulation (I_{geo})								
				Pb	Zn	Mn	Ni	Cu	Cr	Fe	Cd	As
30	Tannery industries		Syria	—	—	—	—	—	—	—	—	—
33	Cement factory	NM	Jordan	0.87	—	—	—	—	—	—	3.47	—
25	Ceramic industries		Bangladesh	—	1.01	—	—	—	—	—	—	—
26	Ceramic industries		Spain	2.93	—	—	—	—	—	—	7.32	—
34	Battery manufacturing	BA	China	3.16	0.25	—	—	—	—	—	—	—
35	Textile, plastic, furniture, industries		India	—	0.42	—	0.85	0.63	1.95	—	—	—
36	Chemicals, dyes, textile, paint industries		India	3.29	—	—	0.42	2.14	0.83	—	—	—
37	Paint, battery, textile, milling and chemical industries		Nigeria	—	—	—	—	0.65	—	—	4.90	—
38	Ceramic, paper products, petrochemical industries		Serbia	0.33	—	—	—	—	—	—	—	—
39	Chemicals, pharmaceutical, batteries plastic products	ND	India	1.12	1.14	—	—	1.52	1.68	—	0.64	0.76
40	Chemicals, furniture, printing, leather, textile		Turkey	1.32	—	—	—	—	—	—	2.49	—
41	Brewery, steel/metal works, paints, pharmaceuticals, packaging, food processing, textiles and plastic products industries		Nigeria	2.25	0.80	—	—	—	—	—	2.69	—
42	Tannery, textile, fertilizer, rerolling and casting, chemicals, paints, plastics		India	0.35	0.17	—	—	—	0.97	—	—	—

[a] No numeric values are shown for the cases with negative values.

survey of metal pollution in different environmental media. Although database which our survey drew upon is limited in terms of number of studies and the range of industry types, it is representative of the diversity and type of studies in the literature and hence the results of our analysis still provide valuable insights into metal pollution in the soil environment. The data obtained in this survey demonstrate that the metal concentrations in soil generally reflect the influence of various local industrial activities which include metal and mining, chemical and petrochemical, textile, leather, cement, and ceramic industries. Observations of generally enhanced metal levels in soils around various industrial facilities are explainable by unregulated, untreated solid and fluid wastes released by the industries to the nearby land. Among the 9 reference trace metals examined, it is seen that Cu, Cd, Pb, and Zn were monitored the most frequently. Evaluation of soil metal data indicates that their maximum values occur in relation to particular industry types, that is, Pb, Zn, Ni, Cu, Fe, and As in smelter and metal industries, Mn and Cd in the textile industry, and Cr in leather industry studies. The observed metal levels in many of the mining and metal industry samples frequently exceeded the guidance levels set by the environmental legislation in the relevant country. The status of soil pollution in this study, if assessed according to I_{geo} values, was classified as moderately-to-extremely contaminated. Most of the samples exhibited the strongest contamination in Pb and Cd. However, the samples were not greatly polluted with respect to Mn and Fe. Considering that there are no regulatory guidelines regarding soil pollution in many developing countries, more efforts should be made to characterize the soil pollution in relation to various industrial activities. This may also help us set proper soil regulation guidelines for sustaining a healthy and balanced environment and protecting human health.

Acknowledgments

This study was supported by a National Research Foundation of Korea (NRF) Grant funded by the Ministry of Education, Science and Technology (MEST) (no. 2010-0007876). The fourth author acknowledges the support made by a grant from the Korea Basic Science Institute (Project no. T31603). The fifth author also acknowledges partial support made by the Human Resources Development of the Korea Institute of Energy Technology Evaluation and Planning (KETEP) Grant funded by the Korea government Ministry of Knowledge Economy (no. 20100092).

References

[1] A. De Bartolomeo, L. Poletti, G. Sanchini, B. Sebastiani, and G. Morozzi, "Relationship among parameters of lake polluted sediments evaluated by multivariate statistical analysis," *Chemosphere*, vol. 55, no. 10, pp. 1323–1329, 2004.

[2] A. Landajo, G. Arana, A. De Diego, N. Etxebarria, O. Zuloaga, and D. Amouroux, "Analysis of heavy metal distribution in superficial estuarine sediments (estuary of Bilbao, Basque Country) by open-focused microwave-assisted extraction and ICP-OES," *Chemosphere*, vol. 56, no. 11, pp. 1033–1041, 2004.

[3] M. Miró, J. M. Estela, and V. Cerdà, "Application of flowing stream techniques to water analysis: part III. Metal ions: alkaline and alkaline-earth metals, elemental and harmful transition metals, and multielemental analysis," *Talanta*, vol. 63, no. 2, pp. 201–223, 2004.

[4] J. C. Amiard, C. Amiard-Triquet, and C. Méayer, "Experimental study of bioaccumulation, toxicity, and regulation of some trace metals in various estuarine and coastal organisms," in *Proceedings of the Symposium on Heavy Metals in Water Organisms*, J. Salanki, Ed., pp. 313–324, Akademiai Kiado, Budapest, Hungary, 1995.

[5] S. Mahimairaja, "An overview of heavy metals: impact and remediation," *Current Science*, vol. 78, no. 7, pp. 34–45, 2000.

[6] S. Dahbi, M. Azzi, N. Saib, M. De la Guardia, R. Faure, and R. Durand, "Removal of trivalent chromium from tannery waste waters using bone charcoal," *Analytical and Bioanalytical Chemistry*, vol. 374, no. 3, pp. 540–546, 2002.

[7] S. Babel and T. A. Kurniawan, "Low-cost adsorbents for heavy metals uptake from contaminated water: a review," *Journal of Hazardous Materials*, vol. 97, no. 1–3, pp. 219–243, 2003.

[8] G. Farabegoli, A. Carucci, M. Majone, and E. Rolle, "Biological treatment of tannery wastewater in the presence of chromium," *Journal of Environmental Management*, vol. 71, no. 4, pp. 345–349, 2004.

[9] J.K. Parikh, V. K. Sharma, U. Gosh, and M. K. Panda, *Trade and Environment Linkages: a case study of India*, Indira Gandhi Institute of Development, 1995.

[10] International Standard Industrial Classification (ISIC), Rev.4 structure (United Nations Statistics Division), 2008.

[11] S. Shallari, C. Schwartz, A. Hasko, and J. L. Morel, "Heavy metals in soils and plants of serpentine and industrial sites of Albania," *Science of the Total Environment*, vol. 209, no. 2-3, pp. 133–142, 1998.

[12] S. Maas, R. Scheifler, M. Benslama et al., "Spatial distribution of heavy metal concentrations in urban, suburban and agricultural soils in a Mediterranean city of Algeria," *Environmental Pollution*, vol. 158, no. 6, pp. 2294–2301, 2010.

[13] R. J. C. Brown, S. L. Goddard, K. C. Blakley, and A. S. Brown, "Improvements to standard methodologies for the analytical determination of metals in stationary-source emissions samples," *Journal of the Air and Waste Management Association*, vol. 61, no. 7, pp. 764–770, 2011.

[14] V. Cappuyns, S. Van Herreweghe, R. Swennen, R. Ottenburgs, and J. Deckers, "Arsenic pollution at the industrial site of Reppel-Bocholt (North Belgium)," *Science of the Total Environment*, vol. 295, no. 1–3, pp. 217–240, 2002.

[15] R. Schulin, F. Curchod, M. Mondeshka, A. Daskalova, and A. Keller, "Heavy metal contamination along a soil transect in the vicinity of the iron smelter of Kremikovtzi (Bulgaria)," *Geoderma*, vol. 140, no. 1-2, pp. 52–61, 2007.

[16] L. Borgna, L. A. Di Lella, F. Nannoni et al., "The high contents of lead in soils of Northern Kosovo," *Journal of Geochemical Exploration*, vol. 101, no. 2, pp. 137–146, 2009.

[17] D. Voutsa, A. Grimanis, and C. Samara, "Trace elements in vegetables grown in an industrial area in relation to soil and air particulate matter," *Environmental Pollution*, vol. 94, no. 3, pp. 325–335, 1996.

[18] L. Medici, J. Bellanova, C. Belviso et al., "Trace metals speciation in sediments of the Basento River (Italy)," *Applied Clay Science*, vol. 53, no. 3, pp. 414–442, 2011.

[19] O. A. Al-Khashman, "Heavy metal distribution in dust, street dust and soils from the work place in Karak Industrial Estate, Jordan," *Atmospheric Environment*, vol. 38, no. 39, pp. 6803–6812, 2004.

[20] J. Bech, C. Poschenrieder, M. Llugany et al., "Arsenic and heavy metal contamination of soil and vegetation around a copper mine in Northern Peru," *Science of the Total Environment*, vol. 203, no. 1, pp. 83–91, 1997.

[21] G. E. Voglar and D. Lestan, "Solidification/stabilisation of metals contaminated industrial soil from former Zn smelter in Celje, Slovenia, using cement as a hydraulic binder," *Journal of Hazardous Materials*, vol. 178, no. 1–3, pp. 926–933, 2010.

[22] B. E. Davies, "Heavy metal contaminated soils in an old industrial area of Wales, Great Britain: source identification through statistical data interpretation," *Water, Air, & Soil Pollution*, vol. 94, no. 1-2, pp. 85–98, 1997.

[23] A. Kashem and B. R. Singh, "Heavy metal contamination of soil and vegetation in the vicinity of industries in Bangladesh," *Water, Air, & Soil Pollution*, vol. 115, no. 1–4, pp. 347–361, 1999.

[24] P. K. Govil, J. E. Sorlie, N. N. Murthy et al., "Soil contamination of heavy metals in the Katedan Industrial Development Area, Hyderabad, India," *Environmental Monitoring and Assessment*, vol. 140, no. 1–3, pp. 313–323, 2008.

[25] M. Nadal, M. Mari, M. Schuhmacher, and J. L. Domingo, "Multi-compartmental environmental surveillance of a petrochemical area: levels of micropollutants," *Environment International*, vol. 35, no. 2, pp. 227–235, 2009.

[26] M. Nadal, M. Schuhmacher, and J. L. Domingo, "Levels of metals, PCBs, PCNs and PAHs in soils of a highly industrialized chemical/petrochemical area: temporal trend," *Chemosphere*, vol. 66, no. 2, pp. 267–276, 2007.

[27] S. Tokahoglu and S. Kartal, "Multivariate analysis of the data and speciation of heavy metals in street dust samples from the Organized Industrial District in Kayseri (Turkey)," *Atmospheric Environment*, vol. 40, no. 16, pp. 2797–2805, 2006.

[28] X. Li and C. Huang, "Environment impact of heavy metals on urban soil in the vicinity of industrial area of Baoji city, P.R. China," *Environmental Geology*, vol. 52, no. 8, pp. 1631–1637, 2007.

[29] O. A. Al-Khashman and R. A. Shawabkeh, "Metals distribution in soils around the cement factory in southern Jordan," *Environmental Pollution*, vol. 140, no. 3, pp. 387–394, 2006.

[30] S. R. Tariq, M. H. Shah, N. Shaheen, A. Khalique, S. Manzoor, and M. Jaffar, "Multivariate analysis of trace metal levels in tannery effluents in relation to soil and water: a case study from Peshawar, Pakistan," *Journal of Environmental Management*, vol. 79, no. 1, pp. 20–29, 2006.

[31] L. Slavkovic, B. Skrbic, and N. Miljevic, "Antonije Onjia, Principal component analysis of trace elements in industrial soils," *Environmental Chemistry Letters*, vol. 2, pp. 105–108, 2004.

[32] A. Möller, H. W. Müller, A. Abdullah, G. Abdelgawad, and J. Utermann, "Urban soil pollution in Damascus, Syria: concentrations and patterns of heavy metals in the soils of the Damascus Ghouta," *Geoderma*, vol. 124, no. 1-2, pp. 63–71, 2005.

[33] E. Martley, B. L. Gulson, and H. R. Pfeifer, "Metal concentrations in soils around the copper smelter and surrounding industrial complex of Port Kembla, NSW, Australia," *Science of the Total Environment*, vol. 325, no. 1–3, pp. 113–127, 2004.

[34] C .P. Dinkins and C. Jones, "Soil sampling strategies. A self learning resource from Montana State University extension," *Agriculture and Natural Resource*, vol. 9, pp. 221–230, 2008.

[35] D. A. Skoog, F. J. Holler, and S. R. Crouch, *Principles of Instrumental Analysis*, Brooks/Cole, 6th edition, 2007.

[36] R. J. C. Brown, R. E. Yardley, A. S. Brown, and M. J. T. Milton, "Sample matrix and critical interference effects on the recovery and accuracy of concentration measurements of arsenic in ambient particulate samples using ICP-MS," *Journal of Analytical Atomic Spectrometry*, vol. 19, no. 5, pp. 703–705, 2004.

[37] J. A. Acosta, A. Faz, S. Martínez-Martínez, and J. M. Arocena, "Enrichment of metals in soils subjected to different land uses in a typical Mediterranean environment (Murcia City, southeast Spain)," *Applied Geochemistry*, vol. 26, no. 3, pp. 405–414, 2011.

[38] E. I. B. Chopin and B. J. Alloway, "Trace element partitioning and soil particle characterisation around mining and smelting areas at Tharsis, Ríotinto and Huelva, SW Spain," *Science of the Total Environment*, vol. 373, no. 2-3, pp. 488–500, 2007.

[39] A. Soriano, S. Pallarés, F. Pardo, A. B. Vicente, T. Sanfeliu, and J. Bech, "Deposition of heavy metals from particulate settleable matter in soils of an industrialised area," *Journal of Geochemical Exploration*, vol. 113, pp. 36–44, 2012.

[40] S. Rodríguez, X. Querol, A. Alastuey et al., "Comparative PM10-PM2.5 source contribution study at rural, urban and industrial sites during PM episodes in Eastern Spain," *Science of the Total Environment*, vol. 328, no. 1–3, pp. 95–113, 2004.

[41] K. K. Deepali and K. Gangwar, "Metals concentration in textile and tannery effluents, associated soils and ground water," *New York Science Journal*, vol. 3, no. 4, pp. 82–89, 2010.

[42] M. M. Jordán, M. A. Montero, S. Pina, and E. García-Sánchez, "Mineralogy and distribution of Cd, Ni, Cr, and Pb in biosolids-amended soils from castellón province (NE, Spain)," *Soil Science*, vol. 174, no. 1, pp. 14–20, 2009.

[43] A. K. Krishna and P. K. Govil, "Heavy metal distribution and contamination in soils of Thane-Belapur industrial development area, Mumbai, Western India," *Environmental Geology*, vol. 47, no. 8, pp. 1054–1061, 2005.

[44] P. Smichowski, J. Marrero, and D. Gómez, "Inductively coupled plasma optical emission spectrometric determination of trace element in PM$_{10}$ airborne particulate matter collected in an industrial area of Argentina," *Microchemical Journal*, vol. 80, no. 1, pp. 9–17, 2005.

[45] S. O. Fakayode and P. C. Onianwa, "Heavy metal contamination of soil, and bioaccumulation in Guinea grass (Panicum maximum) around Ikeja Industrial Estate, Lagos, Nigeria," *Environmental Geology*, vol. 43, no. 1-2, pp. 145–150, 2002.

[46] S. Srinivasa Gowd, M. Ramakrishna Reddy, and P. K. Govil, "Assessment of heavy metal contamination in soils at Jajmau (Kanpur) and Unnao industrial areas of the Ganga Plain, Uttar Pradesh, India," *Journal of Hazardous Materials*, vol. 174, no. 1–3, pp. 113–121, 2010.

[47] S. A. Paulino, J. V. Cirilo, S. L. Quiterio, G. Arbilla, and V. Escaleira, "Analysis of the impact of emission from the peterochemical complex in Campos Eliseos on the air quakity of air basin III. Brazil," *Journal of Petroleum and Gas*, vol. 3, no. 2, pp. 67–74, 2009.

[48] V. E. Toledo, P. B. Almeida, S. L. Quiterio et al., "Evaluation of levels, sources and distribution of toxic elements in PM$_{10}$ in a suburban industrial region, Rio de Janeiro, Brazil," *Environmental Monitoring and Assessment*, vol. 139, no. 1–3, pp. 49–59, 2008.

[49] J. M. Lim, J. H. Lee, J. H. Moon, Y. S. Chung, and K. H. Kim, "Airborne PM$_{10}$ and metals from multifarious sources in an industrial complex area," *Atmospheric Research*, vol. 96, no. 1, pp. 53–64, 2010.

[50] M. K. Kim and W. K. Jo, "Elemental composition and source characterization of airborne PM$_{10}$ at residences with relative proximities to metal-industrial complex," *International*

Archives of Occupational and Environmental Health, vol. 80, no. 1, pp. 40–50, 2006.

[51] N. Manalis, G. Grivas, V. Protonotarios, A. Moutsatsou, C. Samara, and A. Chaloulakou, "Toxic metal content of particulate matter (PM$_{10}$), within the Greater Area of Athens," *Chemosphere*, vol. 60, no. 4, pp. 557–566, 2005.

[52] F. A. Dawodu and K. O. Ajanaku, "Evaluation of the effects of Brewery Effluents disposal on public water bodies in Nigeria," *Aqua Environmental Toxicology*, vol. 31, pp. 25–29, 2008.

[53] A. R. Ipeaiyeda and P. C. Onianwa, "Impact of brewery effluent on water quality of the Olosun river in Ibadan, Nigeria," *Chemistry and Ecology*, vol. 25, no. 3, pp. 189–204, 2009.

[54] R. O. Yusuff and J. A. Sonibare, "Characterization of textile industries' effluents in Kaduna, Nigeria and pollution implications," *GLOBAL NEST: The International Journal*, vol. 6, no. 3, pp. 212–221, 2005.

[55] IUPAC, *Compendium of Chemical Terminology*, The "Gold Book", 2nd edition, 1997.

[56] M. P. Tuchschmid, V. Dietrich, P. Richner et al., *Federal Office of Environment, Forests and Landscape*, Umweltmaterialien no. 32, BUWAL, Berne, Switzerland, 1995.

[57] R. C. Wilmoth, S. J. Hubbard, J. O. Burckle, and J. F. Martin, "Production and processing of metals: their disposal and future risks," in *Metals and Their Compounds in the Environment, Occurrence, Analysis and Biological Relevance*, E. Merian, Ed., vol. 23, pp. 19–65, VCH, Weinheim, Germany, 1991.

[58] S. Dumontet, M. Levesque, and S. P. Mathur, "Limited downward migration of pollutant metals (Cu, Zn, Ni, and Pb) in acidic virgin peat soils near a smelter," *Water, Air, & Soil Pollution*, vol. 49, no. 3-4, pp. 329–342, 1990.

[59] C. Kabala and B. R. Singh, "Fractionation and mobility of copper, lead, and zinc in soil profiles in the vicinity of a copper smelter," *Journal of Environmental Quality*, vol. 30, no. 2, pp. 485–492, 2001.

[60] I. McMartin, P. J. Henderson, A. Plouffe, and R. D. Knight, "Comparison of Cu-Hg-Ni-Pb concentrations in soils adjacent to anthropogenic point sources: examples from four Canadian sites," *Geochemistry*, vol. 2, no. 1, pp. 57–74, 2002.

[61] K. Šichorová, P. Tlustoš, J. Száková, K. Kořínek, and J. Balík, "Horizontal and vertical variability of heavy metals in the soil of a polluted area," *Plant, Soil and Environment*, vol. 50, no. 12, pp. 525–534, 2004.

[62] B. G. Rawlins, R. M. Lark, R. Webster, and K. E. O'Donnell, "The use of soil survey data to determine the magnitude and extent of historic metal deposition related to atmospheric smelter emissions across Humberside, UK," *Environmental Pollution*, vol. 143, no. 3, pp. 416–426, 2006.

[63] J. Susaya, K. H. Kim, and M. C. Jung, "The impact of mining activities in alteration of As levels in the surrounding ecosystems: an encompassing risk assessment and evaluation of remediation strategies," *Journal of Hazardous Materials*, vol. 182, no. 1–3, pp. 427–438, 2010.

[64] C. Pérez-Sirvent, M. L. García-Lorenzo, M. J. Martínez-Sánchez, M. C. Navarro, J. Marimón, and J. Bech, "Metal-contaminated soil remediation by using sludges of the marble industry: toxicological evaluation," *Environment International*, vol. 33, no. 4, pp. 502–504, 2007.

[65] A. Kabata-Pendias and H. Pendias, *Trace Elements in Soils and Plants*, CRC Press, Boca Raton, Fla, USA, 1992.

[66] T. Koljonen, *Geochemical Atlas of Finland, Part 2: Till*, Geological Survey of Finland, Espoo, Finland, 1992.

[67] P. J. Henderson, I. McMartin, G. E. Hall, J. B. Percival, and D. A. Walker, "The chemical and physical characteristics of heavy meals in humus and till in the vicinity of the base metal smelter at Flin Flon, Manitoba, Canada," *Environmental Geology*, vol. 34, no. 1, pp. 39–58, 1998.

[68] I. McMartin, P. J. Henderson, and E. Nielsen, "Impact of a base metal smelter on the geochemistry of soils of the Flin Flon region, Manitoba and Saskatchewan," *Canadian Journal of Earth Sciences*, vol. 36, no. 2, pp. 141–160, 1999.

[69] M. Kaasalainen and M. Yli-Halla, "Use of sequential extraction to assess metal partitioning in soils," *Environmental Pollution*, vol. 126, no. 2, pp. 225–233, 2003.

[70] W. G. Franzin, G. A. McFarlane, and A. Lutz, "Atmospheric fallout in the vicinity of a base metal smelter at Flin Flon, Manitoba, Canada," *Environmental Science and Technology*, vol. 13, no. 12, pp. 1513–1522, 1979.

[71] J. A. Kelley, D. A. Jaffe, A. Baklanov, and A. Mahura, "Heavy metals on the Kola Peninsula: aerosol size distribution," *Science of the Total Environment*, vol. 160-161, pp. 135–138, 1995.

[72] P. E. Rasmussen, "Long-range atmospheric transport of trace metals: the need for geoscience perspectives," *Environmental Geology*, vol. 33, no. 2-3, pp. 96–108, 1998.

[73] G. A. Garmash, "Distribution of heavy metals in soils near metallurgical plants," *Soviet Soil Science*, vol. 17, no. 1, pp. 80–85, 1985.

[74] J. B. Stigter, H. P. M. De Haan, R. Guicherit, C. P. A. Dekkers, and M. L. Daane, "Determination of cadmium, zinc, copper, chromium and arsenic in crude oil cargoes," *Environmental Pollution*, vol. 107, no. 3, pp. 451–464, 2000.

[75] T. C. Hutchinson and F. W. Collins, "Effect of H$^+$ ion activity and Ca^{2+} on the toxicity of metals in the environment," *Water, Air, & Soil Pollution*, vol. 7, pp. 421–429, 1977.

[76] M. Malarkodi, R. Krishnasamy, R. Kumaraperumal, and T. Chitdeshwari, "Characterization of heavy metal contaminated soils of Coimbatore district in Tamil Nadu," *Journal of Agronomy*, vol. 6, no. 1, pp. 147–151, 2007.

[77] M. Bosnic, J. Buljan, and R. P. Daniels, Pollutants in tannery effluents US/RAS/92/120, UNIDO Regional program for pollution control in Tanning industry in South-East, 2000.

[78] D. Alvarez-Bernal, S. M. Contreras-Ramos, N. Trujillo-Tapia, V. Olalde-Portugal, J. T. Frías-Hernández, and L. Dendooven, "Effects of tanneries wastewater on chemical and biological soil characteristics," *Applied Soil Ecology*, vol. 33, no. 3, pp. 269–277, 2006.

[79] K. M. Banat, F. M. Howari, and A. A. Al-Hamad, "Heavy metals in urban soils of central Jordan: should we worry about their environmental risks?" *Environmental Research*, vol. 97, no. 3, pp. 258–273, 2005.

[80] C. Reimann, *Chemical Elements in the Environment*, Springer, Berlin, Germany, 1998.

[81] Australian Soil Resource Information System (ASRIS), 2009, http://www.asris.csiro.au/index_other.html.

[82] Ministry of the Environment, Japan. Soil Contamination Countermeasures, 2003, http://www.env.go.jp/en/water/soil/contami_cm.pdf.

[83] Taiwan EPA, "The regulation of trace elements in soils of Taiwan," in *Soil and Groundwater Pollution Remediation Act*, 2001, http://ww2.epa.gov.tw/SoilGW/english_web/laws.htm.

[84] M. A. Yukselen, "Characterization of heavy metal contaminated soils in Northern Cyprus," *Environmental Geology*, vol. 42, no. 6, pp. 597–603, 2002.

[85] European Community Directive, "Protection of the environment, and in particular of the soil, when sewage sludge is used in agriculture," *Journal European Commission*, vol. 181, pp. 6–12, 1986.

[86] V. Ewers, "Standards, guidelines and legislative regulations concerning metals and their compounds," in *Metals and Their Compounds in the Environment: Occurrence, Analysis and Biological Relevance*, E. Merian, Ed., VCH, Weinheim, Germany, 1991.

[87] BOE, Royal Decree 1310/1990 of 29 October, which regulates the use of sewage sludge in agriculture. BOE No. 262 de 1 de noviembre de 1990, Madrid, Spain, pp. 32339–32340, 1990.

[88] K. Louekari and S. Salminen, "Intake of heavy metals from foods in Finland, West Germany and Japan," *Food Additives and Contaminants*, vol. 3, no. 4, pp. 355–362, 1986.

[89] F. Bourgoin, "Soil protection in French environmental law," *The Journal for European Environmental and Planning Law*, vol. 3, pp. 204–211, 2006.

[90] UK Environment Agency, "Using Soil Guideline Values Better Regulation," Science Programme Science report: SC050021/SGV introduction, 2009.

[91] USEPA. PART 503—Standards for the use or disposal of sweage sludge. Electronic Code of Federal Regulations (e-CFR), 2007, http://www.epa.gov/epacfr40/chapt-I.info/chi-toc.htm.

[92] Ministere de l'Environmement du Quebec, Politique de Protection des Soil et de Rehabilitation des Terrains Contamines. Publications of the MEQ, Collection Terrains Contamines, Quebec, Canada, 2001.

[93] M. Piotrowska, S. Dudka, R. Ponce-Hernandez, and T. Witek, "The spatial distribution of lead concentrations in the agricultural soils and main crop plants in Poland," *Science of the Total Environment*, vol. 158, pp. 147–155, 1994.

[94] G. Yaylali-Abanuz, "Heavy metal contamination of surface soil around Gebze industrial area, Turkey," *Microchemical Journal*, vol. 99, no. 1, pp. 82–92, 2011.

[95] B. Smith, A Workbook for Pollution Prevention by Source Reduction in Textile Wet Processing. Pollution Prevention Pays Program of the North Carolina Division of Environmental Management, 1988.

[96] G. Müller, "Index of geo-accumulation in sediments of the Rhine River," *Geojournal*, vol. 2, pp. 108–118, 1969.

[97] K. Loska, D. Wiechuła, and I. Korus, "Metal contamination of farming soils affected by industry," *Environment International*, vol. 30, no. 2, pp. 159–165, 2004.

[98] H. Ahiamadjie, O. K. Adukpo, J. B. Tandoh et al., "Determination of the elemental contents in soils around Diamond Cement Factory, Aflao," *Research Journal of Environmental and Earth Sciences*, vol. 3, no. 1, pp. 46–50, 2011.

20

Maya and WRB Soil Classification in Yucatan, Mexico: Differences and Similarities

Héctor Estrada-Medina,[1] **Francisco Bautista,**[2] **Juan José María Jiménez-Osornio,**[1] **José Antonio González-Iturbe,**[3] **and Wilian de Jesús Aguilar Cordero**[1]

[1] *Departamento de Manejo y Conservación de Recursos Naturales Tropicales (PROTROPICO),*
Campus de Ciencias Biológicas y Agropecuarias (CCBA), Universidad Autónoma de Yucatán (UADY),
Km 15.5 Carretera Mérida - Xmatkuil, Mérida, Yucatán 97315, Mexico
[2] *Centro de Investigaciones en Geografía Ambiental (CIGA), Universidad Nacional Autónoma de México (UNAM),*
Antigua Carretera a Pátzcuaro No. 8701, Col. Ex-Hacienda de San José de La Huerta, Morelia, Michoacán 58190, Mexico
[3] *Facultad de Arquitectura (UADY), Calle 50 S/N x 57 y 59 Ex-Convento de La Mejorada, Mérida, Yucatán 97000, Mexico*

Correspondence should be addressed to Héctor Estrada-Medina; hector.estrada@uady.mx

Academic Editors: M. B. Adams and C. Martius

Soils of the municipality of Hocabá, Yucatán, México, were identified according to both Mayan farmers' knowledge and the World Reference Base for Soil Resources (WRB). To identify Maya soil classes, field descriptions made by farmers and semistructured interviews were utilized. WRB soils were identified by describing soil profiles and analyzing samples in the laboratory. Mayan farmers identified soils based on topographic position and surface properties such as colour and amount of rock fragments and outcrops. Farmers distinguished two main groups of soils: *K'ankab* or soils of plains and *Boxlu'um* or soils of mounds. *K'ankab* is a group of red soils with two variants (*K'ankab* and *Haylu'um*), whereas *Boxlu'um* is a group of dark soils with five variants (*Tsek'el, Ch'ich'lu'um, Chaltun, Puslu'um,* and *Ch'och'ol*). Soils on the plains were identified as Leptosoils, Cambisols, and Luvisols. Soils identified in mounds were Leptosols and Calcisols. Many soils identified by farmers could be more than one WRB unit of soil and *vice versa*; in these cases no direct relationship between both classification systems was possible. Mayan and WRB soil types are complementary; they should be used together to improve regional soil classifications, help transference of agricultural technologies, and make soil management decisions.

1. Introduction

Local soil classification systems play an important role in many agricultural sites throughout the world but they have not considered to construct scientific classification systems [1]. Opportunities to use traditional systems to improve scientific soil classifications, mapping, and environmental impact monitoring are not fully exploited [2]. In countries like Mexico, indigenous soil knowledge of ancestral groups [3–7] need to be understood to facilitate planning, transmission, and implementation of new agricultural technologies [3, 8].

Local knowledge is restricted geographically, dynamic, collective, diachronic, and holistic; it is the product of a long observation history, analysis, and management of the natural resources, transmitted orally from generation to generation [9]. Traditional soil classification systems, created by the users, have a local importance and are based on properties easily affected by management [10]. This knowledge is enough to understand and manage the soil in a local way to solve short term specific problems [2, 11, 12]. On the other hand, scientific soil classification systems are based in measurable and observable soil characteristics defined in terms of diagnostic properties, materials, and horizons related to the soil morphology [13]. Traditional knowledge is being lost because these new regionally applied scientific schemes do not consider it. Incorporation of both types of knowledge into a more useful scheme requires the development of a

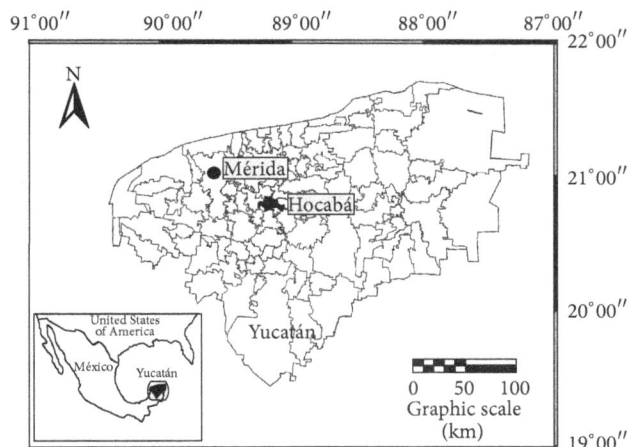

FIGURE 1: Study zone.

TABLE 1: Characterization of interviewed farmers from Hocabá, Yucatán, México.

Farmer age		Years doing *milpa*		Number of *mecates**	
Range (years)	n	Range	n	Range	N
20–29	3	1–9	6	1–9	3
30–39	1	10–19	7	10–19	6
40–49	8	20–29	6	20–29	12
50–59	7	30–39	4	30–39	6
60–69	10	40–49	4	40–49	5
70–80	8	50–59	9	50–59	3
>80	3	>60	4	>60	5

*1 *mecate* = 400 m^2.

common language among farmers, extensionists, technicians, and researchers.

In Yucatán, farmers descended from the old Mayan culture still have a great quantity of knowledge about soils, which they continue using for their agricultural practices [5, 8, 14]; studies concerning this matter are descriptive and only a few have attempted to systematize this knowledge and relate it to scientific soil classification systems [15–17]. In this study the soils of the municipality of Hocabá, Yucatán, México, were identified according to the Mayan farmers' knowledge and the WRB system. The differences and similarities between the two systems were analyzed in order to identify the best correspondence between them.

2. Materials and Methods

2.1. Study Zone. The municipality of Hocabá is located in the central region of the state of Yucatan at 20° 49'N and 89° 15'W within the geomorphologic landscape defined by Lugo [18] as a "structural plain almost horizontal marginal to the coast" with up to 10 m of altitude (Figure 1). Hocabá occupies an area of 81.75 km^2 that represents 0.18% of the state territory [19]. The climate is subhumid tropical with a summer rain season Aw$_1$(i')g [20, 21]. The dominant vegetation is low deciduous forest [22] and the main crops of the land are sisal (*Agave fourcroydes* Lem.) and corn [23]. Two geologic zones converge in this area: a 58 million years ago limestone zone, with fine grain silicated and scarce presence of fossils in the majority of the municipality, and a 13 to 25 million years ago limestone zone, in the southeast part of the municipality, with cream and brownish microcrystalline grey rocks with great amount of fossils [24]. Intercalated zones of plains and mounds compound the topography. Mounds reach diameters of 3 to 10 m and heights up to 3 m; the plains usually have a diameter of 10 to 30 m [14].

Forty semistructured interviews with Mayan farmers were carried out in order to obtain information about the Mayan soil knowledge of the municipality of Hocabá. Interviewed farmers were "milperos" (farmers who grow

milpa—association of corn, bean, and pumpkin) because they are the ones who have more contact and experience using the soil resource. Interviews were conducted directly on the parcel of each farmer where they were asked to mention and show the types of soils they knew, their properties or ways to recognize them, and their abundance and distribution. Farmers were also asked about types of crops they prefer to grow on each kind of soil, type of management, fertilizing, main weeds, and typical problems. The only criterion to select a farmer to be interviewed was the occurrence of their parcel in any of the two main areas of corn production within the municipality [23]. Farmers that only spoke Maya were interviewed with the help of a translator. Based mainly on the predominant responses obtained during the interviews as well as the observations made on the field, the scheme of the Mayan soil classification for this area was built. Once the Mayan soils types were recognized, representative pits for each Mayan soil identified were excavated and profiles were described [25], sampled, analysed, and classified using the WRB classification system [13]. A comparative approach was used to establish similarities and differences between Mayan knowledge and WRB system.

3. Results

The 40 interviewed farmers (4% of the milperos of the municipality) recognized 11 different classes of soils. Most of the interviewed farmers of the study area were older than 40 years, with variable experience on making *milpa* and worked an average area of 1 ha per year (Table 1). Farmers from 60 to 69 years old provided the majority of information about the recognition of the soils, identifying eight classes. Farmers of the three older ranges of age recognized all the types of soils found on the municipality (Table 2). There were only 2 out of 11 classes of soils recognized by all the farmers (*Boxlu'um* and *K'ankab*); the other 9 classes were recognized only by 25% or less of the farmers. The soil properties that Mayan farmers considered to classify their soils are very easy to be observed, these properties included topographic position and colour followed by amount of rock fragments, outcrops, and water retention (Table 3).

Farmers also recognized differences between soils according to the crops they prefer to grow on each class of soil

TABLE 2: Number of soils recognized by the interviewed farmers from Hocabá, Yucatán, México.

Name	Age range							%
	20–29	30–39	40–49	50–59	60–69	70–80	>80	
	$n = 3$	$n = 1$	$n = 8$	$n = 7$	$n = 10$	$n = 8$	$n = 3$	$N = 40$
K'ankab	3	1	8	7	10	8	3	100
Boxlu'um	3	1	8	7	10	8	3	100
Puslu'um				4	3	3		25.0
Ch'ich'lu'um	1		1	2	1			12.5
Muluch buk'tun			2		2		1	12.5
Ek'lu'um	1		1		1			7.5
Ch'och'ol						1		2.5
Tsek'el						1		2.5
Chaltun					1			2.5
Chaklu'um					1			2.5
Haylu'um							1	2.5

TABLE 3: Characteristics of the Mayan soils of the municipality of Hocabá, Yucatán, México.

Name	Visual characteristics				
	Soil color	Topographic position	Superficial rock fragments	Outcrops	Water retention
K'ankab, Chaklu'um	Red	Only on plains	Low or none	Low or none	Good
Haylu'um	Brown or reddish-brown (dark colors)	Base of the mounds	Low amount of fine gravels or none	Hard rock within first 10 cm.	Bad
Chichlu'um	Clear brown or black	On the flat top of the mounds	A lot of fine gravels	Low	Good
Puslu'um	Black	Mounds	Low amount of gravels or none	Hard rock within first 10 cm.	Very bad
Ch'och'ol	Black	Base of the mounds	Piles of cobbles	None	Good
Tsek'el, Yan yan tunichi', Mulu'ch buk'tun	Black	Mounds	High in gravels, stones, and cobbles	High	Bad
Chaltun	Black	Mounds	Low or none	Very high	Bad

and the agronomic problems they perceive (Table 4). They do milpa in any type of soil without discriminating between mound or plain soils. Specifically, they usually prefer sowing varieties of local chilli (Capsicum spp.) in the mound soils called Ch'ich'lu'um. Similarly, vegetable crops and other great diversity of crops are usually sowed in the plain soils free of rock fragments and outcrops called K'ankab. Weeds develop quicker on plain soils because there are more seeds there than in the mound soils and they can germinate at any moment when conditions become favourable. Among the plants that are exclusive or develop quicker on mound soils are Chichibé (Sida acuta Burm.) and Sac kaatzim (Mimosa bahamensis Benth.), while on plain soils Sacchiu (Abutilon permolle (willd.) Sweet) and other grasses, Habín (Piscidia piscipula Sarg.), Tzalam (Lysiloma latisiliquum (L.) Benth), Tsotsk'ab (Mentzelia aspera L.), Kiintal (Desmodium purpureum (Mill.) Fawe), and Tajonal (Viguiera dentata (Cav.)) were also mentioned.

Farmers pointed out the following problems, remarking that they are present with different intensity in each soil class. Generally, mound soils have lower water retention and incidence of gophers, raccoons, and weeds than red soils. The sum of all these factors results, according to farmers, in low yields.

Farmers also use the type of rock associated with soils to classify them. Even more, farmers classify and use those different types of rocks according to their properties and use (Table 5). Farmers recognized five types of rocks; from those, two of them have relevant properties to agriculture, as they appear to have good water retention.

According to the WRB, soil units identified on the plains were Chromic Luvisols (LVcr), characterized by the presence of a Bt horizon and CEC > 24 cmol kg^{-1} through the whole profile; Epileptic Cambisols (lep-CM), Endoleptic Cambisols (len-CM), and Endoskeletic Cambisols (skn-CM) having a Bw horizon but varying in depth and amount of rock fragments; and Lithic Leptosols (li-LP), which are soils up to 10 cm depth (Table 6).

On the mounds, the soil groups were Leptosols (LP) and Calcisols (CL). Both are dark colored (chroma less than 3) and have high organic matter contents from 23 to 50% (Table 7). Both groups have minimal amounts of fine earth due to

TABLE 4: Crops and agronomic problems of the soils of Hocabá, Yucatán, México.

Maya soil name	Preferred crops*	Detected problems
K'ankab	Jamaica (Hibiscus sp.), macal (Xanthosoma yucatanense), jicama (Pachyrhizus erosus), yuca (Manihot esculenta), and sweet potato (Ipomoea batatas).	Weeds grow faster. Tuzas (Dasyprocta mexicana) and raccoons (Procycon spp.) are more frequents.
Haylu'um	Maize (Zea mays), beans (Phaseolus spp., Vigna spp.), and pumpkins (Cucurbita spp.).	As they are shallow soils, maize falls down easily.
Ch'ich'lu'um	Chili pepper (Capsicum spp.) and sometimes sweet potato.	Tuzas (less frequent).
Puslu'um	Maize, beans, and pumpkins	Shallow soils, low water retention.
Ch'och'ol	None	Little surface for planting (too many rock fragments)
Tsek'el, Yan yan tunichi', Mulu'ch buk'tun	Maize, beans, and pumpkins	Little surface for planting (many rock fragments). Presence of weeds.
Chaltun	Maize, beans, and pumpkins	Little surface for planting (too much rock). Very shallow soils.

*Farmers do not have any preference to where to grow Sisal (Agave fourcroydes); they all agreed that the more rock fragments and outcrops in the soil the better the growth of Sisal.

TABLE 5: Types of rock and their characteristics according to the farmers of Hocabá, Yucatán, México.

Maya name	Spanish name*	Use	Characteristics
Saktunich o Sascab	Creta	To build roads	It converts in powder and absorbs much water
Xuxtunich	Roca desgranable	Like sandpaper to cleaning animals, to complete albarradas**	It breaks easily even by hand or when it is burned and absorbs water
Toktunich	Roca fracturable	To build albarradas	Very hard, when it is buried is not broken only turns black and does not absorb water
Sakalbox	Roca soluble	To make hand grinders and albarradas	It is the hardest one and does not absorb water
Haysaltunich	Laja	Pib***, to complete albarradas	Does not absorb water

*Only laja is a common name among farmers, the other 4 names were derived from observations of their properties; **Albarrada is a wall made of rocks; ***Pib means cooking in pits.

the high content of rock fragments. There are three different types of Leptosols: (1) Hyperskeletic Leptosol (LPhsk), having more than 80% by weight of rock fragments; (2) Nudilithic and Lithic Leptosol (LPli), having a depth less than 5 and 10 cm, respectively, and; (3) Calcaric Humic Leptosol (LPcahu), more than 10 cm in depth, high organic matter content, and calcium carbonate content less than 40%. Two types of Calcisols were recognized (CL): (1) Epipetric Skeletic Calcisol (CLptpsk) and (2) Epileptic Skeletic Calcisol (CLlepsk) both of them differing in their depth.

4. Discussion

4.1. Soils Identification. In Hocabá, Yucatán, soil knowledge is being lost because there are less young people interested in making milpa, the main activity that relates farmers to soil. Whit each generation, fewer young people engage in this activity because most of them prefer a salaried work or studying. Moreover, most of the adults younger than 50 years old perform milpa in an intermittent way combining it with a salaried work [26]. The reduction of the available forest area

to make milpa is also a factor in the abandonment of this activity [23]. All these causes are promoting the loss of the traditional soil knowledge; this is supported in this study by the observed relationship between farmers' age and number of soils they recognize. Loss of traditional soil knowledge is occurring similar to other parts of the world [2].

No classification system is static [12] and the Mayan soil classification is not an exception. Synonymies and differences in the descriptions given by the interviewed farmers confirmed this situation. In this study, four cases of possible synonymies were found: Puslu'um and Ch'ich'lu'um, K'ankab and Chaklu'um, Boxlu'um and Eklu'um, and Muluch buk'tun and Tsek'el. The first three cases are reported by [8, 15] as different soil classes.

Soil names and descriptions provided by the farmers were contrasted with those of previous works [15, 17]; all of them presented a similar number of soil classes and the descriptions were highly consistent, although some names varied (Table 8). In those works done at state and regional levels, only three additional soils were reported for the study area (Ya'axhom, Ak'alche, and Kacab), suggesting that Mayan

TABLE 6: Chemical and physical properties of soils located in plains at Hocabá, Yucatán, México.

Soil horizon	Depth cm	Dry color	Structure AS, ASi	PSD Sand %	Clay %	Silt %	TC	Fgr %	Cgr %	CO$_3$ %	pH	OM %	CEC cmol$^+$ kg^{-1}	Exchange cations cmol$^+$ kg^{-1} Ca	Mg	Na	K	BS %
Lithic Leptosol Rhodic																		
A	0–9	5YR 4/4	SBK, VF-M	43.0	19.0	38.0	L	0	0	0.4	7.4	14.0	47.2	23.6	16.6	0.1	2.8	100
Leptic Cambisol																		
A	0–11	5YR 3/3	SBK, F-H	48.8	20.6	31.4	L	0	0	0.1	7.6	15.8	39.7	31.5	2.7	0.3	0.8	89
Bw1	11–23	5YR 3/3	SBK, F-H	48.0	19.6	32.4	L	0	0	0.1	7.6	13.8	47.8	31.0	2.9	0.1	4.4	81
Bw2	23–38	5YR 3/2	ABK, VF-L	46.1	24.5	29.4	L	0	0	0.1	7.5	11.7	29.2	25.8	3.2	0.2	0.4	100
Endoskeletal Cambisol																		
A1	0–9	5YR 3/3	SBK, VF-M	42	26	32	L	0	0	0.1	7.2	17.2	51.2	32.0	21.4	0.1	0.8	100
A2	9–17	5YR 4/4	SBK, VF-L	50	22	28	L	0	0	0.1	7.0	11.0	49.9	27.8	25.3	0.7	0.6	100
Bw1	17–39	5YR 4/6	ABK, VF-L	51	21	28	L	0	0	0.1	6.9	10.0	43.2	30.5	15.9	0.1	0.3	100
Bw2	39–56	5YR 4/6	ABK, VF-L	58	16	26	CL	0	20	0.1	7.0	6.9	44.1	29.2	18.8	0	0.2	100
C	56–100	5YR 4/6	ABK, VF-L	47	19	33	L	23	67	0.1	7.2	7.2	29.2	20.5	12.2	0.1	0.3	100
Endoleptic Cambisol																		
A1	0–4	5YR 4/3	GR, F-H	45.1	21.6	33.3	L	5	0	0.5	7.4	18.8	46.0	39.0	3.6	0.2	2.3	100
A2	4–22	5YR 4/4	SBK, VF-M	49.0	19.6	31.4	L	0	0	0.1	7.3	12.3	33.2	38.5	2.3	0.1	1.0	100
Bw1	22–33	5YR 3/6	ABK, VF-M	54.9	15.7	25.5	CL	5	0	0.1	7.3	8.1	29.0	30.0	2.5	0.1	0.4	100
Bw2	33–55	5YR 4/3	ABK, VF-L	61.8	16.5	22.5	CL	3	0	0.4	7.5	9.1	25.5	19.5	1.2	0.1	0.3	100
Bw3	55–75	5YR 4/3	ABK, VF-L	51.0	17.5	32.5	L	2	0	1.5	7.5	5.5	34.8	35.5	1.0	0.1	0.2	100
Haplic Luvisol Rhodic																		
A1	0–6	5YR 2.5/3	GR, F-H	47.1	25.5	27.5	L	0	0	0.1	7.4	11.4	36.2	21.6	3.6	0.1	1.1	73
A2	6–20	5YR 3/3	SBK, F-M	48.0	26.0	26.0	L	0	0	0.1	6.6	8.1	25.5	12.6	2.5	0.3	0.3	61
Bt1	20–45	2.5YR 3/6	ABK, M-L	40.2	37.7	22.1	CL	0	0	0.1	6.7	3.8	24.9	15.3	0.9	0.2	0.3	67
Bt2	45–85	2.5YR 3/6	ABK, M-L	30.4	47.7	22.5	C	0	0	0.1	7.0	3.1	19.0	16.2	1.8	0.4	0.3	97
Bt3	85–109	2.5YR 4/6	ABK, M-L	36.8	39.2	24.0	CL	0	0	0.1	7.1	3.3	15.1	16.2	1.0	0.3	0.3	100
Bt4	109–150	2.5YR 4/6	ABK, M-L	34.3	37.3	28.4	CL	0	0	0.1	7.2	2.7	10.4	14.8	3.2	0.3	0.3	100

PSD: particle size distribution. AS: aggregates shape (GR: granular, SBK: subangular blocky, ABK: angular blocky). ASi: aggregates size (VF: very fine, F: fine, M: medium). Stability (L: low, M: moderate, H: high). TC: textural class. Fgr: fine gravels (<2 cm), Cgr: coarse gravels (>2 cm); OM: organic matter, CEC: cation exchange capacity, BS: base saturation. L: loam; CL: clay loam; C: clay; TC: textural class.

TABLE 7: Properties of soils located in mounds at Hocabá, Yucatán, México.

Soil horizon	Depth cm	Dry color	Structure AS, ASi	PSD Sand %	Clay %	Silt %	TC	Gr %	St %	CO₃ %	pH	OM %	CEC cmol⁺ kg⁻¹	Exchange Cations cmol⁺ kg⁻ Ca	Mg	Na	K	BS %
Hyperskeletic Leptosol																		
A	0–10	7.5YR 2.5/1	GR, VF-M	70.6	15.7	13.7	SL	50	30	12	8.0	45.0	66.2	54.0	1.8	0.1	3.3	89
Ak/C	10–45	7.5YR 3/1	GR, VF-L	58.8	17.6	23.5	SL	67	25	4	8.0	36.4	19	19.2	5.4	0.4	3.1	100
Rendzic Hyperskeletic Leptosol																		
A	0–7	10YR 2.5/1	GR, F-M	63.7	15.7	20.6	SL	51	40	31	7.8	34.5	54.1	38.4	12.6	0.2	0.9	100
A/C	7–23	10YR 2.5/1	GR, VF-M	71.6	13.7	14.7	SL	51	45	43	7.7	28.6	24.4	39.0	27.0	0.2	1.0	100
Leptic Skeletal Calcisol																		
A	0–1	7.5YR 2.5/1	GR, VF-L	55.0	20.0	25.0	SCL	30	10	41	8.0	30.6	40.7	38.6	0.7	0.2	2.8	100
Ak/Ck	1–15	7.5YR 4/3	GR, VF-M	62.7	13.7	23.5	SL	36	17	34	8.1	19.4	32.2	35.3	1.1	ND	0.6	100
Ck/Ak	15–50	7.5YR 4/3	SBK, VF-L	62.7	15.7	21.6	SL	34	20	42	8.0	19.5	26.7	35.3	0.9	0.5	0.6	100
Epipetric Skeletic Calcisol																		
Ak	0–4	10YR 3/3	GR, F-H	49.0	19.6	31.4	L	22	0	30	8.0	21.3	35.5	25.4	4.0	0.3	1.8	100
Bk1	4–20	10YR 4/3	SBK, F-M	53.9	31.4	14.7	SCL	25	15	35	8.2	11.3	16.7	14.9	1.1	0.3	1.0	100
Bk2	20–35	10YR 4/1	SBK, VF-L	52.9	21.6	25.6	SCL	21	20	36	8.3	12.9	18.2	16.5	1.1	0.3	0.8	100
Ckm	35–40									>50								
IIAk	40–60	10YR 5/1	GR, VF-L	59.8	19.6	20.6	SL	22	15	37	8.6	10.6	14.4	12.1	0.4	0.2	0.5	100
Humic-Hyperskeletal Leptosol																		
Ak1	0–2	7.5YR 2.5/1	GR, VF-M	69.0	11.0	20.0	SL	31	0	36	7.5	49.9	59.9	32.8	34.1	0.1	1.1	100
Ak2	2–22	7.5YR 3.5/1	GR, VF-L	67.0	13.0	20.0	SL	0	90	44	7.7	42.6	37.6	41.8	22.3	0.1	1.2	100
C/A	22–80	7.5YR 4/1	SBK, VF-L	71.0	12.0	17.0	SL	0	95	47	7.8	18.0	28.6	21.1	8.4	0.1	0.2	100

PSD: particle size distribution. AS: aggregates shape (GR: granular, SBK: subangular blocky, ABK: angular blocky). ASi: aggregates size (VF: very fine, F: fine, M: medium). Stability (L: low, M: moderate, H: high). L: loam; CL: clay loam; C: clay; TC: textural class. Gr: gravels, St: stones; OM: organic matter, CEC: cation exchange capacity, BS: base saturation.

TABLE 8: Comparison between the descriptions of the soils found in the municipality of Hocabá, Yucatán, México, and those presented in others studies.

Maya name*	Aguilera (1958) [31]	Duch (1988, 1991) [15, 16]	This study
K'ankab	Light red, deep soils	Reddish brown color. Found it in plain terrains	Red soil
Boxlu'um	—	Black color	Black or dark soil
Puslu'um	—	Remarkable less amount of rock fragments than Boxlu'um	Black soil, soft, with no rock fragments. It dries out quickly
Ch'ich'lu'um	—	Very hard and gravel aggregates	Soil with abundant amount of gravels
Muluch buk'tun	—	—	Soil with a lot of rock fragments
Ek'lu'um	Organic soil on calcareous rock (Ek'lu'um Tsek'el)	—	Black soil
Ch'och'ol	Soil with calcareous rocks along profile	Soil with abundant rock fragments on the surface	Soil under piled rock fragments
Tsek'el	Calcareous rock with a thin layer of soil	Shallow soil with abundant rock fragments	Soil with abundant rock fragments
Chaltun	Soil over laja rock	Soil with calcareous armour exposed	Black shallow soil. with cracked or holed rock
Chaklu'um	—	Soil more red than K'ankab	Dark red soil
Haylu'um	—	—	Very shallow soil. Less than 10 cm depth

*Writing of the Maya names is according to the Porrúa dictionary [32].

knowledge of soils is similar in the whole state. Typically, farmer classifications are highly variable or they have little consistency from region to region [8, 12, 27]; however, it seems that the Mayan soil knowledge is quite homogeneous, even at regional level [8]. This homogeneity can help to facilitate its systematization.

On the other hand, this apparent homogeneity could also indicate a loss of the soil knowledge. This statement is supported by the results of this work in which only two classes of soils were recognized by all the interviewed farmers (Boxlu'um and K'ankab) and these two terms were used to refer to the rest of the soils when they did not know them.

Most of the classification systems reflect the priorities of who propose them [28]. Characteristics that Mayan farmers use to classify soils are mainly visual and intimately related to their agricultural activities.

Farmers recognized colour differences among soils; however, it was observed that there are different tones that farmers do not consider as distinctive elements. A particular case is the Mayan term Box that means black or dark. Many farmers referred the soil called Chichlu'um that is usually light brown to dark brown as simply Boxlu'um. However, apparently this is related to the absence of a Mayan word to designate brown color, although some farmers used the Spanish term "achocolatado" (colored as chocolate) to refer to this colour. Another special case is the soil called K'ankab whose common translation is "yellow place at the bottom" in attention to subsuperficial soils horizons [8]. The Mayan word Chak means red; thus, Chaklu'um are red soils. These soils are darker than other red top soils such as K'ankab.

Soil depth was only an important characteristic for farmers to differentiate between shallow (soils < 10 cm) and deep soils (soils > 10 cm). Farmers recognized these soils empirically during sowing when they insert their sowing stick, by observation of aerial roots on maize plants, or when maize plants fall down due to the wind action because roots lack deep anchorage. Farmers affirmed that mound soils are not very deep but they pointed out that roots always find cracks in the rocks to continue growing. Shallow soils are called Haylu'um in both plains and mounds and correspond to Nudilithic Leptosols. Farmers judge the depth of K'ankab by its surface color, the darker the soil is the shallower it is, and the lighter the soil is the deeper it is. Bautista and Zinck [8] reported these differences in deepness for K'ankab soils.

The microtopographic position, superficial amount of rock fragments, and outcrops are three characteristics that the farmers always consider together. We found five rock types with different uses and it is possible that they influence soil characteristics and soil genesis [14, 26]. For example, soils called Ch'och'ol can only be found at the base of the mounds under stone accumulations (Hyperskeletic Leptosols), while soils designated Chaltun are almost always near the mounds but present bedrock very near to the surface (Nudilithic Leptosols).

Water retention is a characteristic that many farmers recognized but their comments relative to this property were inconsistent. For many farmers soils that do not retain water (they dried out first) were those on the mounds, while others assured that it was the soils on the plains. This disagreement can be due to the farmers not considering the variability in the amount of rock fragments or the depth of the soils as factors that determine the water retention. In fact, this characteristic was only relevant to recognize the soil called Puslu'um, which farmers consistently designated as the soil that dried out first.

Other studies have found that soil texture is an important property for local classifications [11, 29], but it seems that this is not the case of the Mayan classification, because none of their agricultural practices requires a direct physical contact with the deeper layers of soil.

4.2. Soils Uses. Preferences for growing crops were mainly linked to the more availability of workable surface for sowing in plain soils. Farmers like to grow most of their crops on the *K'ankab* soils, because the absence of rock fragments and outcrops makes field operations easier and quicker. The one clear exception is the soil called *Ch'ich'lu'um*, in which farmers prefer to sow peppers and sweet potato, arguing that they only grow well in that class of soil. It is possible that a nutritional reason exists to explain the best development of those crops in that class of soil, but this remains to be confirmed with soil fertility analyses.

Although some farmers said that with good rain the mound soils give better production, most of them assured that crops on mound soils as well as plain soils grow well if it has rained well and on time. On the other hand, when rain is not good, some farmers said that the mound soils produce higher yields while others argue that the plain soils do. The reason for these inconsistencies could be the depth of the soils and the amount of rock fragments and outcrops. Comparatively, shallow soils can store less water than deep soils, but amount of rocks and stones on mound soils help to conserve the humidity better than in shallow plain soils.

Soil water content is a property associated by farmers with the presence of weeds. In this regard, they said that they have many problems to control weeds because as soon as it rains, weeds appear in the soil. They also pointed out differences in weed composition and abundance between mound and plain soils. A farmer making *milpa* in the same place for five years in a row said that every year he had to use more herbicide and dedicate more hours to remove weeds than the earlier year.

In contrast to what the authors of [16, 17] found, the farmers interviewed in this study did not use any terms to refer to soil fertility. This is perhaps because these authors did their studies using a deeper anthropologic approach. The author of [16] developed the hierarchical classification of the soils of the Puuc region of the state of Yucatan and outlined a classification departed from a linguistic point of view, grouping the soil classes according to the meaning of the Mayan names as well as some management aspects. Such research is important to obtain information that may no longer exist among the contemporary inhabitants of a region.

5. Comparison and Systematization

Our results suggest that Mayan soils knowledge is given at three levels (Figure 2). In level one, topographic position mound (*Muluch*) or plain (*K'ankabal*). Mound soils are dark, generally black, grey, or brown, while plain soils are red to red-brown which makes a first division among soil classes resulting in a general designation for soils according to their color. Dark soils are designated as *Box'luum* and red soils as *K'ankab*. The prefix *Box* means literally black but it is used to refer to all classes of dark colours. On the other hand, the prefix *K'an* means literally yellow but is used to designate light red soils on plains. Level two is almost exclusive for

mound soils since in the plains the amount of rock fragments and outcrops are nearly absent. However, in some cases, red mound soils, having abundant amount of rock fragments and outcrops, were observed and recognized by farmers as *Ch'ich-K'ankab* or *Tsek'el-K'ankab*. This was the only case where farmers used a compound name mixing two single names.

In level three, variations of soils from the second level were recognized according to their association with specific topographic position. Here, there were two subdivisions: (1) soil names ending with the Mayan word *lu'um*, which means soil, for example, *Ch'ich'lu'um*—soil with gravels—(Hsk-LP) or *Haylu'um*—very shallow soil—(Nu-LP) and (2) soils that were designated according to the specific microtopographic positions on which they occur, for example, *Ch'och'ol*—soil under piled rock fragments—(Hsk-LP), *Tsek'el*—soil among rock fragments—(Nu-LP), *Chaltun*—soil between outcrops—(Nu-LP). At this level, soils in mounds were recognized as *K'ankab, Haylu'um Ch'ich'lu'um, Tsek'el, Chaltun, Puslu'um, Ch'och'ol,* and so on. In the case of the plain soils, farmers only recognized two variants: *K'ankab* and *Hay'lu'um*. In both cases when farmers were not sure of the specific name of the soil, they designate as *Boxluu'm* to all soils in mounds and *K'ankab* to all soils in plains.

Following this scheme, it can be seen that level one (mound soils and plain soils) is the most studied level so far [5, 14, 15]. It is in the second and third levels that research is needed. It is in those levels where the participation of the farmers is important in order to better understand each one of the elements of the landscape and topography that they recognize and use to identify soils.

Mayan soil types and WRB units cannot be directly related to each other because these systems share few diagnostic properties and assign them different relative importance [8]. Many soils identified by farmers relate with more than one WRB group of soil and vice versa; in these cases, no direct relationship between both classification systems is possible (Table 9).

People's understanding of soils constitutes a complex knowledge system, with some categories similar or complementary to those used by modern soil science [8, 30]. For example, even though hierarchical levels of the WRB system are based on qualitative and quantitative data, they use qualifiers to distinguish soils at secondary levels. Some of those characteristics, that is, gravels or rock fragments percentage, are related to the Mayan approach. For instance, amount of rock fragments is a very important property for building hierarchal levels in the Mayan soil nomenclature (e.g., *Ch'och'ol*) as well as in the WRB classification at the qualifier level (e.g., Hyperskeletic and Skeletic Leptosols).

6. Conclusions

In Hocabá, Yucatán farmers distinguished two main groups of soils: *K'ankab* or soils of plains and *Boxlu'um* or soils of mounds. *K'ankab* is a group of red soils with two variants (*K'ankab* and *Haylu'um*), whereas *Boxlu'um* is a group of dark soils with five variants (*Tsek'el, Ch'ich'lu'um, Chaltun, Puslu'um,* and *Ch'och'ol*). Soils on the plains were identified as Leptosols, Cambisols, Cambisols, and Luvisols. Soils

Level

0   Mayan soil types

1   Soils in plains (*K'ankabal*) (*K'ankab*)   Soils in mounds (Muluch) (*Boxluu'm*)

2   Mounds dominated by rock fragments   Mounds dominated by outcrops

3   Deep soils (less than 10 cm) (color: red or reddish brown)   Shallow soils (less than 10 cm) (color: red or reddish brown)   Soils with abundant fine gravels (color: light brown, grey, or black)   Soils free of rock fragments (color: grey or black)   Soils in between outcrops (color: black)   Soils in piles of stones and cobbles (color: black)   Soils contained in rock holes (color: black)   Shadow soils (less than 10 cm) (color: black)

K'ankab   *Haylu'um*   *Ch'ich'lu'um*   *Puslu'um*   *Tsek'el*   *Ch'och'ol*   *Chaltun*

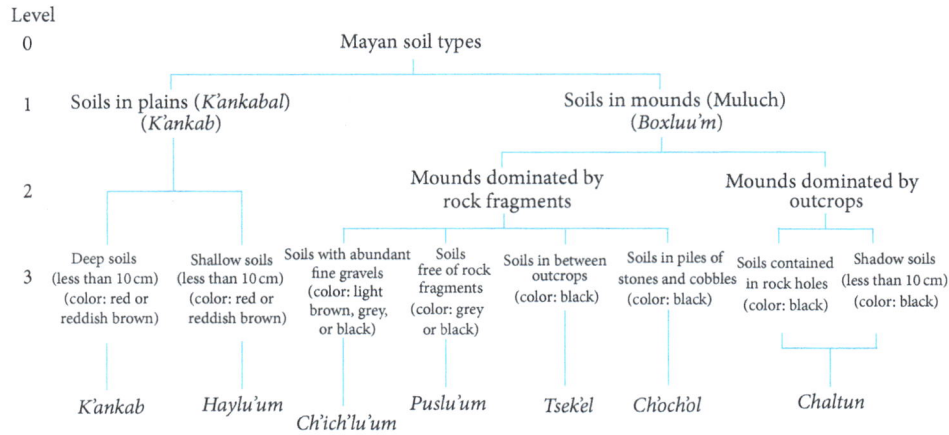

FIGURE 2: Mayan soil types in Hocabá, Yucatán.

TABLE 9: Relationship between WRB soil units and Mayan soil types.

WRB soil group	Dominant kind of rock	Topographic position	Mayan soil class
Chromic Luvisols Eutric Cambisols Mollic Leptosol	Fracturable	Any part of the plains	*K'ankab*
Nudilithic Leptosol	Fracturable	Base of the mounds	*Haylu'um*
Hyperskeletic Leptosol			*Ch'ich'lu'um*
Haplic Calcisol Hyperskeletic Leptosol	Graintable and powderable	Flat tops of the mounds	*Puslu'um* *Ch'ich'lu'um*
Petrocalcic Calcisol	Fracturable and grainable	Top of the mounds	*Puslu'um*
Nudilithic Leptosol Hyperskeletic Leptosol	Fracturable Fracturable Fracturable	Top of the mounds Sides of the mounds Base of the mounds	*Boxlu'um* *Tsek'el* *Ch'och'ol*
Rendzic Leptosol Nudilithic Leptosol	Fracturable	Top of the mounds Sides of the mounds	*Boxlu'um* *Tsek'el*
Nudilithic Leptosol Lithic Leptosol Nudilithic Leptosol	Soluble Soluble Fracturable	Top of the mounds Sides of the mounds Base of the mounds	*Chaltun* *Boxlu'um* *Haylu'um*

identified in mounds were Leptosols and Calcisols. Mayan soil types and WRB groups are complementary; they should be used together in order to improve both soil classifications, to help transference of agricultural technologies, and make soil management decisions. Soil characteristics that should be considered for a local soil classification system are topographic position (plain or mound), colour, amount of rock fragments and outcrops, and soil depth or effective rooting depth.

Acknowledgments

Acknowledgments are due to the Science and Technology National Council (CONACyT) for the scholarship granted to the first author for his master degree studies; to CONACyT project: *Base de datos para la península de Yucatán, incluyendo nomenclatura maya y FAO*, Convenio: R31624-B; to the American Institute for Global Change Research (IAI), project: Biogeochemical cycles under land use change in the semiarid Americas; to Dr. Steve Gliessman, Dr. Robert Graham, and M.Sc. Arturo Caamal Maldonado for their comments and corrections to the paper; to Wendy Huchim and Miguel Huicab for their collaboration on some of the tables and figures. This paper is dedicated to the memory of my friend Mr. Pedro Canché farmer from Hocabá who with his wide knowledge of soils, roads, and people helped me as guide and Mayan-Spanish translator and in contacting other farmers.

References

[1] A. M. G. A. Winklerprins, "Local soil knowledge: a tool for sustainable land management," *Society and Natural Resources*, vol. 12, no. 2, pp. 151–161, 1999.

[2] P. V. Krasilnikov and J. A. Tabor, "Perspectives on utilitarian ethnopedology," *Geoderma*, vol. 111, no. 3-4, pp. 197–215, 2003.

[3] N. Barrera-Bassols and J. A. Zinck, "Ethnopedology: a worldwide view on the soil knowledge of local people," *Geoderma*, vol. 111, no. 3-4, pp. 171–195, 2003.

[4] N. Barrera-Bassols, "Etnoedafología purhépecha," *México Indígena*, vol. 6, no. 24, pp. 47–52, 1988.

[5] F. Bautista, J. Jiménez, J. Navarro, A. Manu, and R. Lozano, "Microrelieve y color del suelo como propiedades de diagnóstico en Leptosoles cársticos," *Terra*, vol. 21, no. 1, pp. 1–12, 2003.

[6] B. J. Williams, "Aztec soil science," *Boletín del Instituto de Geografía*, vol. 7, pp. 115–120, 1975.

[7] B. J. Williams and C. A. Ortiz-Solorio, "Middle American folk soil taxonomy," *Annals, Association of American Geographers*, vol. 71, no. 3, pp. 335–358, 1981.

[8] F. Bautista and J. A. Zinck, "Construction of an Yucatec Maya soil classification and comparison with the WRB framework," *Journal of Ethnobiology and Ethnomedicine*, vol. 6, article 7, 2010.

[9] V. Toledo, "Indigenous knowledge on soils: an ethnoecological conceptualization," in *Ethnopedology in A Worldwide Perspectives: An Annotated Bibliography*, N. Barrera-Bassols and J. A. Zinck, Eds., International Institute for Aerospace Survey and Earth Science, Enschede, The Netherlands, 2000.

[10] A. J. Tabor, "Soil classification systems," in *Soil of Arid Regions of the U.S. and Israel*, 1997, http://cals.arizona.edu/oals/soils/classifsystems.html.

[11] J. A. Sandor and L. Furbee, "Indigenous knowledge and classification of soils in the Andes of Southern Peru," *Soil Science Society of America Journal*, vol. 60, no. 5, pp. 1502–1512, 1996.

[12] A. J. Tabor, "Soil surveys and indigenous soil classification," *Indigenous Knowledge and Development Monitor*, vol. 1, no. 1, 1993.

[13] IUSS-ISRIC-FAO, "2006 World reference base for soil resources 2006: A framework for international classification, correlation and communication," World Soil Resour Report 103, FAO, Rome, Italy.

[14] F. Bautista, H. Estrada-Medina, J. Jiménez-Osornio, and J. González-Iturbe, "Relación entre relieve y suelos en zonas cársticas," *Terra*, vol. 22, no. 3, pp. 243–254, 2004.

[15] G. J. Duch, *La conformación territorial del estado de Yucatán—Los componentes del medio físico*, Centro Regional de la Península de Yucatán CRUPY, Universidad Autónoma de Chapingo, México, Mexico, 1988.

[16] G. J. Duch, *Fisiografía del estado de Yucatán—su relación con la agricultura*, Centro Regional de la Península de Yucatán CRUPY, Universidad Autónoma de Chapingo, México, Mexico, 1991.

[17] P. N. Dunning, "Soils and vegetation," in *Lords of the Hills: Ancient Maya Settement in the Puuc Region, Yucatán, México*, vol. 15 of *Monographs in World Archaeology*, Prehistory Press, Madison, Wis, USA, 1992.

[18] J. Lugo, "Geomorfología," in *Atlas de Procesos Territoriales de Yucatán*, P. P. Chico, Ed., Universidad Autónoma de Yucatán, Yucatán, México, 1999.

[19] Gobierno Constitucional del Estado de Yucatán, "Programa de desarrollo regional de la zona Henequenera de Yucatán 1992–1994," México, 1992.

[20] INEGI, "Anuario estadístico del estado de Yucatán," Gobierno del estado de Yucatán, 1995.

[21] R. Orellana, "Evaluación Climática (Climatología de la Península de Yucatán)," in *Atlas de Procesos Territoriales de Yucatán*, A. García de Fuentes, C. Y. Ordóñez, and P. Chico Ponce de León, Eds., pp. 162–182, Facultad de Arquitectura, Universidad Autónoma de Yucatán, Mérida, Mexico, 2000.

[22] S. J. Flores and C. I. Espejel, "Tipos de vegetación de la península de Yucatán," Fascículo 3, Sostenibilidad Maya. Universidad Autónoma de Yucatán, México, Mexico, 1994.

[23] L. Cano, *Cambio del uso del suelo en el municipio de Hocabá, Yucatán, México [Tesis de maestría en Manejo y Conservación de Recursos Naturales Tropicales]*, Facultad de Medicina Veterinaria y Zootecnia, Universidad Autónoma de Yucatán, 2000.

[24] INEGI, "Instituto Nacional de Estadística, Geografía e Informática," Carta Geológica 1:250000, Secretaría de Programación y Presupuesto, México, Mexico, 1983.

[25] C. Siebe, R. Janh, and K. Stahr, "Manual para la descripción y evaluación ecológica de suelos en el campo.," Publicación Especial 4, Sociedad Mexicana de la Ciencia del Suelo, A. C. Edo. de México, México, Mexico, 1996.

[26] F. Bautista, J. Garcia, and A. Mizrahi, "Diagnóstico campesino de la situación agrícola en Hocabá, Yucatán," *Terra Latinoamericana*, vol. 23, no. 4, 2005.

[27] E. Habarurema and K. G. Steiner, "Soil suitability classification by farmers in southern Rwanda," *Geoderma*, vol. 75, no. 1-2, pp. 75–87, 1997.

[28] M. Corbeels, A. Shiferaw, and M. Haile, "Farmers' knowledge of soil fertility and local managament strategies in Tigray, Ethiopia," Managing Africa's Soils 10, IIED, London, UK, 2000.

[29] L. Pool-Novelo, E. Cervantes-Trejo, and S. Mesa-Díaz, "La clasificación *Tsotsil* de soils en el paisaje cárstico de la subregión San Cristobal de las casas, Chiapas, México," *Terra*, vol. 9, no. 1, pp. 11–23, 1991.

[30] N. Barrera-Bassols and A. J. Zinck, *Ethnopedology in a Worldwide Perspectives: An Annotated Bibliography*, International Institute for Aerospace Survey and Earth Science, Enschede, The Netherlands, 2000.

[31] N. Aguilera Herrera, "Suelos," in *Los recursos naturales del Sureste y su aprovechamiento*, E. Beltrán, Ed., Instituto Mexicano de Recursos Naturales Renovables, México City, Mexico, 1958.

[32] Porrúa, *Diccionario Maya*, Editorial Porrúa, México, Mexico, 2 edition, 1991.

21

The Variations in the Soil Enzyme Activity, Protein Expression, Microbial Biomass, and Community Structure of Soil Contaminated by Heavy Metals

Xi Zhang,[1] Feng Li,[1,2] Tingting Liu,[1] Chen Xu,[1] Dechao Duan,[1] Cheng Peng,[1] Shenhai Zhu,[1] and Jiyan Shi[1]

[1] *Institute of Environmental Science and Technology, Zhejiang University, Hangzhou, Zhejiang 310058, China*
[2] *College of Materials and Environmental Engineering, Hangzhou Dianzi University, Hangzhou, Zhejiang 310018, China*

Correspondence should be addressed to Jiyan Shi; shijiyan@zju.edu.cn

Academic Editors: W. Peijnenburg, D. G. Strawn, and J. Thioulouse

Heavy metals have adverse effects on soil ecology. Given the toxicity of heavy metals, there is an urgent need to select an appropriate indicator that will aid in monitoring their biological effects on soil ecosystems. By combining different monitoring techniques for various aspects of microbiology, the effects of heavy metals on soil microorganisms near a smelter were studied. Our goal was to determine whether proteins could be a proper indicator for soil pollution. This study demonstrated that the activities of acid phosphatase and dehydrogenase, as well as the levels of microbial biomass carbon and proteins, were negatively affected by heavy metals. In addition, significantly negative correlations were observed between these microbial indicators and heavy metals. Denaturing gradient gel electrophoresis analysis was used in this study to demonstrate that heavy metals also have a significantly negative effect on soil microbial diversity and community structure. The soil protein expression was similar across different soils, but a large quantity of presumably low molecular weight protein was observed only in contaminated soil. Based on this research, we determined that the soil protein concentration was more sensitive to heavy metals than acid phosphatase, dehydrogenase, or microbial biomass carbon because it was more dramatically decreased in the contaminated soils. Therefore, we concluded that the soil protein level has great potential to be a sensitive indicator of soil contamination. Further research is essential, particularly to identify the low molecular weight protein that only appears in contaminated soil, so that further insight can be gained into the responses of microbes to heavy metals.

1. Introduction

Heavy metals (e.g., Cd, Cr, Cu, Pb, and Zn) can be introduced into soils from several sources, such as waste from mines and smelters, atmospheric deposition, animal manures, sewage sludge, and, in some circumstances, inorganic fertilizer. Once these elements enter the soil, they can remain for extremely long periods and are difficult to remove [1]. Heavy metals are known to be toxic to most organisms when present in excessive concentrations [2]. Given the toxicity of these elements, there is an urgent need to monitor their biological effects on soil ecosystems.

Soil microbes play vital roles in the recycling of plant nutrients, the maintenance of soil structure, and the detoxification of poisonous chemicals [3]. Therefore, the diversity of microbial communities and the activity of microorganisms are important indicators of soil quality. Alterations in the composition of microbial communities have often been considered effective indicators for soil contamination [4]. Many reports have shown that either long-term or short-term exposure of soil to heavy metals results in a reduction of microbial diversity and microorganism activities [3, 5].

Although soil enzyme activity and microbial biomass are the most tested parameters in soil quality monitoring, these parameters still have many limitations [2, 6]. They need to be supplemented by other valid microbial indicators, working together to fully assess soil contamination. The recent

The Variations in the Soil Enzyme Activity, Protein Expression, Microbial Biomass, and Community Structure of Soil
Contaminated by Heavy Metals

195

emergence of soil proteomics could provide a new perspective on soil microbial activities [7]. Proteins are the chief actors in organism metabolism processes, and they fall into a variety of categories; for example, enzymes are proteins with a catalytic function. Proteins work together to complete various biological functions. Soil proteins primarily come from microorganisms as well as from flora and fauna tissues. The structure of the microbial community, biomass, and microbial status often determine the synthesis levels of soil proteins. Therefore, any change to the microbial composition, which can be caused by environmental changes such as contamination, could reflect in the composition and expressions of soil proteins [8]. Soil proteomics studies aim to investigate the spatial and temporal changes of proteins extracted from soils because proteins are the functional components of the microbial genomic expression products. Such studies more conclusively determine the ecological functions of soil microbes and their roles in soil pollutant transportation and transformation [9–11]. Therefore, proteomics analyses have great potential in soil pollution assessment.

Currently, soil proteomics are predominantly used in studies of the biogeochemical cycle [12–14] and the rhizosphere soil microecosystem [15, 16]. However, there is little literature on the monitoring of contaminated soil, and the existing studies were all carried out under controlled conditions [8, 17]. Unlike a controlled laboratory, natural soil is a complicated ecosystem, and various factors combine to influence the transportation and transformation of pollutants as well as the structure of the microorganism community. Consequently, we were interested in examining the modification of soil proteins under heavy metals stress and determining whether proteins can be a useful indicator of natural soil pollution.

In this study, a typical heavy metal smelter was selected. We intended to study the effects of heavy metals on soil microbes by combining different biochemical analysis techniques. Three compounds were used for heavy metal extraction, $CaCl_2$, EDTA, and DTPA, to assess the bioavailability of trace metals in soil [18, 19]. Soil microbiological and biochemical properties (including enzyme activities, protein contents, and microbial biomass carbon levels) were measured to evaluate the effects of heavy metals. A soil microbial community analysis was carried out using molecular biology techniques. Partial 16S rDNA genes were amplified from soil microorganism community DNA by the polymerase chain reaction (PCR), using primers that bind to evolutionarily conserved regions within the bacteria and actinomycete genes. The diverse PCR amplified products were transformed to genetic fingerprints using denaturing gradient gel electrophoresis (DGGE) [3, 20]. The soil protein composition was analyzed by SDS-polyacrylamide gel electrophoresis (SDS-PAGE) [7].

2. Materials and Methods

2.1. Soil Research Region and Soil Sampling.
The soil area selected for study is downwind of a copper-zinc smelter in Fuyang County, Zhejiang Province, China. The soil in the area has been contaminated by heavy metals, primarily Cu and Zn. This area was formerly used to cultivate crops. Soil samples were taken in April 2012 at nine points along a pollution gradient of Cu and Zn (Table 1). At each point, three replicated samples were taken, and each sample was composed of three soil cores 5 cm in diameter and 20 cm in depth (approximately 500 g). Samples were taken randomly from different areas at each site. Each replicate soil sample was homogenized thoroughly. The site 2,000 m away from the smelter was chosen as the reference site, and the soil collected from it was marked "CK."

2.2. Sample Treatment and Analysis.
Fresh soil samples were sieved ($\Phi \leq 2$ mm) by a nylon sieve to remove stones, large pieces of plant material, and soil animals. Portions of the samples were kept moist in the dark at 4°C to determine the microbial biomass and then stored at −70°C to extract soil DNA. Other portions were freeze-dried and passed through a 0.25 mm nylon sieve prior to soil protein extraction. The remaining soil was air-dried under cool conditions (approximately 20°C), sieved through a 0.25 mm nylon sieve, and stored at 4°C to analyze the enzyme activity, soil pH, organic matter, total heavy metals, and bioavailable heavy metals.

The soil pH was determined with distilled water in a ratio (soil : water) of 1 : 2.5 (w/v) using a pH meter (Orion 5-Star, Thermo). The soil was heated at 105°C for 12 h and then combusted at 550°C overnight to measure the organic matter content [21]. The samples were digested using the method described by Tang et al. [22]. The dry sample (0.2 g) was weighed and digested with a mixture of nitric acid (HNO_3), hydrofluoric acid (HF), and perchloric acid ($HClO_4$). The total concentrations of Cd, Cr, Cu, Pb, and Zn were measured by flame atomic absorption spectrometry (MKIL-M6, Thermo).

The soil bioavailable heavy metal fractions were estimated by extraction with 0.01 M $CaCl_2$ (soil : solution, 1 : 5), 0.05 M EDTA (soil : solution, 1 : 5), and DTPA (0.005 M DTPA, 0.01 M $CaCl_2$, and 0.1 M TEA at pH 7.3) (soil : solution, 1 : 2.5). The soil suspensions were shaken at 200 rpm for 2 h, centrifuged at 4,000 ×g for 10 min, and filtered [23–25]. The Cd, Cr, Cu, Pb, and Zn contents of the bioavailable fraction in each filtrate were determined by flame atomic absorption spectrometry.

2.3. Enzyme Activity Determination.
We chose acid phosphatase and dehydrogenase as the subjects for our investigation of enzyme changes under heavy metal stress because they are closely related to the soil phosphorous cycle and to microbial metabolic activities [26, 27]. To determine acid phosphatase activity, dry soil was incubated in disodium phenyl phosphate for 2 h, and the results are expressed as the micrograms of P_2O_5 released per 100 g of dry soil using 4-aminoantipyrine colorimetry at 510 nm. To measure dehydrogenase activity, dry soil was incubated in triphenyltetrazolium chloride (TTC) for 24 h, and the results are expressed as the micrograms of triphenyl formazan (TPF) released per 1 g of dry soil. The concentration of TPF in the extract was measured using a colorimeter at 485 nm [27, 28].

TABLE 1: Selected soil physicochemical properties and the soil concentrations of heavy metals.

| Soil number | CK | | | Distance from the smelter | | | | | | | | |
| | | | | 800 m | | | 400 m | | | 200 m | | |
	1	2	3	1	2	3	1	2	3	1	2	3
pH	5.94	5.76	5.72	7.57	7.34	6.23	6.61	8.00	6.56	6.52	6.95	7.10
Organic matter (%)	4.76	5.36	4.69	4.06	4.64	3.23	6.84	5.42	6.21	7.65	6.78	6.13
Cu (mg·kg^{-1})												
$CaCl_2$	0.10	0.05	0.05	0.17	0.28	0.17	0.68	0.68	0.33	1.54	0.81	0.65
DTPA	4.45	4.78	5.16	33.41	48.67	21.59	124.29	85.39	102.08	138.29	129.54	110.12
EDTA	8.13	8.05	8.72	83.51	107.06	36.40	200.03	143.65	154.78	293.92	238.30	200.66
Total	**52**	**46**	**46**	**314**	**414**	**175**	**711**	**746**	**511**	**1147**	**794**	**624**
Zn (mg·kg^{-1})												
$CaCl_2$	0.96	0.93	1.14	0.46	0.44	5.93	12.32	8.44	8.74	46.02	20.69	13.48
DTPA	3.44	3.75	4.46	18.36	25.56	17.83	55.86	62.40	41.71	179.49	148.50	139.94
EDTA	5.33	5.69	6.77	43.45	63.33	30.84	75.43	121.80	63.59	680.38	394.31	468.81
Total	**116**	**113**	**111**	**260**	**364**	**187**	**599**	**596**	**545**	**2472**	**1456**	**1421**
Pb (mg·kg^{-1})												
$CaCl_2$	0.60	0.65	0.57	0.59	0.59	0.57	0.58	0.59	0.63	0.70	0.61	0.60
DTPA	3.38	4.21	4.25	6.46	13.06	5.58	25.79	89.79	23.99	71.09	101.70	78.98
EDTA	8.60	9.73	9.56	16.79	29.97	10.68	71.45	169.41	56.51	360.61	233.78	174.38
Total	**83**	**76**	**75**	**94**	**176**	**77**	**281**	**616**	**230**	**1199**	**770**	**557**
Cd (mg·kg^{-1})												
$CaCl_2$	0.04	0.05	0.05	0.03	0.04	0.28	0.31	0.24	0.25	0.55	0.30	0.19
DTPA	0.22	0.21	0.21	0.35	0.69	0.57	1.63	1.13	1.41	1.73	1.15	0.99
EDTA	0.23	0.24	0.24	0.50	0.99	0.78	1.92	1.79	1.65	2.32	1.45	1.25
Total	**1.29**	**1.52**	**1.14**	**2.54**	**4.17**	**3.16**	**5.82**	**7.26**	**5.29**	**7.15**	**4.79**	**4.11**
Cr (mg·kg^{-1})												
$CaCl_2$	N.D.	N.D.	N.D.	N.D.	N.D.	N.D.	N.D.	N.D.	N.D.	N.D.	N.D.	N.D.
DTPA	N.D.	N.D.	N.D.	N.D.	N.D.	N.D.	N.D.	N.D.	N.D.	N.D.	N.D.	N.D.
EDTA	N.D.	N.D.	N.D.	N.D.	N.D.	N.D.	N.D.	N.D.	N.D.	N.D.	N.D.	N.D.
Total	**96**	**115**	**115**	**108**	**154**	**116**	**67**	**63**	**75**	**44**	**73**	**87**

Note: values presented in the table are the arithmetic mean of three replicates. N.D.: not detected.

2.4. Microbial Biomass Carbon. The amount of microbial biomass carbon (biomass C) in the soil was measured by a modified fumigation extraction (FE) method [29]. Soil samples were divided into two parts: one part was fumigated with ethanol-free chloroform for 24 h at 25°C in the dark, and the second part was stored in the dark for 24 h at 25°C but not fumigated. The organic carbon was extracted from both soils using 0.5 M K_2SO_4 (soil : solution, 1 : 4). Soil extractions were shaken at 150 rpm for 30 min and filtered, and then the total organic carbon was determined using a TOC analyzer (TOC-102A, Analytik Jena). The microbial biomass carbon (MBC) was calculated using the following equation:

$$MBC = [TOC \text{ (fumigated soil)}$$
$$- TOC \text{ (nonfumigated soil)}] \times 2.22. \quad (1)$$

The microbial biomass was expressed as milligrams of biomass C per kilogram of soil.

2.5. Soil Protein Extraction. The citrate-SDS sequential extraction method [15, 16] was used to extract the soil proteins. Specifically, 3.0 g of dried soil was mixed with 15 mL of 0.25 M citrate buffer (pH 8.0), and the homogenate was shaken at 1,200 rpm in room temperature for 4 h. Then, the suspension was centrifuged for 15 min at 15,000 ×g and 4°C and filtered through filter paper (approximately 30 to 50 μm). Next, the soil was extracted using 15 mL of SDS buffer, which contained 1.25% (w/v) SDS, 0.1 M Tris-HCl (pH 6.8), and 20 mM dithiothreitol (DTT). This SDS soil mixture was shaken for 1 h at 1,200 rpm and room temperature, and then it was centrifuged for 15 min at 15,000 ×g and 4°C. For protein recovery, both the citrate extract and the SDS extract were mixed with buffered phenol (pH 8.0) at a volume ratio (extract : phenol) of 3 : 1 and centrifuged at 15,000 ×g and 4°C for 30 min. After centrifugation, the phenol phase was precipitated at −20°C overnight with five volumes of cold 0.1 M ammonium acetate dissolved in methanol. The proteins were recovered by centrifugation at 20,000 ×g and 4°C for 20 min. The pellets were washed once with cold methanol and twice with cold acetone, and then they were air-dried for further use.

2.6. Protein Content Determination and Protein Separation. Protein pellets were solubilized in 500 μL of lysis buffer, which contained 9 M urea, 4% w/v CHAPS, 1% w/v DTT, 0.5% ampholyte, and 1 mM PMSF. The concentration of protein in the supernatant was determined by the Bradford method [30], and the protein was stored at −70°C.

The Variations in the Soil Enzyme Activity, Protein Expression, Microbial Biomass, and Community Structure of Soil
Contaminated by Heavy Metals

197

Samples of the extracted proteins were added to an equal volume of loading buffer, which contained 100 mM Tris-HCl (pH 6.8), 4% (w/v) SDS, 20% glycerol, 0.5% (w/v) bromophenol blue, and 100 mM DTT, and then they were heated in water at 95°C for 5 min prior to SDS-PAGE [8]. Subsequently, discontinuous SDS-PAGE was performed using the Mini-PROTEAN 3 Electrophoresis Cell (Bio-Rad) with a 4% stacking gel and a 12% separating gel. The process was run at a constant 75 V/gel through the stacking gel and a constant 150 V/gel through the separating gel. A prestained protein ladder (approximately 10 to 170 kDa, Fermentas) was loaded as a molecular weight marker, and each lane was loaded with the same quantity of protein. After separation, the gels were stained by silver staining [31] for further comparisons.

2.7. DNA Extraction. The extraction of the total soil DNA was accomplished using a Sangon DNA isolation kit, according to the manufacturer's protocol (SK8233).

2.8. PCR-DGGE Microbial Community Analysis. The primers F357 and R518, 5′-CCT ACG GGA GGC AGC AGC-3′ and 5′-ATT ACC GCG GCT GCT GG-3′ [20], respectively, were used in this study for the amplification of bacterial 16S rDNA genes. The primers Com2xf and Ac1186r-(pH), 5′-AAA CTC AAA GGA ATT GAC GG-3′ and 5′-CTT CCT CCG AGT TGA CCC-3′ [32], respectively, were used for the amplification of actinomycete 16S rDNA genes. A GC clamp (CGC CCG CCG CGC CCC GCG CCC GGC CCG CCG CCC CCG CCC C) was added to the forward primers to facilitate the DGGE [20]. The PCR reaction mixture (Takara) contained 5 μL of 10x reaction buffer (0.1 M Tris-HCl at pH 8.3 and 0.5 M KCl), 6 μL of MgCl$_2$ (25 μM), 1 μL of each primer (20 μM), 4 μL of each dNTP (2.5 mM), 0.5 μL of Taq DNA polymerase (5 U·μL^{-1}), and 2 μL of DNA; milli-Q water was added to reach a total reaction volume of 50 μL. The PCR protocol used to amplify the soil bacterial and actinomycete 16S rDNA gene fragments was as follows: a 5 min initial denaturation step at 94°C, followed by 20 cycles of 94°C for 1 min, 65°C for 1 min, and 72°C for 1 min, then, 10 cycles of 94°C for 1 min, 55°C for 1 min, and 72°C for 1 min, with a final extension at 72°C for 7 min (S1000 Thermal Cycler, Bio-Rad). Prior to DGGE, the PCR products were checked by electrophoresis in 1.0% (w/v) agarose gels stained with ethidium bromide.

For the DGGE analysis, the PCR products generated from each sample were separated on an 8% acrylamide gel using the Bio-Rad Dcode System with a linear denaturant gradient range from 30% to 60%. DGGE was performed in 1x TAE buffer at 60°C and 150 V for 8 h. Gels were stained with SYBR-Green I, and the gels were scanned (Bio-Rad).

2.9. Analysis of DGGE Profiles. The digitized DGGE images were analyzed by Quantity One image analysis software (Version 4.62, Bio-Rad). The Shannon index (H) was used to estimate the soil microbial diversity based on the intensity and number of bands using the following equation:

$$\text{Shannon index } (H) = - \sum \frac{n_i}{N} \ln \frac{n_i}{N}, \qquad (2)$$

where i is the number of bands in each lane of the DGGE gel, n_i is the peak height of band i, and N is the sum of the peak heights in a given lane of the DGGE gel [33].

Principal component analysis (PCA) was performed by SPSS 16.0 [34], and WPGAMA cluster analysis was performed by Quantity One 4.62 to calculate the similarities of the gel patterns.

2.10. Data Analysis. All measurements of soil pH, enzyme activity, and levels of organic matter, heavy metals, microbial biomass carbon, and proteins were performed in triplicate. All of the values reported are the average of three determinations. All of the statistical analyses were performed using SPSS software version 16.0. One-way analysis of variance (ANOVA) was used for statistical comparisons, and the Pearson coefficient was used for correlation analysis. A value of $P < 0.05$ was considered statistically significant.

3. Results

3.1. Soil Properties. Selected soil physicochemical properties, the total heavy metal contents, and the bioavailable fraction of heavy metals are shown in Table 1. Compared with the contaminated area, soil samples from the reference site were mildly more acidic, with a pH range from 5.72 to 5.94, and had a lower organic matter content between 4.69% and 5.36%. In the field contaminated by heavy metals, the soil pH varied from 6.23 to 8.00, and the range of soil organic matter was from 3.23% to 7.65%. The amount of organic matter in each sample was lower the further the sample was from the smelter.

In all of the soil samples, the Cr concentration was either below or only slightly above the natural background concentration of 90 mg·kg^{-1}, as defined by the Chinese Environmental Quality Standards for Soil. In addition, the bioavailable Cr fraction was not detected by the apparatus. Therefore, the effects of Cr stress on soil microorganisms were not considered.

The soils varied greatly in their concentrations of total Cu (approximately 175 to 1,147 mg·kg^{-1}), Zn (approximately 187 to 2,472 mg·kg^{-1}), Pb (approximately 77 to 1,199 mg·kg^{-1}), and Cd (approximately 2.54 to 7.26 mg·kg^{-1}) with their distances from the smelter. The highest concentrations of Cu, Zn, and Pb were found in the samples nearest to smelter (200 m), but the highest Cd level was found in the sample 400 m away (Table 1).

3.2. Concentration of Bioavailable Heavy Metals. The fraction of metal, that is, in its bioavailable form, is crucial to understanding metal ecological toxicity. The bioavailability of heavy metals is related to their chemical forms in soil. Several fractions or compartments of the soil act as reservoirs for bioavailable metals [3]. Extractions by CaCl$_2$, DTPA, and EDTA are widely used methods of soil analysis, providing an operationally defined soil compartment that is characterized by its mobility and bioavailability. CaCl$_2$ is used to estimate the soluble and exchangeable metals. DTPA and EDTA are predominantly used to study the metals that have bonded

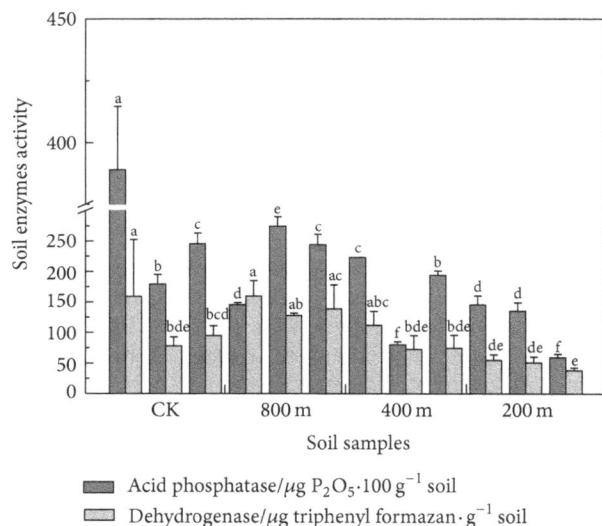

FIGURE 1: The effect of heavy metals on the activities of soil enzymes. Note: different letters in same enzyme indicate significant differences ($P < 0.05$).

with organic matter and the overall phytoavailability of heavy metals [18, 19].

The concentrations of bioavailable Cu, Zn, Pb, and Cd in the soil samples varied significantly (Table 1). The amount of bioavailable heavy metals in the soil decreased with increasing distance from the smelter, and the bioavailable fractions of heavy metals were significantly positively correlated with the total heavy metals ($P < 0.05$). In all of the soil samples, the concentration of $CaCl_2$-extractable metals < DTPA-extractable metals < EDTA-extractable metals. These results correlated with the extraction abilities of the reagents [35]. The concentration of $CaCl_2$-extractable heavy metals was far less than those of the DTPA- and EDTA-extractable heavy metals. This result indicates that heavy metals bonded with organic matter and were predominantly in their bioavailable form, agreeing with the conclusion that heavy metals in soil are primarily associated with soil organic matter [36, 37]. It is this bioavailable fraction of heavy metals that greatly affects soil microorganisms [1].

3.3. Soil Enzyme Activities.

The variations in the activities of acid phosphatase and dehydrogenase in the soil samples are shown in Figure 1. Good correlations were observed between enzyme activity and distance from the smelter. Compared with the reference area (CK), acid phosphatase activity was inhibited in the sites contaminated by heavy metals. Soil acid phosphatase activity tended to decrease with increasing heavy metals. A significant decrease was observed in the samples only 200 m away from the smelter. Conversely, dehydrogenase activity was slightly increased in the soil samples 800 m away from the smelter and decreased as the distance from the smelter decreased. This result indicates that a low level of heavy metals will stimulate soil microbes to synthesize dehydrogenase, promoting microbial metabolic activity.

TABLE 2: The Pearson correlation coefficients between the microbial parameters and the concentrations of heavy metals.

	Acid phosphatase	Dehydrogenase	Microbial biomass C	Protein
Cu				
CaCl$_2$	−0.500	−0.597*	−0.632*	−0.705*
EDTA	−0.576*	−0.628*	−0.834**	−0.884**
DTPA	−0.568	−0.649*	−0.839**	−0.874**
Total	**−0.586***	**−0.594***	**−0.729****	**−0.866****
Zn				
CaCl$_2$	−0.400	−0.584*	−0.604*	−0.543
EDTA	−0.554	−0.685*	−0.683*	−0.596*
DTPA	−0.613*	−0.734**	−0.748**	−0.698*
Total	**−0.531**	**−0.680***	**−0.716****	**−0.653***
Pb				
CaCl$_2$	−0.239	−0.485	−0.303	−0.219
EDTA	−0.586*	−0.709**	−0.624*	−0.643*
DTPA	−0.719**	−0.760**	−0.583*	−0.669*
Total	**−0.569**	**−0.693***	**−0.604***	**−0.645***
Cd				
CaCl$_2$	−0.379	−0.515	−0.551	−0.542
EDTA	−0.486	−0.544	−0.679*	−0.814**
DTPA	−0.425	−0.539	−0.726**	−0.794**
Total	**−0.528**	**−0.494**	**−0.576***	**−0.817****

Note: ** correlation is significant at the 0.01 level (2-tailed).
* Correlation is significant at the 0.05 level (2-tailed).

Soil acid phosphatase activity was significantly negatively correlated with EDTA-extractable Cu and Pb; DTPA-extractable Zn and Pb; and total Cu. Similarly, dehydrogenase activity was significantly negatively correlated with $CaCl_2$-extractable Cu and Zn; EDTA-extractable Cu, Zn, and Pb; DTPA-extractable Cu, Zn, and Pb; and total Cu, Zn, and Pb (Table 2).

3.4. Microbial Biomass Carbon in Soil.

The amount of microbial biomass carbon (MBC) in the contaminated soil samples ranged from 44.6 to 105.0 mg·kg^{-1}, as measured by the FE method, which are lower values than those of the reference site (Figure 2). A significant decrease was observed in the soil samples only 200 m away from the smelter, with the MBC increasing from the plot nearest to the plot farthest away (800 m). The amount of soil MBC was significantly negatively correlated with $CaCl_2$-extractable Cu and Zn; EDTA-extractable Cu, Zn, Pb, and Cd; DTPA-extractable Cu, Zn, Pb, and Cd; and total Cu, Zn, Pb, and Cd (Table 2). These results indicate that the concentrations of DTPA- and EDTA-extractable heavy metals are better predictors of the effects that heavy metals have on MBC compared with the concentrations of $CaCl_2$-extractable heavy metals.

3.5. Soil Microorganism Community Composition.

The idea of using a DGGE analysis of soil microbial communities to

The Variations in the Soil Enzyme Activity, Protein Expression, Microbial Biomass, and Community Structure of Soil Contaminated by Heavy Metals

199

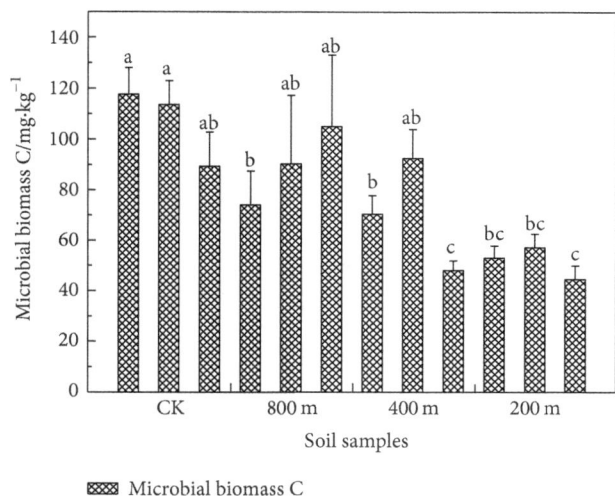

FIGURE 2: The effect of heavy metals on the microbial biomass carbon in soil. Note: the different letters indicate significant differences ($P < 0.05$).

investigate soil microbial diversity was first introduced by Muyzer et al. (1993). DGGE proved to be capable of accurately measuring soil microbial diversity, so it has become the predominant technique in soil microbial ecology for profiling soil microbial communities and monitoring shifts in microbial community composition due to environmental changes [38]. The DGGE profiles of the soil bacteria and actinomycetes are shown in Figures 3(a) and 4(a). Although many bands were found in all lanes, obvious differences in the intensities of the soil microbial communities were clearly observed (Figures 3(a) and 4(a)). As shown in Figure 3(a), the number of bands in the soil was significantly increased with an increasing distance from the smelter; in other words, the number of bands in the DGGE profile increased as the heavy metals presence decreased. A similar result was also found in the soil actinomycete community composition. Compared with the soils from the reference site and from 800 m away, the number of bands in the plots situated at 400 m and 200 m were significantly decreased (Figure 4(a)).

Subsequently, the DGGE gels were interpreted by the Shannon index, principal component analysis (PCA), and cluster analysis to examine the number, presence, and relative intensity of the bands. The Shannon index indicated that the diversity of the soil microorganisms (bacteria and actinomycetes) was decreased significantly with an increase in the heavy metals concentration (Table 3). PCA and the cluster analysis are also useful methods for analyzing DGGE profiles. The PCA plots show a clear separation due to the different distances from the smelter, demonstrating the altered structure and diversity of the microbial community. Using the presence of the bands as input data, the first two principal components (PC1 and PC2) were sufficient to explain 79.5% and 78.0% of the variance in the soil bacteria and the actinomycetes, respectively (Figures 3(b) and 4(b)). The similarity between the soil bacterial and actinomycete communities of the reference site and those of the soil taken

around the smelter was obtained from a cluster analysis of the DGGE profile and was determined to be less than 50% (Figures 3(c) and 4(c)). The cluster analysis indicated that the soil microorganism community structure was greatly affected by the presence of heavy metals.

3.6. Soil Protein Expression. The amount of protein in the contaminated soil samples ranged from 15.6 to 104 μg·g^{-1}, far less than the amount found at the reference site (Figure 5(a)). The concentration of protein decreased drastically when the soil was slightly contaminated by heavy metals at a distance of 800 m from the smelter. The quantity of soil proteins was significantly negatively correlated with soil CaCl$_2$-extractable Cu; EDTA-extractable Cu, Zn, Pb, and Cd; DTPA-extractable Cu, Zn, Pb, and Cd; and total Cu, Zn, Pb, and Cd (Table 2).

The protein profiles are shown in Figure 5(b). The results indicate that no discernible differences exist between the composition and location of the citrate-extracted protein band of the reference site and those of the soil samples from 800 m and 400 m. The bands from these three sample sets were almost entirely located between approximately 55 and 70 kD. However, some type of protein, that is, 40 kD in size, was only found in the soil 200 m from the smelter. Additionally, although the SDS buffer extracts showed similar protein patterns between different soil samples, the expression of some proteins was significantly different. Compared with the contaminated soil samples, the uncontaminated samples had more abundant proteins overall. In the contaminated soil, with its increased concentration of heavy metals, there were fewer large molecular weight (>35 kD) proteins and more low molecular weight proteins (approximately 15 kD). At the base of each contaminated profile was a large agglomeration of presumably low molecular weight proteinaceous material that produced dark areas on the gels. Therefore, these low molecular weight proteins were largely aggregated.

4. Discussion

The activity and community composition of soil microbes are closely related to soil fertility and environmental quality. Additionally, soil biological parameters may have potential for use as early and sensitive indicators of soil ecological stress and restoration. Currently, the amount of microbial biomass carbon and enzyme activity are the metrics predominantly used to provide information about the biochemical processes occurring in the soil [3, 39]. The soil microbial biomass is the total mass of microorganisms living in the soil, which is the living part of the soil organic matter [1]. Soil phosphatase plays an essential role in the mineralization of organic phosphorus, while dehydrogenase is an intracellular enzyme that is involved in microbial oxidoreductase metabolism [3, 39].

In our study, the amount of microbial biomass C in the soil decreased when the concentration of heavy metals increased, although the difference was not significant when the heavy metals were at a low level. This finding is not surprising. Microorganisms differ in their sensitivity to metal toxicity, and the development of metal-tolerant strains, as well as shifts in the community structure, could compensate for

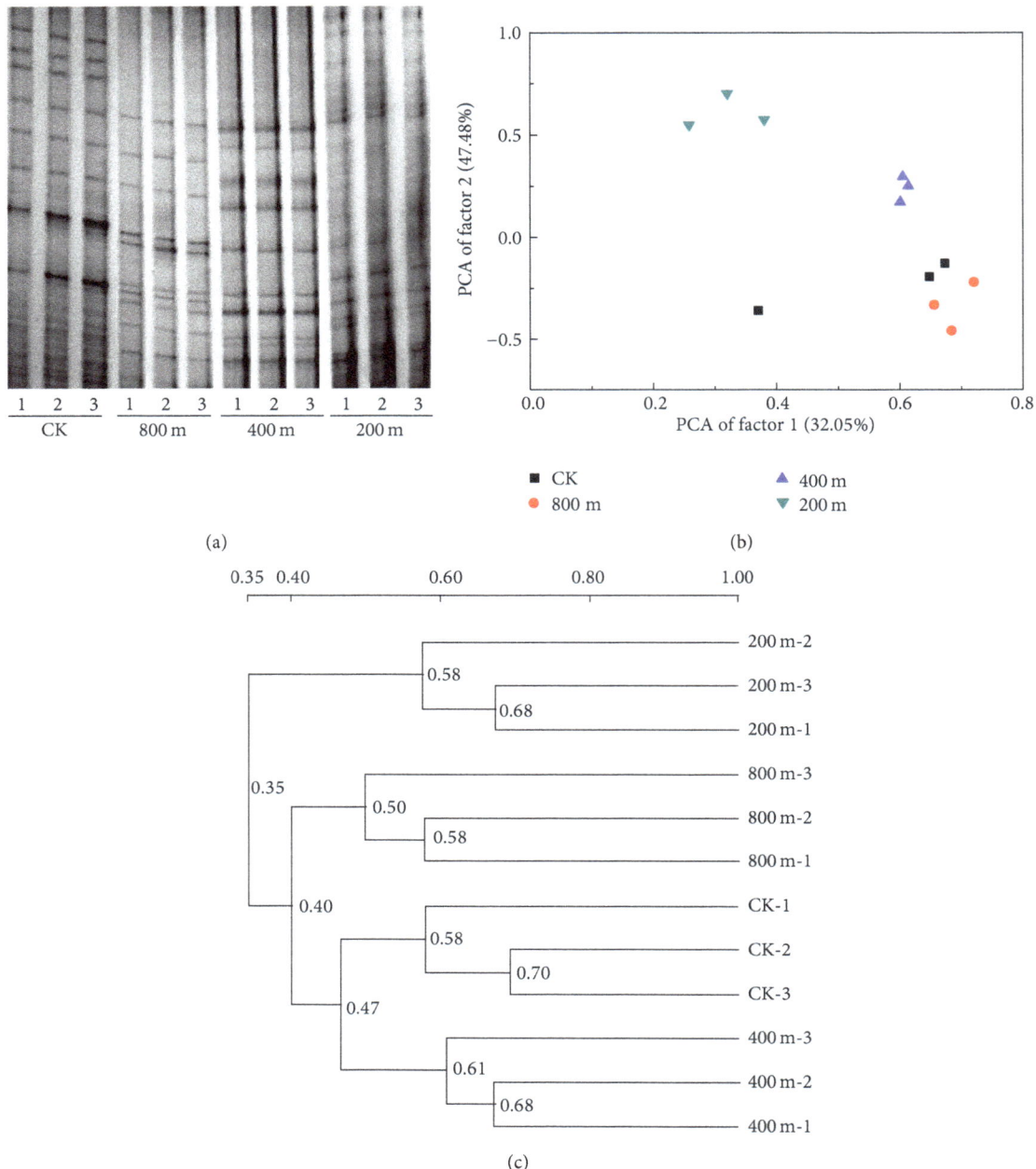

(a)

(b)

(c)

FIGURE 3: The effect of heavy metals on soil bacterial communities. (a) The DGGE profiles of the bacterial communities inhabiting each soil sample. (b) A principle component analysis based on the DGGE profiles of soil bacteria. (c) A cluster analysis based on the DGGE profiles of soil bacteria.

the loss of more sensitive populations [2]. The reduction in soil microbial biomass C might also explain the inhibition of enzyme activity observed in the heavy metal contaminated soil.

It is important to differentiate between extracellular and intracellular enzymes when we study the changes of enzymes under pollutant stress. The dehydrogenase activity, which is only present in viable cells and essentially depends on the metabolic state of the soil biota, may therefore be considered a direct measure of soil microbial activity. In contrast, phosphatase activity can occur extracellularly as well as within a

living cell [1, 3, 6, 26, 39]. For example, Brookes et al. [40] reported less dehydrogenase activity in metal-contaminated soil than in similar uncontaminated soil, while the soil phosphatase was unaffected. This might explain why, in our study, the dehydrogenase activity was also more sensitive to the heavy metal stress than the acid phosphatase activity.

Furthermore, the dehydrogenase activity was significantly correlated with soil microbial biomass C ($r = 0.611$, $P < 0.05$). These results suggest that microbial biomass C and dehydrogenase can be useful measures of the level of heavy metal contamination in a soil sample [26, 39].

The Variations in the Soil Enzyme Activity, Protein Expression, Microbial Biomass, and Community Structure of Soil
Contaminated by Heavy Metals

201

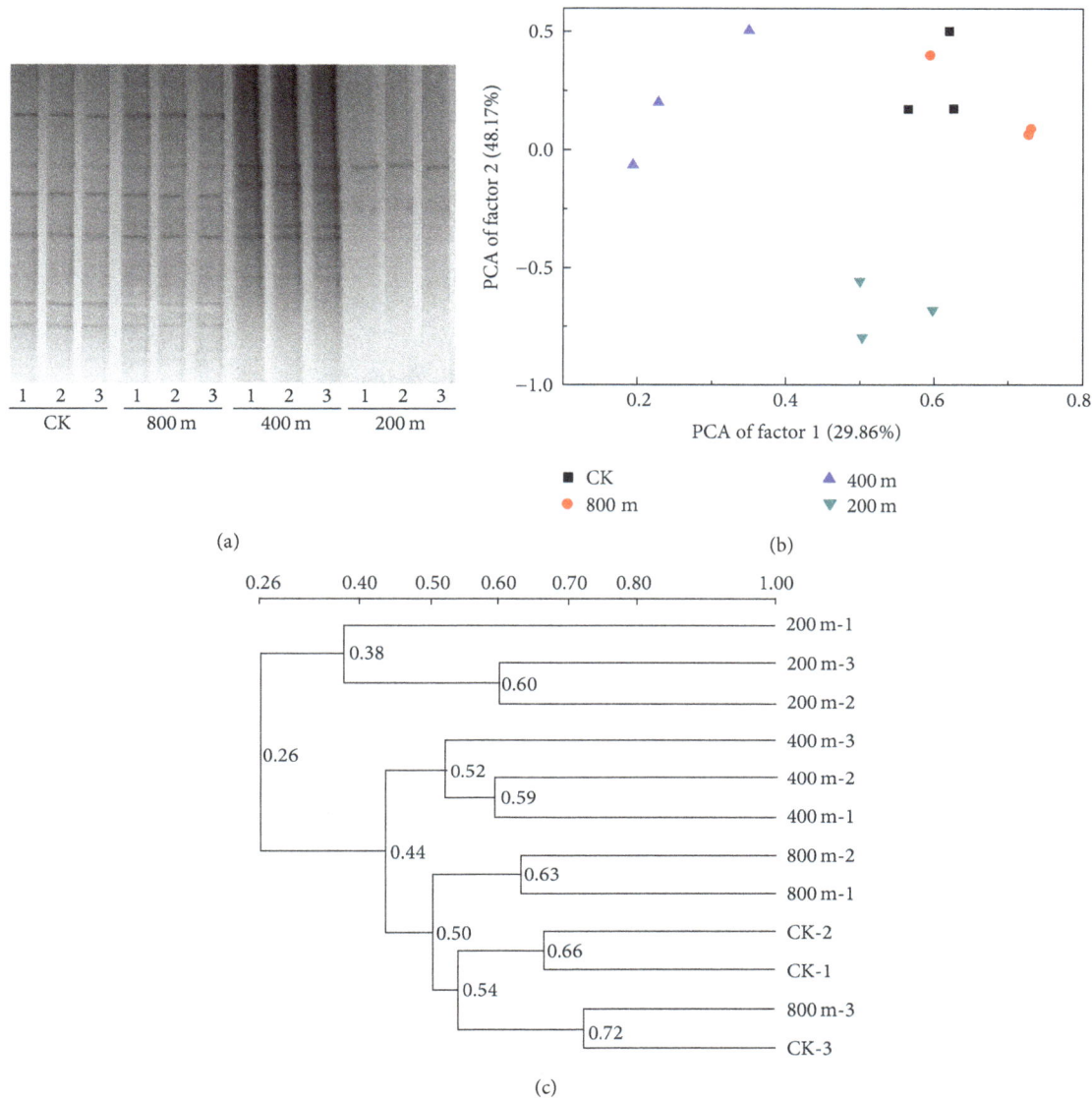

FIGURE 4: The effect of heavy metals on soil actinomycetes communities. (a) The DGGE profiles of the actinomycete communities inhabiting each soil sample. (b) A principle component analysis based on the DGGE profiles of soil actinomycetes. (c) A cluster analysis based on the DGGE profiles of soil actinomycetes.

The microbial biomass and enzyme activity measurements have their limitations in soil pollution studies, however. For example, the changes were not notable when heavy metal concentrations were low and highly soluble Cu generated chromogen interference to dehydrogenase activity measurements [41]. In addition, the changes in the microorganism community structure cannot be evaluated by such techniques. Therefore, the use of soil enzyme and microbial biomass as indicators for soil contaminated by heavy metals still needs further discussions.

The changes in the metabolic profiles indicate the possibility that heavy metal contamination results in a community that is more variable and less stable [3]. In our study, the genetic structure and protein expression of the indigenous soil microbial communities were evaluated by an advanced molecular technique that is based on PCR-DGGE and SDS-PAGE. The PCR-DGGE method has become a predominant molecular technique to study the changes in soil microorganisms due to contamination and agricultural practices [3, 20, 42]. In contrast, the use of SDS-PAGE to monitor contaminated soil is still in its infancy [7, 8, 17]. The results of the DGGE profiles, as analyzed by principal component analysis (PCA) and cluster analysis in this study, make it clear that the number and intensity of DNA bands in the different samples changed significantly. This significant difference between the samples indicates that the soil microbial community structure changed greatly in response to heavy metals contamination (Table 3, Figures 3-4). We also detected an increased relative abundance in the microbial populations with increasing distance from the

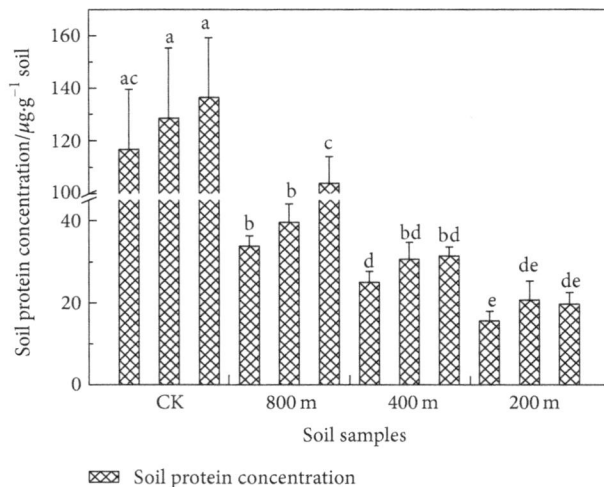

(a)

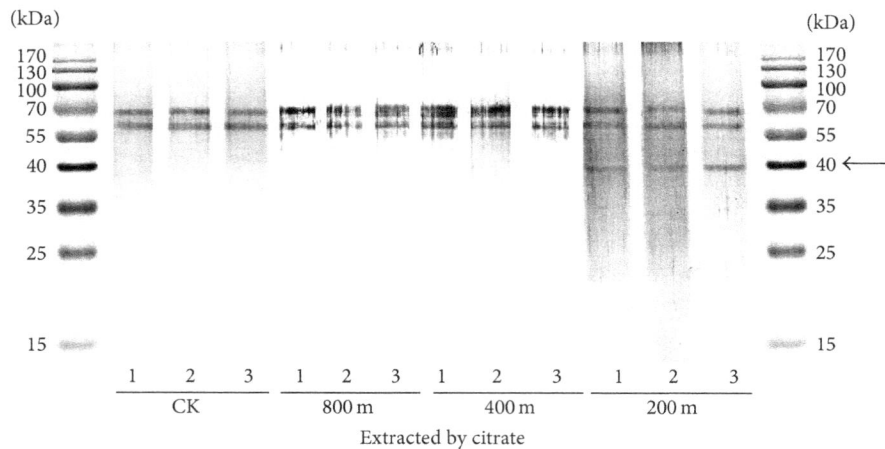

Extracted by citrate

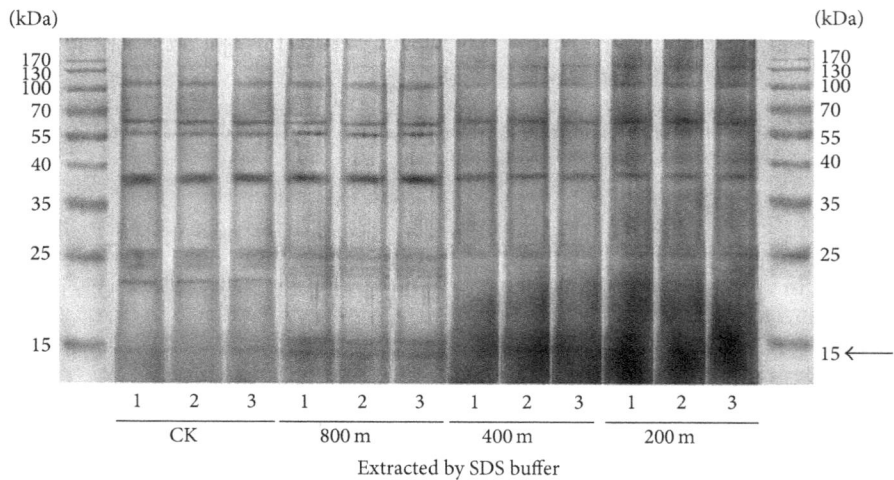

Extracted by SDS buffer

(b)

FIGURE 5: The effect of the concentration of heavy metals on soil protein. (a) The effect of the concentration of heavy metals on the amount of soil proteins. Note: Different letters indicate significant differences ($P < 0.05$). (b) The effect of the heavy metals concentration on soil protein expressions.

The Variations in the Soil Enzyme Activity, Protein Expression, Microbial Biomass, and Community Structure of Soil
Contaminated by Heavy Metals

203

TABLE 3: The Shannon diversity index for the soil samples contaminated by heavy metals.

Distance	Soil bacteria	Soil actinomycetes
CK	3.11 ± 0.11^{a}	0.52 ± 0.02^{a}
800 m	2.91 ± 0.12^{b}	0.40 ± 0.03^{b}
400 m	2.90 ± 0.07^{b}	0.35 ± 0.01^{c}
200 m	2.48 ± 0.08^{c}	0.23 ± 0.02^{d}

Note: values presented in the table are the mean $\pm$ standard deviation ($n = 3$).

Values in each column followed by different letters are significantly different ($P < 0.05$).

smelter, demonstrating that heavy metals exerted a major effect on bacterial diversity by promoting changes in species composition (represented by the position of the bands) and in species richness (represented by the number of occurring bands). Namely, contamination inhibited certain bacterial groups and stimulated others, and it changed the overall microbial diversity [2, 3, 38, 43, 44].

Subsequently, the content and composition of the soil proteins were studied to investigate their changes under heavy metals stress. Compared with the reference site, the soil protein levels decreased significantly due to heavy metals contamination, and the percentage decrease in total soil proteins was larger than the decrease detected in soil enzyme activity or microbial biomass carbon (Figure 5(a)). The soil protein concentrations were significantly negatively correlated with heavy metals (Table 2). These results indicate that proteins are more sensitive to the changes caused by pollutants, making protein concentration a great potential indicator of soil pollution. The variation in soil proteins may be due to modifications in the microbial community composition [7, 8]. This hypothesis was confirmed by the DGGE profiles in our study (Figures 3(a) and 4(a)). Although protein has potential to be an indictor for soil contamination, its reliability still needs more studies since Bradford method is easily affected by humic substances present in protein solution [45].

The collected soil proteins were separated by SDS-PAGE, and we found that several bands were common to all soils. We also found more presumably low molecular weight proteins in the contaminated soil than in the soil of the reference site (Figure 5(b)). This protein could be related to the presence of heavy metals and could be a microbial response to the metals [8, 46]. For example, many eukaryotic microbes are known to produce metallothioneins (a low molecular weight protein) in response to metal (particularly Cd) exposure [47, 48]. Another explanation may be that sufficient metal exposure will result in the immediate death of cells due to a disruption of their essential functions; thus, the cells are lysed, and their proteins are released into the environment, where they are quickly degraded into low molecular weight materials [2, 8]. Obviously, more work is needed to identify this low molecular weight protein and understand its role in the microbial responses to heavy metals stress. This protein should also be investigated as an indicator for soil pollution.

Although PCR-DGGE and SDS-PAGE are useful for monitoring the changes in microbial community structures, they still have drawbacks. These approaches are easily interfered with by impurities that are coextracted with the soil

DNA and proteins, primarily humic substances [7, 38]. The DNA and proteins that are mixed in with the organic matter are not easily separated by DGGE or SDS-PAGE. In addition, relatively small populations of microbes and low abundances of proteins may not be identified by DGGE and SDS-PAGE, respectively [8, 38]. In this study, for example, both the DGGE and SDS-PAGE profiles have a dark background and smeared dark areas at the bottom of SDS-PAGE gel, which make the accuracy and reliability of the analysis questionable. Consequently, the molecular techniques used in this study have much room for improvement, and more work is required to develop effective extraction methods for soil DNA and proteins.

No single technique can provide a comprehensive depiction of the soil microbial situation. Therefore, soil microbial activity and diversity are difficult to elucidate. To obtain a better understanding of soil microbial ecology, we need to integrate different methods to create a comprehensive analysis.

5. Conclusions

By combining different monitoring techniques for different aspects of microbiology, we can obtain a better understanding of the soil microorganisms that live in soils contaminated by heavy metals. Our study demonstrated that heavy metals have a significant negative effect on the activities of soil acid phosphatase and dehydrogenase as well as on the levels of proteins and microbial biomass carbon. Compared with the other potential contamination markers tested, the protein level showed the most dramatic decrease in slightly contaminated soil, indicating that it may be more sensitive to heavy metals. Subsequently, the denaturing gradient gel electrophoresis analysis used in this study demonstrated that heavy metals had a significant negative effect on soil microbial diversity and community structure. Furthermore, the soil protein expression was similar in different soils, but a large quantity of presumably low molecular weight protein was observed only in heavy metal contaminated soil. Based on the research described in this paper, we can conclude that the soil protein level has great potential to be a sensitive indicator of soil contamination. Further research is essential to identify the low molecular weight protein that only appears in contaminated soil.

Acknowledgments

This work was financially supported by the National Natural Science Foundation of China (11179025), National Natural Science Foundation for the Youth of China (21007055), Science Foundation for Postdoctor of China (20100471714, 201104718), and Program for New Century Excellent Talents in University (NCET-11-0455).

References

[1] P. C. Brookes, "The use of microbial parameters in monitoring soil pollution by heavy metals," *Biology and Fertility of Soils*, vol. 19, no. 4, pp. 269–279, 1995.

[2] K. E. Giller, E. Witter, and S. P. Mcgrath, "Toxicity of heavy metals to microorganisms and microbial processes in agricultural soils: a review," *Soil Biology and Biochemistry*, vol. 30, no. 10-11, pp. 1389–1414, 1998.

[3] Y. Wang, J. Shi, H. Wang, Q. Lin, X. Chen, and Y. Chen, "The influence of soil heavy metals pollution on soil microbial biomass, enzyme activity, and community composition near a copper smelter," *Ecotoxicology and Environmental Safety*, vol. 67, no. 1, pp. 75–81, 2007.

[4] G. Renella, M. Mench, L. Landi, and P. Nannipieri, "Microbial activity and hydrolase synthesis in long-term Cd-contaminated soils," *Soil Biology and Biochemistry*, vol. 37, no. 1, pp. 133–139, 2005.

[5] A. K. Müller, K. Westergaard, S. Christensen, and S. J. Sørensen, "The effect of long-term mercury pollution on the soil microbial community," *FEMS Microbiology Ecology*, vol. 36, no. 1, pp. 11–19, 2001.

[6] C. Trasar-Cepeda, M. C. Leirós, S. Seoane, and F. Gil-Sotres, "Limitations of soil enzymes as indicators of soil pollution," *Soil Biology and Biochemistry*, vol. 32, no. 13, pp. 1867–1875, 2000.

[7] F. Bastida, J. L. Moreno, C. Nicolás, T. Hernández, and C. García, "Soil metaproteomics: a review of an emerging environmental science. Significance, methodology and perspectives," *European Journal of Soil Science*, vol. 60, no. 6, pp. 845–859, 2009.

[8] I. Singleton, G. Merrington, S. Colvan, and J. S. Delahunty, "The potential of soil protein-based methods to indicate metal contamination," *Applied Soil Ecology*, vol. 23, no. 1, pp. 25–32, 2003.

[9] P.-A. Maron, L. Ranjard, C. Mougel, and P. Lemanceau, "Metaproteomics: a new approach for studying functional microbial ecology," *Microbial Ecology*, vol. 53, no. 3, pp. 486–493, 2007.

[10] P. Wilmes and P. L. Bond, "The application of two-dimensional polyacrylamide gel electrophoresis and downstream analyses to a mixed community of prokaryotic microorganisms," *Environmental Microbiology*, vol. 6, no. 9, pp. 911–920, 2004.

[11] P. Wilmes and P. L. Bond, "Metaproteomics: studying functional gene expression in microbial ecosystems," *Trends in Microbiology*, vol. 14, no. 2, pp. 92–97, 2006.

[12] S. Criquet, A. M. Farnet, and E. Ferre, "Protein measurement in forest litter," *Biology and Fertility of Soils*, vol. 35, no. 5, pp. 307–313, 2002.

[13] W. X. Schulze, G. Gleixner, K. Kaiser, G. Guggenberger, M. Mann, and E.-D. Schulze, "A proteomic fingerprint of dissolved organic carbon and of soil particles," *Oecologia*, vol. 142, no. 3, pp. 335–343, 2005.

[14] E. B. Taylor and M. A. Williams, "Microbial protein in soil: influence of extraction method and C amendment on extraction and recovery," *Microbial Ecology*, vol. 59, no. 2, pp. 390–399, 2010.

[15] S. Chen, M. C. Rillig, and W. Wang, "Improving soil protein extraction for metaproteome analysis and glomalin-related soil protein detection," *Proteomics*, vol. 9, no. 21, pp. 4970–4973, 2009.

[16] H.-B. Wang, Z.-X. Zhang, H. Li et al., "Characterization of metaproteomics in crop rhizospheric soil," *Journal of Proteome Research*, vol. 10, no. 3, pp. 932–940, 2011.

[17] D. Benndorf, G. U. Balcke, H. Harms, and M. Von Bergen, "Functional metaproteome analysis of protein extracts from contaminated soil and groundwater," *ISME Journal*, vol. 1, no. 3, pp. 224–234, 2007.

[18] L. A. Brun, J. Maillet, P. Hinsinger, and M. Pépin, "Evaluation of copper availability to plants in copper-contaminated vineyard soils," *Environmental Pollution*, vol. 111, no. 2, pp. 293–302, 2001.

[19] E. Meers, R. Samson, F. M. G. Tack et al., "Phytoavailability assessment of heavy metals in soils by single extractions and accumulation by *Phaseolus vulgaris*," *Environmental and Experimental Botany*, vol. 60, no. 3, pp. 385–396, 2007.

[20] G. Muyzer, E. C. de Waal, and A. G. Uitterlinden, "Profiling of complex microbial populations by denaturing gradient gel electrophoresis analysis of polymerase chain reaction-amplified genes coding for 16S rRNA," *Applied and Environmental Microbiology*, vol. 59, no. 3, pp. 695–700, 1993.

[21] E. Baath, A. Frostegaard, T. Pennanen, and H. Fritze, "Microbial community structure and pH response in relation to soil organic matter quality in wood-ash fertilized, clear-cut or burned coniferous forest soils," *Soil Biology and Biochemistry*, vol. 27, no. 2, pp. 229–240, 1995.

[22] X. Tang, C. Shen, D. Shi et al., "Heavy metal and persistent organic compound contamination in soil from Wenling: an emerging e-waste recycling city in Taizhou area, China," *Journal of Hazardous Materials*, vol. 173, no. 1–3, pp. 653–660, 2010.

[23] E. Lakanen and R. Ervio, "A comparison of eight extractants for determination of plant available micronutrients in soil," *Acta Agraria Fennica*, vol. 123, pp. 223–232, 1971.

[24] W. L. Lindsay and W. A. Norvell, "Development of a DTPA soil test for zinc, iron, manganese, and copper," *Soil Science Society of America Journal*, vol. 42, pp. 421–428, 1978.

[25] P. Quevauviller, "Operationally defined extraction procedures for soil and sediment analysis I. Standardization," *Trends in Analytical Chemistry*, vol. 17, no. 5, pp. 289–298, 1998.

[26] C. Garcia, T. Hernandez, F. Costa, and B. Ceccanti, "Biochemical parameters in soils regenerated by the addition of organic wastes," *Waste Management and Research*, vol. 12, no. 6, pp. 457–466, 1994.

[27] S. Y. Guan, *Soil Enzyme and Its Research Methods*, Agricultural Press, Beijing, China, 1986.

[28] X. Lin, *Principles and Methods of Soil Microbiology Research*, Higher Education Press, Beijing, China, 2010.

[29] E. D. Vance, P. C. Brookes, and D. S. Jenkinson, "An extraction method for measuring soil microbial biomass C," *Soil Biology and Biochemistry*, vol. 19, no. 6, pp. 703–707, 1987.

[30] M. M. Bradford, "A rapid and sensitive method for the quantitation of microgram quantities of protein utilizing the principle of protein dye binding," *Analytical Biochemistry*, vol. 72, no. 1-2, pp. 248–254, 1976.

[31] J. X. Yan, R. Wait, T. Berkelman et al., "A modified silver staining protocol for visualization of proteins compatible with matrix-assisted laser desorption/ionization and electrospray ionization- mass spectrometry," *Electrophoresis*, vol. 21, pp. 3666–3672, 2000.

[32] J. Schäfer, U. Jäckel, and P. Kämpfer, "Development of a new PCR primer system for selective amplification of Actinobacteria," *FEMS Microbiology Letters*, vol. 311, pp. 103–112, 2010.

[33] G. W. Yeates, H. J. Percival, and A. Parshotam, "Soil nematode responses to year-to-year variation of low levels of heavy metals," *Australian Journal of Soil Research*, vol. 41, no. 3, pp. 613–625, 2003.

[34] C. D. Clegg, R. D. L. Lovell, and P. J. Hobbs, "The impact of grassland management regime on the community structure of selected bacterial groups in soils," *FEMS Microbiology Ecology*, vol. 43, no. 2, pp. 263–270, 2003.

[35] C. R. M. Rao, A. Sahuquillo, and J. F. Lopez Sanchez, "A review of the different methods applied in environmental geochemistry for single and sequential extraction of trace elements in soils

The Variations in the Soil Enzyme Activity, Protein Expression, Microbial Biomass, and Community Structure of Soil
Contaminated by Heavy Metals

205

and related materials," *Water, Air, and Soil Pollution*, vol. 189, no. 1–4, pp. 291–333, 2008.

[36] A. Manceau, M.-C. Boisset, G. Sarret et al., "Direct determination of lead speciation in contaminated soils by EXAFS spectroscopy," *Environmental Science and Technology*, vol. 30, no. 5, pp. 1540–1552, 1996.

[37] D. G. Strawn and L. L. Baker, "Molecular characterization of copper in soils using X-ray absorption spectroscopy," *Environmental Pollution*, vol. 157, no. 10, pp. 2813–2821, 2009.

[38] J. Zhou, X. Sun, J. Jiao, M. Liu, F. Hu, and H. Li, "Dynamic changes of bacterial community under the influence of bacterial-feeding nematodes grazing in prometryne contaminated soil," *Applied Soil Ecology*, vol. 64, pp. 70–76, 2013.

[39] J. C. García-Gil, C. Plaza, P. Soler-Rovira, and A. Polo, "Long-term effects of municipal solid waste compost application on soil enzyme activities and microbial biomass," *Soil Biology and Biochemistry*, vol. 32, no. 13, pp. 1907–1913, 2000.

[40] P. Brookes, S. McGrath, D. Klein, and E. Elliott, "Effects of heavy metals on microbial activity and biomass in field soils treated with sewage sludge," in *Environmental Contamination*, pp. 574–583, CEP Consultants, Edinburgh, Scotland, 1984.

[41] K. Chander and P. C. Brookes, "Is the dehydrogenase assay invalid as a method to estimate microbial activity in copper-contaminated soils?" *Soil Biology and Biochemistry*, vol. 23, no. 10, pp. 909–915, 1991.

[42] M. Cea, M. Jorquera, O. Rubilar, H. Langer, G. Tortella, and M. C. Diez, "Bioremediation of soil contaminated with pentachlorophenol by Anthracophyllum discolor and its effect on soil microbial community," *Journal of Hazardous Materials*, vol. 181, no. 1–3, pp. 315–323, 2010.

[43] K. Arnebrant, E. Bååth, and A. Nordgren, "Copper tolerance of microfungi isolated from polluted and unpolluted forest soil," *Mycologia*, vol. 79, no. 6, pp. 890–895, 1987.

[44] E. Baath, "Effects of heavy metals in soil on microbial processes and populations (a review)," *Water, Air, and Soil Pollution*, vol. 47, no. 3-4, pp. 335–379, 1989.

[45] P. Roberts and D. L. Jones, "Critical evaluation of methods for determining total protein in soil solution," *Soil Biology and Biochemistry*, vol. 40, no. 6, pp. 1485–1495, 2008.

[46] M. Hodson, "Effects of heavy metals and metalloids on soil organisms," in *Heavy Metals in Soils*, B. J. Alloway, Ed., pp. 141–160, Springer, Amsterdam, The Netherlands, 2013.

[47] C. A. Blindauer, "Bacterial metallothioneins: past, present, and questions for the future," *Journal of Biological Inorganic Chemistry*, vol. 16, no. 7, pp. 1011–1024, 2011.

[48] M. Mejáre and L. Bülow, "Metal-binding proteins and peptides in bioremediation and phytoremediation of heavy metals," *Trends in Biotechnology*, vol. 19, no. 2, pp. 67–73, 2001.

Permissions

The contributors of this book come from diverse backgrounds, making this book a truly international effort. This book will bring forth new frontiers with its revolutionizing research information and detailed analysis of the nascent developments around the world.

We would like to thank all the contributing authors for lending their expertise to make the book truly unique. They have played a crucial role in the development of this book. Without their invaluable contributions this book wouldn't have been possible. They have made vital efforts to compile up to date information on the varied aspects of this subject to make this book a valuable addition to the collection of many professionals and students.

This book was conceptualized with the vision of imparting up-to-date information and advanced data in this field. To ensure the same, a matchless editorial board was set up. Every individual on the board went through rigorous rounds of assessment to prove their worth. After which they invested a large part of their time researching and compiling the most relevant data for our readers. Conferences and sessions were held from time to time between the editorial board and the contributing authors to present the data in the most comprehensible form. The editorial team has worked tirelessly to provide valuable and valid information to help people across the globe.

Every chapter published in this book has been scrutinized by our experts. Their significance has been extensively debated. The topics covered herein carry significant findings which will fuel the growth of the discipline. They may even be implemented as practical applications or may be referred to as a beginning point for another development. Chapters in this book were first published by Hindawi Publishing Corporation; hereby published with permission under the Creative Commons Attribution License or equivalent.

The editorial board has been involved in producing this book since its inception. They have spent rigorous hours researching and exploring the diverse topics which have resulted in the successful publishing of this book. They have passed on their knowledge of decades through this book. To expedite this challenging task, the publisher supported the team at every step. A small team of assistant editors was also appointed to further simplify the editing procedure and attain best results for the readers.

Our editorial team has been hand-picked from every corner of the world. Their multi-ethnicity adds dynamic inputs to the discussions which result in innovative outcomes. These outcomes are then further discussed with the researchers and contributors who give their valuable feedback and opinion regarding the same. The feedback is then collaborated with the researches and they are edited in a comprehensive manner to aid the understanding of the subject.

Apart from the editorial board, the designing team has also invested a significant amount of their time in understanding the subject and creating the most relevant covers. They scrutinized every image to scout for the most suitable representation of the subject and create an appropriate cover for the book.

The publishing team has been involved in this book since its early stages. They were actively engaged in every process, be it collecting the data, connecting with the contributors or procuring relevant information. The team has been an ardent support to the editorial, designing and production team. Their endless efforts to recruit the best for this project, has resulted in the accomplishment of this book. They are a veteran in the field of academics and their pool of knowledge is as vast as their experience in printing. Their expertise and guidance has proved useful at every step. Their uncompromising quality standards have made this book an exceptional effort. Their encouragement from time to time has been an inspiration for everyone.

The publisher and the editorial board hope that this book will prove to be a valuable piece of knowledge for researchers, students, practitioners and scholars across the globe.

List of Contributors

Tekin Öztekin
Department of Biosystem Engineering, Agricultural Faculty, Gaziosmanpasa University, Tasliciftlik Campus, 60250 Tokat, Turkey

Klement Rejsek, Valerie Vranova, and Pavel Formanek
Department of Geology and Soil Science, Mendel University in Brno, Zemedelska 3, 613 00, Brno, Czech Republic

Song Chen, Dangying Wang, Liping Chen, Chunmei Xu and Xiufu Zhang
China National Rice Research Institute, Chinese Academy of Agricultural Sciences, Zhejiang, Hangzhou 310006, China

Xi Zheng
985-Institute of Agrobiology and Environmental Sciences, Zhejiang University, Hangzhou 310029, China

Wang Xing-run, Zhang Yan-xia, Wang Qi and Shu Jian-min
State Key Laboratory of Environmental Criteria and Risk Assessment, Chinese Research Academy of Environmental Sciences, Beijing 100012, China

Zhang Yan-xia
College of Science, Northwest A&F University, Shaanxi 712100, China

Ligang Xu and Xiaolong Wang
State Key Laboratory of Lake Science and Environment, Nanjing Institute of Geography and Limnology, Chinese Academy of Sciences, Nanjing, China

Hailin Niu
College of Ecological and Environmental Engineering, Qinghai University, Xining, China

Jin Xu
Department of Environmental Engineering, Nanjing Institute of Technology, Nanjing, China

Jana Albrechtova
Department of Experimental Plant Biology, Faculty of Science, Charles University in Prague, 12844 Vinicna 5, Czech Republic

Jana Albrechtova and Miroslav Vosatk
Institute of Botany of the Academy of Sciences of the Czech Republic, 25243 Pruhonice, Czech Republic

Ales Latr and Miroslav Vosatka
Symbiom Ltd., Sazava 170, 56301 Lanskroun, Czech Republic

Ludovit Nedorost and Robert Pokluda
Department of Vegetable Sciences and Floriculture, Mendel University in Brno, Valticka 337, 69144 Lednice, Czech Republic

Katalin Posta
Microbiology and Environmental Toxicology Group, Plant Protection Institute, Szent István University, 2100 Gödöllő, Hungary

Jaqueline Aparecida Ribaski Borges and Luiz Fernando Pires
Laboratory of Soil Physics and Environmental Sciences, Department of Physics, State University of Ponta Grossa (UEPG), Avenue Carlos Cavalcanti 4748, 84030-900 Ponta Grossa, PR, Brazil

André Belmont Pereira
Department of Soil Science, State University of Ponta Grossa (UEPG), Avenue Carlos Cavalcanti 4748, 84030-900 Ponta Grossa, PR, Brazil

Mário A. Camargo, Paulo C. Facin, and Luiz F. Pires
Laboratory of Soil Physics and Environmental Sciences, Department of Physics, State University of Ponta Grossa (UEPG), 84.030-900 Ponta Grossa, PR, Brazil

Junzeng Xu, Shihong Yang, Shizhang Peng, Qi Wei and Xiaoli Gao
State Key Laboratory of Hydrology-Water Resources and Hydraulic Engineering, Hohai University, Nanjing 210098, China

Junzeng Xu and Xiaoli Gao
College of Water Conservancy and Hydropower Engineering, Hohai University, Nanjing 210098, China

Songhao Shang
State Key Laboratory of Hydroscience and Engineering, Tsinghua University, Beijing 100084, China

Songhao Shang
Department of Hydraulic Engineering, Tsinghua University, Beijing 100084, China

Weidong Li and Chuanrong Zhang
Department of Geography and Center for Environmental Sciences & Engineering, University of Connecticut, Storrs, CT 06269, USA

Dipak K. Dey
Department of Statistics, University of Connecticut, Storrs, CT 06269, USA

Michael R. Willig
Center for Environmental Sciences & Engineering and Department of Ecology & Evolutionary Biology, University of Connecticut, Storrs, CT 06269, USA

Renáta Sándor and Nándor Fodor
Centre for Agricultural Research, Hungarian Academy of Sciences, 2462 Martonvásár, Hungary

P. Karimi, R. A. Khavari-Nejad, F. Ghahremaninejad and F. Najafi
Faculty of Biological Sciences, Kharazmi University, Tehran 15719-14911, Iran

R. A. Khavari-Nejad
Department of Biology, Faculty of Science, Islamic Azad University, Science and Research Branch, Tehran 14778-93855, Iran

V. Niknam
School of Biology and Center of Excellence in Phylogeny of Living Organisms, College of Science, University of Tehran, Tehran 14115-154, Iran

Małgorzata Brzezińska, Magdalena Nosalewicz, Marek Pasztelan and Teresa Włodarczyk
Institute of Agrophysics, Polish Academy of Sciences, Ulica Doświadczalna 4, 20290 Lublin, Poland

D. S. Karam, A. Arifin, N. M. Majid, A. H. Hazandy and T. X. Rui
Department of Forest Production, Faculty of Forestry, Universiti Putra Malaysia, 43400 Serdang, Selangor, Malaysia

A. Arifin and A. H. Hazandy
Laboratory of Sustainable Bioresource Management, Institute of Tropical Forestry and Forest Products, Universiti Putra Malaysia, 43400 Serdang, Selangor, Malaysia

O. Radziah and J. Shamshuddin
Department of Land Management, Faculty of Agriculture, Universiti Putra Malaysia, 43400 Serdang, Selangor, Malaysia

O. Radziah
Laboratory of Food Crops and Floriculture, Institute of Tropical Agriculture, Universiti Putra Malaysia, 43400 Serdang, Selangor, Malaysia

I. Zahari and A. H. Nor Halizah
Forestry Department Peninsular Malaysia, Jalan Sultan Salahuddin, 50660 Kuala Lumpur, Malaysia

Yanhong Li, Huimei Wang, Wenjie Wang, Lei Yang and Yuangang Zu
The Key Laboratory of Forest Plant Ecology Ministry of Education, Harbin, Heilongjiang 150040, China

Ofelia Andrea Valdés-Rodríguez, Arturo Pérez-Vázquez
Colegio de Postgraduados, Campus Veracruz 421, 91690 Veracruz, VER, Mexico

Odilón Sánchez-Sánchez
Centro de Investigaciones Tropicales, UV 91110 Xalapa, VER, Mexico

Joshua S. Caplan
Department of Ecology, Evolution and Natural Resources, Rutgers, The State University of New Jersey, New Brunswick, NJ 08901, USA

Frédéric Danjon
INRA, UMR1202 BIOGECO, 33610 Cestas, France

Frédéric Danjon
Universitié de Bordeaux, UMR1202 BIOGECO, 33610 Cestas, France

Huimei Wang, Wei Liu, Wenjie Wang and Yuangang Zu
Key Laboratory of Forest Plant Ecology, Ministry of Education, Northeast Forestry University, Harbin 150040, China

Wei Liu
College of Art and Landscape, Jiangxi Agricultural University, Nanchang 330045, China

Ehsanul Kabir, Sharmila Ray, Ki-Hyun Kim and Eui-Chan Jeon
Department of Environment and Energy, Sejong University, Seoul 143-747, Republic of Korea

Ehsanul Kabir
Department of Farm Power and Machinery, Bangladesh Agricultural University, Mymensingh, Bangladesh

Hye-On Yoon
Korea Basic Science Institute, Seoul Center, Anamdong, Seoul 136-713, Republic of Korea

Yoon Shin Kim and Yong-Sung Cho
Institute of Environmental and Industrial Medicine, Hanyang University, Seoul 133-791, Republic of Korea

Seong-Taek Yun
Department of Earth and Environmental Sciences, Korea University, Seoul 136-701, Republic of Korea

Richard J. C. Brown
Analytical Science Division, National Physical Laboratory, Hampton Road, Teddington, TW11 0LW, UK

Héctor Estrada-Medina, Juan José María Jiménez-Osornio and Wilian de Jesús Aguilar Cordero
Departamento de Manejo y Conservación de Recursos Naturales Tropicales (PROTROPICO),
Campus de Ciencias Biológicas y Agropecuarias (CCBA), Universidad Autónoma de Yucatán (UADY),
Km 15.5 Carretera Mérida - Xmatkuil, Mérida, Yucatán 97315, Mexico

Francisco Bautista
Centro de Investigaciones en Geografía Ambiental (CIGA), Universidad Nacional Autónoma de México (UNAM),
Antigua Carretera a Pátzcuaro No. 8701, Col. Ex-Hacienda de San José de La Huerta, Morelia, Michoacán 58190,
Mexico

José Antonio González-Iturbe
Facultad de Arquitectura (UADY), Calle 50 S/N x 57 y 59 Ex-Convento de La Mejorada, Mérida, Yucatán 97000,
Mexico

Xi Zhang, Feng Li, Tingting Liu, Chen Xu, Dechao Duan, Cheng Peng, Shenhai Zhu and Jiyan Shi
Institute of Environmental Science and Technology, Zhejiang University, Hangzhou, Zhejiang 310058, China

Feng Li
College of Materials and Environmental Engineering, Hangzhou Dianzi University, Hangzhou, Zhejiang 310018, China